Biotechnology & Genetic Engineering Reviews

Volume 1

BIOTECHNOLOGY & GENETIC ENGINEERING REVIEWS

Biotechnology & Genetic Engineering Reviews

Volume 1

Executive Editor:

GORDON E. RUSSELL

Emeritus Professor of Agricultural Biology, University of Newcastle upon Tyne

Intercept
Newcastle upon Tyne

British Library Cataloguing in Publication Data
Biotechnology & genetic engineering reviews.—
Vol. 1
1. Biochemical engineering—Periodicals
660'.6'05 TP248.3

ISBN 0-946707-01-4
ISSN 0264-8725

Published in February 1984 by Intercept Limited,
P.O. Box 2, Ponteland, Newcastle upon Tyne NE20 9EB, England.

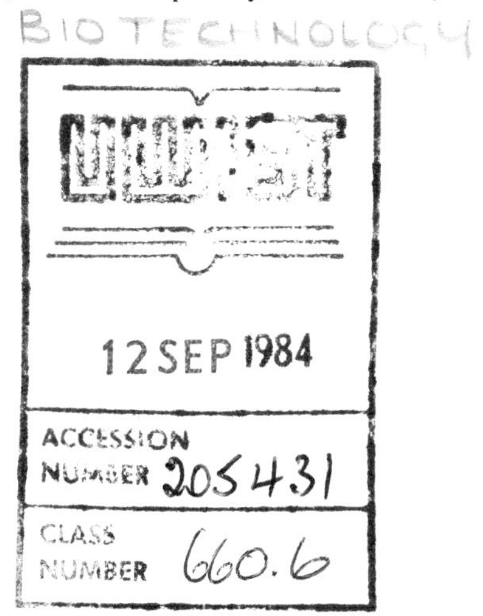

Filmset in 'Monophoto' Times by
Northumberland Press Ltd, Gateshead
Printed by Richard Clay (The Chaucer Press) Ltd, Bungay, Suffolk

Preface

Biotechnology is not a new subject. Micro-organisms have been used for centuries in brewing, wine-making and baking, although without much understanding of the processes concerned. What *is* new is the great explosion of knowledge about the micro-organisms themselves and about the biochemical reactions involved in these and other biological systems. This has already resulted in the more efficient exploitation of microbes and the scaling-up of operations in the 'industrial' use of biology, which is nowadays usually referred to as 'biotechnology'. So many different disciplines and approaches are covered by the term 'biotechnology' that it is almost impossible to arrive at a definition which would be universally acceptable. In this series of volumes, biotechnology will be interpreted in a very broad sense, including industrial activities (such as the production and application of biological compounds), agriculture, veterinary and human medicine, and applied environmental science.

The ability to manipulate the genes of micro-organisms, higher plants and animals, has revolutionized the future of biotechnology and, for this reason, biotechnology and genetic engineering will become increasingly more closely interlinked. These two areas of research and development are so large and involve so many different disciplines, that it is becoming more and more difficult for specialists in one aspect to keep abreast of important developments in others; many of these developments may be very relevant to work outside these specialist areas. It is hoped that these reviews will enable specialists to follow recent work in areas other than their own, and that these volumes will help to improve communication between disciplines and between those working on different applications of similar techniques.

Volume 1 of *Biotechnology and Genetic Engineering Reviews* comprises 13 up-to-the-minute, in-depth review articles on a wide range of topics by leading experts from many parts of the world. Although at first sight these topics may appear to be unrelated and distinct from each other, they are closely interlinked in many ways. For example, genetic manipulation of microbes is mentioned in no less than seven of the 13 contributions. The subjects of improved production methods for sugars and amino acids; biosensors; immobilized enzymes; the treatment of sewage and industrial wastes; and the control of agricultural pests, are each referred to in several chapters. These linkages emphasize the interdisciplinary nature of these reviews.

All the chapters give authoritative views on the current status and future development of different aspects of biotechnology. In Chapter 1 it is pointed out that there is considerable scope for the improvement of enzyme technology, because only about 10% of all the available enzymes have so far been characterized and exploited by man. Work on the genetic manipulation of nitrogen-fixing bacteria will not only result in improved crop production, but could also, it has been suggested, perhaps lead to increased livestock production, through the genetic engineering of rumen microbes (Chapter 3). Valuable products, or energy, can be produced by biotechnological processes from surplus or waste materials (Chapters 5, 10, 11 and 12), and new techniques employing biosensors and immobilized enzymes will play an increasingly important part in manufacturing and waste disposal industries (Chapters 1, 2, 4, 5 and 7). The more efficient use of micro-organisms, often using new, genetically engineered strains, will improve the recovery of oil from reservoirs (Chapter 8), create new possibilities for the production of vaccines (Chapter 9) and will provide alternative biological pest-control methods in agriculture (Chapter 13). It is important that progress in biotechnology and genetic manipulation is not restricted or delayed by problems, either real or theoretical, regarding the hazards of handling micro-organisms in the laboratory; this important topic is appraised in Chapter 6.

Future volumes in this series will include major review articles covering all the main aspects of biotechnology (in the widest sense of that word), with particular emphasis on the interface between biological systems in industry on the one hand, and genetic manipulation of the organisms concerned on the other. These comprehensive reviews are intended to complement the many excellent research journals, newsletters and other publications which publish short reviews and news items concerning biotechnology.

Finally, I would like to acknowledge the help and encouragement which I have received from members of the Advisory Board in the launching of this series of reviews. The guidance and advice of the Editorial Consultant, Dr W. F. J. Cuthbertson O.B.E., based on many years of practical experience in industry, has been invaluable.

GORDON E. RUSSELL
October 1983

Contributors

W.J. ASTON, *Biotechnology Centre, Cranfield Institute of Technology, Cranfield, Bedfordshire MK43 0AL, UK*

J.E. BERINGER, *Soil Microbiology Department, Rothamsted Experimental Station, Harpenden, Hertfordshire AL5 2JQ, UK*

J.B. CARTER, *Department of Biology, Liverpool Polytechnic, Byrom Street, Liverpool L3 3AF, UK*

ANDREW K. CHEUNG, *Biogen SA, 46 Route des Acacias, Carouge, Geneva, Switzerland*

C.H. COLLINS, *Public Health Laboratory, Dulwich Hospital, East Dulwich Grove, London SE22 8QF, UK*

J. COOMBS, *Bio-Services, King's College, University of London, 68 Half Moon Lane, London SE24 9JF, UK*

P. GALZY, *Chaire de Génétique et Microbiologie, ENSA–INRA, 34060 Montpellier Cedex, France*

TOKUYA HARADA, *Kobe Women's University, Suma, Kobe 654, Japan*

P.R. HIRSCH, *Soil Microbiology Department, Rothamsted Experimental Station, Harpenden, Hertfordshire AL5 2JQ, UK*

HANS KÜPPER, *Biogen SA, 46 Route des Acacias, Carouge, Geneva, Switzerland*

G. MOULIN, *Chaire de Génétique et Microbiologie, ENSA–INRA, 34060 Montpellier Cedex, France*

SAUL L. NEIDLEMAN, *Cetus Corporation, 1400 Fifty-Third Street, Emeryville, California USA 94608*

HALINA Y. NEUJAHR, *Department of Biochemistry and Biotechnology, The Royal Institute of Technology, S-100 44 Stockholm, Sweden*

POUL BØRGE POULSEN, *Novo Industri A/S, Novo Allé, DK-2880 Bagsvaerd, Denmark*

DEREK G. SPRINGHAM, *School of Biological Sciences, Queen Mary College, University of London, Mile End Road, London E1 4NS, UK*

A.P.F. TURNER, *Biotechnology Centre, Cranfield Institute of Technology, Cranfield, Bedfordshire MK43 0AL, UK*

A.D. WHEATLEY, *Environmental Biotechnology Group, Department of Chemical Engineering, University of Manchester Institute of Science and Technology, PO Box 88, Manchester M60 1QD, UK*

Contents

1
Applications of Biocatalysis to Biotechnology

SAUL L. NEIDLEMAN

Cetus Corporation, 1400 Fifty-Third Street, Emeryville, California USA 94608

Introduction

Enzymes are protein catalysts synthesized by living systems. The major emphasis in this review will be on the activities of single enzymes, either purified or confined within cells, living and dead. There will be less consideration of the biocatalytic activity of whole, living cells.

Before proceeding to technical detail, it is appropriate to indicate what is to be covered and what is not to be covered in this chapter.

This is not an enzyme textbook: therefore, only brief attention will be given to enzyme structure, properties and kinetics. General enzyme classification will be outlined to indicate the considerable breadth of biocatalytic reactions, and a few specific examples, subjectively selected, will be given.

Enzymes are important as synthetic and degradative catalysts. This variety of effects will become clear during the discussion of present applications of enzymes in the food, chemical, medical, detergent and textile industries.

Enzyme technology has several new techniques under development that are

Table 1. Some key references to biocatalysis in biotechnology.

Topic	Reference
Industrial enzymes—broad coverage	Aunstrup (1978), Dunnill (1980), Fogarty (1983a), Godfrey and Reichelt (1983a), Katchalski-Katzir and Freeman (1982), Rose (1980), Schmid (1979a), Scott (1980), Yamamoto (1978).
Enzymes in the food industry	Birch, Blakebrough and Parker (1981), Ruttloff (1982), Schwimmer (1981), Vanbelle, Meurens and Crichton (1982).
Enzyme production	Lambert and Meers (1983).
Enzyme purification	Bruton (1983).
Enzyme immobilization	Keyes (1980), Klibanov (1983), Trevan (1980).
Enzyme stabilization	Barker (1983), Klibanov (1979).
Pharmaceutical enzymes	Cooney, Stergis and Jayaram (1980), Ruyssen and Lauwers (1978).
Enzyme structure/chemistry	Dixon and Webb (1979), Scrimgeour (1977), Walsh (1979).
Enzymes in organic synthesis	Jones (1980), Kieslich (1976, 1982), Sebek and Kieslich (1977), Sih, Abushanab and Jones (1977).

Biotechnology and Genetic Engineering Reviews – Vol. 1, February 1984
0264–8725/84/01/01–38$10.00 + $0.00

Table 2. Enzyme classification.

Class	General description	Selected examples	EC numbers
1. Oxidoreductases	Enzymes of this group catalyse oxidation–reduction reactions involving oxygenation or overall removal or addition of hydrogen atom equivalents.	Alcohol dehydrogenase	1.1.1.2
		Glucose oxidase	1.1.3.4
		Amino acid oxidase	1.4.3.2
		Cytochrome oxidase	1.9.3.1
		Catalase	1.11.1.6
		Peroxidase	1.11.1.7
		Steroid 11β-monooxygenase	1.14.15.4
2. Transferases	These enzymes mediate the transfer of a group, such as aldehydic or ketonic, acyl, sugar, phosphoryl, methyl or a sulphur-containing one, from one molecule to another.	Homocysteine methyltransferase	2.1.1.10
		Hexokinase	2.7.1.1.
		Aryl sulphotransferase	2.8.2.1
		Transketolase	2.2.1.1
		Transaldolase	2.2.1.2
		Alanine aminotransferase	2.6.1.2
3. Hydrolases	The range of functional groups hydrolysed by such enzymes is very broad. It includes esters, anhydrides peptides and others. C–O, C–N and C–C bonds may be cleared as well as some others.	Triacylglycerol lipase	3.1.1.3
		Pectinesterase	3.1.1.11
		Alkaline phosphatase	3.1.3.1
		Ribonuclease II	3.1.13.1
		Deoxyribonuclease I	3.1.21.1
		α-Amylase	3.2.1.1
		Cellulase	3.2.1.4
		α-D-Glucosidase	3.2.1.20
		Aminopeptidase	3.4.11.11
		Chymosin (rennin)	3.4.23.4
		Trypsin	3.4.21.4
		Papain	3.4.22.2

4. Lyases	The types of reactions catalysed are additions to, or formation of, double bonds such as C=C, C=O, C=N.	Pyruvate decarboxylase	4.1.1.1
		Citrate (*pro*-3*S*)-lyase	4.1.3.6
		Carbonate dehydratase	4.2.1.1
		Enolase	4.2.1.11
		Phenylalanine ammonia-lyase	4.3.1.5
5. Isomerases	A variety of isomerizations, including racemization, can be effected	Methionine racemase	5.1.1.2
		Glutamate racemase	5.1.1.3
		Glucosaminephosphate isomerase (glutamine-forming)	5.3.1.19
		Xylose(glucose) isomerase	5.3.1.5
		Alanine racemase	5.1.1.1
		UDPglucose 4-epimerase	5.1.3.2
		S-Methylmalonyl-CoA mutase	5.4.99.2
6. Ligases	These are often called synthetases, and catalyse the formation of C—O, C—S, C—N and C—C bonds with accompanying adenosine triphosphate (ATP) or other nucleoside triphosphate cleavage.	Glutathione synthetase	6.3.2.3
		D-Alanylalanine synthetase	6.3.2.4
		Arginyl-tRNA synthetase	6.1.1.19
		5,10-Methenyltetrahydrofolate synthetase	6.3.3.2
		Carbamoyl-phosphate synthetase (ammonia)	6.3.4.16
		Pyruvate carboxylase	6.4.1.1

certain to change the face of applied biocatalysis. Some of these will be discussed to indicate new directions that are imminent: enzyme reactions in solvents, metal exchange at the active site of the enzyme, and chemical and genetic modification of enzyme structure and activity, for example.

Finally, consideration will be given to deficiencies in existent biocatalysts in the sweetener, detergent, diagnostic, food and other industries and what may be done to improve their suboptimal performance.

Many useful books and reviews are available in the literature that cover in detail many topics of direct or indirect relevance to this review. A selection of these is shown in *Table 1.*

General properties of enzymes

For nearly every chemically catalysed reaction, there is an enzyme-catalysed equivalent (Sih, Abushanab and Jones, 1977). The question that must be asked, then, is: For any given catalytic application, which—the chemical or the biochemical catalyst—solves the problem most economically and acceptably? For any given case, there may be one or more reasons why an enzyme might offer an effective solution. Some of the general properties of enzymes that can contribute are:

1. High catalytic power, up to 10^9–10^{12} the rate of the non-enzymatic reaction, is provided.
2. A broad range of reactions can be catalysed.
3. Reactions can be run under mild conditions of temperature, pH and pressure.
4. Often, high specificity is offered with regard to regioselectivity and stereospecificity.
5. A chiral centre may be created at a prochiral centre.
6. Nature provides a vast reservoir of available enzyme variants for special requirements.

Enzyme classification

At present, more than 2100 enzymes have been recognized by the International Union of Biochemistry (International Union of Biochemistry, 1979; Lowe, 1983). It has been speculated that 25 000 natural enzymes exist (Kindel, 1981). If this is true, about 90% of the vast reservoir of biocatalysts still remains to be discovered and characterized. As will be discussed below, through methods of chemical modification, random mutation, and genetic and protein engineering, a large number of modified protein catalysts will be available for industrial consideration as well.

Table 2 provides a glimpse at enzyme classification (International Union of Biochemistry, 1979). Six classes of enzymes have been designated by official taxonomic dictum. *Table 3* indicates the number of representatives in each of the six enzyme classes (Schmid, 1979b), while *Table 4* clearly indicates which of these enzyme classes is the major source of industrial enzymes (Godfrey and Reichelt, 1973a): about 85% are hydrolases, with the remaining 15% divided

Table 3. Number of enzymes in each class.

Enzyme class	Number in class
Oxidoreductases	537
Transferases	559
Hydrolases	490
Lyases	231
Ligases	83
Isomerases	98
Total	1998

Data from Schmid (1979b)

Table 4. Industrial use of enzyme classes

Class	Major enzymes used
Oxidoreductases	Glucose oxidase†
	Catalase†
	Peroxidase†
	Steroid hydroxylases, e.g. EC 1.14.15.4, EC 1.14.99.9, EC 1.14.99.10
Transferases	—
Hydrolases	Protease (acid, neutral, alkaline)
	α-Amylase†
	β-Amylase, EC 3.2.1.2
	Cellulase†
	Amyloglucosidase, EC 3.2.1.3
	Invertase (β-D-Fructofuranosidase), EC 3.2.1.26
	Pectinase (polygalacturonase), EC 3.2.1.15
	Lactase (β-D-Galactosidase), EC 3.2.1.23
	Naringinase
	Anthocyanase
	Lipase (triacylglycerol lipase)†
	Aminoacylase, EC 3.5.1.14
	Penicillinase, EC 3.5.2.6
Lyases	—
Ligases	—
Isomerases	Xylose (glucose) isomerase†

† For EC numbers see Table 2

among oxidoreductases and isomerases. Of the hydrolases, some 70% hydrolyse proteins, 26% hydrolyse carbohydrates and 4% hydrolyse lipids. Relatively few oxidoreductases and no transferases are commercially important, despite their large numbers.

The fact that such an overwhelming percentage of industrial enzymes are hydrolytic in nature is of particular interest in that many of these same hydrolytic reactions are reversible under certain reaction conditions discussed below. The synthetic use of hydrolytic enzymes will be a new direction for industrial enzymology.

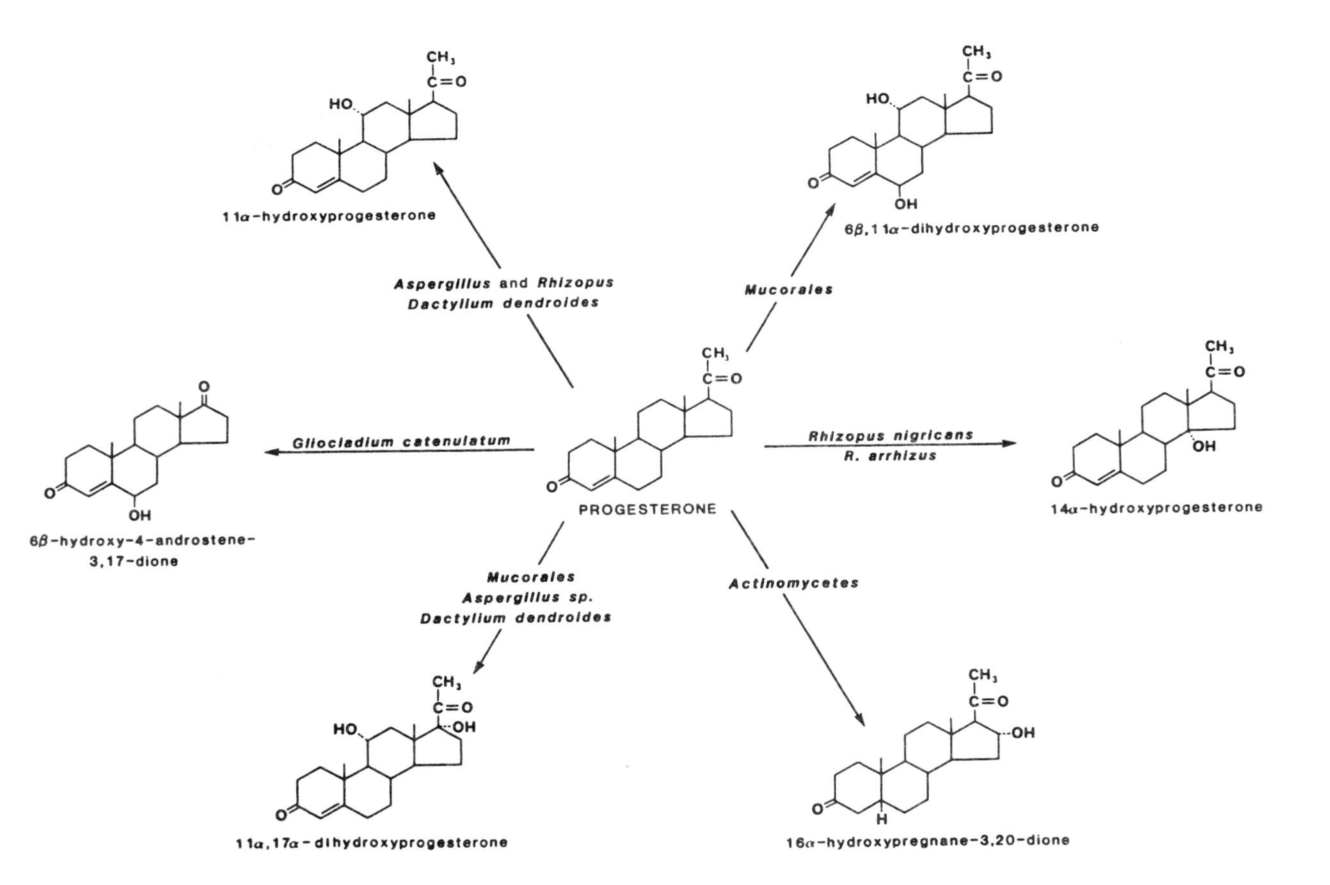

Figure 1. Selected specific steroid hydroxylations.

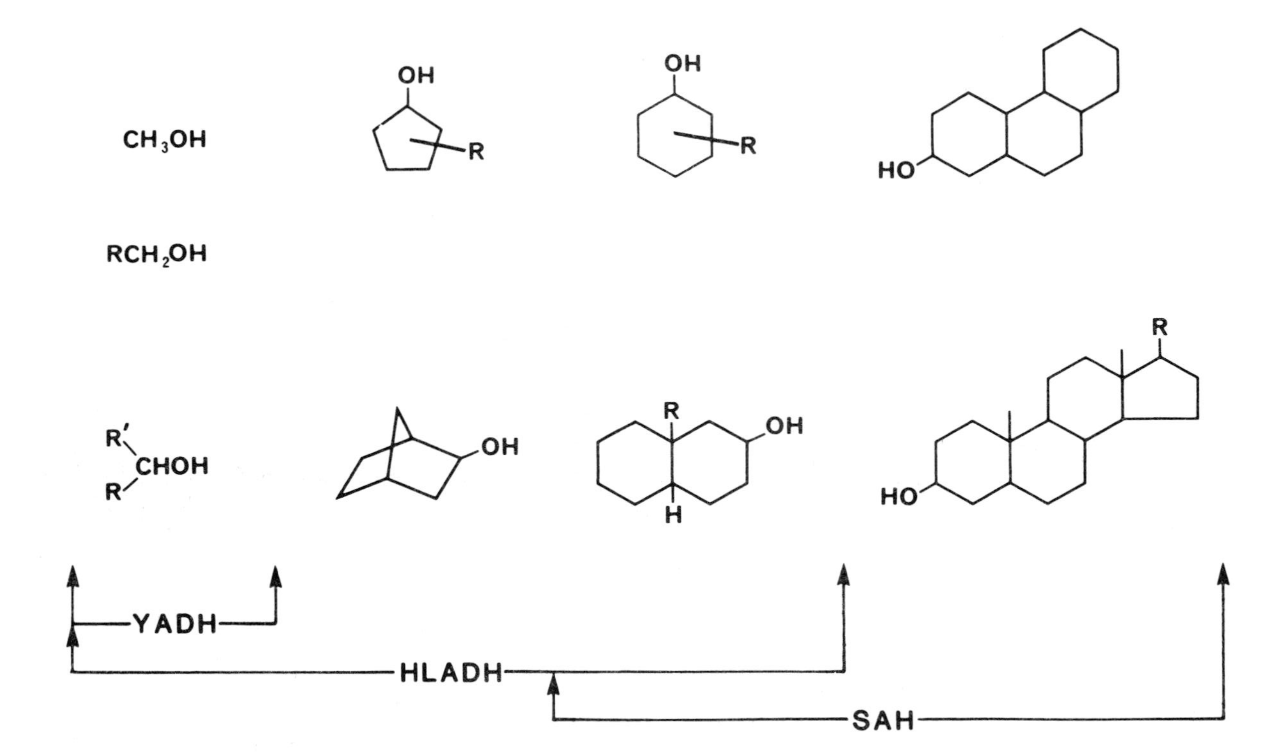

Figure 2. Specificity overlap of alcohol substrates. YADH, yeast alcohol dehydrogenase; HLADH, horse liver alcohol dehydrogenase; SAH, steroid alcohol dehydrogenase.

Many of the specific industrial applications of these and other enzymes will be discussed below. Preceding this, a brief illustration of enzyme talent will be given in the form of some personal favourites.

Some examples of enzyme activity

The use of microbial enzymes to catalyse specific and useful reactions in academic and commercial organic chemistry was stimulated by the success in steroid transformations (Charney and Herzog, 1967; Kieslich, 1980). *Figure* 1 offers just a superficial view of the variety of micro-organisms and hydroxylations that have been studied. In many of these cases, chemical catalysis would be difficult, if not impossible.

It is often the case that an enzyme recognizes only a small region of a complex molecule and can also interact with a less complex molecule having the same or a similar region. A practical outcome of this regiospecificity is that a broad structural range of substrates is often accessible using a very limited number of enzymes. For example, CH(OH)→C=O oxidations can be effected on substrates ranging in complexity from the simple aliphatic alcohols to complex polycyclic alcohols, using only three alcohol dehydrogenases with overlapping specificities. This is shown in *Figure 2* (Jones, 1980).

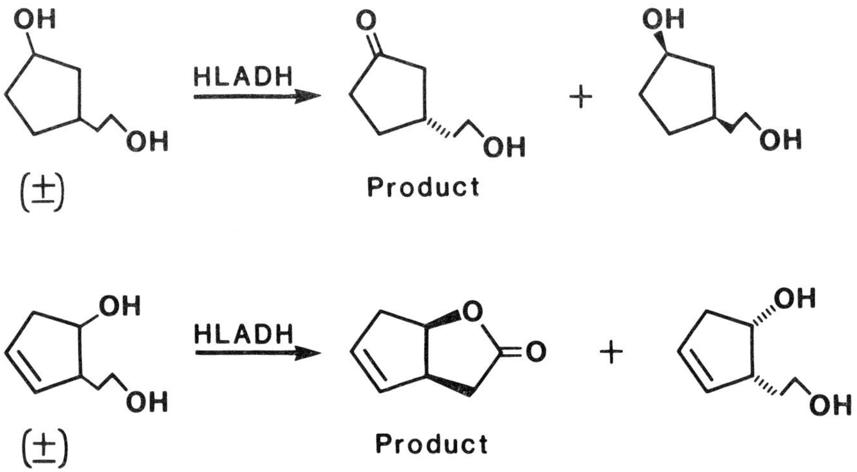

OH OH (±) HLADH O OH Product + OH OH

OH OH (±) HLADH O =O Product + OH OH

Figure 3. Enantiomeric selectivity and regiospecificity. HLADH, horse liver alcohol dehydrogenase.

In *Figure 3*, it is indicated that an enzyme may be able to achieve enantiomeric selectivity as well as regiospecificity as in a single oxidative reaction catalysed by horse liver dehydrogenase (Jones, 1980). *Figure 4* shows how this exquisite specificity is carried further, by illustrating enantiomeric selectivity combined with enantioface specificity in reduction reactions catalysed by horse liver alcohol dehydrogenase (Jones, 1980).

Enzymes have very subtle stereochemical distinctions that can be applied to particular synthetic problems. This is indicated in *Figure 5*, where three different

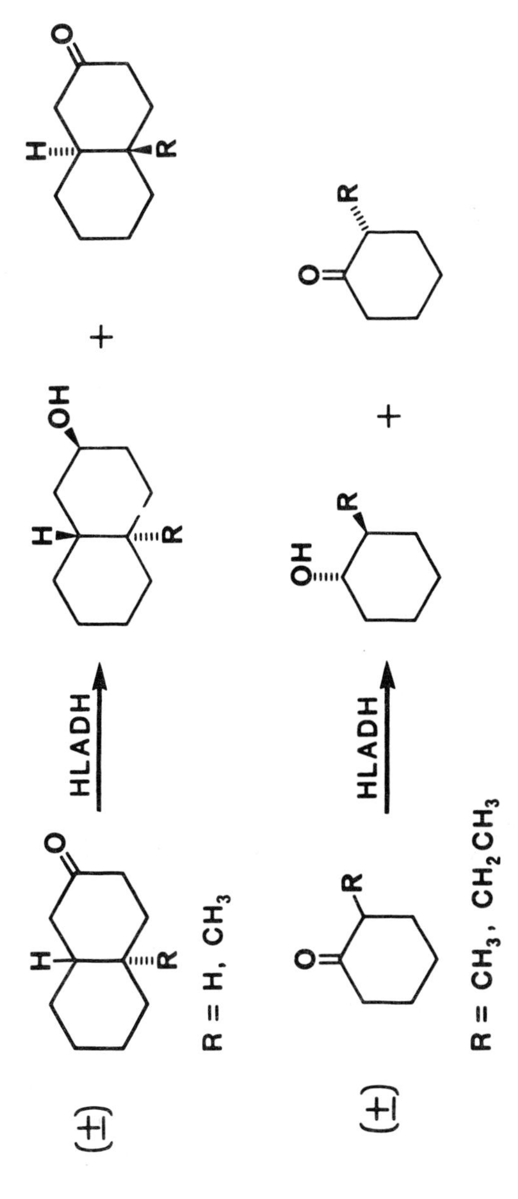

Figure 4. Enantioface specificity and enantiomeric selectivity. HLADH, horse liver alcohol dehydrogenase.

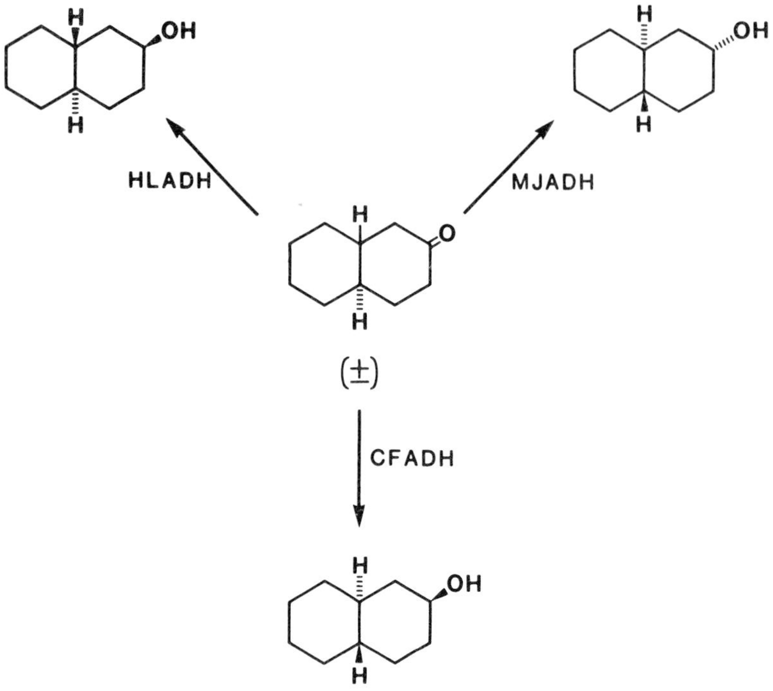

Figure 5. Diastereoisomer formation with different enzymes. HLADH, horse liver alcohol dehydrogenase; MJADH, *Mucor javanicus* alcohol dehydrogenase; CFADH, *Curvularia falcata* alcohol dehydrogenase.

diastereoisomers are formed, depending upon which alcohol dehydrogenase is used (Jones, 1980).

Figure 6 illustrates the extreme specificity of enzymatic oxidations in the carbohydrate field. With two enzymes (D-glucose-1-oxidase, EC 1.1.3.4, and pyranose-2-oxidase, EC 1.1.3.10) alone and in sequence, three useful derivatives of D-glucose can be prepared in excellent yields: (1) D-glucosone, which can be chemically reduced to D-fructose (Geigert, Neidleman and Hirano, 1983); (2) D-glucono-1,5-lactone, which converts to D-gluconic acid in water; and (3) 2-D-ketogluconic acid (Geigert *et al.*, 1983d). Distinguishing C1 and C2 of D-glucose to this degree using chemical catalysis is not possible.

A prime example of selective catalyst design in nature is that of enzymes that catalyse halogen incorporation into a wide spectrum of substrates. Among the known enzymes that catalyse halogenation are those indicated in *Table 5* (Neidleman and Geigert, 1983). They are widespread in nature, and they are all haem-containing peroxidases that vary in their ability to activate particular halide ions. None of these enzymes are able to activate F^- (Morrison and Schonbaum, 1976), and all the marine haloperoxidases that have been carefully studied are bromoperoxidases; that is, they cannot activate Cl^- (Hager, 1982). These facts have interesting consequences, as discussed below.

These enzymes can cause halide incorporation into sulphur- and nitrogen-containing compounds, as well as β-diketones and β-ketoacids (Neidleman,

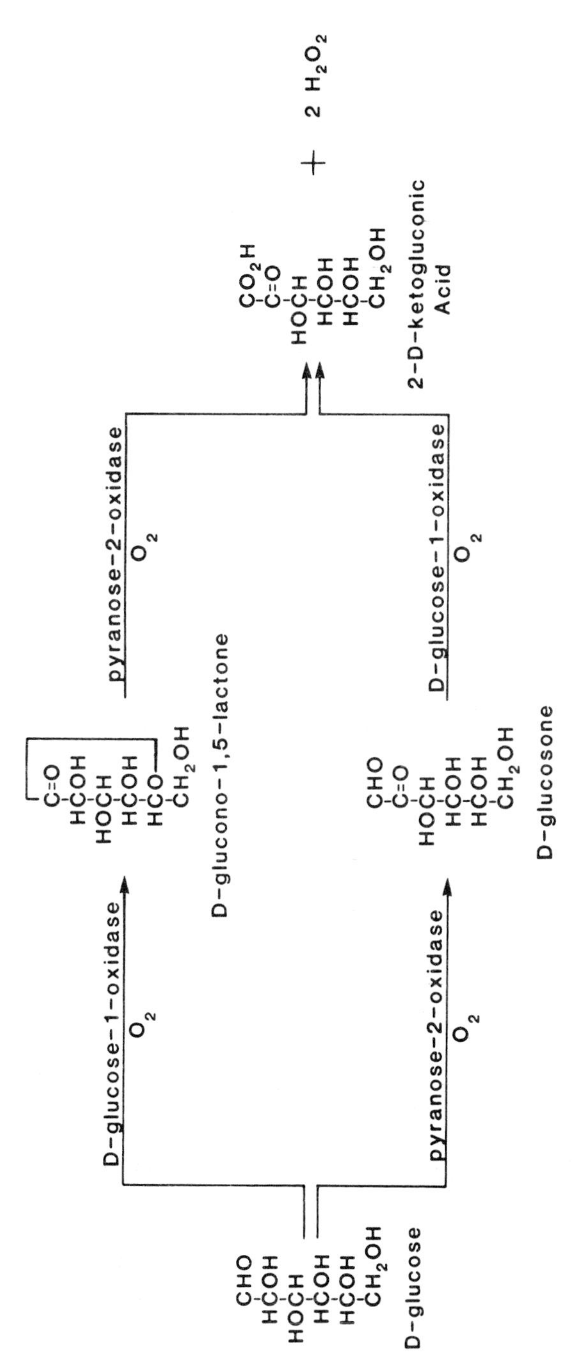

Figure 6. Oxidation of D-glucose to 2-D-ketogluconic acid.

Table 5. Haloperoxidase variants

Enzyme	Source	Approximate optimum pH	Halide activated
Chloroperoxidase (CPO) (chloride peroxidase), EC 1.11.1.10	*Caldariomyces fumago*	3	Cl^- Br^- I^-
Myeloperoxidase (MPO), EC 1.11.1.7	Leucocytes	6–7	Cl^- Br^- I^-
Lactoperoxidase (LPO), EC 1.11.1.7	Milk	6–7	Br^- I^-
Bromoperoxidase (BPO)	Seaweed	6–7	Br^- I^-
Thyroid (iodide) peroxidase (TPO), EC 1.11.1.8	Thyroid	6–7	I^-
Horseradish peroxidase (HRPO), EC 1.11.1.7	Horseradish	6–7	I^-

1975). In addition, recent work has shown that various alkenes, alkynes and cyclopropanes are good substrates for halogenation (Geigert, Neidleman and Dalietos, 1983; Geigert *et al.*, 1983b, c, e). *Figure 7* shows a variety of products that can be synthesized from allyl alcohol in the presence of chloroperoxidase (CPO) (Geigert *et al.*, 1983a; Neidleman and Geigert, 1983). It should be noted that the nature of the halogenated product is dictated by the concentration of the halide ion and by whether one or two different halides are present. This effect will be discussed in a later section. It is also of interest that, in the absence of halide ion, CPO oxidizes allyl alcohol to acrolein.

Figure 8 illustrates the catalytic incorporation of Cl^- into the alkyne, methyl acetylene, and the cyclopropane, methylcyclopropane (Geigert, Neidleman and Dalietos, 1983). The products obtained with methyl acetylene are very responsive

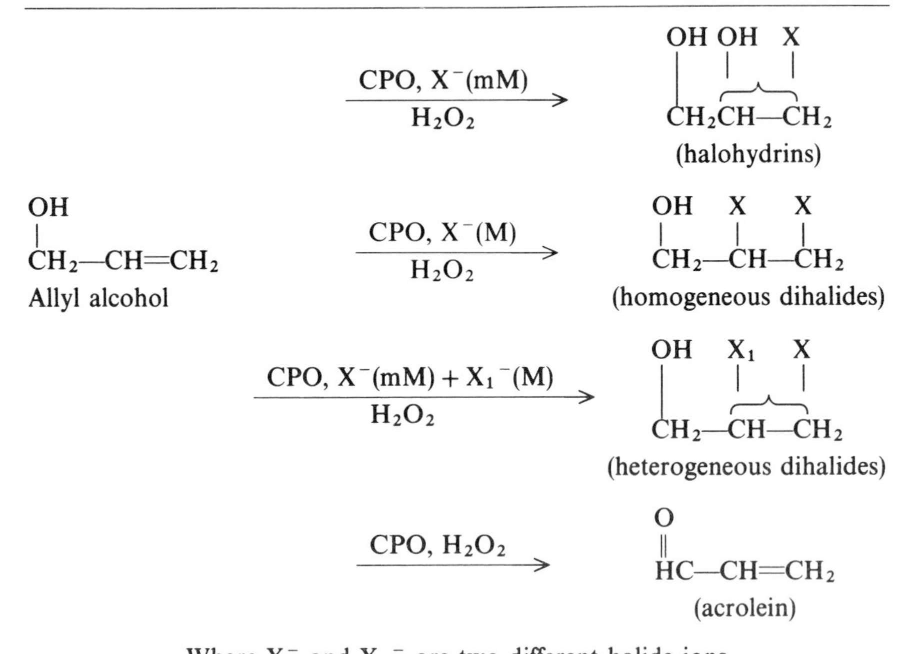

Figure 7. Various reactions between allyl alcohol and chloroperoxidase (CPO), EC 1.11.1.10.

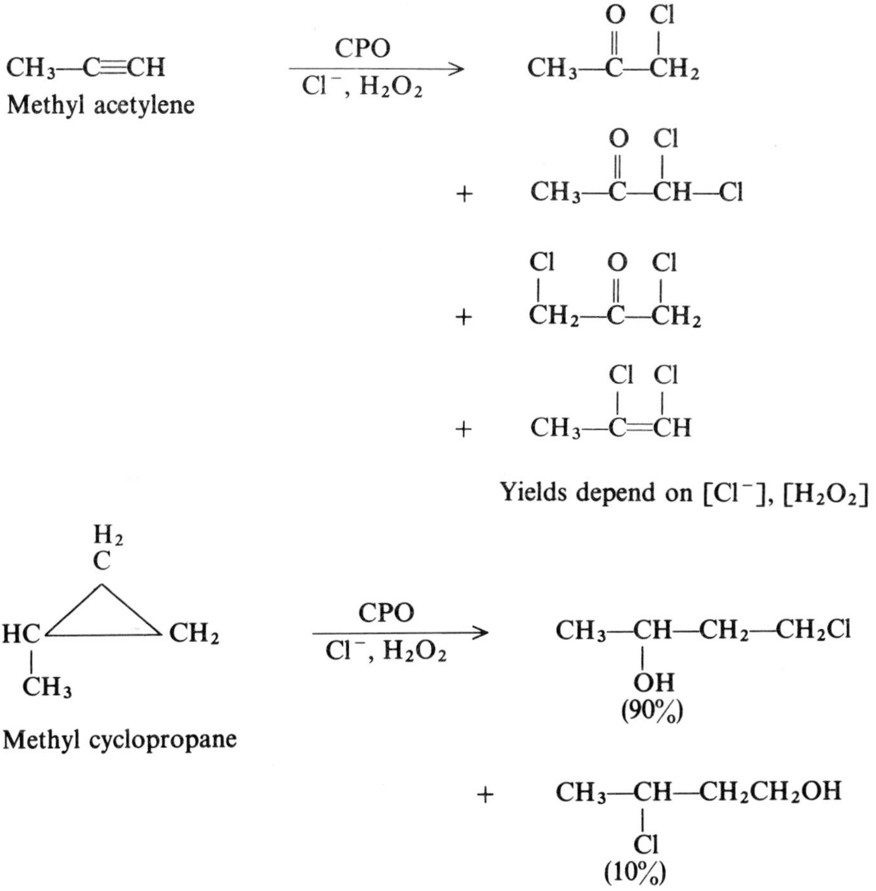

$$CH_3{-}C{\equiv}CH \xrightarrow[Cl^-,\ H_2O_2]{CPO} CH_3{-}C({=}O){-}CH_2Cl$$

Methyl acetylene

$$+\ CH_3{-}C({=}O){-}CHCl{-}Cl$$

$$+\ ClCH_2{-}C({=}O){-}CH_2Cl$$

$$+\ CH_3{-}CCl{=}CHCl$$

Yields depend on $[Cl^-]$, $[H_2O_2]$

$$\text{Methyl cyclopropane} \xrightarrow[Cl^-,\ H_2O_2]{CPO} CH_3{-}CH(OH){-}CH_2{-}CH_2Cl\ (90\%)$$

Methyl cyclopropane

$$+\ CH_3{-}CHCl{-}CH_2CH_2OH\ (10\%)$$

Figure 8. Reactions of chloroperoxidase (CPO) on methyl acetylene and methyl cyclopropane.

to the concentrations of halide ion and hydrogen peroxide—this would be equally true in nature as it is in the laboratory.

Another useful reaction in the synthesis of chemicals is the specific insertion of double bonds into lipids. Depending upon the properties of the desaturating enzyme, double bonds can be placed at different positions within the lipid chain (Fulco, 1977; Weete, 1980). Selectively unsaturated esters and glycerides are of value as flavours, fragrances and as components of various food oils and baked products (Heath, 1981; Macrae, 1983a, b; Strobel, 1983; Strobel *et al.*, 1983). As an example of the production of unsaturated wax esters (monoesters), *Table 6* shows the capacity of *Acinetobacter* sp. HO1-N to form such compounds from ethanol and *n*-eicosane (C_{20}-*n*-alkane) (DeWitt *et al.*, 1982; Ervin *et al.*, 1983). The wax esters contain 0, 1 or 2 double bonds. In the case of a monoenic wax ester, the unsaturation may be in either the acyl or alkoxy moiety of the ester. In the case of a dienic wax ester, one double bond is in the acyl moiety, and the other is in the alkoxy moiety. The ability to control the level of unsaturation in these wax esters with temperature will be illustrated below.

Table 6. Formation of wax esters by *Acinetobacter* sp. HO1-N from *n*-eicosane and ethanol

Substrate →	Fatty acids	+ Fatty alcohols →	Wax esters		
	RCOOH	+ R′OH →	RC(=O)—OR′		
n-eicosane	16:0, 16:1*	20:0, 20:1	36:0	38:0	40:0
	18:0, 18:1		36:1	38:1	40:1
	20:0, 20:1		36:2	38:2	40:2
Ethanol	16:0, 16:1	16:0, 16:1	32:0	34:0	36:0
	18:0, 18:1	18:0, 18:1	32:1	34:1	36:1
			32:2	34:2	36:2

* 16, 18, 20, 32, 34, 36, 38, 40 = number of carbon atoms: 0, 1, 2 = number of double bonds, located at Δ^9 or Δ^{11} in either the acyl or alkoxy fragment of the wax ester.

Industrial applications

FOODS

The use of biocatalysts in the food industry involves a wide range of effects, including the production of food components such as flavours and fragrances, and the control of colour, texture, appearance and nutritive value. In many cases, the subtleties and nuances of these effects require exquisite control bordering more on art than science. *Table 7* shows various enzymes and their application to particular aspects of the food industry (Shahani *et al.*, 1976; Crouzet, 1977; Yamada, 1977; Aunstrup, 1978; Yamamoto, 1978; Schmid, 1979b; Ter Haseborg, 1981; Marschall *et al.*, 1982; Norman, 1982; Pilnik, 1982; Schindler and Schmid, 1982; Stewart, 1982; Vanbelle, Meurens and Crichton, 1982; Atkinson, 1983; Felix and Villetaz, 1983; Godfrey, 1983a, c; Hollo, Laszlo and Hoschke, 1983; Janda, 1983).

DETERGENTS

One of the major industrial applications of enzymes is in the detergent area. More than 95% of the enzymes sold for laundry detergents are alkaline serine proteases from *Bacillus* sp., especially the *Bacillus subtilis* group. The enzymes must have the following important properties to be acceptable:

1. Stable and active at pH 9·0–10·5.
2. Thermotolerant in the range 55–100°C.
3. Compatibility with perborate, surfactants, chelators, etc.

There is a real possibility that future detergents will contain enzymes other than alkaline serine proteases: lipases for oily or fatty soiling, amylases for starch soiling, and perhaps hydrolases or oxidases (Starace & Barfoed, 1980; Barfoed, 1983; Starace, 1983).

Table 7. Applications of enzymes in the food industry.

Enzymes	Sources	Substrates	Applications
α-Amylase†	Fungal, bacterial	Starch	Liquefaction to dextrins, alcohol production, proper volume in baked goods
β-Amylase‡	Plant	Starch	Maltose production, proper volume of baked goods
Anthocyanase	Fungal	Anthocyanine Glycoside	Decolourization of juice/wine
Catalase†	Fungal, mammals	Hydrogen peroxide	Milk sterilization, cheese-making
Cellobiase (β-D-glucosidase), EC 3.2.1.21	Fungal	Cellobiose	Ethanol production
Cellulase†	Fungal	Cellulose	Ethanol production
Glucoamylase (exo-1,4-α-D-glucosidase), EC 3.2.1.3	Fungal	Dextrins	Dextrin degradation to glucose
Glucose isomerase†	Bacterial	Glucose	High fructose syrup
Glucose oxidase† (± catalase)	Fungal	Glucose, oxygen	Flavour and colour preservation
Hemicellulase*	Fungal, bacterial	Hemicellulose	Clarification of plant extracts
Hesperidinase	Fungal	Hesperidin glycoside	Clarification of juice
Invertase‡	Yeast	Sucrose	Production of invert sugar, sugar confectionery
Lactase‡	Fungal, yeast	Lactose	Lactose hydrolysis in cheese whey
Lipase†	Fungal, bacterial, goat, calf, lamb throat	Lipid	Cheese ripening
	Fungal, bacterial	Lipid	Modify milk fat for butter
	Bacterial	Lipid	Sausage curing
Lipoxidase (lipoxygenase), EC 1.13.11.12	Plant	Carotene	Bleaching agent in baking
Melibiase (α-D-galactosidase), EC 3.2.1.22	Fungal	Raffinose	Improve sucrose production from sugar beets
Naringinase	Fungal	Naringin glycoside	Debittering of juice
Pectinase (polygalacturonase), EC 3.2.1.15	Fungal	Pectin	Wine/fruit juice clarification, viscosity reduction in fruit processing
Proteases, e.g. EC 3.4.22.2, 3.4.22.4, 3.4.23.4	Fungal, plant (papain, bromelain)	Protein	Meat tenderizer
	Fungal	Protein	Condensed fish solids
	Fungal, bacterial, calf stomach (rennin)	Casein	Cheese production
	Fungal	Gluten	Dough conditioner
	Bacterial	Protein	Sausage curing
	Plant (papain)	Protein	Beer haze removal
	Pancreas	Protein	Peptone manufacture
	Bacterial	Protein	Soy sauce preparation
Pullulanase, EC 3.2.1.41	Bacterial	Amylopectin	Beer production, improve glucose and maltose processes

* Mixture † For EC numbers see Table 2 ‡ For EC numbers see Table 4

Table 8. Application of enzymes in the medical field.

Enzyme	Source	Substrate	Effect
Amylase, EC 3.2.1.1, 3.2.1.2	Fungal, bacterial	Starch	Digestive aid
L-Asparaginase, EC 3.5.1.1	Bacteria	L-Asparagine	Antitumour
Bromelain, EC 3.4.22.4	Pineapple	Protein	Digestive aid, reduction of oedema
Cellulase, EC 3.2.1.4	Fungal	Cellulose	Digestive aid
Chymopapain, EC 3.4.22.6	Papaya	Protein	Treatment of herniated discs
Chymotrypsin, EC 3.4.21.1	Pancreas	Protein	Treatment of inflammatory conditions
Diphtheria toxin	Bacterial	Elongation factor 2	Antitumour
Ficin, EC 3.4.22.3	Fig	Protein	Digestive aid
Hyaluronidase, EC 3.2.1.35 and EC 3.2.1.36	Beef seminal vesicles	Hyaluronic acid	In conjunction with anaesthetics (facilitates absorption)
Kallikrein, EC 3.4.21.8	Pancreas	Protein	Reduce blood pressure
Lactase (β-D-galactosidase)	Yeast, fungal	Lactose	Increase milk digestibility
Lipase (triacylglycerol lipase), EC 3.1.1.3	Fungal, pancreas	Lipids	Digestive aid
Lysozyme, EC 3.2.1.17	Chicken egg white	Mucopolysaccharide	Antiviral, antibacterial
Pancreatin (mixture)	Pancreas	Starch, lipid, protein	Digestive aid
Papain, EC 3.4.22.2	Papaya	Protein	Digestive aid, reduction of oedema, treatment of herniated discs
Pepsin, EC 3.4.23.1	Mammalian stomach	Protein	Digestive aid
Plasma, EC 3.4.21.7	Human blood	Protein	Dissolve blood clots
Protease	Bacterial	Protein	Digestive aid
Streptodornase (deoxyribonuclease I), EC 3.1.21.1	Bacterial	DNA	Wound débridement
Streptokinase (plasmin, EC 3.4.21.7)	Human urine	Protein	Dissolve blood clots
Trypsin, EC 3.4.21.4	Pancreas	Protein	Digestive aid, treat athletic injuries
Urokinase, EC 3.4.21.31	Human urine	Protein	Dissolve blood clots

MEDICAL FIELD

Table 8 indicates that the major use of enzymes in the medical field is in digestive aids. However, it is also clear that a range of other applications is existent: use as antitumour or antimicrobial agents, and in the treatment of blood clots and herniated discs, for example. There is considerable research in this area at present, and there is no doubt that more sophisticated applications, such as the treatment of enzyme deficiencies, will be forthcoming (Ruyssen and Lauwers, 1978; Cooney, Stergis and Jayaram, 1980; Anonymous, 1982).

ANALYTICAL APPLICATIONS

The use of enzymes in various aspects of chemical analysis is a rapidly growing area of interest. One of the major applications is in the diagnostic field, as indicated in *Table 9*. These uses have been aided by the development of highly sophisticated solid-support systems and instrumentation. The most common

Table 9. Analytical applications of enzymes.

Enzyme	Source	Substance analysed
Alcohol dehydrogenase, EC 1.1.1.1	Yeast, horse liver	Ethanol
Cholesterol oxidase, EC 1.1.3.6	Bacterial	Cholesterol
Creatinase, EC 3.5.2.10	Bacterial	Creatinine
Galactose oxidase, EC 1.1.3.9	Fungal	Galactose
Glucose oxidase, EC 1.1.3.4	Fungal	Glucose
Glycerol kinase, EC 2.7.1.30	Bacterial, yeast	Triglycerides
L-Glycerol-3-phosphate dehydrogenase, EC 1.1.99.5	Rabbit	Triglycerides
Lactate dehydrogenase, EC 1.1.1.27	Mammalian	Triglycerides, transaminase
Lipase (triacylglycerol lipase), EC 3.1.1.3	Fungal, yeast, wheat germ, pig	Triglycerides
Luciferase (luciferin sulphotransferase) EC 2.8.2.10	Firefly	ATP
Peroxidase, EC 1.11.1.7	Horseradish	Hydrogen peroxide from various reactions
Urease, EC 3.5.1.5	Jack bean	Urea
Uricase (urate oxidase), EC 1.7.3.3	Yeast, pig	Uric acid

tests are carried out on human serum, a liquid of reasonably consistent characteristics. This fact has allowed for the development of regularized, miniaturized and highly automated analytical systems. Other industrial applications are primarily aimed at monitoring of various industrial processes for such compounds as acetic acid, citric acid, galactose and glucose. Another use is in the analysis of pesticides: organophosphates and carbamates (Keyes, 1980; Langley, 1983).

PRODUCTION OF CHEMICALS

The production of the chemicals listed in *Table 10* is usually carried out with enzymes immobilized within living or dead cells, or by fermentation with live cells. The wide diversity of compounds that have been produced is obvious.

Table 10. Chemicals

Compound	Producing organism
I. Organic Acids, Alcohols, Aldehydes	
Acetic acid	*Clostridium* sp., *Gluconobacter* sp.
Acetone	*Clostridium* sp.
Butanol	*Clostridium* sp.
Citric acid	*Aspergillus niger*, *Candida lipolytica*
Erythorbic acid	*Penicillium notatum*
Ethanol	Various yeasts
Fumaric acid	*Rhizopus delemar*
Gluconic acid	*Aspergillus niger*
Itaconic acid	*Aspergillus terreus*
2-Ketogluconic acid	*Pseudomonas meldenbergii*
5-Ketogluconic acid	*Gluconobacter suboxydans*
α-Ketoglutaric acid	*C. hydrocarbofumarica*
Kojic acid	*Aspergillus oryzae*
Lactic acid	*Lactobacillus delbrukii*
Malic acid	*Lactobacillus brevis*
Propionic acid	*Propionibacterium shermanii*
Succinic acid	*Bacillus succinicum*
II. Amino Acids	
L-Alanine	*Pseudomonas dacunhae*
DL-Alanine	*Corynebacterium gelatinosum*
L-Arginine	*Brevibacterium flavum*
L-Aspartic acid	Aspartase from *Escherichia coli* and *Erwinia herbicola*
L-Citrulline	*Brevibacterium flavum*, *Pseudomonas putida*
L-Glutamic acid	*Corynebacterium glutamicum*, *Brevibacterium flavum*
L-Histidine	*Brevibacterium flavum*
L-Isoleucine	*Brevibacterium flavum*
L-Leucine	*Bacillus lactofermentum*
L-Lysine	*Corynebacterium glutamicum*, *Brevibacterium flavum*, H-enzyme (*Cryptococcus laurentii*) plus R-enzyme (*Achromobacter obae*)
L-Ornithine	*Corynebacterium glutamicum*
L-Phenylalanine	*Bacillus subtilis*, *Brevibacterium flavum*
L-Proline	*Brevibacterium flavum*
L-Serine	*Corynebacterium glycinophilum*
L-Threonine	*Brevibacterium flavum*
L-Tryptophan	*Bacillus subtilis*, *Brevibacterium flavum*, tryptophanase from *Proteus rettgeri*
L-Tyrosine	L-Tyrosinase from *Erwinia herbicola*
L-Valine	*Bacillus lactofermentum*
III. Miscellaneous Chemicals	
Adenosine	*Bacillus* sp.
Carotenoids	Algae, fungi
Gibberellins	*Gibberella fujikuroi*
Guanosine	*Bacillus subtilis*
Inosine	*Bacillus subtilis*
Inosine-5′-phosphate	*Brevibacterium ammoniagenes*
Riboflavin	*Clostridium* sp., *Ascomycetes* sp., *Candida* sp.
Vitamin B_{12}	*Propionobacter shermanii*, *Bacillus* sp., *Pseudomonas* sp.
Xanthan gum	*Xanthomonas campestris*

IV. Possible Chemicals of the Future

Acrylic acid	Propylene glycol
Adipic acid	Phthalic anhydride
Ethylene oxide	Tartaric acid
Ethylene glycol	Propylene oxide
Glycerol	

V. Chemicals Not Detailed

Antibiotics	Steroid derivatives
Enzyme inhibitors	Flavours
Alkaloids	Fragrances
Lipid derivatives	Surfactants

The list is not exhaustive, and not all examples are currently commercially significant. Usually, the use of purified enzymes is precluded by the economic disadvantage of the high cost of preparing the enzyme. In other cases, the starting material is D-glucose and its conversion to the final product requires multiple enzymatic activities with associated cofactors; this is most feasible with whole cells. There is considerable research and progress, however, in the application of coenzyme-dependent enzymes in industry (Lowe, 1983). Many references may be consulted for further information on production of chemicals (Kieslich, 1976; Arima, 1977; Perlman, 1977; Sebek and Kieslich, 1977; Yamada and Kumagai, 1978; Gray and Tribe, 1979; Tribe and Gray, 1979; Janshekar and Fiechter, 1982; Rosazza, 1982; Tong, 1982; Chibata, Tosa and Sato, 1983; Linden and Moreira, 1983).

WASTE TREATMENT

The use of enzymes for waste treatment is widespread. Among the enzymes that are used are amylase, amyloglucosidase, cellulase, glucoamylase, lipase, pectinase and protease. They are used to digest their normal substrates, the carbohydrates, proteins, fats and oils. Often, the goal is to 'recycle' the waste for reuse: for example, to convert starch to sugar, to convert whey to various useful products, and for the recovery of additional oil from oil-seeds (Callely, Forster and Stafford, 1976; Godfrey, 1983b; Moo-Young, 1983). A novel procedure for the removal of phenols and aromatic amines, which occur widely in industrial waste waters, has been reported (Klibanov, 1982; Klibanov, Tu and Scott, 1983) and utilizes horseradish peroxidase. The recovery and treatment of lignocellulosic substances is one of the greatest challenges of waste—or, rather, resource—recycling. Economic processes for cellulose and lignin upgrading await improvements in the enzymes involved in their degradation.

MISCELLANEOUS APPLICATIONS

An effort has been made in this review to give a broad view of the use of enzymes in industry. However, the field is so complex and diverse that only partial coverage could be given. *Table 11* gives additional examples of areas where biocatalysis has been applied, which did not fall gracefully into other classifications. Not all are in present use, but the possibility of commercial application exists.

New techniques and concepts

The purpose of this section is to illustrate in several ways that the application of biocatalysis to industrial problems is approaching an explosion of sophistication at least equal to that of genetic engineering. Among the elements in this surge of biocatalysis will be the following:

1. Enzyme reactions in organic solvents.
2. Metal exchange at the active site of enzymes.

Table 11. Miscellaneous applications of enzymes

Enzyme	Substrate	Use	Reference
Aminoacylase EC 3.5.1.14	L-Amino acids	Production of L-amino acids	Chibata, Tosa and Sato (1983)
α-Amylase, EC 3.2.1.1	Starch	Textile desizing, dental hygiene	Godfrey and Reichelt, 1983c, d
Dextranase, EC 3.2.1.11	Dextran	Dental hygiene	Godfrey and Reichelt, 1983d
Glucose oxidase, EC 1.1.3.4	Glucose	Dental hygiene	Godfrey and Reichelt, 1983d; Schmid, 1979b
Papain, EC 3.4.22.2	Protein	Dental hygiene	Godfrey and Reichelt, 1983d
Pectinase, EC 3.2.1.15	Pectins	Wood preservation, retting of textile fibres	Fogarty and Kelly, 1983
Pencillin amidase, EC 3.5.1.11	Penicillins	Synthesis of antibiotics	Arima, 1977
Protease	Protein	Dental hygiene	Godfrey and Reichelt, 1983d
		Leather tanning, photographic silver recovery	Godfrey and Reichelt, 1983; Ward, 1983; Cowan, 1983

3. Chemical modification of enzymes.
4. Enzyme variants from nature, mutation, and genetic (protein) engineering.
5. Effects of temperature on lipid unsaturation.
6. Effects of reactant concentration on reaction product chemistry.
7. Effects of enzyme immobilization.

EFFECTS OF ORGANIC SOLVENTS ON ENZYME REACTIONS

Organic solvents can have deleterious effects on enzyme reactions by causing protein denaturation and, therefore, a loss in enzymatic activity. However, as in many other instances, there is another side of the coin: organic solvents can have a positive effect on certain enzyme reactions. Examples are finding their way into the literature with increasing frequency. Several mechanisms appear to be involved in these advantageous cases, including:

1. Increasing substrate solubility in the aqueous phase and therefore increasing substrate availability for enzymatic transformation.
2. Reducing hydrolytic reactions by making water a limiting reactant.
3. Altering the conformation of the active site of the enzyme.

Each of these mechanisms will be illustrated by examples from the literature.

Two basic types of systems have been studied in pursuing enzyme reactions in solvents. One is a monophasic system in which a water-miscible solvent is employed. A second version of a monophasic system is that in which the only solvent is an organic one. In contrast to these monophasic systems, there are biphasic systems in which water and an immiscible organic solvent are used. As not all enzymes can retain activity in the presence of high levels of organic solvent, the biphasic systems have the advantage that the enzyme in the aqueous phase is not necessarily contending with elevated and inhibitory solvent concentrations (Lilly, 1983).

Table 12. Improvement by solvents of cortisone reduction by 20β-hydroxysteroid dehydrogenase (*20β-HSDH*)†

Organic solvent	Activity of 20β-HSDH (% inhibition)	Cortisone solubility in solvent (g/100 ml)	Solvent solubility in water (%; w/v)	Cortisone reduction (%)
n-hexane	0	0·002	0·014	<5
Carbon tetrachloride	0	0·004	0·08	<5
Chlorobenzene	0	0·030	0·048	15
Diethyl ether	62	0·017	7·5	10
Butyl acetate	52	0·160	0·5	100
Ethyl acetate	71	0·270	8·6	90

† EC 1.1.1.53.

Some of these principles are illustrated in *Table 12.* In aqueous medium, the percentage conversion of cortisone to preg-4-en-17,20,21-triol-3,11-dione is <5%. The Table clearly indicates that butyl and ethyl acetate markedly improve the transformation (Antonini, Arrea and Cremonesi, 1981). A number of points should be noted. These solvents do inhibit the activity of 20β-hydroxysteroid dehydrogenase (EC 1.1.1.53). The logical reaction to this inhibitory effect might be to eliminate these solvents as candidates to improve the reaction in favour of a solvent such as chlorobenzene, which does not inhibit activity of the enzyme. However, it is seen that the inhibition by butyl and ethyl acetate is compensated for by the increased solubility of cortisone in the aqueous phase, resulting from a balance of cortisone solubility in the organic solvent and solubility of the organic phase in the aqueous phase. It is necessary to assess carefully the net worth of both positive and negative factors.

Another principle to be illustrated is that certain hydrolytic reactions can be reversed to afford synthetic reactions. Esterases and lipases that normally hydrolyse their ester and glyceride substrates in an aqueous milieu can often synthesize these substances when water is made a limiting reagent. Thus, a reaction which is thermodynamically at a disadvantage in water is thermodynamically favoured in organic solvent. The proportion of organic solvent can approach 100%.

Table 13 shows an example of ester synthesis by α-chymotrypsin in the presence of chloroform (Klibanov *et al.*, 1977). With water as the solvent, the ester

Table 13. Synthesis of esters by 'hydrolytic' enzymes in solvents

Acid	Alcohol		Ester
N-acetyl L-tryptophan	+ethyl	EC 3.4.21.1 α-chymotrypsin → $CHCl_3$	*N*-acetyl L-tryptophan ethyl ester (100%)
Linoleic	+cetyl		Cetyl linoleate (~90%)
Palmitic	+octyl		Octyl palmitate (~90%)
Pentanoic	+pentyl	Dried mycelia of → *Rhizopus arrhizus*	Pentyl pentanoate
Butyric	+benzyl		Benzyl butyrate
Acetic	+geranyl		Geranyl acetate (30–70%)

is not synthesized. *Table 13* also illustrates the capacity of one of the reactants, the alcohol, to serve as the solvent for ester synthesis with dried mycelia of *Rhizopus arrhizus* (Bell *et al.*, 1978; Patterson *et al.*, 1979; Strobel, 1983; Strobel *et al.*, 1983). Methods such as these will be employed to produce flavours and fragrances for industrial use.

These techniques can be used to prepare mono-, di- and triglycerides, as well as simple esters. This is shown in *Table 14*. Glycerol and fatty acid are dissolved

Table 14. Synthesis of glycerides in solvent by dried *Rhizopus arrhizus* mycelia

$$\underset{(0{\cdot}2\%\ \text{w/v})}{\text{Glycerol}} + \underset{(10\%\ \text{w/v})}{\text{Oleic acid}} \xrightarrow[\text{Acetone}]{R.\ arrhizus\ \text{(dried)}} \begin{matrix}\text{1-Monoglyceride} \\ \text{1,2/1,3-Diglycerides} \\ \text{1,2,3-Triglyceride (trace)}\end{matrix}$$

70% yield based on conversion of hydroxyl groups to ester.

in a solvent and, in the presence of dried mycelia of *R. arrhizus*, glycerides are synthesized (Bell *et al.*, 1978). A similar process with *Corynebacterium* sp. 5–401, in which glycerol and oleic acid are dissolved in *n*-hexane, yielded triolein, rather than mono- or diolein (Seo, Yamada and Okada, 1982). Methods such as these will synthesize tailor-made glycerides for industrial applications (Tanaka *et al.*, 1981; Fukui and Tanaka, 1982).

It was shown in *Table 12* that the nature of the organic solvent can have a marked influence on the efficiency with which an enzymatic transformation of a steroid can be performed. The same holds true for enzyme-catalysed ester synthesis: *Table 15* illustrates that chloroform is the solvent of choice for ester synthesis with α-chymotrypsin (Martinek and Semenov, 1981; Martinek, Semenov and Berezin, 1981).

Table 15. Synthesis of *N*-benzoyl-L-phenylalanine ethyl ester by α-chymotrypsin in the presence of various solvents

Solvent	Yield (%)
Chloroform	80
Benzene	64
Carbon tetrachloride	63
Diethyl ether	26
Water	0

The process of interesterification is another example of a potential industrial application of biocatalysis in biphasic systems. Typically, lipase-coated inorganic particles are activated with 10% water, and stirred in a reactor with substrates dissolved in a solvent such as petroleum ether (Macrae, 1983a, b). Such a method can use a cheap feedstock such as olive oil, and, under appropriate conditions, a reaction product closely resembling the more valuable cocoa butter can be obtained. Reactions such as these offer a clear indication of the greater specificity that can be achieved with biocatalysis, compared with traditional chemical catalysis.

Table 16. Interesterification of olive and coconut oils (1 : 1; w/w)

Carbon no. in triglyceride fatty acids	Starting mixture (wt %)	Percentage change from starting mixture in interesterified oils treated with:	
		C. cylindracae lipase†	Alkali metal
26–28	29·2	−13·0	−14·0
40–48	12·0	+48·3	+48·9
50–56	58·7	−35·2	−34·2

†Triacylglycerol lipase, EC 3.1.1.3.

Table 16 illustrates that both chemically catalysed and enzyme-catalysed interesterification can result in random acyl migration and exchange (Macrae, 1983a, b). In this case, sodium or sodium alkoxide and a non-specific lipase offer little to choose from with respect to the end product. However, when enzymes with intrinsic catalytic specificity are used, the situation is quite different. For example, in *Table 17* (Macrae, 1983a, b) the use of a 1,3-specific lipase in the presence of olive oil and stearic acid enriches the 1,3-positions of olive oil with stearic acid. In *Table 18* a linoleic-specific lipase results in the enrichment of olive oil with linoleic acid, but not stearic acid (Macrae, 1983a, b). Remarkable reaction specificity such as this can yield novel products with selected properties not achievable through random chemical or enzymatic interesterification.

The final example of organic solvent effects on biocatalysis is the most intriguing because it has basic implications for both industry and nature. It is the regulation of enzyme activity and specificity through solvent-mediated conformational changes in the active site of enzymes. α-Thrombin is an enzyme with both esterase and amidase activity. *Table 19* shows the influence of dimethylsulphoxide (DMSO) on this dual activity: increasing DMSO concentration increases esterase activity and decreases amidase activity (Pal and Gertler, 1983). This is an example wherein the catalytic capacity of the enzyme in an aqueous reaction mixture can be increased by addition of an organic solvent.

The effects of solvents on biocatalysis are complex, but promise exciting developments in industrial applications.

METAL EXCHANGE AT ACTIVE SITES OF ENZYMES

In chemical catalysis, it is legend to believe that to change the catalyst, you change the metal. In biocatalysis, this concept is in its infancy. There are provocative data in the literature to show that this underdeveloped approach to enzyme modification and improvement deserves more attention. Examples will be given to illustrate metal exchange as a means of altering enzyme specificity, stability and inhibition.

Like α-thrombin, carboxypeptidase A is an enzyme with two catalytic activities —esterase and peptidase. In the case of α-thrombin, esterase activity is increased and amidase activity decreased in the presence of increasing concentrations of DMSO. In the case of carboxypeptidase A, esterase activity may be increased, and peptidase activity decreased by metal exchange. *Table 20* shows that when zinc, the natural metal component, is replaced by cobalt, both esterase and

Table 17. Olive oil triglyceride interesterification with a 1,3-specific lipase (EC 3.1.1.3) of *Rhizopus delemar* and stearic acid (5:1, w/w)

Fatty acid	Total triglycerides		Positions 1, 3		Position 2	
	Olive oil (wt %)	Interesterified oil (% change)	Olive oil (wt %)	Interesterified oil (% change)	Olive oil (wt %)	Interesterified oil (% change)
16:0	16·6	−2·9	23·2	−4·3	3·5	−0·3
16:1	1·8	−0·2	2·0	−0·4	1·3	+0·3
18:0	2·0	+13·6	2·5	+20·5	1·0	−0·3
18:1	66·8	−10·2	64·2	−15·4	72·0	+0·2
18·2	12·8	−0·2	8·1	−0·4	22·2	+0·1

Table 18. Olive oil triglyceride interesterification using the linoleic acid specific lipase† of *Geotrichum candidum* with stearic acid, and linoleic acid (1·0:0·15:0·15, w/w/w).

Fatty acid	Fatty acid content of olive oil (%)	Change in interesterified oil product (%)
16:0	11·6	−0·1
16:1	0·8	+0·3
18:0	3·6	+0·9
18:1	72·8	−8·0
18:2	10·6	+7·7
20:1	0·6	−0·2

† Triacylglycerol lipase, EC 3.1.1.3.

Table 19. DMSO effects on the enzyme activity of bovine α-thrombin†

	Activity (%)	
DMSO (% v/v)	Amidase‡	Esterase§
0	100	100
5	60	135
10	30	175
15	30	190
20	9	200

† EC 3.4.21.5.
‡ 0·2 mM *N*-Benzoyl-L-phenylalanyl-L-valyl-arginine-*p*-nitroanilide HCl.
§ 1 mM *p*-Tosyl-L-arginine methyl ester HCl.

Table 20. Effects of metal exchange on the enzyme activity of bovine carboxypeptidase A†

	Relative rates	
Metal	Esterase‡	Peptidase§
Apo (−metal)	0	0
Zinc	100	100
Cobalt	114	200
Nickel	43	47
Manganese	156	27
Cadmium	143	0
Mercury	86	0
Rhodium	71	0
Lead	57	0
Copper	0	0

† EC 3.4.17.1.
‡ 0·01 M-benzoylglycyl-DL-phenyllactate, pH 7·5, 25°C.
§ 0·02 M-benzyloxycarbonylglycyl-L-phenylalanine, pH 7·5, 0°C.

peptidase activities are increased. When zinc is replaced by manganese or cadmium, esterase activity is increased, while peptidase activity is markedly reduced or vanishes completely (Vallee, 1980).

A somewhat more subtle effect of metal exchange is illustrated in *Table 21.* The cases of α-thrombin and carboxypeptidase A involve an alteration in general classes of enzyme activity: esterase, peptidase and amidase. In the case of the aminoacylase, substitution of zinc by cobalt has an effect on substrate specificity in one type of hydrolytic activity, namely deacylation. The data show that

Table 21. Metal exchange effects on substrate specificity of the aminoacylase of *Aspergillus oryzae*

Substrates	Relative activity (%)	
	Co^{2+}	Zn^{2+}
N-Chloro-acetyl-Ala	127	100
N-Chloro-acetyl-Met	163	286
N-Chloro-acetyl-Norleu	191	326
N-Chloro-acetyl-Leu	153	60
N-Chloro-acetyl-Phe	195	218
N-Acetyl-Glu	2·5	1
N-Acetyl-Gln	59	12
N-Acetyl-Ala	39	16·5
N-Acetyl-Lys	6·5	1

substituting cobalt for zinc increases activity against six substrates and decreases activity versus three (Gilles, Loffler and Schneider, 1981); depending on the specific substrate, therefore, either the zinc or the cobalt enzyme might be the choice. Furthermore, the pH optimum of the zinc enzyme towards *N*-chloroacetyl-alanine is 8·5, whereas that of the cobalt enzyme is 7·0.

Table 22. Metal exchange effects in iodosylbenzene-supported oxygenation with cytochrome $P450_{cam}$

Cyt $P450_{cam}$	Reactions	Substrates	Relative activity (%)
Fe(III)	5-hydroxylation	Camphor	100
Fe(III)	5,6-epoxidation	Dehydrocamphor	100
Mn(III)	5-hydroxylation	Camphor	$<0{\cdot}5$
Mn(III)	5,6-epoxidation	Dehydrocamphor	53

The examples just discussed are concerned with hydrolytic enzymes. The effects of metal exchange are not limited to this enzyme class. In *Table 22* two different oxidative activities of cytochrome $P450_{cam}$ are considered, namely 5-hydroxylation of camphor and 5,6-epoxidation of dehydrocamphor. Whereas the Fe(III) enzyme (the natural enzyme) carries out both reactions, the Mn(III) enzyme catalyses only the latter (Gelb, Toscano and Sligan, 1982). In a hypothetical case, if one had a substrate that could be hydroxylated and epoxidized but only the epoxide product was desired, the side-reaction (the hydroxylation) could be eliminated by constructing and using the Mn(III) enzyme.

The case of substituting manganese for iron in a superoxide dismutase of *Bacteroides fragilis* (*Table 23*) illustrates a remarkable alteration in enzyme

Table 23. Metal exchange effects on the superoxide dismutase of *Bacteroides fragilis*.

Superoxide dismutase (SOD)	Effect of 5 mM H_2O_2 on half-life (min)	Inhibition by mM NaN_3 (%)		
		0·2	1·0	20·0
Native Fe SOD	4	56	89	—
Reconstituted Fe SOD	5	51	84	—
Reconstituted Mn SOD	>20	—	0	45

stability and inhibition. The natural and reconstituted iron-containing enzymes are considerably more sensitive to sodium azide (NaN_3) inhibition and hydrogen peroxide instability than the reconstituted manganese-containing enzyme. This effect mirrors the difference noted in naturally occurring iron and manganese variants of superoxide dismutase (EC 1.15.1.1): a satisfactory agreement between two 'laboratories', the natural and the academic (Gregory and Dapper, 1983).

It is clear from these dramatic alterations in biocatalytic properties by metal exchange at the active site that more attention should be given to the technique.

EFFECT OF TEMPERATURE ON ENZYMATIC UNSATURATION OF LIPIDS

The production of saturated and unsaturated wax esters by *Acinetobacter* sp. HO1-N was discussed above. The capacity of the micro-organism to vary dramatically the level of unsaturation can be enhanced by imposing a temperature stress on the biosynthetic process. There is in nature a common and inverse relationship between lipid unsaturation and temperature: high temperature leads to low unsaturation; low temperature leads to high unsaturation. The major function of this relationship is to ensure membrane function and fluidity over a range of temperature conditions. *Table 24* illustrates that the wax esters

Table 24. Effect of temperature on unsaturation of wax esters produced from ethanol and *n*-eicosane by *Acinetobacter* sp. HO1-N

Temperature (°C)	Di-ene fractions (Total)	Mono-ene fractions (Total)	Saturated fractions (Total)
Ethanol	(32:2, 34:2, 36:2)	(32:1, 34:1, 36:1)	(32:0, 34:0, 36:0)
17	72%	18%	10%
24	29%	40%	31%
30	9%	25%	66%
n-Eicosane	(36:2, 38:2, 40:2)	(36:1, 38:1, 40:1)	(36:0, 38:0, 40:0)
17	71%	24%	5%
24	41%	37%	22%
30	35%	40%	25%

produced by *Acinetobacter* sp. HO1-N from *n*-eicosane and ethanol at 17°C are much more unsaturated than those produced at 30°C (Ervin *et al.*, 1983). This response may be mediated through temperature-sensitive desaturases, but the mechanism is not yet clarified. However, even in the absence of a biochemical explanation, the technique can, in appropriate situations, give very high levels of specifically unsaturated lipid derivatives.

EFFECTS OF SUBSTRATE (HALIDE ION) CONCENTRATION ON PRODUCT CHEMISTRY

It was noted in *Figures 7* and *8* that the nature of the halogenated products synthesized by haloperoxidases (E.C. 1.11.1.10) from allyl alcohol depends upon halide ion concentration, among other things. At low millimolar levels of activatable halide ion, the major products are halohydrin, whereas at molar levels the product is the homogeneous dihalide. It was further indicated that in the presence of two different halide ions, heterogeneous (mixed) dihalide derivatives could be formed. The requirements for this reaction include the following: (1) at least one of the halide ions must be activatable by the chosen haloperoxidase; (2) the most readily activated halide ion should be present at millimolar concentration with the less or non-activatable halide ion at molar concentrations; and (3) the ratio of the lesser to the more activated halide ion should be > 100. Halide ions can be arranged in the following order according to their ease of activation: I > Br > Cl ⋙ F^-. In fact, F^- is not activated by any known haloperoxidase.

Taking advantage of these various factors, some unusual heterogeneous dihalide derivatives of allyl alcohol can be synthesized, as illustrated in *Table 25*. Some of these compounds are new compositions of matter, showing that biocatalysis can yield truly novel products. The fluorinated derivative represents the first C–F bond synthesized in an enzyme-associated reaction with a purified biocatalyst (Neidleman and Geigert, 1983).

There will be many examples in the future wherein the nature of the end-product of the enzyme-catalysed reaction will depend upon the substrate preference of the enzyme and the relative concentrations of 'competing' substrates.

Table 25. Various halogenated products formed from allyl alcohol† by haloperoxides as a function of halide concentrations and ratios

Enzyme	mM KBr	mM KCl	mM KI	mM KF	Major product
Chloroperoxidase	17	—	—	—	$CH_2(OH)$—CH(Br)—$CH_2(OH)$ ‡
Chloroperoxidase	3389	—	—	—	$CH_2(OH)$—CH(Br)—$CH_2(Br)$
Chloroperoxidase	—	200	—	—	$CH_2(OH)$—CH(Cl)—$CH_2(OH)$ ‡
Chloroperoxidase	—	2000	—	—	$CH_2(OH)$—CH(Cl)—$CH_2(Cl)$
Chloro- or lacto-peroxidase	20	2000	—	—	$CH_2(OH)$—CH(Br)—$CH_2(Cl)$ ‡
Chloroperoxidase	—	2000	20	—	$CH_2(OH)$—CH(I)—$CH_2(Cl)$ ‡
Horseradish peroxidase	2000	—	20	—	$CH_2(OH)$—CH(I)—$CH_2(Br)$ ‡
Horseradish peroxidase	—	—	20	2000	$CH_2(OH)$—CH(I)—$CH_2(F)$ ‡,§

† $CH_2(OH)$—CH=CH_2 $\xrightarrow[H_2O_2,\ halides]{Haloperoxidase}$ Various products

‡ Positional isomers in 1 : 1 ratio.

§ Actually minor product (10% yield); major product: $CH_2(OH)$—CH(I)—$CH_2(OH)$

ENZYME VARIANTS FROM NATURE, RANDOM MUTATION, GENETIC ENGINEERING AND CHEMICAL MODIFICATION

The preparation of enzyme variants by metal exchange at the active site has already been discussed. Other ways to isolate or prepare modified enzymes exist, including: (1) screening for enzyme variants in nature; (2) creating new enzymes by random mutation; (3) constructing alternative biocatalysts through genetic engineering and site-specific mutation; and (4) chemically modifying the enzyme.

Nature has an enormous reservoir of variants for many classes of enzymes. *Table 5* shows variant haloperoxidases. *Table 26* illustrates variants of superoxide dismutases (Neidleman, 1983). Other enzymes that show variability in structure and properties include α-amylase (Fogarty, 1983b), glucose isomerase (Bucke, 1983), pectinolytic enzymes (Fogarty and Kelly, 1983), cellulases (Enari, 1983), lipases (Macrae, 1983a, b) and proteinases (Ward, 1983). A particularly interesting example of an enzyme variant is that of a non-haem, manganese-containing catalase of *Lactobacillus plantarum*. Traditional catalase is an iron–haem enzyme. In this case, two markedly different enzymes catalyse the same reaction—the degradation of hydrogen peroxide (Kono and Fridovich, 1983a, b).

Table 26. Variants of superoxide dismutases.

Metal	Source
Fe	*Rhizobium japonicum*
Mn	Pea leaf
	Escherichia coli
Mn/Fe	*Rhodococcus bronchialis*
Cu/Zn	Vertebrates
	Green pea
	Neurospora crassa
	Saccharomyces cerevisiae
Fe/Zn	*Thermoplasma acidophilum*

$O_2^- + O_2^- + 2H^+ \longrightarrow O_2 + H_2O_2$.

In addition to the fact that the search for, and the discovery and characterization of, variant enzymes may offer solutions to problems in industrial biocatalysis, information derived from these studies will supply structure–activity data that will aid in the directed synthesis of tailored enzymes by protein or genetic engineering as well as by chemical modification (Ulmer, 1983). The design of optimized enzymes will include consideration of thermostability, turnover number, substrate specificity, product inhibition, cofactor requirements, pH optimum, size and stability to pH, salts and other enzymes.

Table 27 gives some examples of enzyme alteration by random mutation and site-specific genetic engineering. It should be noted that the effect is not always in a positive direction, but even the negative effects on activity afford useful insights for future studies.

The effects of chemical modification on the catalytic activity of various enzymes are illustrated in *Table 28*. In each of these four cases, a positive change is affected either by increasing the activity of the biocatalyst or by initiating a new catalytic function.

Table 27. Enzyme modifications by mutation and genetic engineering.

Enzyme	Source	Modification†	Effect	Reference
Aspartate transcarbamoylase (aspartate carbamoyltransferase), EC 2.1.3.2	*E. coli* HS513	Change of glycine^{125} to aspartic acid at position other than active site by mutation	Loss of activity, binding of catalytic and regulatory subunits weakened	Kim *et al.*, 1981
β-Lactamase (penicillinase), EC 3.5.2.6	*E. coli* K12	(1) Serine^{70} to cysteine by recombinant DNA	Decreased enzyme activity	Sigal, Harwood and Arentzen, 1982.
		(2) Serine^{70} to threonine and threonine^{71} to serine	Inactive enzyme	Dalbadie-McFarland *et al.*, 1982
Tyrosyl-tRNA-synthetase, EC 6.1.1.1	*Bacillusstearothermophilus* cloned into *E. coli*	Cysteine^{35} to serine	K_m for ATP lowered	Winter *et al.*, 1982.
Xanthine dehydrogenase (purine hydroxylase I), EC 1.2.1.37	*A. nidulans*	Alteration in relative positions of catalytic and orienting sites through mutation	Change in substrate specificity	Scazzocchio and Sealy-Lewis, 1978

† Superscripts indicate the positions of the amino acids in the protein.

Table 28. Effect of chemical modification of enzymes on catalytic activity.

Enzyme	Chemical modification	Effect	Reference
α-Amylase (*B. subtilis*), EC 3.2.1.1	Tryptophan at active site modified	Greater % of maltose and glucose in product	Hollo, Laszlo and Hoschke, 1983
α-Chymotrypsin, EC 3.4.21.1	Pyridoxal added to the reaction mixture	Allows for hydrolysis of D-aromatic amino esters with a free amino group; decreases similar activity against L-enantiomers (the 'normal' substrate)	Kraicsovits, Previero and Otvos, 1981
Papain (pineapple), EC 3.4.22.2	Alkylation of cysteine at active site with 7α-bromoacetyl-10-methylisoalloxazine	Initiates oxidation of dihydronicotinamides	Levine and Kaiser, 1978
Rennet (chymosin) (*Mucor pusillus*), EC 3.4.23.4	Acylation with various anhydrides	>100% increase in milk coagulating activity	Cornelius, Asmus and Sternberg, 1982

The few examples given as illustrations serve to indicate that both the discovery and the synthesis of variant enzymes will be rich sources of novel biocatalysts, differing in properties from the currently available selection of enzymes. One can anticipate that, with an increasing knowledge of structure–activity relationships in enzymes, there will come an era of 'enzymes-to-order', optimized to the needs of specific processes.

Further, it is challenging to realize that 'non-enzymes' can also carry out catalytic reactions under certain conditions. Bovine serum albumin has been shown to cause hydrolysis of *p*-nitrophenyl α-methoxyphenylacetates (Kokubo *et al.*, 1982). The L-enantiomer is hydrolysed three times faster than the D-enantiomer. The level of biocatalytic activity is considerably less than that of hydrolytic enzymes. Another instance of 'non-enzymatic' catalysis is the hydroxylation of aniline by immobilized haemoglobin in the presence of riboflavin and NADH. Activity similar to that with cytochrome P450 was obtained (Guillochon *et al.*, 1982).

Is it possible that chemical or genetic modification of common proteins such as these can produce economically and commercially significant catalysts?

ENZYME IMMOBILIZATION

Discussing the diverse advantages and techniques of immobilizing enzymes or their producing cells on solid supports for industrial applications is beyond the scope of this review. Numerous useful references are available (Keyes, 1980; Trevan, 1980; Klibanov, 1983). Five general methods for immobilization are available: these are covalent attachment, adsorption, encapsulation, entrapment and cross-linking. Each method may have desirable advantages over other procedures, dependent on whether an enzyme or a cell is being immobilized, and on the cost–benefit picture for the process under consideration.

The major advantages that may result from immobilization are: (1) the biocatalytic system may be readily separated from the products and reused; and (2) immobilization often affords improved enzyme stability against temperature, pH and other environmental denaturants. These effects allow for an increase in the useful lifetime for enzymes (*see also* Chapter 5).

Future developments in immobilization will allow for a more interactive role between the enzyme and the immobilization matrix. The chemical and physical nature of the solid matrix will be modified to afford additional stabilization effects, for example, hydrogen peroxide degradation to protect sensitive oxidases, and microenvironments to favour enzyme reactions in solvents. Heterogeneous matrices, allowing a mosaic of hydrophilic and hydrophobic environments, will allow for the simultaneous or sequential immobilization of multi-enzyme systems to produce a variety of chemicals. Sophisticated immobilization methods will serve as an adjunct in optimizing the use of the enzyme variants discussed above.

Desirable enzyme improvements

As nothing is perfect, it is illogical to expect that for each process there exists an enzyme which is ideal in the eyes of those concerned. Some enzymes approach

Table 29. Some suggested enzyme improvements from the literature

Enzyme	Applications	Useful improvement	Reference
α-Amylase	Detergent	Alkali and bleach resistance	Barfoed, 1983.
Amyloglucosidase	High fructose corn syrup	Immobilized, higher productivity	Reichelt, 1983
Esterases, lipases, proteases	Flavour development	More specificity in flavour development	Godfrey, 1983b.
Glucose isomerase	High fructose corn syrup	Increased thermostability, lower pH optimum	Hollo, Laszlo and Hoschke, 1983.
Glucose-1-oxidase and other oxido-reductases	Diagnostic assay, food quality, dentifrice	Increased stability to hydrogen peroxide	Schmid, 1979b; Richter, 1983
Limoninase (limonin-D-ring lactonase), EC 3.1.1.36	Fruit juice debittering	More complete degration of limonin	Janda, 1983.
Protease	Detergent	Bleach resistance	Starace, 1983; Bucke, 1983.
Protease	Chill-proofing beer	More specific enzymes	Godfrey, 1983a.
Pullulanase, debranching enzyme	High fructose corn syrup	Thermostable	Reichelt, 1983.

the ideal more closely than others. *Table 29* offers a selection of desirable enzyme improvements drawn from recent literature. The Table suggests that the methods discussed above for generating enzyme variants can be applied to numerous examples, and that much remains to be accomplished in biocatalysis.

Conclusion

This review has attempted to present an overview of many of the contact points between industry and biocatalysis, and it seems clear that these are both many and varied. The range is from the subtlety of food aroma and texture to the extraordinary specificity of enzymes in the synthesis of complex organic chemicals, and from the much-acclaimed benefits of therapeutic enzymes to the more down-to-earth necessities of waste disposal.

A most significant development, growing more pervasive in biocatalytic research and applications each year, is the realization that enzymes need not be confined to aqueous reaction conditions, room temperature, and atmospheric pressure. They can tolerate more hostile environments: solvents, temperatures greater than 100°C, and pressures far above atmospheric. Note the recent report (Baross and Deming, 1983) of micro-organisms functioning at 250°C, at a pressure of 265 atmospheres (26·5 MPa) and at the salt concentrations of sea water. Enzymes are not as fragile as has been thought, and will often do new 'tricks' in adverse surroundings.

It is also clear that there is a vast reservoir of untapped, undiscovered and unsynthesized enzyme variants with properties to match most process requirements. A search in nature, the use of chemical modification, and the application of the techniques of mutation and genetic and protein engineering will find or construct the answer to a process engineer's dream.

Finally, it is heartening to note that existent enzymes are not perfect, not all problems have been solved, and that an abundance of exciting and profitable research remains to challenge the willing scientist.

References

ANONYMOUS (1982). New help for slipped discs. *Time, 6 December* p. 86.

ANTONINI, E., ARREA, G. AND CREMONESI, P. (1981). Enzyme catalyzed reactions in water-organic solvent two-phase systems. *Enzyme and Microbial Technology* **3,** 291–296.

ARIMA, K. (1977). Recent developments and future direction of fermentations in Japan. *Developments in Industrial Microbiology* **18,** 78–117.

ATKINSON, B. (1983). Requirements for innovatory research in support of malting and brewing. *Journal of the Institute of Brewing* **89,** 160–163.

AUNSTRUP, K. (1978). Enzymes of industrial interest: traditional products. In *Annual Reports on Fermentation Processes* (D. Perlman and G. T. Tsao, Eds), pp. 125–154. Academic Press, New York.

BARFOED, H.C. (1983). Detergents. In *Industrial Enzymology* (T. Godfrey and J. Reichelt, Eds), pp. 284–293. The Nature Press, New York.

BARKER, S.A. (1983). New approaches to enzyme stabilisation. In *Topics in Enzyme and Fermentation Biotechnology* (A. Wiseman, Ed.), volume 7, pp. 68–78. Ellis Horwood, Chichester.

BAROSS, J.A. AND DEMING, J.W. (1983). Growth of 'black smoker' bacteria at temperatures of at least 250°C. *Nature* **303**, 423–426.

BELL, G., BLAIN, J.A., PATTERSON, J.D.E., SHAW, C.E.L. AND TODD, R. (1978). Ester and glyceride synthesis by *Rhizopus arrhizus* mycelia. *FEMS Microbiology Letters* **3**, 223–225.

BIRCH, G.G., BLAKEBROUGH, N. AND PARKER, K.J., EDS (1981). *Enzymes and Food Processing*. Applied Science, London.

BRUTON, C.J. (1983). Large-scale purification of enzymes. *Philosophical Transactions of the Royal Society of London* **B300**, 249–261.

BUCKE, C. (1983). Glucose-transforming enzymes. In *Microbial Enzymes and Biotechnology* (W.M. Fogarty, Ed.), pp. 93–129. Applied Science, New York.

CALLELY, A.G., FORSTER, C.F. AND STAFFORD, D.A. (1976). *Treatment of Industrial Effluents*. John Wiley & Sons, New York.

CHARNEY, W. AND HERZOG, H.L. (1967). *Microbial Transformations of Steroids*. Academic Press, New York.

CHIBATA, I., TOSA, T. AND SATO, T. (1983). Immobilized cells in preparation of fine chemicals. *Advances in Biotechnological Processes* **1**, 203–222.

COONEY, D.A., STERGIS, G. AND JAYARAM, H.M. (1980). Enzymes, Therapeutic. In *Kirk-Othmer Encyclopedia of Chemical Technology*, volume 9, pp. 225–240. John Wiley & Sons, New York.

CORNELIUS, D.A., ASMUS, C.V. AND STERNBERG, M.M. (1982). Acylation of *Mucor pusillus* microbial rennet enzyme. United States Patent No. 4,362,818.

COWAN, D. (1983). Proteins. In *Industrial Enzymology* (T. Godfrey and J. Reichelt, Eds), pp. 352–374. The Nature Press, New York.

CROUZET, J. (1977). La regeneration enzymatique des aromes. *Bios* **8**, 29–35.

DALBADIE-MCFARLAND, G., COHEN, L.W., RIGGS, A.D., MORIN, C., ITAKURA, K. AND RICHARDS, J.H. (1982). Oligonucleotide-directed mutagenesis as a general and powerful method for studies of protein function. *Proceedings of the National Academy of Sciences of the United States of America* **79**, 6409–6413.

DEWITT, S., ERVIN, J.L., HOWES-ORCHISON, D., DALIETOS, D. AND NEIDLEMAN, S.L. (1982). Saturated and unsaturated wax esters produced by *Acinetobacter* sp. HO1-N grown on C_{16}–C_{20} *n*-alkanes. *Journal of the American Oil Chemists' Society* **59**, 69–74.

DIXON, M. AND WEBB, E.C. (1979). *Enzymes*. Academic Press, New York.

DUNNILL, P. (1980). The current status of enzyme technology. In *Enzymatic and Non-enzymatic Catalysis* (P. Dunnill, A. Wiseman and N. Blakebrough, Eds), pp. 28–53. Ellis Horwood, Chichester.

ENARI, T.M. (1983). Microbial cellulases. In *Microbial Enzymes and Biotechnology* (W.M. Fogarty, Ed.), pp. 183–223. Applied Science, New York.

ERVIN, J.L., GEIGERT, J., NEIDLEMAN, S.L. AND WADSWORTH, J. (1983). Substrate-dependent and growth temperature-dependent changes in the wax ester compositions produced by *Acinetobacter* sp. HO1-N. *Journal of the American Oil Chemists' Society* (in press).

FELIX, R. AND VILLETAZ, J.C. (1983). Wine. In *Industrial Enzymology* (T. Godfrey and J. Reichelt, Eds), pp. 410–421. The Nature Press, New York.

FOGARTY, W.M. (1983a). *Microbial Enzymes and Biotechnology*. Applied Science, New York.

FOGARTY, W.M. (1983b). Microbial amylases. In *Microbial Enzymes and Biotechnology* (W.M. Fogarty, Ed.), pp. 1–92. Applied Science, New York.

FOGARTY, W.M. AND KELLY, C.T. (1983). Pectic enzymes. In *Microbial Enzymes and Biotechnology* (W.M. Fogarty, Ed.), pp. 131–182. Applied Science, New York.

FUKUI, S. AND TANAKA, A. (1982). Bioconversion of lipophilic or water-insoluble compounds by immobilized biocatalysts in organic solvent systems. In *Enzyme Engineering* (I. Chibata, S. Fukui and L.B. Winegard, Jr, Eds), volume 6, pp. 191–200. Plenum Press, New York.

FULCO, A.J. (1977). Fatty acid desaturation in microorganisms. In *Polyunsaturated Fatty Acids* (W.H. Kunau and R.T. Holman, Eds), pp. 19–36. American Oil Chemists' Society, Champaign.

GEIGERT, J., NEIDLEMAN, S.L. AND DALIETOS, D.J. (1983). Novel haloperoxidase substrates: alkynes and cyclopropanes. *Journal of Biological Chemistry* **258,** 2273–2277.

GEIGERT, J., NEIDLEMAN, S.L. AND HIRANO, D.S. (1983). Convenient laboratory procedure for reducing D-glucosone to D-fructose. *Carbohydrate Research* **13,** 159–162.

GEIGERT, J., DALIETOS, D.J., NEIDLEMAN, S.L., LEE, T.D. AND WADSWORTH, J. (1983a). Peroxide oxidation of primary alcohols to aldehydes by chloroperoxidase catalysis. *Biochemical and Biophysical Research Communications* **114,** 1104–1108.

GEIGERT, J., NEIDLEMAN, S.L., DALIETOS, D.J. AND DEWITT, S.K. (1983b). Haloperoxidases: Enzymatic synthesis of α,β-halohydrins from gaseous alkenes. *Applied and Environmental Microbiology* **45,** 366–374.

GEIGERT, J., NEIDLEMAN, S.L., DALIETOS, D.J. AND DEWITT, S.K. (1983c). Novel haloperoxidase reaction: synthesis of dihalogenated products. *Applied and Environmental Microbiology* **45,** 1575–1581.

GEIGERT, J., NEIDLEMAN, S.L., HIRANO, D.S., WOLF, B. AND PANSCHAR, B.M. (1983d). Enzymatic oxidation of D-arabino-hexos-2-ulose (D-glucosone) to D-arabino-2-hexulosonic acid ('2-keto-D-gluconic acid'). *Carbohydrate Research* **113,** 163–165.

GEIGERT, J., NEIDLEMAN, S.L., LIU, T.E., DEWITT, S.K., PANSCHAR, B.M., DALIETOS, D.J. AND SIEGEL, E.R. (1983e). Production of epoxides from α, β-halohydrins by *Flavobacterium* sp. *Applied and Environmental Microbiology* **45,** 1148–1149.

GELB, M.H., TOSCANO, W.A., JR. AND SLIGAN, S.G. (1982). Preparation and properties of manganese-substituted cytochrome $P\text{-}450_{cam}$. In *Cytochrome P-450, Biochemistry, Biophysics and Environmental Implications* (E. Hietanen, M. Laitenen and O. Hanninen, Eds), pp. 573–580. Elsevier Biomedical Press, Amsterdam.

GILLES, I., LOFFLER, H.G. AND SCHNEIDER, F. (1981). Co^{+2}-substituted acylamino acid amido hydrolase from *Aspergillus oryzae. Zeitschrift für Naturforschung* **36C,** 751–754.

GODFREY, T. (1983a). Brewing. In *Industrial Enzymology* (T. Godfrey and J. Reichelt, Eds). pp. 221–259. The Nature Press, New York.

GODFREY, T. (1983b). Effluent, by-product and biogas. *Ibid*, pp. 294–304.

GODFREY, T. (1983c). Flavouring and colouring. *Ibid*, pp. 305–314.

GODFREY, T. AND REICHELT, J. (1983a). *Industrial Enzymology.* The Nature Press, New York.

GODFREY, T. AND REICHELT, J. (1983b). Leather. *Ibid*, pp. 321–330.

GODFREY, T. AND REICHELT, J. (1983c). Textiles. *Ibid*, pp. 397–409.

GODFREY, T. AND REICHELT, J. (1983d). A guide to the use of enzymes in the more common industrial areas. *Ibid*, pp. 549–551.

GRAY, P.P. AND TRIBE, D.E. (1979). The commercial production of microbial enzymes. Part 2. Production processes. *Food Technology in Australia* **31,** 254–260.

GREGORY, E.M. AND DAPPER, C.H. (1983). Isolation of iron-containing superoxide dismutase from *Bacteroides fragilis:* Reconstitution as a Mn-containing enzyme. *Archives of Biochemistry and Biophysics* **220,** 293–300.

GUILLOCHON, D., LUDOT, J.M., ESCLADE, L., CAMBOU, B. AND THOMAS, D. (1982). Hydroxylase activity of immobilized haemoglobin. *Enzyme and Microbial Technology* **4,** 96–98.

HAGER, L.P. (1982). Mother nature likes some halogenated compounds. *Basic Life Sciences* **19**, 415–428.

HEATH, H.B. (1981). Flavours in food products—art or science? *Journal of the Royal Society of Health* **101,** 6–12.

HOLLO, J., LASZLO, E. AND HOSCHKE, A. (1983). Enzyme engineering in starch industry. *Stärke* **35,** 169–175.

INTERNATIONAL UNION OF BIOCHEMISTRY (1979). *Enzyme Nomenclature. Recommendations (1978) of the Nomenclature Committee of the International Union of Biochemistry.* Academic Press, New York and London.

JANDA, W. (1983). Fruit juice. In *Industrial Enzymology* (T. Godfrey and J. Reichelt, Eds), pp. 315–320. The Nature Press, New York.

JANSHEKAR, H. AND FIECHTER, A. (1982). Biochemical Engineering. *Topics in Current Chemistry* **100,** 97–126.

JONES, J.B. (1980). Enzymes in synthetic organic chemistry. In *Enzymatic and Non-enzymatic Catalysis* (P. Dunnill, A. Wiseman and N. Blakebrough, Eds), pp. 54–83. Ellis Horwood, Chichester.

KATCHALSKI-KATZIR, E. AND FREEMAN, A. (1982). Enzyme engineering reaching maturity. *Trends in Biochemical Sciences* **7,** 427–431.

KEYES, M.H. (1980). Enzymes, Immobilized. In *Kirk-Othmer Encyclopedia of Chemical Technology*, volume 9, pp. 148–172. John Wiley & Sons, New York.

KIESLICH, K. (1976). *Microbial Transformations of Non-Steroid Cyclic Compounds.* John Wiley & Sons, New York.

KIESLICH, K. (1980). Steroid conversions. In *Microbial Enzymes and Bioconversions* (A.H. Rose, Ed.), pp. 370–465. Academic Press, New York.

KIESLICH, K. (1982) Practical applications of microbial transformations in the synthesis of natural compounds or analogues. In *Overproduction of Microbial Products* (V. Kreumphanzyl, B. Sikyta and Z. Vanek, Eds), pp. 311–324. Academic Press, New York.

KIM, R., YOUNG, T.S., SCHACHMAN, H.K. AND KIM, S.H. (1981). Crystallization and preliminary x-ray diffraction studies of an inactive mutant aspartate trans-carbamoylase from *Escherichia coli. Journal of Biological Chemistry* **256,** 4691–4692.

KINDEL, S. (1981). Enzymes, the bio-industrial revolution. *Technology* **1,** 62–74.

KLIBANOV, A.M. (1979). Enzyme stabilization by immobilization. *Analytical Biochemistry* **93,** 1–25.

KLIBANOV, A.M. (1982). Enzymatic removal of hazardous pollutants from industrial aqueous effluents. In *Enzyme Engineering* (I. Chibata, S. Fukui and B. Winegard, Jr, Eds), volume 6, pp. 319–324. Plenum Press, New York.

KLIBANOV, A.M. (1983). Immobilized enzymes and cells as practical catalysts. *Science* **219,** 722–727.

KLIBANOV, A.M., SAMOKHIN, G.P., MARTINEK, K. AND BEREZIN, I.V. (1977). A new approach to preparative enzymatic synthesis. *Biotechnology and Bioengineering* **19,** 1351–1361.

KLIBANOV, A.M., TU, T.M. AND SCOTT, K.P. (1983). Peroxidase-catalyzed removal of phenols from coal-conversion waste waters. *Science* **221,** 259–261.

KOKUBO, T., UCHIDA, T., TANIMOTO, S., OKANO, M. AND SUGIMOTO, T. (1982). A hydrolytic enzymelike behavior of bovine serum albumin in hydrolysis of *p*-nitrophenyl esters. *Tetrahedron Letters* **23,** 1593–1596.

KONO, Y. AND FRIDOVICH, I. (1983a). Isolation and characterization of the pseudo-catalase of *Lactobacillus plantarum.* A new manganese-containing enzyme. *Journal of Biological Chemistry* **258,** 6015–6019.

KONO, Y. AND FRIDOVICH, I. (1983b). Functional significance of manganese catalase in *Lactobacillus plantarum. Journal of Bacteriology* **155,** 742–746.

KRAICSOVITS, F., PREVIERO, A. AND OTVOS, L. (1981). The pyridoxal effect on the specificity of α-chymotrypsin. *Bulgarian Academy of Sciences* **32,** 386–390.

LAMBERT, P.W. AND MEERS, J.L. (1983). The production of industrial enzymes. *Philosophical Transactions of the Royal Society of London* **B300,** 263–282.

LANGLEY, T.J. (1983). Analytical applications. In *Industrial Enzymology* (T. Godfrey and J. Reichelt, Eds), pp. 194–209. The Nature Press, New York.

LEVINE, H.L. AND KAISER, E.T. (1978). Oxidation of dihydronicotinamides by flavopapain. *Journal of the American Chemical Society* **100,** 7670–7767.

LILLY, M.D. (1983). Two-liquid phase biocatalytic reactors. *Philosophical Transactions of the Royal Society of London* **B300,** 391–398.

LINDEN, J.C. AND MOREIRA, A. (1983). Anaerobic production of chemicals. In *Basic Biology of New Developments in Biotechnology* (A. Hollaender, A.I. Laskin and P. Rogers, Eds), pp. 377–403. Plenum Press, New York.

LOWE, C.R. (1983). The application of coenzyme-dependent enzymes in biotechnology. *Philosophical Transactions of the Royal Society of London* **B300,** 335–353.

MACRAE, A.R. (1983a). Lipase-catalyzed interesterification of oils and fats. *Journal of the American Oil Chemists Society* **60,** 291–294.

MACRAE, A.R. (1983b). Extracellular microbial lipases. In *Microbial Enzymes and Biotechnology* (W.M. Fogarty, Ed.), pp. 225–250. Applied Science, New York.

MARSCHALL, J.J., ALLEN, W.G., DENAULT, L.J., GENISTER, P.R. AND POWER, J. (1982). Enzymes in brewing. *Brewers Digest* **57,** 14–22.

MARTINEK, K. AND SEMENOV, A.N. (1981). Enzymes in organic synthesis: Physicochemical means of increasing the yield of end products in biocatalysis. *Journal of Applied Biochemistry* **3,** 93–126.

MARTINEK, K., SEMENOV, A.N. AND BEREZIN, I.V. (1981). Enzymatic synthesis in biphasic aqueous-organic systems 1. Chemical equilibrium shift. *Biochimica et biophysica acta* **658,** 76–89.

MOO-YOUNG, M. (1983). *Biotechnology and Waste Treatment.* Pergamon Press, Oxford.

MORRISON, M. AND SCHONBAUM, G.R. (1976). Peroxidase-catalyzed halogenation. *Annual Reviews of Biochemistry* **45,** 861–888.

NEIDLEMAN, S.L. (1975). Microbial halogenation. In *CRC Critical Reviews in Microbiology* (A.I. Laskin, Ed.), pp. 333–358. CRC Press, Cleveland.

NEIDLEMAN, S.L. AND GEIGERT, J. (1983). The enzymatic synthesis of heterogeneous dihalide derivatives: A unique biocatalytic discovery. *Trends in Biotechnology* **1,** 21–25.

NEIDLEMAN, S.L. (1983). Aspects of enzyme catalysis. In *Proceedings, Ninth Conference on Catalysis of Organic Reactions, Charleston, SC, 1982* (in press).

NORMAN, B.E. (1982). A novel debranching enzyme for application in the glucose syrup industry. *Stärke* **34,** 340–346.

PAL, P.K. AND GERTLER, M.M. (1983). The catalytic activity and physical properties of bovine thrombin in the presence of dimethyl sulfoxide. *Thrombosis Research* **29,** 175–185.

PATTERSON, J.D.E., BLAIN, J.A., SHAW, C.E.L., TODD, R. AND BELL, G. (1979). Synthesis of glycerides and esters by fungal cell-bound enzymes in continuous reactor systems. *Biotechnology Letters* **1,** 211–216.

PERLMAN, D. (1977). The fermentation industries—1977. *American Society for Microbiology News* **43,** 82–89.

PILNIK, W. (1982). Utilisation des enzymes dans l'industrie des foissons (jus de fruit, nectars, vins, spiritueux et bieres). *Bios* **13,** 7–19.

REICHELT, J. (1983). Starch. In *Industrial Enzymology* (T. Godfrey and J. Reichelt, Eds), pp. 375–396. The Nature Press, New York.

RICHTER, G. (1983). Glucose oxidase. In *Industrial Enzymology* (T. Godfrey and A. Reichelt, Eds), pp. 428–436. The Nature Press, New York.

ROSAZZA, J.P. (1982). *Microbial Transformation of Bioactive Compounds.* CRC Press, Boca Raton.

ROSE, A.H. (1980). *Microbial Enzymes and Bioconversions.* Academic Press, New York.

RUTTLOFF, H. (1982). Biotechnology and aroma production. *Nährung* **26,** 575–589.

RUYSSEN, R. AND LAUWERS, A., EDS (1978). *Pharmaceutical Enzymes.* E-Story-Scienta, Gent.

SCAZZOCCHIO, C. AND SEALY-LEWIS, H.M. (1978). A mutation in the xanthine dehydrogenase (purine hydroxylase I) of *Aspergillus nidulans* resulting in altered specificity. *European Journal of Biochemistry* **91,** 99–109.

SCHINDLER, J. AND SCHMID, R.D. (1982). Fragrance or aroma chemicals—microbial synthesis and enzymatic transformation—a review. *Process Biochemistry, Sept/Oct.* pp. 2–8.

SCHMID, R.D. (1979a). Stabilized soluble enzymes. *Advances in Biochemical Engineering* **12,** 41–118.

SCHMID R.D. (1979b). Oxidoreductases—present and potential applications in technology. *Process Biochemistry, May* pp. 2–8, 35.

SCHWIMMER, S. (1981). *Source Book of Food Enzymology.* Avi, Westport, Conn.

SCOTT, D. (1980). Enzymes, Industrial. In *Kirk-Othmer Encyclopedia of Chemical Technology*, volume 9, pp. 173–224. John Wiley & Sons, New York.

SCRIMGEOUR, K.G. (1977). *Chemistry and Control of Enzyme Reactions*. Academic Press, New York.

SEBEK, O.K. AND KIESLICH, K. (1977). Microbial transformations of organic compounds. In *Annual Reports on Fermentation Processes* (D. Perlman and G.T. Tsao, Eds), volume 1, pp. 267–297, Academic Press, New York.

SEO, C.W., YAMADA, Y. AND OKADA, H. (1982). Synthesis of fatty acid esters by *Corynebacterium* sp. S-401. *Agricultural and Biological Chemistry* **46,** 405–409.

SHAHANI, K.M., ARNOLD, R.G., KILARA, A. AND DEVIVEDI, B.K. (1976). Role of microbial enzymes in flavor development in foods. *Biotechnology and Bioengineering* **18,** 891–907.

SIGAL, I.S., HARWOOD, B.G. AND ARENTZEN, R. (1982). Thiol-β-lactamase: Replacement of the active-site serine of RTEM β-lactamase by a cysteine residue. *Proceedings of the National Academy of Sciences of the United States of America* **79,** 7157–7160.

SIH, C.J., ABUSHANAB, E. AND JONES, J.B. (1977). Biochemical procedures in organic synthesis. *Annual Reports in Medicinal Chemistry* **12,** 298–308.

STEWART, G.G. (1982). Biotechnology in the 1980's. *Bakers Digest, Feb.* pp. 24–30.

STARACE, C.A. (1983). Detergent enzyme—past, present and future. *Journal of the American Oil Chemists Society* **60,** 1025–1027.

STARACE, C. AND BARFOED, H.C. (1980). Enzyme Detergents. In *Kirk-Othmer Encyclopedia of Chemical Technology*, volume 9, pp. 138–148. John Wiley & Sons, New York.

STROBEL, R. (1983). Enzymatic esterification: An effective alternative to chemical synthesis. *Biotechnology News* **3,** (7), p. 5.

STROBEL, R.J., JR, CIAVARELLI, L.M., STARNES, R.L. AND LANZILOTTA, R.P. (1983). Biocatalytic synthesis of esters using dried *Rhizopus arrhizus* mycelium as a source of enzyme. *Abstract 053 of the Annual American Society for Microbiology Meeting* p. 248.

TANAKA, T., ONO, E., ISHIHARA, M., YAMANAKA, S. AND TAKINAMI, K. (1981). Enzymatic acyl exchange of triglyceride in *N*-hexane. *Agricultural and Biological Chemistry* **45,** 2387–2389.

TER HASEBORG, E. (1981). Enzymes in flour and baking applications, especially waffle batters. *Process Biochemistry, Aug/Sept.* pp. 16–19.

TONG, G.E. (1982). Industrial chemicals from fermentation. *Enzyme and Microbial Technology* **1,** 173–179.

TREVAN, M.D. (1980). *Immobilized Enzymes*. John Wiley & Sons, New York.

TRIBE, D.E. AND GRAY, P.P. (1979). The commercial production of microbial enzymes. Part 1. Biochemical background. *Food Technology in Australia* **31,** 190–197.

ULMER, K.M. (1983). Protein engineering. *Science* **219,** 666–671.

VALLEE, B.L. (1980). Zinc and other active site metals as probes of local conformation and function of enzymes. *Carlsberg Research Communications* **45,** 423–441.

VANBELLE, M., MEURENS, M. AND CRICHTON, R.R. (1982). Enzymes in foods and feeds. *Revue des Fermentations et des Industries Alimentaires* **37,** 124–135.

WALSH, C. (1979). *Enzymatic Reaction Mechanisms*. Freeman, San Francisco.

WARD, O.P. (1983). Proteinases. In *Microbial Enzymes and Biotechnology* (W.M. Fogarty, Ed.), pp. 251–317. Applied Science, New York.

WEETE, J.D. (1980). *Lipid Biochemistry of Fungi and Other Organisms*. Plenum Press, New York.

WINTER, G., FERSHT, A.R., WILKINSON, A.J., ZOLLER, M. AND SMITH, M. (1982). Redesigning enzyme structure by site-directed mutagenesis: tyrosyl tRNA synthetase and ATP binding. *Nature* **299,** 756–758.

YAMADA, K. (1977). Recent advances in industrial fermentations in Japan. *Biotechnology and Bioengineering* **19,** 1563–1621.

YAMADA, H. AND KUMAGAI, H. (1978). Microbial and enzymatic processes for amino acid production. *Pure and Applied Chemistry* **50,** 1117–1127.

YAMAMOTO, Y. (1978). Industrial applications of enzymes. *Hakkokogaku Kaishi* **56,** 656–661.

2
Isoamylase and its Industrial Significance in the Production of Sugars from Starch

TOKUYA HARADA

Kobe Women's University, Suma, Kobe 654, Japan

Introduction

Debranching enzymes are important in the basic and applied science of starch. The usual composition of starch is about 80% amylopectin and 20% amylose. Amylopectin is partially hydrolysed by α-amylase because its branch points with (1,6-α)-glucosidic linkages are resistant to attack by the usual α-amylase; only α-amylase (EC 3.2.1.1) from *Thermoactinomyces vulgaris* has been shown to hydrolyse both (1,6-α)-glucosidic linkages and (1,4-α)-glucosidic linkages (Sakano *et al*., 1982). Glucoamylase (exo-1,4-α-D-glucosidase, EC 3.2.1.3), which is capable of producing glucose from starch, can hydrolyse (1,6-α)-glucosidic linkages in amylopectin, although slowly. β-Amylase (EC 3.2.1.2) forms maltose from the non-reducing point of the chain in amylopectin but the hydrolysis stops near the branching points of the chain, leaving β-amylase limit dextrin (β-limit dextrin). Thus, the ability to produce maltose or glucose from starch can be improved by using a debranching enzyme.

Debranching enzymes are classified as direct or indirect (Lee and Whelan, 1971). The former, which can attack amylopectin and glycogen directly, are principally divided into isoamylase (EC 3.2.1.68) and pullulanase (EC 3.2.1.41). Isoamylase can split all the branching points of glycogen but not those of pullulan whereas pullulanase can split pullulan completely, but has limited hydrolytic activity on glycogen. Pullulan produced by *Aureobasidium pullulans* is a linear polymer of α-maltotriose joined endwise through (1,6-α) bonds. Isoamylase is particularly useful for elucidating the structures of glucans such as glycogen, whereas pullulanase was first used to clarify the structure of pullulan. There are several differences in the modes of decomposition of starch by isoamylase and pullulanase and these enzymes are useful for elucidation of the structure of starch. Isoamylase is produced in high yield by a mutant strain of *Pseudomonas amyloderamosa* SB15 and shows very high specific activity towards starch. This isoamylase is very useful for the industrial production of glucose or maltose from starch.

The major sugars produced from starch are glucose (dextrose), maltose, and

Biotechnology and Genetic Engineering Reviews—Vol. 1, February 1984
0264–8725/84/01/39–25$10.00 + $0.00 © Intercept Ltd.

isomerized sugar (isomerose). Isomerose is a mixture of glucose and fructose approaching the 50/50 composition of invert sugar, and which is increasingly being used as a substitute for sucrose, especially in soft drinks. (In acid drinks, sucrose is hydrolysed to invert sugar anyway, after a few days). Isomerose resembles invert sugar in being much sweeter than glucose and nearly as sweet as sucrose itself. There are now methods of obtaining pure fructose from isomerose but these are not dealt with in this review. The efficiency of production of isomerose from glucose depends very much on the purity of the glucose used as raw material: even a very small percentage of non-glucose sugars can have a disastrous effect on yields.

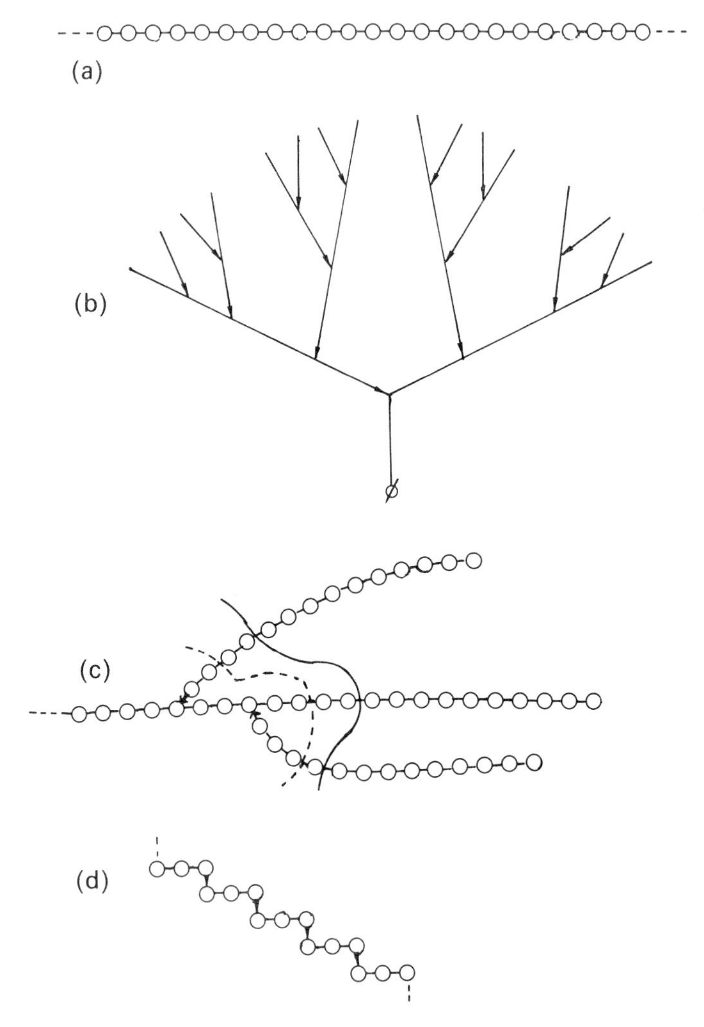

Figure 1. Structures of amylose, amylopectin and their related compounds. (1,4-α)-Linkages and (1,6-α)-linkages are represented by solid lines and by arrow heads, respectively. The position of the reducing end of the glucose unit is indicated by ⌀. (a) Amylose; (b) Meyer's model of amylopectin and glycogen (this model is no longer recognized). The ratios of (1,6-α)-linkage to (1,4-α)-linkage are about 1 : 24 for amylopectin and about 1 : 12 for glycogen; (c) Amylopectin, amylopectin β-limit dextrin_ _ _ _ _and phosphorylase______limit dextrin; (d) Pullulan.

The major problems of conversion of starch into the desired sugars involve maximizing the yield of the target substance(s) with minimum contamination with undesirable by-products such as saccharides with (1,6-α)-links, retrograded starch and other higher polysaccharides. The reactions should be at the highest temperatures and reagent concentrations possible in order to economize in time, in enzymes and in the cost of removal of water in the preparation of stable syrups or pure solid products.

Structures of amylose, amylopectin and their related compounds are shown in *Figure 1*.

Occurrence of isoamylase

The name isoamylase was first given by Maruo and Kobayashi (1951) to an enzyme found in yeast cells and able to debranch intracellular glycogen. The enzymatic activity in these cells was too weak to allow detailed studies on the properties of the enzyme, which previously had been known as amylosynthetase. Lee, Nielsen and Fisher (1967) reported, however, that the combined actions of glucosyl transferase (4-α-D-glucanotransferase, EC 2.4.1.25) and amylo-1,6-glucosidase (EC 3.2.1.33) in an autolysate, debranch amylopectin to form a linear chain with a much longer chain length than the inner chain stubs of the original amylopectin. In 1968, Harada, Yokobayashi and Misaki selectively isolated colonies of *Pseudomonas* which develop a blue coloration with iodo-reagent in a medium containing amylopectin as the sole carbon source. The organism produced extracellular isoamylase and was named *Pseudomonas amyloderamosa* SB15. After this finding, many workers tried to obtain useful micro-organisms capable of producing isoamylase. In 1970, Gunja-Smith, Marshall and Whelan (1970) found the enzyme in a crude preparation of *Cytophaga*. Evans, Manners and Stark (1979) partially purified the enzyme and studied its properties, and Whelan's group (Gunja-Smith *et al.*, 1970; Marshall and Whelan, 1974) and Manners's group (Manners and Matheson, 1981) used the enzyme to examine the structures of glycogen and starch, respectively. Urlaub and Wöber (1975) detected isoamylase activity in an enzyme preparation without α-amylase extracted from cells of *Bacillus amyloliquefaciens* which produces α-amylase. Harada and Yokobayashi (unpublished data) could not detect the enzyme activity in the culture broth of 42 stock cultures of *Pseudomonas* tested in the Institute of Fermentation, Osaka: no strain other than the one that we had isolated was of industrial value, because other strains produced much lower amounts of enzyme.

Harada, Amemura and Konishi (unpublished data) showed that antiserum to *Pseudomonas* isoamylase formed a precipitin line with *Pseudomonas* isoamylase, but not with *Cytophaga* isoamylase. Thus, the two enzymes differ immunologically.

Occurrence of pullulanase

In 1961, pullulanase was discovered in *Aerobacter aerogenes* by Bender and Wallenfels (1961). Later, Walker (1968) detected pullulanase activity in cells of some strains of *Streptococcus mitis* and (1,6-α)-glucosidase activity in place of pullulanase in three other strains of this species. The former strains did not accumulate glycogen, but the latter accumulated glycogen in the cells. A strain of *Streptomyces* (No. 280) was also reported to produce pullulanase (Yagisawa *et al.*, 1972). Griffin and Forgarty (1973) found a strain of *Bacillus polymyxa* that was able to produce pullulanase and β-amylase. A strain of *Bacillus cereus* var. *mycoides* was also shown to produce pullulanase and β-amylase extracellularly (Takasaki, 1976a). This enzyme preparation was reported to be useful for the production of maltose from starch (Takasaki, 1976b); Yamanobe and Takasaki, 1979). Morgan, Adams and Priest (1979) detected the enzyme activity in many strains of *Bacillus*. Recently, Novo Company (Norman, 1982a) found a strain of *Bacillus* capable of producing an acidic and thermophilic pullulanase and attempted to use it in conjunction with glucoamylase from *Aspergillus niger* for the production of glucose from starch. Wöber (1973) showed that *Pseudomonas stutzeri* produced pullulanase extracellularly.

Strains of *Aerobacter aerogenes* are classified as strains of *Klebsiella pneumoniae* and of *Enterobacter aerogenes*. Pullulanase was reported to occur in a strain of *Escherichia coli* (Palmer, Wöber and Whelan, 1973). Thus we (Konishi *et al.*, 1979) investigated the distribution of pullulanase in the family Enterobacteriaceae. First, we examined the growth of the organisms on a medium containing starch or pullulan, and their ability to produce pullulanase, finding that only species of *Klebsiella pneumoniae* can grow on such media and produce pullulanase: nine strains of *Enterobacter aerogenes* tested were unable to produce pullulanase. The presence of pullulanase was also examined by immunodiffusion. Culture filtrates and cell-free extracts of all eight strains of *Klebsiella pneumoniae* reacted immunologically with antiserum prepared with crystalline pullulanase from *Klebsiella pneumoniae* IFO 3321. Their precipitation lines fused completely with the homologous lines, except in the cases of *Klebsiella pneumoniae* IFO 3317 and ATCC 21073, the precipitation lines of which fused completely with each other, however. In contrast, nine strains of *Enterobacter aerogenes* did not cross-react with the antiserum. Thus, pullulanase formation may be a character common to all *Klebsiella* and, if so, it may be a key character in differentiating strains of *Klebsiella* from other strains of Enterobacteriaceae.

Palmer, Wöber and Whelan (1973) suggested that *E. coli* uses pullulanase for debranching of extracellular carbohydrate, whereas isoamylase is presumed to function intracellularly in the degradation of glycogen. Dessein and Schwartz (1974) reported, however, that no pullulanase activity was detectable in 11 different strains of *E. coli* grown on maltose and that neither these strains, nor 40 other strains of the same species, would grow on pullulan. We also could not detect pullulanase activity in cultures of two strains of *E. coli* tested. On the other hand, Jeannigros *et al.* (1976) detected activity of a debranching enzyme in a strain of *E. coli*. The purified enzyme hydrolysed (1,6-α)-glucosidic linkages in phosphorylase and β-amylase limit dextrins prepared from glycogen and amylopectin. It also completely hydrolysed amylopectin, but showed only very low activity on glycogen and no activity on pullulan. Thus, this enzyme cannot be classified as pullulanase or an isoamylase. Pullulanase has been shown to be produced by strains from *Nocardia*, *Lactobacillus*, *Micrococcus* and other bacteria (Sakai, 1981). Pullulanase appears to be widely distributed in micro-organisms, in contrast to the limited distribution of isoamylase.

High activity of *Pseudomonas* isoamylase towards amylopectin and glycogen

The specific activities on starch and glycogen of our *Pseudomonas* isoamylase may be the highest reported for debranching enzymes. The debranching actions of crystalline isoamylase of *Pseudomonas* on amylopectin and glycogen have been investigated by Harada *et al.* (1972). The increase in reducing power was monitored and digests were fractionated on a Sephadex column. The results were compared with those obtained with crystalline pullulanase from *Klebsiella* (*Figure 2*). On incubation, 0·01 mg of the isoamylase hydrolysed the branching

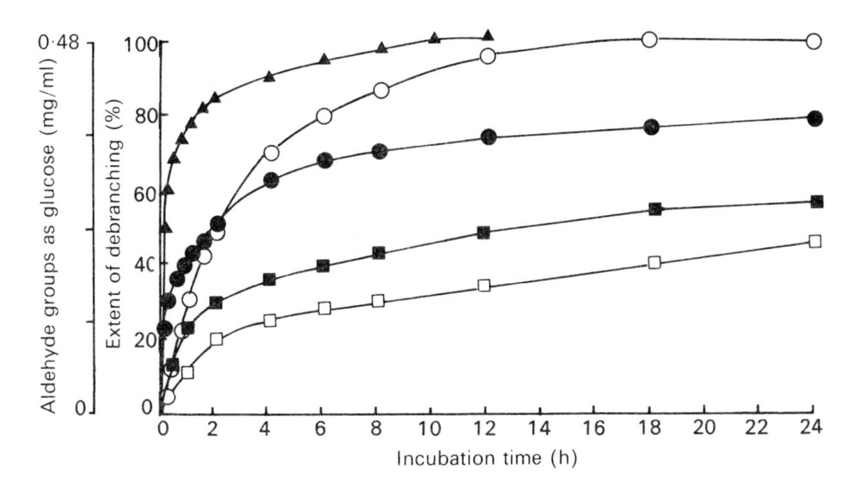

Figure 2. Debranching actions of purified enzymes on waxy-maize amylopectin (Harada *et al.*, 1972). One gram of amylopectin was treated with 0·001 mg (□) or 0·01 mg (○) of isoamylase, pH 3·5 or with 0·13 mg (■), 1·3 mg (●) or 13 mg (▲) of pullulanase, pH 5·5 in a volume of 100 ml.

linkages in 1 g of waxy-maize amylopectin in 20 h, whereas 1·3 mg of the pullulanase did not completely hydrolyse them even in 24 h, although 13 mg of the enzyme caused complete hydrolysis in 10 h. The distribution pattern of the linear (1,4-α)-linked unit chains in potato amylopectin is illustrated in *Figure 3* (Akai *et al.*, 1971b). Similar distribution patterns were obtained with waxy-maize amylopectin using *Pseudomonas* isoamylase and *Klebsiella* pullulanase (Lee, Mercier and Whelan, 1968), with waxy-rice amylopectin using *Pseudomonas* isoamylase (Akai *et al.*, 1971b) and with wheat amylopectin using *Cytophaga* isoamylase (Atwell, Hoseney and Lineback, 1980). There were two characteristic peaks with average degrees of polymerization (DP_n values) at the apex of about 50 and 20, respectively. Mercier and Kainuma (1975) showed that *Pseudomonas* isoamylase and *Klebsiella* pullulanase can be used in dimethylsulphoxide solution to study the fine structure of water-insoluble branched polysaccharides.

On incubation with oyster glycogen, 0·027 mg of the isoamylase caused com-

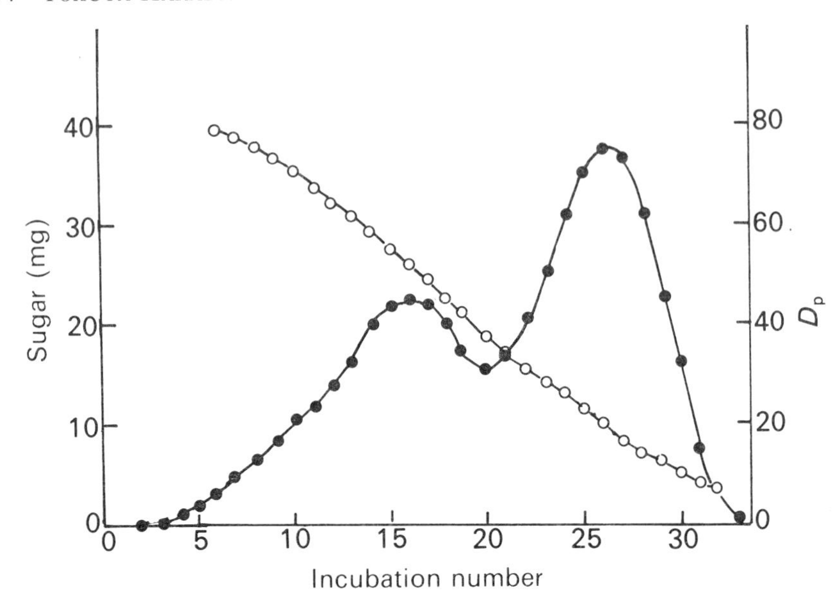

Figure 3. Fractionation of potato amylopectin (500 mg) on a Sephadex G-75 column after treatment for 20 h with *Pseudomonas* isoamylase (Akai *et al.*, 1971b). ○—○, D_p (degrees of polymerization); ●—●, mg sugar.

plete hydrolysis of the branching linkages in 24 h, while 29 mg of the pullulanase caused about 30% hydrolysis of the linkages in 24 h (Harada *et al.*, 1972). *Pseudomonas* isoamylase cleaved (1,6-α)-glucosidic linkages extensively to form linear (1,4-α)-linked unit chains of DP_n 15 from rabbit liver glycogen and DP_n 11 from oyster glycogen (Akai *et al.*, 1971a): the products gave one peak with a fairly broad distribution of chain lengths ranging from 3 to 50 glucose units; such patterns from glycogen were not obtained with pullulanase. *Pseudomonas* isoamylase has been used to clarify the structure of glycogens from various different sources, such as from the blue-green alga *Anacytis nidulans* (Weber and Wöber, 1975), and from shellfish such as scallops and abalone (Hata *et al.*, 1983) in addition to that from oysters (Akai *et al.*, 1971a; Umeki and Yamamoto, 1977).

The specific activities of *Pseudomonas* isoamylase were determined by Yokobayashi *et al.* (1972). The activities were measured as amounts of cleavage of glucosidic bonds per minute per milligram of protein at substrate concentrations of 2%, in comparison with those of *Klebsiella* pullulanase. The activities of *Pseudomonas* isoamylase were 110–280 μmol for amylopectins and glycogens and 1·1 μmol for pullulan, while those of *Klebsiella* pullulanase were 3–5 μmol for amylopectins, 0·5–1·2 μmol for glycogens and 53 μmol for pullulan. The K_m values (g/ml) of the isoamylase were about 1×10^{-4}–2×10^{-4} for amylopectins and glycogens and 2×10^{-3} for pullulan, whereas those of the pullulanase were about 8×10^{-3}–1×10^{-2} for amylopectins, about 2×10^{-2}–5×10^{-2} for glycogens and $1{\cdot}7 \times 10^{-5}$ for pullulan. Thus, the debranching activities of isoamylase towards glycogen and amylopectin are much higher than those of pullulanase: the molecular activities of the isoamylase are about 9500–26 700 for amylopectins and glycogens, but only about 110 for pullulan whereas those of the pullulanase with amylopectin and glycogen were only 130–1400 and that with pullulan was 7700. These molecular activities were defined as the number of equivalents of a bond cleaved per minute per molecule of enzyme.

Comparison of substrate specificities of *Pseudomonas* isoamylase and *Klebsiella* pullulanase

The activities of *Pseudomonas* isoamylase and *Klebsiella* pullulanase for β-amylolysis of amylopectin, glycogen and related dextrins are shown in *Table 1* (Yokobayashi, Misaki and Harada, 1969, 1970). Isoamylase, like pullulanase, caused the complete breakdown of amylopectin, both when added before and also when added at the same time as β-amylase. Similarly isoamylase hydrolysed glycogen completely when acting either before or together with β-amylase. However, although pullulanase caused the complete degradation of glycogen when acting simultaneously with β-amylase, it caused only a slight increase in the β-amylolysis of glycogen when added before β-amylase. The incomplete degradation of glycogen by pullulanase is attributable to its inability to penetrate to the interior of the glycogen molecule. On the other hand, *Pseudomonas* isoamylase can penetrate the interior of the compact glycogen molecule and can hydrolyse all (1,6-α)-glucosidic inter-chain linkages. Isoamylase thus can be used to elucidate the structure of glycogen.

With glycogen β-limit dextrin, isoamylase caused extensive β-amylolysis (79%) when added before β-amylase, whereas pullulanase caused 31% degradation under the same conditions (*Table 1*), showing that isoamylase can split interior branch points in glycogen. Although the combined actions of *Pseudomonas* isoamylase and β-amylase caused degradation of β-limit dextrin, 20–30% of amylopectin β-limit dextrin and 20–25% of glycogen β-limit dextrin, respectively, were resistant to enzyme action. Pullulanase caused complete degradation of amylopectin β-limit dextrin both when added before and also when added at the same time as β-amylase; however, pullulanase caused only 32% degradation of glycogen when allowed to act before β-amylase.

The difference in the degradation rates of β-limit and phosphorylase limit dextrins by isoamylase can also be seen in *Table 1*. This difference is due to difference in the side chains in β-amylase limit dextrins and phosphorylase limit dextrins, the former being two or three glucose units long, and the latter four glucose units long (*see Figure 1c*). The action of pullulanase on waxy-maize amylopectin and oyster glycogen phosphorylase β-amylase limit dextrins, which have two glucose unit stubs, is similar to that on amylopectin, glycogen and their phosphorylase or β-amylase limit dextrin. However, the combined action of *Pseudomonas* isoamylase and β-amylase cause hydrolysis of 48% of amylopectin phosphorylase β-amylase limit dextrin and of 44% of the glycogen phosphorylase β-amylase limit dextrin. Gunja-Smith *et al.* (1970) noted in their report using *Cytophaga* isoamylase that 'if the ideal Meyer structure would be true,

Table 1. Effects of *Pseudomonas* isoamylase and *Klebsiella* pullulanase on β-amylolysis of amylopectin, glycogen and related dextrins (Yokobayashi, Misaki and Harada, 1969, 1970)

Substrate	Conversion to maltose (%)				
	β-Amylase alone	Action before β-amylase		Action simultaneous with that of β-amylase	
		Pseudomonas isoamylase	*Klebsiella* pullulanase	*Pseudomonas* isoamylase	*Klebsiella* pullulanase
Waxy-maize amylopectin	50	99	95	95	103
Potato amylopectin	47	96	98	97	103
Oyster glycogen	38	102	46	100	99
Rabbit liver glycogen	42	100	51	99	97
Waxy-maize amylopectin β-limit dextrin	0	80	97	72	97
Oyster glycogen β-limit dextrin	0	79	31	76	99
Waxy-maize amylopectin phosphorylase limit dextrin	21	95	97	98	101
Rabbit liver glycogen phosphorylase limit dextrin	28	94	32	97	99
Waxy-maize amylopectin phosphorylase β-limit dextrin	0	48	96	56	100
Rabbit liver glycogen phosphorylase β-limit dextrin	0	44	29	50	97

every B chain liberated by isoamylase from the Meyer phosphorylase β-amylase limit dextrin would be branched near its non-reducing end by a maltosyl A chain. Thus, the debranched molecule should remain inert to the action of β-amylase'. However, the extent of amylolysis of isoamylase-treated phosphorylase β-limit dextrin is significantly high, as shown in our experiments and theirs. There are several possible explanations for this. One possibility is that B chains do not carry an A chain. This was postulated by Gunja-Smith *et al.* (1970) who then revised the Meyer branched model of glycogen and amylopectin (the Meyer structure is no longer recognized). The second possibility is that there may be 'buried chains' in the structure: that is, that some of the A chains are not subject to the action of phosphorylase and β-amylase. There is a third possibility, that the preparations of phosphorylase β-limit dextrin used by Yokobayashi, Misaki and Harada (1970) and by Gunja-Smith *et al.* (1970) have branched points with maltotriose stubs as pointed out by Nakamura (1977). Another possibility is that isoamylase can split some of branching points with maltose stubs in them. We (Yokobayashi, Misaki and Harada, 1970) have shown that *Pseudomonas* isoamylase liberates maltose and maltotriose in a molar ratio of 1 : 13·4 with amylopectin β-limit dextrin and 1 : 9·7 with glycogen β-limit dextrin under our conditions. Evans, Manners and Stark (1979) also detected the presence of maltose in a reaction mixture containing a purified *Cytophaga* isoamylase and amylopectin. Maltose production was also clearly observed after a longer incubation by Manners and Matheson (1981). Isoamylase can slowly split branching points with maltose stubs in β-limit dextrin, and this may explain in part why phosphorylase β-limit dextrin with maltose stubs is degraded by the action of isoamylase. The comparatively poor ability of isoamylase to hydrolyse such branching points with maltose stubs does not affect its efficiency in the industrial production of sugars from starch.

The detailed structural requirements of the substrate of *Pseudomonas* isoamylase were compared qualitatively and quantitatively with those of *Klebsiella* pullulanase (Kainuma, Kobayashi and Harada, 1978) and some of the data are shown in *Figure 4* and *Table 2*. The best substrates for the isoamylase are polymers of higher molecular weight, such as amylopectin and glycogen, whereas those of the pullulanase are branched oligosaccharides derived from amylopectin. An important difference between isoamylase and pullulanase is that the former only slowly hydrolyses pullulan whereas pullulanase has a similarly slow action on glycogen. Maltosyl branches in the oligosaccharides are hydrolysed

Table 2. Relative reaction rates of *Pseudomonas* isoamylase and *Klebsiella* pullulanase on various branched oligosaccharides and polysaccharides (Kainuma, Kobayashi and Harada, 1978)

Substrate	*Pseudomonas* isoamylase	*Klebsiella* pullulanase
Potato amylopectin	100	15
Oyster glycogen	124	1
Pullulan	1	100
6^3-*O*-α-Maltosyl-maltotriose	3	22
6^3-*O*-α-Maltotriosyl-maltotriose	10	162
6^2-*O*-α-Maltotetraosyl-maltotriose	7	26
6^3-*O*-α-Maltotriosyl-maltotetraose	33	146

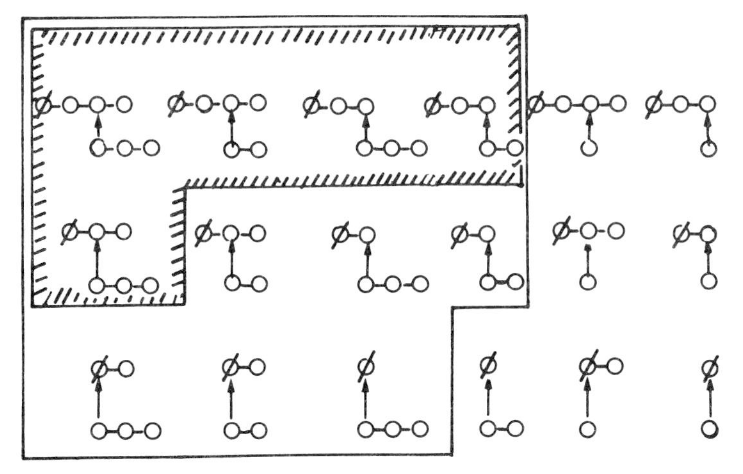

Figure 4. Susceptibility of branched oligosaccharides to *Pseudomonas* isoamylase and *Klebsiella* pullulanase in relation to their structures (Kainuma, Kobayashi and Harada, 1978). Boxed area susceptible to pullulanase and isoamylase. (1,4-α)-Linkages and (1,6-α)-linkages are represented by solid lines and by arrow heads, respectively. The position of the reducing end of the glucose unit is indicated by ⌀.

by *Pseudomonas* isoamylase much more slowly than the maltotriosyl branches. *Pseudomonas* isoamylase requires a minimum of three D-glucose residues in the chains at the reducing end of the branched oligosaccharides.

Endo-type and exo-type cleavage of amylopectin and glycogen, respectively, by *Pseudomonas* isoamylase

In order to determine whether isoamylase has an endolytic or exolytic action on branching points with (1,6-α)-glucosidic linkages in amylopectin and glycogen, the hydrolysis products with isoamylase, obtained after appropriate periods, were fractionated on Sephadex G-75 and the sugars in each fraction were measured (Harada *et al.*, 1972), as shown in *Figures 5* and *6*. For comparison, the action of pullulanase was also examined. After about 50% or 70% hydrolysis, the products obtained with isoamylase clearly contained less residual polymer and more large glucans with branching linkages than the products of pullulanase action. Amounts of 0·01 mg of isoamylase or 1·3 mg of pullulanase caused about 50% hydrolysis of 1 g of waxy-maize amylopectin in 2 h. After 80–85% hydrolysis with isoamylase, the amounts of residual polymer had diminished and two peaks appeared, one peak being that of linear glucan and the other peak probably being that of linear glucans and glucans containing branching linkages. Isoamylase differs from pullulanase in that it can hydrolyse higher-molecular-weight branched glucans more readily than branched oligosaccharides of lower molecular weight. The reaction mixture became white due to precipitation at an early stage of the reaction with isoamylase. This precipitation could have been caused by the retrodegradation of the short-chain amylose and branched oligosaccharides produced. After 24 h incubation with isoamylase a typical pattern of amylopectin products with linear linkages was seen, but after 24 h incubation with pullulanase there was still a great deal of residual glucan. Incubation with

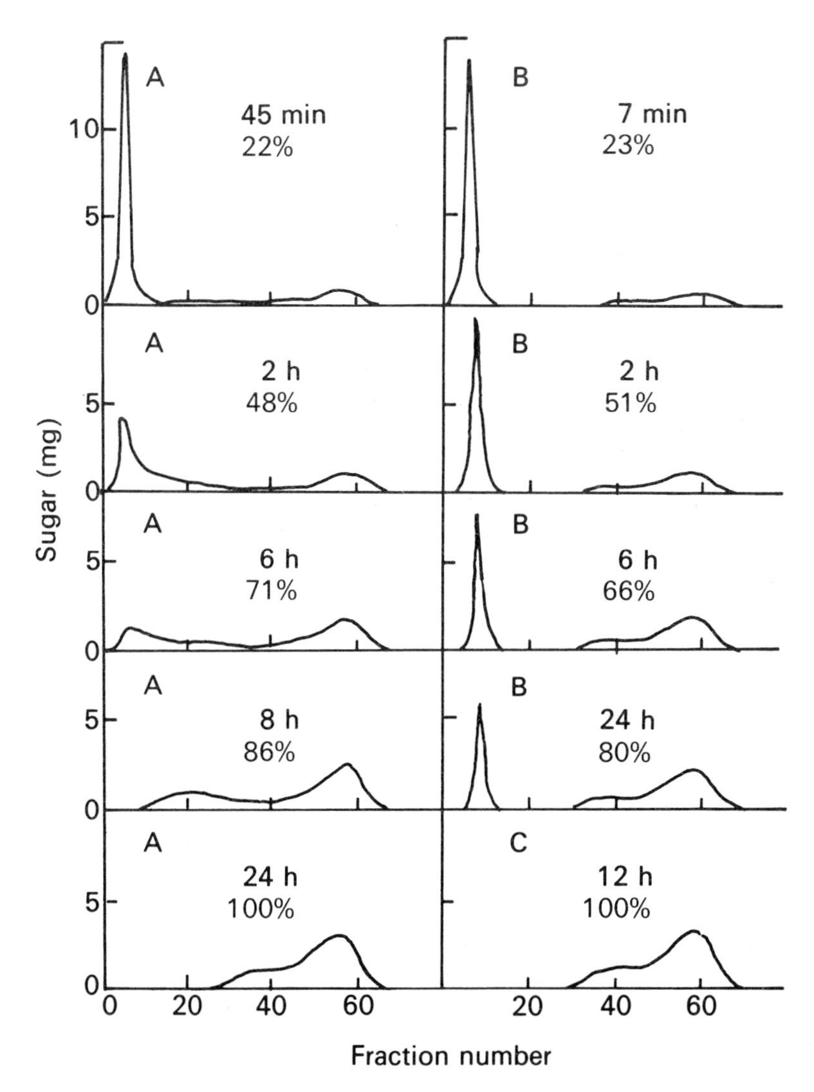

Figure 5. Fractionation of waxy-maize amylopectin on a Sephadex G-75 column after enzyme treatment (Harada *et al.*, 1972). A = 0·01 mg isoamylase; B = 1·3 mg pullulanase; C = 13 mg pullulanase. Incubation time and degree of hydrolysis is shown for each fractionation pattern.

a tenfold increase in pullulanase (13 mg) for 12 h resulted in complete scission of the branching linkages of amylopectin, giving the same pattern of linear linkages as that obtained with isoamylase. These observations suggest that the isoamylase hydrolyses both inner and outer branching linkages of amylopectin, whereas the pullulanase hydrolyses outer linkages of amylopectin efficiently, but affects inner linkages only slowly.

Significant differences between the actions of the two enzymes on glycogen were observed: the isoamylase hydrolysed all the branching linkages comparatively well, whereas the pullulanase hydrolysed relatively few. The isoamylase

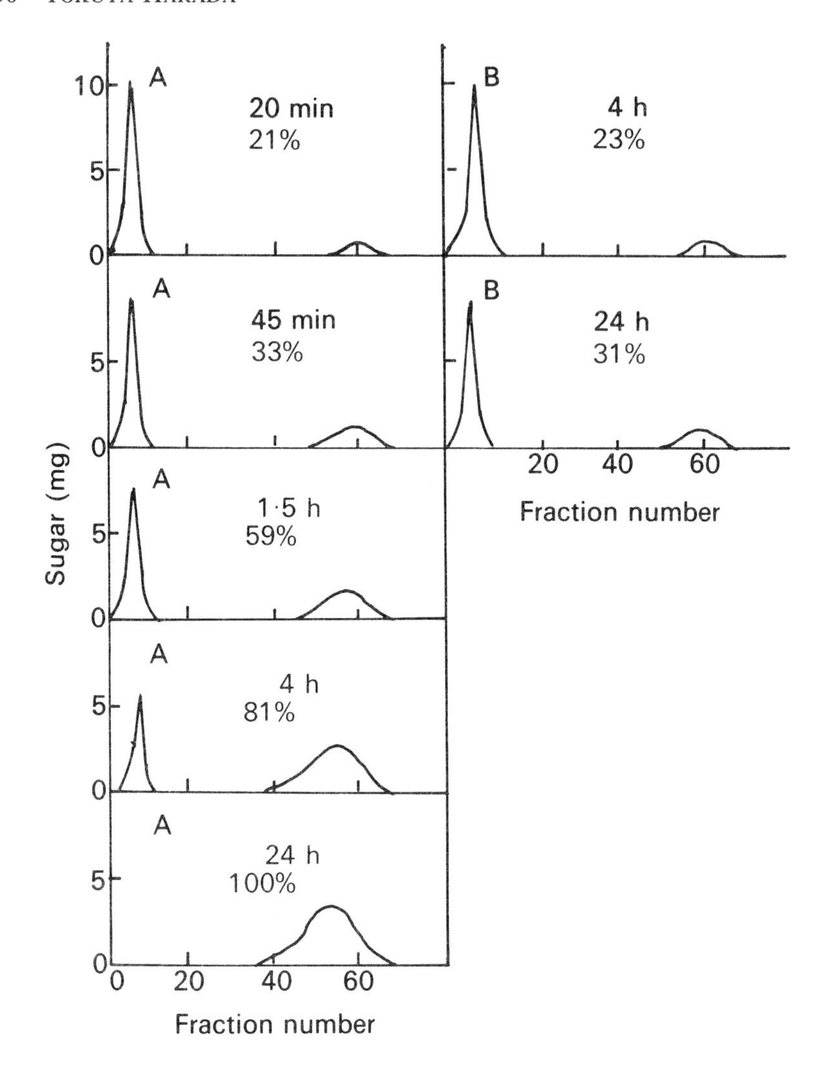

Figure 6. Fractionation of oyster glycogen on a Sephadex G-75 column after enzyme treatment (Harada *et al.*, 1972). A = 0·027 mg isoamylase; B = 29 mg pullulanase. Incubation time and degree of hydrolysis shown for each fractionation pattern.

hydrolysed glycogen exolytically: this action is attributable to the structure of the polysaccharide, not to the specificity of binding of the active site of the enzyme. Palmer, Macakie and Grewal (1983) extended these experiments using *Cytophaga* isoamylase and suggested that, during hydrolysis by the enzyme, branched oligosaccharides are not produced in measurable amounts from glycogen. These findings suggest that isoamylase from *Pseudomonas* and *Cytophaga* have exolytic actions on glycogen; i.e. they catalyse the ordered and sequential hydrolysis of (1,6-α)-glucosidic linkages proceeding from the non-reducing ends to the interior of the chain.

Enhanced production of isoamylase by *Pseudomonas amyloderamosa*

Pseudomonas amyloderamosa SB15 produces isoamylase extracellularly, and no other amylase or α-glycosidase apart from isoamylase can be detected in the culture filtrate of this organism, thus providing a very convenient method of obtaining pure isoamylase, whereas contaminating enzymes usually impede enzyme purification.

We have used an iodine colour test to detect and isolate organisms that produce only isoamylase. In the test, filter paper is dipped in a 1% solution of waxy amylopectin containing 0·1 M acetate buffer, pH 3·5, and the wet paper is placed on the colonies formed on agar medium and incubated at 40°C for 30 min. The paper is then exposed to iodine vapour: blue spots indicate the position of colonies that produce isoamylase constitutively. These colonies were picked up by the replica method.

Pseudomonas amyloderamosa SB15 produces isoamylase inducibly (i.e. it can be induced to do so) in the presence of glycogen, starch or maltose. The most satisfactory medium consists of 2% maltose or starch, 0·4% sodium glutamate, 0·3% diammonium hydrogen phosphate, and other inorganic salts. This organism requires glutamate or aspartate for growth. The optimum pH for enzyme production is 5–6; the enzyme is stable at pH 2–6, but is extremely labile above pH 7. The maximum yield of the enzyme is obtained after 2 days in the above culture medium (Harada, Yokobayashi and Misaki, 1968).

We used mutation procedures to increase the enzyme production by this organism over 500-fold. Sugimoto, Amemura and Harada (1974), using the test paper method (Harada, 1965), by the following procedure, isolated a constitutive mutant strain MS1 that produces isoamylase: strain SB15 cells treated with *N*-methyl-*N*′-nitro-*N*-nitrosoguanidine, were examined by the iodine colour test paper method already described; we then selected much better producers from the original strain by the iodine test paper method, and finally obtained the best mutant strain 1168, which is now used industrially.

Short-chain amyloses formed by the action of extracellular isoamylase can enter the cells in which amyloses are degraded by α-amylase (Kato *et al.*, 1975) and α-glucosidase (Amemura, Sugimoto and Harada, 1974) to form glucose as a final product. The α-amylase attacks amylose much faster than it does amylopectin and glycogen: this enzyme may therefore have a role in degrading short-chain amylose entering the cells after formation by the action of extracellular isoamylase. This α-amylase produces almost equimolar amounts of glucose and maltose as final products and the resultant maltose is hydrolysed by the action of the α-glucosidase. The parent strain SB15 can produce the two intracellular enzymes inducibly, as well as isoamylase, but the mutant strain MS1 produces them constitutively (i.e. as a normal process). The extracellular isoamylase and two intracellular enzymes together permit utilization of extracellular branched α-glucans such as glycogen and amylopectin.

An isoamylase-negative mutant strain, K1, was derived from MS1, and K1C was obtained as a revertant of K1 (Kuswanto *et al.*, 1976). Strain K1 grew as poorly on glycogen or amylopectin medium as on a non-carbohydrate medium. It has no isoamylase activity, and α-amylase and α-glucosidase activities similar to those of the parent strain MS1. Norrman and Wöber (1975) reported that strain SB15 showed a preference for glycogen to amylopectin for growth, but we have not been able to confirm this.

Isoamylase can easily be purified with cross-linked amylose gel as an absorbent (Kato *et al.*, 1977). The enzyme can be eluted from the gel with 5% maltotriose or linear maltodextrin solution. Crystalline enzyme was prepared by dropwise addition of ammonium sulphate to a solution of the purified enzyme (*Figure 7*). By this procedure 96 mg of the enzyme was purified to homogeneity from 20 litres of culture broth of strain K1C with a yield of about 70%. Strain 1168 produces about 500 mg of isoamylase as the purified enzyme per litre of culture broth.

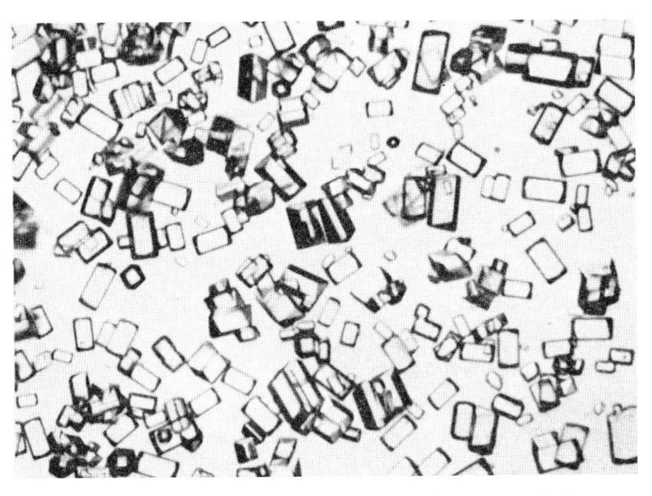

Figure 7. Photograph of crystalline isoamylase from *Pseudomonas amyloderamosa* SB15.

Molecular properties of *Pseudomonas* isoamylase

Crystalline isoamylase (*Figure 7*) was found to be contaminated with a trace of proteolytic enzyme (Amemura, Konishi and Harada, 1980). This contaminant digested the isoamylase under neutral or alkaline conditions, especially in the presence of sodium dodecyl sulphate. A reliable molecular weight of the enzyme was obtained by SDS-polyacrylamide gel electrophoresis and by gel filtration on Sepharose-6B in 6 M guanidine hydrochloride after heat inactivation of the proteolytic contaminant. The molecular weight of the undegraded polypeptide chain of the isoamylase was about 90 000: that of *Klebsiella* pullulanase is 143 000 (Eisele, Rashed and Wallenfels, 1972). Comparison of the amino acid compositions of isoamylase and pullulanase has shown that isoamylase has a much lower lysine content (Kitagawa, Amemura and Harada, 1975). Isoamylase has no sugar component, but the pullulanase has a sugar content of about 2%. The effects of inhibitors on the two enzymes were compared: an SH group did not seem to be related to the activity of either enzyme, but isoamylase is inhibited by iodo-2,4-nitrofluorobenzene and Hg ion. These results and the pK_a value suggest that an imidazole radical may be important for the activity of isoamylase.

Tryptophan groups may not be essential for the activity of isoamylase, although, like pullulanase, the enzyme was inhibited by *N*-bromosuccinimide and 2-hydroxy-5-nitro-benzylbromide. A tryptophan group may be important for the activity of pullulanase (Amemura, Kitagawa and Harada, 1975). Isoamylase was inhibited by maltotriose but not by maltose, which would explain why it does not attack 6-α-maltosyl maltodextrin. The fact that maltose is not inhibitory is very useful in the production of maltose by the combined actions of isoamylase and β-amylase, as shown later. Isoamylase was not inhibited by cyclodextrin, whereas pullulanase was (Marshall, 1973). Isoamylase was inactivated more than pululanase by photo-oxidation in the presence of 0·001% rose bengal (Kato and Harada, unpublished).

Preliminary X-ray studies were carried out on isoamylase in the crystalline state and in aqueous solution (Sato *et al.*, 1982). The diffraction patterns of the ho1 and ok1 zones were recorded with a Buerger precession camera. The crystal is orthorhombic, the space group is $p2_12_12_1$, and the unit-cell dimensions are $a = 137{\cdot}9$, $b = 52{\cdot}9$, $c = 151{\cdot}2$ Å ($V = 1{\cdot}1 \times 10^6$ Å^3). If we assume that the asymmetric unit contains one enzyme molecule of molecular weight 95 000, the V_m and V_{prot} values are within the range usually found for proteins. Solutions at concentrations of 0·4, 0·8, 1·0 and 1·5% were prepared for the X-ray generator. Guinier plots for these four solutions showed that the size and shape of the enzyme molecule are uniform. The four plots were essentially parallel.

The radii of gyration of this enzyme molecule, obtained from these plots, showed a good coincidence, giving an R_g value of 27·5 Å (2·75 nm) after slit error correction. The approximate maximum dimension of the enzyme can be obtained by Fourier transformation of the small-angle scattering intensity observed, the so-called $p(r)$ function, 84 Å (8.4 nm). Referring to the results obtained for Takaamylase A (from *Aspergillus oryzae*) (molecular weight 45 000), we finally concluded that the molecule is prolate.

Glucoamylase, glucose isomerase and β-amylase; the principal enzymes for production of glucose, isomerized sugar and maltose

In 1958, Tsujisaka, Fukumoto and Yamamoto developed a method for the commercial production of glucose using an enzyme from a strain of *Rhizopus delemar*. Subsequently, good producers of glucoamylase, such as strains of *Aspergillus niger* have been discovered (Underkofler, Denault and Hou, 1965; Freedberg *et al.*, 1974). The enzyme from *Aspergillus niger* has been used for the production of glucose from starch in many countries. In Japan the enzymes from *Rhizopus delemar* or *Rhizopus niveus* also have been used industrially. A strain of *Endomycopsis fibuligera* (old name, *Endomyces fibuliger*) (Hattori and Takeuchi, 1961) was used in some Japanese companies at an early stage of the industry. The optimum pH and temperature of enzymes from *Aspergillus niger* are 4·5 and 60°C, whereas those from *Rhizopus* are 5·0 and 56°C. Enzymes from *Aspergillus niger* can be produced by liquid culture, but enzymes from *Rhizopus* are produced by culture on solids. Thus, in general, the former enzyme is preferred to the latter enzyme for industrial purposes. *Rhizopus* sp. (Ueda, Ohba and Kano, 1975) and *Aspergillus niger* (Smiley *et al.*, 1971) have multiple glucoamylases, whereas *Endomycopsis fibuligera* has a single enzyme (Fukui and Nikuni, 1969; Sukhumavasi, Kato and Harada, 1975; Kato *et al.*, 1976). A strain of *Rhizopus* produces glucoamylases I and II; the former, which is capable of hydrolysing raw starch, can hydrolyse the (1,6-α)-glucosidic linkage, but the latter cannot (Ueda, 1980). Many strains of *Rhizopus*, *Aspergillus niger* and *Endomycopsis fibuligera* have been used to produce the traditional fermented foods in South-East Asia: *Rhizopus* is used for Chinese liquor in China and for Tempe in Indonesia, *Aspergillus niger* for the special Japanese beverage Shochu, and *Endomycopsis fibuligera* for the production of sugar syrups from starch—in Thailand as Look Pang, in Indonesia as Ragi, and in the Philippines as Busbad. The glucoamylases from *Aspergillus awamori* (Ueda, 1957; Ueda, Ohba and Kano, 1974; Gasdorf *et al.*, 1975; Hayashida *et al.*, 1982) and *Aspergillus oryzae* (Morita *et al.*, 1966; Miah and Ueda, 1977a, b) which have been used for the production of the Japanese beverages Awamori and Sake, respectively, also consist of two or three forms. *Penicillium oxalicum* also has two forms (Yamasaki, Suzuki and Ozawa, 1977). The glucoamylase preparation of *Aspergillus niger*, Glucozyme XL-128 (Nagase Co.) contains a phosphatase (EC 3.1.3.1), the concomitant action of which was shown by Abe, Takeda and Hizukuri (1982) to be necessary for the complete hydrolysis of the starch. These workers mentioned that the idea that glucoamylases of various origins are capable of the complete hydrolysis of starch should be reconsidered, in the light of the presence of phosphatase, because starch can bind considerable amounts of phosphate covalently.

Glucoamylase can polymerize glucose by a reaction that is the reverse of hydrolysis (Fukumoto, 1969; Hehre, Okada and Genghof, 1969). The main products of this reverse reaction are maltose and isomaltose, although on prolonged incubation at high substrate concentrations other disaccharides can be detected. Polymerization can also occur by another mechanism, namely glucose transfer, which is catalysed by α-glucosidase (EC 3.2.1.20); this reaction was also shown to be catalysed by a transglucosylase (Pazur and Ando, 1961), which is often present as an impurity in crude glucoamylase preparations (Maher, 1968). Currently used commercial preparations of glucoamylase of many different origins do not now contain glucose transferase activity: Novo Company isolated a mutant strain of *Aspergillus niger* which was devoid of such transferase activity. Most α-glucosidase catalyses the formation of non-fermentable glucose oligomers by transferring a glucosyl moiety from the (1,4-α) to a (1,6-α) position. If transferase activity is present, considerable amounts of pannose and isomaltose are formed during saccharification and so the final yield of glucose is considerably reduced.

In 1957, Marshall and Kooi found that *Pseudomonas hydrophila* accumulates glucose isomerase in the cells when grown on xylose. Subsequently, numerous producers of the enzyme have been discovered by many research workers. In 1964 and 1965, D-xylose ketol-isomerase (xylose isomerase, EC 5.3.1.5) was discovered in strains of *Streptomyces* independently by two groups in Japan (Takasaki and Tanabe, 1964; Sato and Tsumura, 1964; Tsumura and Sato, 1965; Takasaki, 1966). This enzyme is called expediently glucose isomerase: it can convert glucose to fructose efficiently at a temperature of about 70°C. Thus, this enzyme was developed for commercial production of isomerized sugar composed of about equal amounts of glucose and fructose. This enzyme is induced by xylose, but Takasaki, Kosugi and Kanbayashi (1969) isolated a mutant strain of *Streptomyces albus* capable of using xylan as a much less expensive inducer than xylose to produce the enzyme. Later, Sanchez and Quinto (1975) isolated a D-xylose isomerase constitutive mutant strain that was insensitive to D-glucose repression, and Novo Company found a strain of *Bacillus coagulans* capable of producing glucose isomerase and developed the enzyme on an industrial scale.

The isomerase is immobilized by a chemical process in which the enzyme is bound to a solid support. Many companies such as Denki Kagaku Company and Novo Company have developed immobilized glucose isomerase to make isomerized sugar. Isomerized sugar being a mixture of glucose and fructose, comparable with invert sugar, is used in large amounts in food processing, particularly in soft-drink bottling; for this purpose it is used as a substitute for sucrose. In Japan about 500 000 t of isomerized sugar currently are produced commercially (*see* Chapter 5).

β-Amylase (EC 3.2.1.2) occurs in higher plants, such as sweet potato, soya bean, wheat, barley, oats, maize and rye. Robyt and French (1964) found an enzyme producing maltose in the culture broth of *Bacillus polymyxa* and later Higashihara and Okada (1974) discovered β-amylase in *Bacillus megatherium*. β-Amylase was also found in many strains of bacteria, such as *Bacillus* sp. (Shinke, Kunimi and Nishira, 1975b), *Streptomyces* sp. (Shinke, Nishira and Mugibayashi, 1974) and *Pseudomonas* sp. (Shinke, Kunimi and Nishira, 1975a). Some of these strains produce β-amylase with pullulanase. However, crude enzyme from soya bean has been used for commercial production of maltose in Japan, because much soya bean protein is produced industrially in Japan and crude β-amylase can be obtained fairly easily from it. The properties of soya bean β-amylase have been studied by Mikami, Aibara and Morita (1982).

Use of isoamylase in glucose production from starch

Glucose syrups are used for the production of glucose and as the starting material for the production of isomerized sugar consisting of equal amounts of glucose and fructose. Such syrups are mainly produced from starch by the action of glucoamylase, instead of the previously used (and now discontinued) acid treatment, which gave a lower glucose yield and an undesirable colour, derived from the sugar. Glucoamylase can hydrolyse (1,6-α)-glucosidic linkages, but at a much slower rate than (1,4-α)-glucosidic linkages. Industrially, it is desirable to obtain syrups with a high glucose content from starch: this can be achieved using a debranching enzyme with glucoamylase for the saccharification of starch (Harada *et al.*, 1982; Norman, 1982a, b).

One example of our experiments on the possible commercial production of glucose from starch was as follows. A 33% suspension of corn starch was liquefied by the action of Termamyl, a thermostable α-amylase (Madsen, Norman and Scott, 1973) at 98°C for 30 min. The optimum pH of the isoamylase is 3·0–4·5 at 30°C and about 4·0–4·5 at 55°C, whereas that of glucoamylase is 4·0–4·5. A pH of 4·5 was employed for the combined actions of glucoamylase from *Asper-*

gillus niger (amyloglucosidase Novo 150) and *Pseudomonas* isoamylase. The optimum temperature for the action of isoamylase is 50–55°C. Norman (1982b) reported that the limiting temperature for practical application of the glucoamylase is about 60°C. We examined the effect of temperatures of 50°C, 55°C and 60°C on the yield of glucose and found that 55°C was by far the best. The yield of glucose on hydrolysis at 55°C for 40 h was 95·9% whereas it was 94·2% when only glucoamylase was used.

Norman (1982a, b) examined the use of a debranching enzyme, *Pseudomonas* isoamylase or *Bacillus* pullulanase, in glucose syrup production and showed that only the *Bacillus* pullulanase can act at 60°C, a temperature which is useful for industrial operation because it prevents microbial contamination during saccharification. However, *Pseudomonas* isoamylase is much more acidophilic than *Bacillus* pullulanase and *Klebsiella* pullulanase, the optimum pH of *Bacillus* pullulanase being 4·5–5·5 whereas that of *Klebsiella* pullulanase is 5·5–6·0. Combined activity of glucoamylase and *Bacillus* pullulanase can be achieved at pH 4·5–5·0, whereas that of glucoamylase and isoamylase is achieved at pH 4·0–4·5. *Klebsiella* pullulanase is unsuitable for such combined activity because its optimum pH is much higher than that of glucoamylase. Thus, a much lower pH can be used for isoamylase than for *Bacillus* pullulanase. Use of the isoamylase which is capable of acting at such a low pH is advantageous for avoiding microbial contamination. If isoamylase or pullulanase is used with glucoamylase, the amount of glucoamylase required to obtain a high level of glucose can be reduced: the amount of glucoamylase required decreases as the amount of added isoamylase is increased. Glucoamylase reversion reaction products can thus be reduced, leading to an increase of glucose yield. Use of debranching enzyme and the glucoamylase system not only leads to a higher glucose level but also to a much shorter incubation time. The yield of glucose increases greatly with increase in the concentration of isoamylase used.

The ability of a particular glucoamylase to digest raw starch depends not only on the amylase activity, but also on the debranching activity of the enzymes. *Pseudomonas* isoamylase effectively enhances the production of glucose from raw starch by glucoamylase, as shown by Ueda, Ohba and Kano (1974). This process is important in the production of ethanol from starch.

Use of isoamylase in maltose production from starch

The action of β-amylase alone on amylopectin results in the formation of maltose (50–60% yield), leaving β-limit dextrin. On addition of isoamylase or pullulanase, the branching point of amylopectin is cleaved and the yield of maltose greatly increases. The following experiment was used to assess the commercial production of maltose from starch with *Pseudomonas* isoamylase and soya bean β-amylase (Harada *et al.*, 1982). An aqueous suspension of 20% corn starch was liquefied by the action of Termamyl and was then treated simultaneously (at various pH values of 3·5–6·0 at 50°C for 48 h) with β-amylase from soya bean and with *Pseudomonas* isoamylase. Bacterial α-amylase, which can cleave maltotriose to form maltose and glucose, was added to the reaction mixture to enhance maltose production. A pH of 4·5 was best for maltose production: this is under-

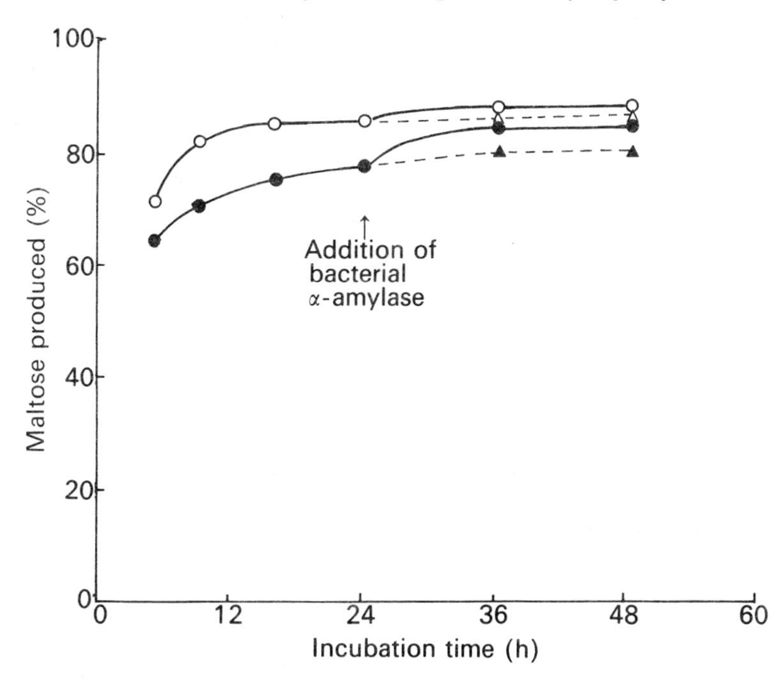

Figure 8. Production of maltose from corn starch at concentration of 21% (DE, 2·9) at 50°C by soya bean β-amylase (20 μg/g starch) with isoamylase (○) (6 μg/g starch) at pH 4·7 or pullulanase (●) (17 μg/g starch) at pH 6·0 (Harada *et al.*, 1982). The amounts of the enzyme used were expressed as those of purified enzyme. DE (dextrose equivalent), reducing sugars expressed as D-glucose and calculated as a % of dry substance. The curves with isoamylase (△) or pullulanase (▲) with β-amylase without bacterial α-amylase are shown as broken lines.

standable, bearing in mind that the optimum pH of the β-amylase is 5·0–6·0 and that of the isoamylase is 4·0–4·5 at 55°C. *Pseudomonas* isoamylase was compared with *Klebsiella* pullulanase (*Figure 8*). A pH of 4·7 was used for the isoamylase, but pullulanase was tested at pH 6·0 because the optimum pH of pullulanase is 5·5–6·0. The yield of maltose is higher with *Pseudomonas* isoamylase (87·6%) than with *Klebsiella* pullulanase (85·6%). In these experiments three times more pullulanase than isoamylase was added. High concentrations of substrate, such as the 20% used in this experiment, give a much lower maltose yield than do low concentrations such as 0·1% (*Table 1*). Chain-stubs of starch are composed of odd or even numbers of glucose residues: small amounts of glucose and maltotriose therefore are produced with maltose from starch by the simultaneous actions of β-amylase and isoamylase. To enhance the yield of maltose, the development of a precipitate due to retrogradation of the products formed by isoamylase should be minimized, by adding isoamylase at the most appropriate time and by adding isoamylase and glucoamylase or β-amylase in the correct proportions.

Some maltose produced by the Hayashibara Company (over 1000 t/year) is sold by the Ootsuka Pharmaceutical Company under the trade name of Martos 10. Maltose may be preferred to glucose for intravenous injection because of its lower osmotic pressure and slower release of glucose so that the time required for injection of glucose may be shortened when maltose is injected at an equivalent level to glucose. The Hayashibara Company succeeded in producing crystalline maltose and now produces over twenty thousand tons per year as a food sweetener. Maltose is a much better sweetener than sucrose for use in Japanese confectionery because it does not easily become crystalline when kept for a long time, whereas sucrose readily crystallizes in confectionery. This company has also succeeded in making crystalline maltitol, which can easily be obtained from maltose by chemical reduction; Ohno, Hirao and Kido (1982) have studied it by X-ray diffraction methods. Purified maltitol is not hygroscopic and therefore has numerous applications. In co-operation with other companies, Hayashibara Company produces now over 3000 t of maltitol per year as a low-calorie sweetener. Atsuji *et al.* (1972) and Oku, Him and Hosoya (1981) reported that maltitol is a low-calorie sweetener and Rennhard and Bianchine (1976) reported that maltitol is readily metabolized in rat, dog and man.

Maltose is almost as sweet as sucrose (Oda, 1974). Japan has the greatest commercial production of maltose and *Pseudomonas* isoamylase and soya bean β-amylase are mainly used for its production. *Pseudomonas* isoamylase is preferable to *Klebsiella* pullulanase for this purpose for the following reasons: (1) isoamylase splits over 20 times more (1,6-α)-glucosidic linkages in amylopectin than pullulanase per unit of enzyme protein; (2) because the action of isoamylase is not inhibited by maltose, a higher concentration of starch can be used as starting material; (3) the reaction of the isoamylase is non-reversible, whereas that of pullulanase is reversible. Pullulanase polymerizes maltose, forming a tetrasaccharide, and links maltose with amylose (Abdullah and French, 1970); the yield of maltose is therefore much higher with isoamylase than with pullulanase.

An account of the uses of debranching enzymes in the brewing industry has been given by Enevoldsen (1975).

Acknowledgement

The author wishes to thank a number of workers in Hayashibara Biochemical Laboratories, Inc. in Okayama for many helpful comments on the manuscript.

References

ABDULLAH, M. AND FRENCH, D. (1970). Substrate specificity of pullulanase. *Archives of Biochemistry and Biophysics* **137,** 483–493.

ABE, J., TAKEDA, Y. AND HIZUKURI, S. (1982). Action of glucoamylase from *Aspergillus niger* on phosphorylated substrate. *Biochimica et biophysica acta* **703,** 26–33.

AKAI, H., YOKOBAYASHI, K., MISAKI, A. AND HARADA, T. (1971a). Complete hydrolysis of branching linkages in glycogen by *Pseudomonas* isoamylase: Distribution of linear chains. *Biochimica et biophysica acta* **237,** 422–429.

AKAI, H., YOKOBAYASHI, K., MISAKI, A. AND HARADA, T. (1971b). Structural analysis of amylopectin using *Pseudomonas* isoamylase. *Biochimica et biophysica acta* **252,** 427–431.

AMEMURA, A., KITAGAWA, H. AND HARADA, T. (1975). Role of the tryptophan group in the action of pullulanase of *Aerobacter aerogenes*. *Journal of Biochemistry* **77,** 575–578.

AMEMURA, A., KONISHI, Y. AND HARADA, T. (1980). Molecular weight of the undegraded polypeptide chain of *Pseudomonas* isoamylase. *Biochimica et biophysica acta* **611,** 390–393.

AMEMURA, A., SUGIMOTO, T. AND HARADA, T. (1974). Characterization of intracellular α-glucosidase of *Pseudomonas* SB15. *Journal of Fermentation Technology* **52,** 778–780.

ATSUJI, H., ASANO, S., SUZUKI, S., KATAOKA, K., NAKAJIMA, M., SUZUKI, Y., TAKAHASHI, Y. AND MARUYAMA, H. (1972). Metabolism of maltitol. *Rinsho Eiyo* **41,** 200–208 (in Japanese).

ATWELL, W.A., HOSENEY, R.C. AND LINEBACK, D.R. (1980). Debranching of wheat amylopectin. *Cereal Chemistry* **57,** 12–16.

BENDER, H. AND WALLENFELS, K. (1961). Pullulan 2. Specific decomposition by a bacterial enzyme. *Biochemische Zeitschrift* **334,** 79–95.

DESSEIN, A. AND SCHWARTZ, M. (1974). Is there a pullulanase in *Escherichia coli*? *European Journal of Biochemistry* **45,** 363–366.

EISELE, E., RASHED, I.R. AND WALLENFELS, K. (1972). Molecular characterization of pullulanase from *Aerobacter aerogenes. European Journal of Biochemistry* **26,** 62–67.

ENEVOLDSEN, B.S. (1975). Debranching enzymes in brewing. In *Proceedings of the 15th Congress of the European Brewery Convention, Nice*, pp. 684–697. Elsevier Scientific Publishing Company, Amsterdam.

EVANS, R.M., MANNERS, D.J. AND STARK, J.R. (1979). Partial purification and properties of bacterial isoamylase. *Carbohydrate Research* **76,** 203–213.

FREEDBERG, I.M., LEVEN, Y., KAI, C.M., MCCUBBIN, W.D. AND KATCHALSKI-KATZIR, E. (1974). Purification and characterization of *Aspergillus niger* exo-1,4-glucosidase. *Biochimica et biophysica acta* **391,** 361–381.

FUKUI, T. AND NIKUNI, Z. (1969). Preparation and properties of crystalline glucoamylase from *Endomyces* species IFO 0111. *Journal of Agricultural and Biological Chemistry* **33,** 884–891.

FUKUMOTO, J. (1969). Industrial production of enzymes and its application. In *Biological Industrial Chemistry* (J. Fukumoto *et al.*, Eds), pp. 1–116. Asakura Publisher, Tokyo (In Japanese).

GASDORF, H.J., ATTHASAMPUNNA, P., DAN, V., HENSLEY, D.E. and SMILEY, K.L. (1975). Patterns of action of glucoamylase isozymes from *Aspergillus* species on glycogen. *Carbohydrate Research* **42,** 147–156.

GRIFFIN, P.J. AND FORGARTY, W.M. (1973). Preliminary observations on the starch-degrading system elaborated by *Bacillus polymyxa. Biochemical Society Transactions* **1,** 397–400.

GUNJA-SMITH, Z., MARSHALL, J.J. AND WHELAN, W.J. (1970). A glycogen-debranching enzyme from *Cytophaga. FEBS Letters* **12,** 96–100.

GUNJA-SMITH, Z., MARSHALL, J.J., MERCIER, C., SMITH, E.E. AND WHELAN, W.J. (1970). A revision of the Meyer-branched model of glycogen and amylopectin. *FEBS Letters* **12,** 101–104.

HARADA, T. (1965). A new method using test paper containing chromogenic substrates of hydrolases for the isolation and identification of microorganisms. *First International Congress of Food Science and Technology*, volume 2. pp. 629–634. Research Science Publishers, London.

HARADA, T., YOKOBAYASHI, K. AND MISAKI, A. (1968). Formation of isoamylase by *Pseudomonas. Applied Microbiology* **10,** 1939–1944.

HARADA, T., AMEMURA, A., KASAI, N. AND SAKAI, N. (1982). *Pseudomonas* isoamylase and its industrial application. In *Proceedings, XIth International Carbohydrate Symposium, Vancouver*, (G.G.S. Dutton, Ed.), pp. V7

HARADA, T., MISAKI, A., AKAI, H., YOKOBAYASHI, K. AND SUGIMOTO, K. (1972). Characterization of *Pseudomonas* isoamylase by its actions on amylopectin and glycogen: Comparison with *Aerobacter* pullulanase. *Biochimica et biophysica acta* **268,** 497–505.

HATA, K., HATA, MA, HATA, MI AND MATSUDA, K. (1983). The structures of shellfish glycogens. 2. A comparative study of the structures of glycogens from oyster, scallop and abalone. *Journal of Japanese Society of Starch Science* **30,** 95–101.

HATTORI, Y. AND TAKEUCHI, I. (1961). Studies on amylolytic enzymes produced by *Endomyces* sp. 2. Purification and general properties of amyloglucosidase. *Journal of Agricultural and Biological Chemistry* **25,** 895–901.

HAYASHIDA, S., KUNISAKI, S., NAKAO, M. AND FLOR, P.O. (1982). Evidence for raw starch-affinity site on *Aspergillus awamori* glucoamylase 1. *Journal of Agricultural and Biological Chemistry* **46,** 83–89.

HEHRE, E.J., OKADA, G. AND GENGHOF, D.S. (1969). Configurational specificity: Unappreciated key to understanding enzymic reversions and de novo glycosidic bond synthesis. *Archives of Biochemistry and Biophysics* **135,** 75–89.

HIGASHIHARA, M. AND OKADA, S. (1974). Studies on β-amylase of *Bacillus megatherium* strain No. 32. *Journal of Agricultural and Biological Chemistry* **38,** 1023–1029.

JEANNINGROS, R., CREUZET-SIGAL, N., FRIXON, C. AND CATTANEO, J. (1976). Purification and properties of a debranching enzyme from *Escherichia coli. Biochimica et biophysica acta* **438,** 186–199.

KAINUMA, K., KOBAYASHI, S. AND HARADA, T. (1978). Action of *Pseudomonas* isoamylase on various branched oligo and polysaccharides. *Carbohydrate Research* **61,** 345–357.

KATO, K., KONISHI, Y., AMEMURA, A. AND HARADA, T. (1977). Affinity chromatography of *Pseudomonas* isoamylase on cross-linked amylose gel. *Journal of Agricultural and Biological Chemistry* **41**, 2077–2080.

KATO, K., KUSWANTO, K., BANNO, I. AND HARADA, T. (1976). Identification of *Endomycopsis fibuligera* isolated from Ragi in Indonesia and properties of its crystalline glucoamylase. *Journal of Fermentation Technology* **54,** 831–837.

KATO, K., SUGIMOTO, T., AMEMURA, A. AND HARADA, T. (1975). A *Pseudomonas* intracellular amylase with high activity on maltodextrins and cyclodextrins. *Biochimica et biophysica acta* **391,** 96–108.

KITAGAWA, H., AMEMURA, A. AND HARADA, T. (1975). Studies on the inhibition and molecular properties of crystalline *Pseudomonas* isoamylase. *Journal of Agricultural and Biological Chemistry* **39,** 989–994.

KONISHI, Y., AMEMURA, A., TANABE, S. AND HARADA, T. (1979). Immunological study of pullulanase from *Klebsiella* strains and the occurrence of this enzyme in the Enterobacteriaceae. *International Journal of Systematic Bacteriology* **29,** 13–18.

KUSWANTO, K.R., KATO, K., AMEMURA, A. AND HARADA, T. (1976). Isoamylase production and growth on glucose, maltose and α-glucans of *Pseudomonas amyloderamosa* SB15 and its mutant strains. *Journal of Fermentation Technology* **54,** 192–196.

LEE, E.Y.C. AND WHELAN, W.J. (1971). Glycogen and starch debranching enzymes. In *The Enzymes* (P. D. Boyer, Ed.) volume 5, pp. 191–234. Academic Press, New York.

LEE, E.Y.C., MERCIER, C. AND WHELAN, W.J. (1968). A method for the investigation of the fine structure of amylopectin. *Archives of Biochemistry and Biophysics* **125,** 1028–1029.

LEE, E.Y.C., NIELSEN, L.D. AND FISHER, E.H. (1967). A new glycogen-debranching enzyme system in yeast. *Archives of Biochemistry and Biophysics* **121,** 245–246.

MADSEN, G.B., NORMAN, B.E. AND SCOTT, S. (1973). A new, heat stable bacterial amylase and its use in high temperature liquefaction. *Stärke* **25,** 304–308.

MAHER, G.G. (1968). Inactivation of transglucosidase in enzyme preparations from *Aspergillus niger. Stärke* **20,** 228–232.

MANNERS, D.J. AND MATHESON, N.K. (1981). The fine structure of amylopectin. *Carbohydrate Research* **90**, 99–110.

MARSHALL, J.J. (1973). Inhibition of pullulanase by Schardinger dextrins. *FEBS Letters* **37,** 269–273.

MARSHALL, J.J. AND WHELAN, W.J. (1974). Multiple branching in glycogen and amylopectin. *Archives of Biochemistry and Biophysics* **161,** 234–238.

MARSHALL, R.O. AND KOOI, E.R. (1957). Enzymatic conversion of D-glucose to D-fructose as well as D-xylose to xylulose. *Science* **125,** 648–649.

MARUO, B. AND KOBAYASHI, T. (1951). Enzymatic scission of the branch links in amylopectin. *Nature* **167,** 606–607.

MERCIER, C. AND KAINUMA, K. (1975). Enzymatic debranching of starch from maize of various genotypes in high concentration of dimethylsulphoxides. *Stärke* **9,** 289–292.

MIAH, M.N.N. AND UEDA, S. (1977a). Multiplicity of glucoamylase of *Aspergillus oryzae.* 1. Separation and purification of three forms of glucoamylase. *Stärke* **29,** 191–196.

MIAH, M.N.N. AND UEDA, S. (1977b). Multiplicity of glucoamylase of *Aspergillus oryzae.* 2. Enzymatic and physiochemical properties of three forms of glucoamylase. *Stärke* **29,** 235–239.

MIKAMI, B., AIBARA, S. AND MORITA, Y. (1982). Distribution and properties of soya bean β-amylase isozyme. *Journal of Agricultural and Biological Chemistry* **46,** 943–953.

MORGAN, F.J., ADAMS, K.R. AND PRIEST, F.G. (1979). A cultural method for the detection of pullulan degrading enzymes in bacteria and its application to the Genus *Bacillus. Journal of Applied Bacteriology* **46,** 291–294.

MORITA, Y., SHIMIZU, K., OHGA, M. AND KORENAGA, T. (1966). Studies on amylases of *Aspergillus oryzae* cultured on rice. 1. Isolation and purification of glucoamylase. *Journal of Agricultural and Biological Chemistry* **30,** 114–121.

NAKAMURA, M. (1977). Structure of amylopectin. In *Denpun-Kagaku (Starch Science) Handbook* (M. Nakamura and S. Suzuki, Eds), pp. 12–19. Asakura Publisher, Tokyo (in Japanese).

NORMAN, B.E. (1982a). A novel debranching enzyme for application in the glucose syrup industry. *Stärke* **34,** 340–346.

NORMAN, B.E. (1982b). The use of debranching enzymes in dextrose syrup production. In *Maize: Recent Progress in Chemistry and Technology,* pp. 157–179. Academic Press, Inc., New York.

NORRMAN, J. AND WÖBER, G. (1975). Comparative biochemistry of α-glucan utilization in *Pseudomonas amyloderamosa* and *Pseudomonas saccharophila.* Physiological significance of variations in the pathways. *Archives of Microbiology* **102,** 253–260.

ODA, T. (1974). Studies on the character and application of maltitol and other sugar alcohols. *Denpun Kagaku Kaishi (Starch Science)* **21,** 322–327 (in Japanese).

OHNO, S., HIRAO, M. AND KIDO, M. (1982). X-ray crystal structure of maltitol (4-*O*-α-D-glucopyranosyl-D-glucitol). *Carbohydrate Research* **108,** 163–171.

OKU, T., HIM, S.H. AND HOSOYA, N. (1981). Effect of maltose and diet containing starch on maltitol hydrolysis in rats. *Eiyo to Schokuryo* **34,** 145–157 (in Japanese).

PALMER, T.N., MACAKIE, L.E. AND GREWAL, K.K. (1983). Spatial distribution of unit chains is glycogen. *Carbohydrate Research* **115,** 139–150.

PALMER, T.N., WÖBER, G. AND WHELAN, W.J. (1973). The pathway of exogenous and endogenous carbohydrate utilization in *E. coli*: A dual function for the enzymes of the maltose operon. *European Journal of Biochemistry* **39,** 601–612.

PAZUR, J.H. AND ANDO, T. (1961). The isolation and mode of action of a fungal transglucosylase. *Archives of Biochemistry and Biophysics* **93,** 43–49.

RENNHARD, H.H. AND BIANCHINE, J.R. (1976). Metabolism and caloric utilization of orally administered maltitol-^{14}C in rat, dog and man. *Journal of Agriculture and Food Chemistry* **24,** 287–291.

ROBYT, J. AND FRENCH, D. (1964). Purification and action pattern of an amylase from *Bacillus polymyxa. Archives of Biochemistry and Biophysics* **104,** 338–345.

SAKAI, S. (1981). Production and usage of maltose. *Denpun Kagaku Kaishi* **28,** 72–78 (in Japanese).

SAKANO, Y., HIRAIWA, S., FUKUSHIMA, J. AND KOBAYASHI, T. (1982). Enzymatic properties and action patterns of *Thermoactinomyces vulgaris* α-amylase. *Journal of Agricultural and Biological Chemistry* **46,** 1121–1129.

SANCHEZ, S. AND QUINTO, C.M. (1975). D-Glucose isomerase: Constitutive and catabolic repression-resistant mutants of *Streptomyces phaeochromogenes. Journal of Applied Microbiology* **30,** 750–754.

SATO, M., HATA, Y., II, Y., MIKI, K., KASAI, N. AND HARADA, T. (1982). Preliminary X-ray studies on *Pseudomonas* isoamylase. *Journal of Molecular Biology* **160,** 669–671.

SATO, T. AND TSUMURA, N. (1964). Japanese Patent 489867.

SHINKE, R., KUNIMI, Y. AND NISHIRA, H. (1975a). Isolation and characterization of β-amylase-producing microorganisms. *Journal of Fermentation Technology* **53,** 687–692.

SHINKE, R., KUNIMI, Y. AND NISHIRA, H. (1975b). Production and some properties of β-amylase of *Bacillus* sp. BQ10. *Journal of Fermentation Technology* **53,** 693–697.

SHINKE, R., NISHIRA, H. AND MUGIBAYASHI, N. (1974). Isolation of β-amylase-producing microorganisms. *Journal of Agricultural and Biological Chemistry* **38,** 665–666.

SMILEY, K.L., HENSLEY, D.E., SMILEY, M.J. AND GASDORI, H.J. (1971). Kinetic patterns of glucoamylase isozymes isolated from *Aspergillus* species. *Archives of Biochemistry and Biophysics* **144,** 694–699.

SUGIMOTO, T., AMEMURA, A., AND HARADA, T. (1974). Formations of extracellular isoamylase and innercellular α-glucosidase and amylase(s) by *Pseudomonas* SB15 and a mutant strain. *Journal of Applied Microbiology* **28,** 336–339.

SUKHUMAVASI, S., KATO, K. AND HARADA, T. (1975). Glucoamylase of a strain of *Endomycopsis fibuligera* isolated from mould bran (Look Pang) of Thailand. *Journal of Fermentation Technology* **53,** 559–565.

TAKASAKI, T. (1966). Studies on sugar-isomerizing enzyme. Production and utilization of glucose isomerase from *Streptomyces* sp. *Journal of Agricultural and Biological Chemistry* **30**, 1247–1252.

TAKASAKI, T. AND TANABE, N. (1964). Japanese Patent 484472.

TAKASAKI, T., KOSUGI, Y. AND KANBAYASHI, A. (1969). *Streptomyces* glucose isomerase. In *Fermentation Advances* (D. Perman, Ed.), pp. 561–570. Academic Press, New York.

TAKASAKI, Y. (1974). Formation of glucose isomerase by *Streptomyces* sp, *Journal of Agricultural and Biological Chemistry* **38,** 667–668.

TAKASAKI, Y. (1976a). Productions and utilizations of β-amylase and pullulanase from *Bacillus cereus* var. *mycoides*. *Journal of Agricultural and Biological Chemistry* **40,** 1515–1530.

TAKASAKI, Y. (1976b). Purification and enzymatic properties of β-amylase and pullulanase from *Bacillus cereus* var. *mycoides*. *Journal of Agricultural and Biological Chemistry* **40,** 1523–1530.

TSUJISAKA, Y., FUKUMOTO, J. AND YAMAMOTO, T. (1958). Specificity of crystalline saccharogenic amylase of molds. *Nature* **181,** 770–771.

TSUMURA, N. AND SATO, T. (1965). Enzymatic conversion of D-glucose to D-fructose. Properties of the enzyme from *Streptomyces phaeochromogenes*. *Journal of Agricultural and Biological Chemistry* **29**, 1129–1134.

UEDA, S. (1957). Studies on the amylolytic system of black-koji-molds. 5. Debranching activity of the saccharogenic amylase fraction. *Bulletin of Agricultural Chemical Society of Japan* **21,** 379–385.

UEDA, S. (1980). Raw starch digestion. In *Mechanisms of Saccharide Polymerization and Depolymerization* (J.J. Marshall, Ed.), pp. 55–72. Academic Press, New York.

UEDA, S., OHBA, R. AND KANO, S. (1974). Fractionation of the glucoamylase system from black-koji mold and the effects of adding isoamylase and alpha-amylase on amylolysis by the glucoamylase fractions. *Stärke* **26,** 374–378.

UEDA, S., OHBA, R. AND KANO, S. (1975). Multiple forms of glucoamylase of *Rhizopus* species. *Stärke* **27,** 123–128.

UMEKI, K. AND YAMAMOTO, T. (1977). Determination of A and B chains of shellfish glycogen. *Journal of Agricultural and Biological Chemistry* **41,** 1515–1517.

UNDERKOFLER, L.A., DENAULT, L.J. AND HOU, E.F. (1965). Enzymes in the starch industry. *Stärke* **17,** 179–184.

URLAUB, H. AND WÖBER, G. (1975). Identification of isoamylase, a glycogen debranching enzyme, from *Bacillus amyloliquefaciens*. *FEBS Letters* **57,** 1–4.

WALKER, G.J. (1968). Metabolism of the reserve polysaccharide of *Streptococcus mitis*, some properties of pullulanase. *Biochemical Journal* **108,** 33–40.

WEBER, M. AND WÖBER, G. (1979). The fine structure of the branched α-glucan from the blue-green alga *Anacystis nidulans*: Comparison with other bacterial glycogens and phytoglycogen. *Carbohydrate Research* **39,** 295–302.

WÖBER, B. (1973). The pathway of maltodextrin metabolism in *Pseudomonas stutzeri. Hoppe Seyler's Zeitschrift für physiologische Chemie* **354,** 75–82.

YAGISAWA, M., KATO, K., KOBA, Y. AND UEDA, S. (1972). Pullulanase of *Streptomyces* sp. 280. *Journal of Fermentation Technology* **50,** 572–579.

YAMANOBE, T. AND TAKASAKI, Y. (1979). Production of maltose from starch of various origins by β-amylase and pullulanase of *Bacillus cereus* var. *mycoides. Nippon Nogeikagaku Kaishi* **53,** 77–80 (in Japanese).

YAMASAKI, Y., SUZUKI, Y. AND OZAWA, J. (1977). Purification and properties of two forms of glucoamylase from *Penicillium oxalicum. Journal of Agricultural and Biological Chemistry* **41,** 755–762.

YOKOBAYASHI, K., MISAKI, A. AND HARADA, T. (1969). Specificity of *Pseudomonas* isoamylase. *Journal of Agricultural and Biological Chemistry* **33,** 625–627.

YOKOBAYASHI, K., MISAKI, A. AND HARADA, T. (1970). Purification and properties of *Pseudomonas* isoamylase. *Biochimica et biophysica acta* **212,** 458–469.

YOKOBAYASHI, K., AKAI, H., SUGIMOTO, T., HIRAO, M., SUGIMOTO, K. AND HARADA, T. (1972). Comparison of the kinetic parameters of *Pseudomonas* isoamylase and *Aerobacter* pullulanase. *Biochimica et biophysica acta* **293,** 197–202.

3
Genetic Engineering and Nitrogen Fixation

J. E. BERINGER AND P. R. HIRSCH

Soil Microbiology Department, Rothamsted Experimental Station, Harpenden, Hertfordshire AL5 2JQ, UK

Introduction

Nitrogen is extremely important in agriculture because it is a constituent of proteins, nucleic acids and other essential molecules in all organisms. Most of this nitrogen is derived from reduced or oxidized forms of N in the soil by growing plants, because plants and animals are unable to utilize N_2, which is abundant in the atmosphere. Under most cropping conditions N is limiting for growth and is provided in fertilizers, usually at rates of between 50 and 300 kg of N per ha per year (Anonymous, 1979). The only other sources available to plants are from decomposing organic matter, soil reserves, biological nitrogen fixation, the little that is deposited in rainfall and from other sources such as automobile exhausts.

Biological nitrogen fixation, the enzymic conversion of N_2 gas to ammonia, is much the most important source of fixed nitrogen entering those soils which receive less than about 5 kg N per ha per year from fertilizers. The reduction of N_2 is catalysed by the nitrogenase system, which is very similar in composition and function in all prokaryotes which produce it. Indeed, subunits of nitrogenase obtained from different nitrogen-fixing species can often be mixed to produce a functional system (Emerich and Burris, 1978). In addition, DNA coding for the structural proteins is so highly conserved in sequence that this coding has been used in hybridization experiments to demonstrate the presence of these genes in all nitrogen-fixing species of prokaryotes tested (Mazur, Rice and Haselkorn, 1980; Ruvkun and Ausubel, 1980). Nitrogenase is found only in prokaryotic micro-organisms and thus eukaryotes, such as plants, can benefit from N_2 fixation only if they interact with N_2-fixing species of micro-organism or obtain the fixed N after the death of the organisms.

Nitrogenase functions only under anaerobic conditions because it is irreversibly inactivated by oxygen. The fixation of N_2 requires large amounts of energy, about 30 moles of ATP per mole N_2 reduced (Hill, 1976; Schubert and Wolk, 1982), and thus can act as a major drain for energy produced by N_2-fixing micro-organisms. The requirement for an anaerobic environment and large amounts of energy presents problems to the micro-organisms that fix N_2 and

Biotechnology and Genetic Engineering Reviews—Vol. 1, February 1984
0264–8725/84/01/65–24$10.00 + $0.00

to the geneticists who wish to extend the range of N_2-fixing organisms. Many micro-organisms fix N_2 anaerobically and thus avoid the oxygen problem. However, energy production from organic compounds is usually much more efficient when they are metabolized by oxidative phosphorylation. Thus, in general, nitrogen fixation under aerobic or microaerobic conditions should be more efficient, unless too much energy is lost in protecting the enzyme from oxygen or replacing oxygen-damaged proteins.

An important consequence of the large energy cost for biological nitrogen fixation is that the activity of nitrogenase needs to be regulated very carefully to ensure that only the required amount of fixed N is produced. We discuss the regulation of N_2 fixation in *Klebsiella pneumoniae* in some detail in this chapter because a full understanding of how nitrogenase is regulated will be necessary if the transfer of N_2 fixation genes (*nif*) into other species, or even plants, is to be beneficial to the recipient organism.

The preceding remarks about the energy requirement and oxygen stability of nitrogenase point to two of the most important problems that will be faced in transferring *nif* genes to new hosts. In this review we will discuss other potential problems and show how our knowledge of the genetics of nitrogen fixation might be exploited in future.

N_2-fixing organisms and associations

The ability to fix N_2 is found in a wide range of prokaryotic micro-organisms, some of which are listed in *Table 1*. Genetic studies showing that the structural genes are highly conserved in all known N_2-fixing species (Mazur, Rice and Haselkorn, 1980; Ruvkun and Ausubel, 1980) suggest that the *nif* genes must have evolved and then spread throughout virtually all groups of prokaryotic micro-organisms. The very great similarities in DNA sequence imply that nitrogenase probably evolved after most of the major groups of prokaryotes diverged and thus that the distribution of the genes occurred as a result of gene transfer. If the evolution had been convergent, one might expect to see similar proteins, but much less similarity in DNA sequence. If this evolutionary argument is correct there is good historical precedence for the potential for

Table 1. Nitrogen-fixing prokaryotes of particular interest for genetic manipulation.

Genus	Properties of interest
Azotobacter	Aerobic N_2 fixer; genetic work in progress.
Azospirillum	Commonly associated with plant roots; genetic work in progress.
Rhizobium	Nodulates some species of leguminous plants; genetic studies of fast-growing species well developed.
Klebsiella	Genetics of *nif* well understood, particularly after transfer to *Escherichia coli*.
Frankia	Nodulates a range of unrelated shrubby dicotyledonous plants; some species can be cultured.
Rhodospirillum, *Rhodopseudomonas*	Photosynthetic bacteria; genetic work in progress.
Anabaena, *Nostoc*	Form symbiotic associations with a range of plants, fungi etc.; photosynthetic, genetic work in progress.

expanding the range of N_2-fixing organisms in the future. However, it should be pointed out that no eukaryotic organism has yet been shown to code for nitrogenase and thus the barriers to transfer discussed in this chapter may indeed present very real obstacles to geneticists, who even now are mere amateurs in relation to the processes that have directed evolution.

It should be made clear also that the amount of N_2 fixed by micro-organisms in nature is related almost directly to their access to available sources of energy. Thus for free-living micro-organisms levels of N_2 fixation seldom approach 5 kg N per ha per year (*see* Beringer, 1983), although free-living blue-green algae (cyanobacteria) can fix larger amounts when adequate moisture is available because they are photosynthetic and thus are able to meet their own needs for photosynthetically produced organic compounds. In general, with the possible exception of blue-green algae in rice paddies, it is unlikely that gene manipulation to improve N_2 fixation in agriculture will be worth contemplating for free-living micro-organisms.

A number of N_2-fixing micro-organisms grow in the rhizosphere of a range of crop plants, particularly grasses (Dobereiner and Boddey, 1981) and, under these conditions where root exudates are fairly abundant, have the potential to fix quite large amounts of N_2. Indeed, rates as high as 100 kg N per ha per year have been reported (von Bülow and Dobereiner, 1975). Unfortunately, experience has shown that the amounts of N_2 fixed are relatively low (probably around 10–20 kg N per ha per year; *see* Beringer, 1983) and that only a small part of this is available directly to the plant. While the potential sources of energy in the rhizosphere are greater than those in the bulk of the soil it must be remembered that many other micro-organisms are attracted to the exudates and will also compete with the plant for fixed N released by the N_2-fixing species. Whether such associations can be exploited in the future to improve nutrient exchange, and hence the benefit to the plant, remains to be determined. Until we know much more about these associations it is difficult to see how important gene manipulation could be in improving associative N_2 fixation for crop production.

Much the most important N_2-fixing associations that occur in nature, both in terms of the amounts of N_2 fixed and the benefits to particular crops, involve symbiotic associations in which plants provide specific structures in which the micro-organisms are contained. The best-known examples are root nodules which are induced by species of *Rhizobium* on leguminous plants, by the actinomycete *Frankia* on a range of woody dicotyledonous plants, by the blue-green alga *Nostoc* on cycads, and the special pockets in the leaves of the water fern *Azolla* containing the blue-green alga *Anabaena.*

In all of these symbioses the micro-organisms live inside the plant as monocultures. They receive carbon compounds directly from the host and the fixed N is made available directly to the host. Indeed, it is becoming clear now that the micro-organisms are modified in the host so that when nitrogenase is functioning they are no longer able to assimilate the fixed N and thus they release NH_4^+ directly to be assimilated by the plant (O'Gara and Shanmugam, 1976; Peters *et al.*, 1982). Furthermore, the plant can regulate N_2 fixation by limiting the amount of infected tissue (e.g. the number of nodules) and by regulating carbon flow to the micro-organisms.

In relation to the comments made earlier in this chapter about the efficient use of carbon compounds, it is clear (particularly for *Rhizobium*, which is an aerobe) that N_2 fixation will be most efficient in a micro-aerobic environment. In the legume root nodule this is facilitated by producing haemoglobin which ensures that O_2 is transported to the centre of the nodule, which otherwise could become anaerobic, because of the large number of bacteria present (probably about 10^6–10^7 per nodule), at a sufficiently low P_{O_2} to prevent inactivation of nitrogenase.

Thus in our estimation the symbioses provide an example of the extent to which plants and microbes have had to adapt themselves in order for nitrogenase to function in a way to provide the maximum benefit to the host. We feel that it is important to bear this sophistication in mind when considering possible approaches towards the production of novel N_2-fixing organisms.

Genetics of free-living nitrogen fixation

Some of the earliest studies on the biochemistry of nitrogenase were made using enzyme isolated from the anaerobe *Clostridium pasteurianum*, and there has been much interest in aerobic nitrogen-fixing bacteria such as *Azotobacter*. However, the greatest advances in our understanding of biological nitrogen fixation have involved *Klebsiella pneumoniae*, which is closely related to the non-nitrogen-fixing bacterium *Escherichia coli*, the standard organism for bacterial geneticists. Methods for genetic analysis in *E. coli* could be applied to *K. pneumoniae* and genes could be transferred between the two species and expressed in either of the two organisms. 'Classical' genetics with mutants defective in nitrogen fixation (Nif^-) showed the *nif* genes of *K. pneumoniae* to be located on the chromosome, between genes for histidine biosynthesis (*his*) and shikimic acid uptake (*shiA*). The proximity of the *his* and *nif* regions was used by Dixon and Postgate (1972) to transfer these regions entire from a strain of *K. pneumoniae*, carrying a conjugative gene-mobilizing plasmid, to an *E. coli* strain which required histidine. *E. coli* recipients no longer requiring histidine for growth were found to have acquired the ability to fix nitrogen. Subsequently, a conjugative plasmid, pRDI, with a wide host-range and which had picked up the *K. pneumoniae nif* and *his* genes, was selected and transferred to several different bacterial genera (Dixon, Cannon and Kondorosi, 1976).

Complementation analysis of different *nif* mutations, transductional and other forms of genetic mapping, and the physical purification of DNA fragments by cloning into small plasmids have led to the construction of a physical map over a region of about 24 kb which is the *nif* DNA (Cannon, Riedel and Ausubel, 1979; Riedel, Ausubel and Cannon, 1979) and to the recognition and mapping of 17 separate genes within the *nif* cluster. Mutagenesis using DNA elements such as transposons and bacteriophage Mu also facilitated the identification of eight operons (groups of genes transcribed as a block from one promotor) within the *nif* gene cluster (Dixon *et al.*, 1977; Kennedy, 1977; MacNeil *et al.*, 1978; Merrick *et al.*, 1980; Pühler and Klipp, 1981).

More recently, RNA transcripts have been mapped and the promotor regions from the *nif* gene cluster have been cloned and sequenced so that the initiation point for each operon is now known (Beynon *et al.*, 1983; Drummond *et al.*, 1983; Sundaresan *et al.*, 1983). One method used to identify promotors is to fuse a fragment lacking a promotor but carrying the structural gene for β-galactosidase (*lacZ*), to putative promotor regions. Promotor activity can then be measured by assaying β-galactosidase activity, and this has contributed towards understanding the control of expression of the *nif* operons (Dixon *et al.*, 1980; Hill *et al.*, 1981; MacNeil, Zhu and Brill, 1981; Merrick *et al.*, 1982; Drummond *et al.*, 1983; Ow and Ausubel, 1983). The genetic map, gene functions, organization and regulation are summarized in *Figures 1–3* and discussed below.

Nitrogenase has two soluble protein components, I and II. Component I consists of two copies of two subunits Iα and Iβ, which are encoded by *nifK* and *nifD* respectively. It contains both Mo and Fe atoms and is often referred to as the iron–molybdenum protein. Component II has two identical subunits encoded by *nifH*. Because it contains Fe atoms it is sometimes referred to as the Fe protein. These three genes, together with *nif Y* (the function of the gene product is not yet known), are arranged in one operon. It is suggested that *nifF* encodes a flavodoxin which transfers electrons to nitrogenase, and that *nif J* encodes a pyruvate oxidoreductase which transfers electrons from pyruvate to the flavodoxin. Products of two other genes, *nifM* and *nifS*, are required for processing the *nifH* gene product, component II of nitrogenase (Roberts *et al.*, 1978). Functional nitrogenase requires an inorganic cofactor containing Fe and Mo atoms, known as FeMoCo and four genes, *nifB*, *nifN*, *nifE* and *nifV*, are involved in its synthesis and maturation (Roberts *et al.*, 1978). Nitrogen fixation requires transfer of an electron from component II to component I with concomitant hydrolysis by component II of MgATP to MgADP. The main reductant for *in vivo* nitrogen fixation by *K. pneumoniae* is probably pyruvate, and *nifF* and *nifJ* gene products are required for transport of electrons from pyruvate via a flavodoxin to component II (Hill and Kavanagh, 1980). The functions of three other *nif* genes (*nifQ*, *nifU*, *nifX*) are not yet known.

All the *nif* operons (shown in *Figure 1*) are regulated by products of *nifA* and *nifL*. The *nifA* gene product is required for expression of all the *nif* operons except *nifLA*, and the *nifL* gene product represses expression, probably by inactivating the *nifA* gene product.

Nitrogenase is expressed only under certain physiological conditions: the enzyme is rapidly and irreversibly inactivated by exposure to oxygen and *K. pneumoniae* fixes nitrogen only in anaerobic or extremely micro-aerobic conditions. It is a facultative aerobe and when exposed to O_2, *nif* expression rapidly ceases (Eady *et al.*, 1978), probably because of *nifL*-mediated repression (Hill *et al.*, 1981). Nitrogen fixation uses energy, and when fixed nitrogen (nitrate, ammonia or amino acids) is available *nifA* expression is repressed, as are other genes involved in nitrogen metabolism. The regulation of genes under 'nitrogen control' is therefore of critical importance to expression of *nif*.

Early models for regulation of genes under nitrogen control in *K. pneumoniae* proposed that glutamine synthetase (GS) was responsible, but recent work with *E. coli*, *Salmonella typhimurium* and *K. pneumoniae* has shown them to be incorrect. Merrick (1982) has reviewed the evidence and presented a new model for nitrogen control in these bacteria. The structural gene for GS, *glnA* has

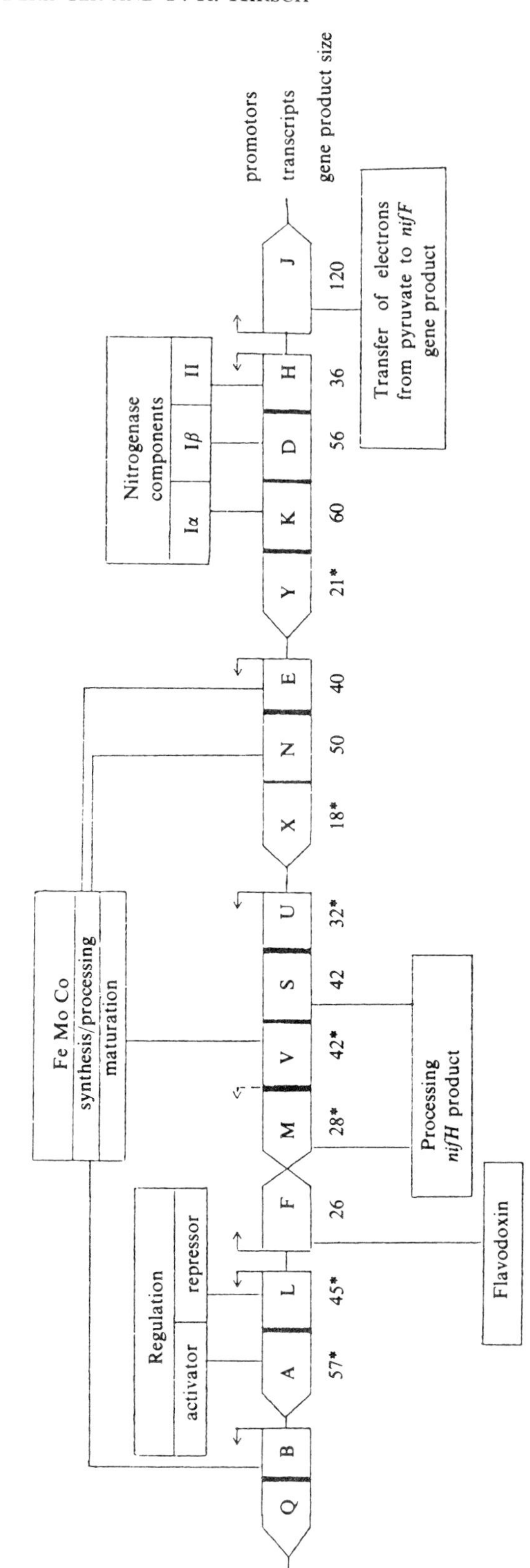

Figure 1. Summary of *nif* gene cluster organization, transcription and gene products. Transcripts and promotors according to Beynon *et al.* (1983) and M. Cannon (personal communication. *nifM* can also be transcribed off its own less efficient promotor, Gene product size: molecular weight × 10^{-3}. * Ausubel and Cannon (1980), Roberts and Brill (1980).

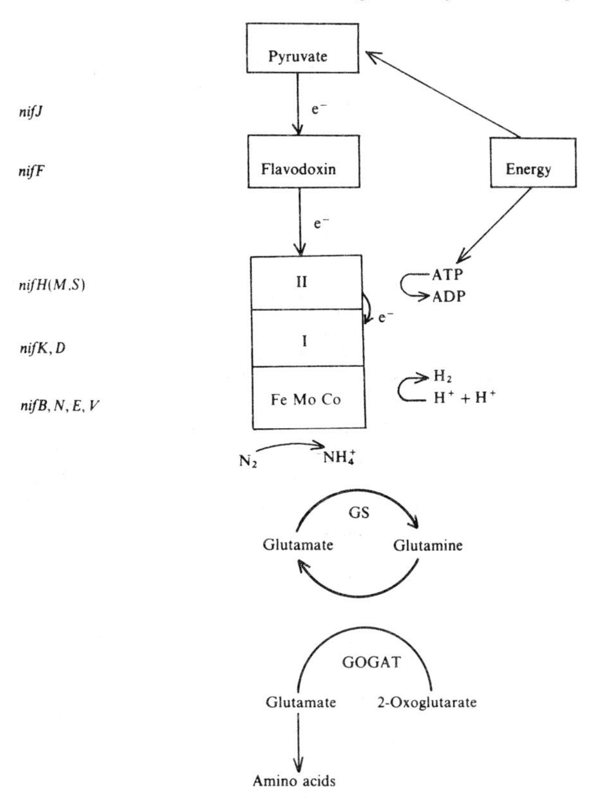

Figure 2. Functions of *nif* gene products in nitrogen fixation.

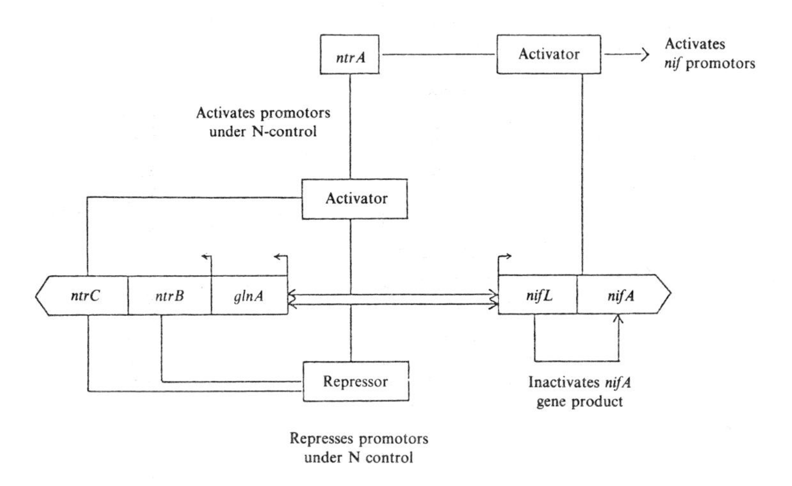

Figure 3. Nitrogen control and *nif* expression in *K. pneumoniae.*

been found to be the first in a transcriptional unit. The second gene, *ntrB* (formerly *glnL*) and the third, *ntrC* (*glnG*) are implicated in nitrogen control in conjunction with a separate gene, *ntrA* (*glnF*). Genes *ntrB* and *ntrC* also have their own weak promotor within the *glnA* operon. The *glnA* operon is transcribed constitutively at a low level, and the *ntrB* and *ntrC* gene products together repress operons under nitrogen control including *glnA* but not the internal *ntrB* promotor. This allows the constitutive expression of the *ntrB* and *ntrC* gene products. However, when the *ntrA* gene product is expressed, it associates with the *ntrC* gene product to form an activator for the *glnA* promotor and others under nitrogen control. It is proposed that *ntrC* encodes a DNA-binding protein which activates transcription in conjunction with the *ntrA* gene product, and that the *ntrB* gene product is either a competitor for DNA-binding sites with the *ntrC-ntrA* activator or inactivates it. In this model, (summarized in *Figure 3*), *ntrA* expression is essential for derepression of genes under nitrogen control (including the *nif* regulator *nifLA*) and must therefore be expressed when fixed nitrogen becomes deficient, in effect switching on the appropriate genes.

It has recently been reported (Drummond *et al.*, 1983; Ow and Ausubel, 1983) that the *nifA* gene product can substitute for *ntrC* in activating operons under nitrogen control in conjunction with the *ntrA* gene product. In addition, the *ntrC* or *nifA* gene products are necessary for transcription of the *nifLA* operon (Drummond *et al.*, 1983; Ow and Ausubel, 1983). The similarity between the *nifA* and *ntrC* gene products and the *nifLA* and *ntrBC* operons has led to the suggestion that *nifA* evolved from *ntrC* (Merrick *et al.*, 1982; Drummond *et al.*, 1983; Ow and Ausubel, 1983). However, they have diverged sufficiently to prevent *ntrC* from replacing *nifA* in activating the *nifH* promotor. The DNA sequences for all the *nif* operon promotors have recently been published (Beynon *et al.*, 1983; Drummond *et al.*, 1983) and have been shown to differ from other promotor sequences known in enteric bacteria. Beynon and co-workers (1983) report that the *nif* promotors contain similar DNA sequences arranged in two conserved regions. One of the regions is suggested to be specific for *nifA* protein binding, and the *nifLA* promotor alone has a slightly different sequence which may be because the primary activator is the *ntrC* rather than the *nifA* gene product. The other conserved region was found in all the *nif* promotors and may be specific to promotors expressed under conditions of nitrogen starvation. It is possible that the *ntrA* gene product interacts with RNA polymerase to produce a form which recognizes promotors of genes to be switched on under these conditions.

In summary, the *nifLA* operon is not transcribed when high levels of fixed nitrogen are present, but when this drops the *ntrA* gene product is active and in conjunction with the *ntrC* gene product de-represses transcription of *nifLA*. However, the *nifL* gene product prevents the *nifA* product from activating other *nif* promotors until the fixed-nitrogen level drops and nitrogen starvation begins. The *nifA* gene product is then active (the *nifL* product inactive) and, in association with the *ntrC* gene product, allows the *nif* operons to be expressed. Mutants with constitutive expression of *nifA* and no *nifL* gene products can transcribe the *nif* operons in the presence of fixed N and O_2 (Buchanan-Wollaston *et al.*, 1981; Merrick *et al.*, 1982). It is probable that the *nifL* gene product is in an

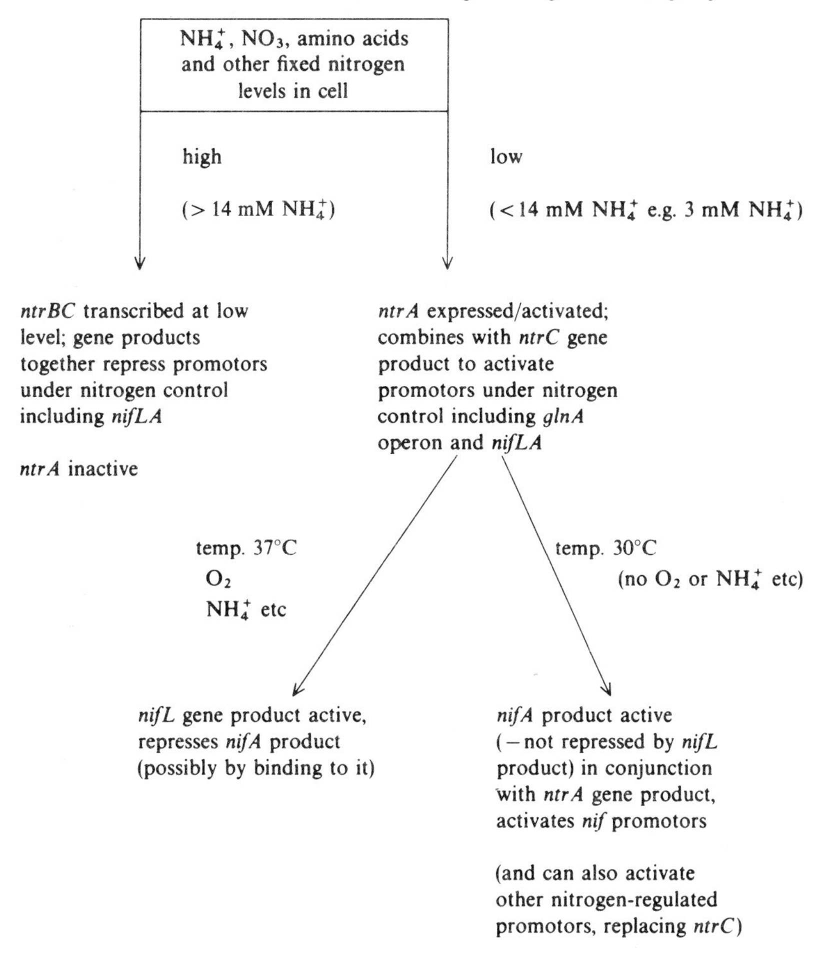

Figure 4. Regulation of *nif* expression in *K. pneumoniae* (after Merrick, 1982; Merrick *et al.*, 1982; Ow and Ausubel, 1983; Drummond *et al.*, 1983).

active form at high temperature or in the presence of fixed N and oxygen, but inactive at lower temperatures or in the absence of O_2 and fixed N (*Figure 4*).

Transfer of *nif* genes from *K. pneumoniae* to other organisms

The description of the *K. pneumoniae nif* genes and their regulation illuminates some of the problems which will be encountered when attempting to transfer *nif* to organisms other than the closely related enteric bacteria. Repression of *nif* expression by temperature, O_2 or fixed N can be overcome by deleting the *nifL* gene, which allows constitutive expression of the *nif* promotors mediated by the *nifA* and *ntrC* gene products. However, recognition of the *nif* promotors by DNA-dependent RNA polymerase is also required, and the ability of the host to translate the *nif* RNA correctly: ribosome-binding sites and codon usage (i.e. which of the several possible triplet codons is generally used to code for a particular amino acid) must be compatible. For functional expression, still more factors are required. The nitrogenase must be protected from oxygen, but an energy source must be available to provide MgATP, a reductant (such as pyruvate) and a suitable electron-transfer system, for nitrogenase activity (although the *nifJ* and *F* gene products are implicated in electron transfer, other cellular components are certainly required). In addition, fixed N must be removed from the vicinity of the nitrogenase either by incorporation into amino acids or by excretion.

Nitrogenase activity in *E. coli* carrying the *nif* plasmid pRDI was found to be equivalent to that in the parent *K. pneumoniae* strain under suitable conditions (Dixon, Cannon and Kondorosi, 1976). Nitrogen fixation has also been reported in *Salmonella typhimurium* carrying *K. pneumoniae nif* genes (Cannon, Dixon and Postgate, 1976; Postgate and Krishnapillai, 1977). These enteric bacteria are all closely related and thus it is not surprising that expression was observed. Transfer of pRDI to *Agrobacterium tumefaciens* did not result in nitrogen-fixing recombinants. This species, which is an obligate aerobe, is not closely related to *Klebsiella* although it is a close relative of *Rhizobium* spp. which are symbiotic nitrogen-fixing bacteria. However, when *A. tumefaciens* carrying pRDI were grown with very low levels of O_2 and combined N, the presence of a protein cross-reacting antigenically to *K. pneumoniae* nitrogenase component I was detected (Dixon, Cannon and Kondorosi, 1976). These authors found no evidence for the expression of nitrogenase in *R. meliloti* carrying pRDI under similar conditions, but this species also has its own *nif* genes which are known to be repressed in free-living cultures. Mutants of *Azotobacter vinelandii* lacking either nitrogenase component I or II regained these activities when pRDI was transferred to them (Cannon and Postgate, 1976). This is significant because *Azotobacter* sp. are obligate aerobes which fix nitrogen in the presence of oxygen. The nitrogenase is thought to be protected from O_2 by intracellular compartmentalization and a very high respiration rate. A secondary mechanism to protect nitrogenase from oxygen inactivation is a protein which binds to the enzyme complex, first described by Shethna, Dervartanian and Beinert (1968) in *A. vinelandii*. Subsequently, purified protein was added to extracts from *Azotobacter* and shown to protect nitrogenase from oxidation. The O_2-protective proteins have been shown to contain Fe and S atoms and to bind to nitrogenase in the presence of O_2, preventing both N_2 fixation and irreversible O_2-inactivation (Robson, 1979). The O_2-protective protein of *A. vinelandii* has a molecular weight of 23000 (Shethna, Dervartanian and Beinert, 1968) and that of *A. chroococcum* is 14000 (Robson, 1979). It would be of great interest to isolate the genes which encode them, since it might be possible to transfer the O_2 protection to other nitrogen-fixing systems, and to new hosts together with the *K. pneumoniae nif* genes. However, it should be recognized that these proteins protect nitrogenase by binding to it and that the enzyme bound to these proteins is inactive. Thus N_2 fixation can re-start only when the oyxgen tension is sufficiently low to enable the proteins to be released without the enzyme becoming inactivated by oxygen.

The O_2 sensitivity of nitrogenase appears to be an intrinsic feature which cannot be overcome by altering the proteins: the irreversible O_2 inactivation is due to the oxidation of non-haem Fe–S groups in both nitrogenase components (Robson, 1979). Another property of nitrogen fixation appears to be the evolution of hydrogen. Nitrogenase acts as a powerful reducing agent, reducing N_2 to NH_4^+, and the protons (H^+) which are abundant in living cells, to H_2. In the absence of N_2, H_2 is still evolved, with concomitant ATP hydrolysis. The proportion of N_2 fixed to H_2 evolved can vary (Andersen and Shanmugan, 1977). The process of N_2 fixation would require less energy if none were 'wasted' on H_2 evolution, but the two functions may be inseparable. However, some energy can be saved by recycling H_2 to H_2O, releasing electrons, and many N_2-fixing bacteria have 'uptake hydrogenases' (Evans *et al.*, 1981). It would therefore seem to be an advantage to introduce an uptake hydrogenase together with *nif* genes into new hosts which do not possess this function.

Any attempts to get *nif* genes expressed in eukaryotic cells will have to overcome considerable problems. There are basic differences in the organization, transcription and translation of genes in prokaryotes (i.e. bacteria) and the nuclei of eukaryotes (higher organisms). For instance, the promotor sequences recognized by the DNA-dependent RNA polymerases are different, as are the sites on the mRNA to which ribosomes bind and where translation is initiated. Furthermore, many bacteria have several genes transcribed from one promotor (as in the *nif* operons) resulting in polycistronic mRNA which has internal ribosome-binding sites for each gene, whereas eukaryotes almost invariably have monocistronic mRNAs with a binding site at the 5′ end. No internal ribosome-binding or re-initiation of protein synthesis occurs, unlike prokaryotes which can re-initiate translation within the mRNA (Kozak, 1983). Therefore, if *nif* were to be expressed as nuclear genes in a eukaryote, it would be necessary to fuse the coding sequence of each gene to eukaryotic promotors in such a way as to produce a suitable ribosome-binding site at the 5′ end of the transcript. Another potential problem with eukaryotic DNA is the 'switching off' of gene expression due to methylation of the DNA. This has been observed in some plant tumour tissues carrying integrated T-DNA from *A. tumefaciens* (Clarke, Pearson and Hepburn, 1983; Hepburn *et al.*, 1983).

It is perhaps surprising that expression of some prokaryotic genes has been reported in a eukaryote, the yeast *Saccharomyces cerevisiae*. These were antibiotic resistance genes derived from bacterial plasmids which are expressed in a wide variety of prokaryotic backgrounds. However, it was not proved that the genes (β-lactamase (Hollenberg, 1982); chloramphenicol acetyltransferase (CAT) (Cohen *et al.*, 1980); and neomycin phosphotransferase (NPT) (Jiminez and Davies, 1980)) were transcribed from their own promotors. Some β-lactamase has also been detected in the fungus *Podospora anserina* carrying a plasmid with a β-lactamase gene (Stahl *et al.*, 1982). No expression of CAT or NPT was detected in the fungus *Neurospora crassa* (Hughes *et al.*, 1983). To get expression of CAT and NPT in plant cells using the T-DNA as a vector, it was necessary to fuse the coding sequence to a eukaryotic promotor (Bevan, Flavell and Chilton, 1983; Herrera-Estrella *et al.*, 1983).

Although the *nif* gene cluster from *K. pneumoniae* has been transferred to *S. cerevisiae* (Elmerich *et al.*, 1981; Zamir *et al.*, 1981), expression of the genes has not been reported, and it is highly unlikely that they could be transcribed correctly.

The organization of the *nif* genes in operons with a common regulator in *K. pneumoniae* is probably important. This will result in the required amount of each product being synthesized at the correct time. There may be slight differences between the amount of the products from the first and last gene in an operon; the extra promotor observed in front of *nifM*, the last gene in the *nifUSVM* operon (Beynon *et al.*, 1983), may be relevant in this respect. Co-ordinated regulation of *nif* expression would be necessary in a new host. For instance, the *nifM* and *nifS* products are required to process the *nifH* product, and equal amounts of nitrogenase components I and II must be present for efficient N_2 fixation. For example, an excess of component I results in more ATP hydrolysed by component II for each electron transferred to I (Ljones and Burris, 1972). Normally two ATP molecules are hydrolysed for each electron transferred, and in *K. pneumoniae* 29 ATP molecules (i.e. 58 electrons) are required for each molecule of N_2 fixed (Hill, 1976); excess component I would therefore lead to ATP (and energy) being wasted during N_2 fixation.

Cloning, in the field of molecular biology, has come to mean the isolation and propagation of individual DNA fragments (often encoding a specific gene) by insertion *in vitro* into small plasmid vectors which can be transformed into, and selected for, in *E. coli.* For successful transformation certain factors are required. First, the organisms in question must be able to take up DNA: many animal cells can take up exogenous DNA, and this has been achieved with some fungi and plants by incubating protoplasts with DNA and then allowing regeneration of the cell wall. Fusion of protoplasts with artificial liposomes containing DNA has also been successful (Fraley and Papahadjopoulos, 1982). Other methods include micro-injection, which has been used to introduce DNA into *Drosophila* embryos (Germeraad, 1976) and the potential use of naturally infectious molecules such as the cauliflower mosaic virus (Hohn, Richards and Lebeurier, 1982). The phytopathogenic bacterium *A. tumefaciens* carries a large Ti (tumour inducing) plasmid, part of which becomes integrated in infected plant cells, and is known as the T-DNA. The mechanism of transformation is not yet understood and normally involves an interaction between the bacterium and plant cell; purified Ti plasmid DNA is also infective (Davey *et al.*, 1980; Krens *et al.*, 1982). The T-DNA can be used as a vector to introduce other genes into plant chromosomal DNA if the genes have been introduced previously into T-DNA which can be transferred to the plant by *Agrobacterium.* There are certain limitations to the system: (1) only certain dicotyledonous plants can be transformed; (2) the T-DNA produces a highly abnormal tumorous phenotype due to altered phytohormone levels; (3) deletions in the T-DNA are common. New vectors derived from T-DNA are being developed which produce apparently normal cells that can develop into plants capable of setting viable seed containing the transferred DNA. Some of these problems and advances are discussed by Cocking *et al.* (1981), Chilton (1983) and Schell and Van Montagu (1983). One disadvantage is that without the tumorous phenotype it is difficult to detect transformants. In bacterial and fungal transformation systems, the vector carries genes for enzyme activities which the host lacks: either the gene replaces a defective one in host auxotrophic mutants, or specifies a new property such as antibiotic resistance. In such systems the presence of the vector can be maintained by constant selection for the genetic marker. The third requirement is the ability of the vector either to replicate autonomously in the host cell or to integrate into the host genomic DNA. In the former case it may be possible to have many copies of the vector; in the latter, only one or two (unless integration into a multiple-copy target site is possible).

Integration can be directed into specific sites if the vector carries a chromosomal DNA segment giving a region of homology. Such vectors can be made readily by introducing cloned host DNA into the vector. If the cloned DNA is for a known gene the site of integration of DNA cloned on the vector will be in or beside that gene in the host genome. To maintain genes carried on an autonomously replicating vector it is necessary to have a selectable marker. In the absence of selection, autonomously replicating plasmids may be lost rapidly as the host cells divide, and especially during meiosis. This problem may be overcome by cloning a chromosomal centromere into the plasmid: in *S. cerevisiae* such meiotically stable vectors have been constructed (Clarke and Carbon, 1980).

The organelles of eukaryotes are possible sites for the integration of foreign DNA. They resemble prokaryotes in some respects of their DNA organization: they contain their own RNA and protein-synthetic systems and their DNA genomes are circular, like bacterial chromosomes and plasmids. Plant mitochondrial genomes investigated so far are larger and more complex than those of animals (which are about 15 kb), ranging from 330 kb to 2400 kb (Ward, Anderson and Bendich, 1981) and smaller circular DNA molecules have been observed in some species. The ribosomes in plant mitochondria have features in common with both prokaryotic and eukaryotic cytoplasmic ribosomes (Levings, 1983). No promotor sequences for plant mitochondrial genes have been described, to our knowledge. The genome sizes of chloroplasts which have been reported ranged from 120 to 180 kb (Palmer, Singh and Pillay, 1983; Spielman, Oritz and Stutz, 1983) and all have an unusual structure with inverted repeats that can recombine to form two alternative orientations of the unique regions (Palmer, 1983).

The chloroplast genes for the large subunit of ribulose biphosphate carboxylase from maize and wheat have been shown to be transcribed and translated in *E. coli* and have prokaryotic-type promotor sequences (Gatenby, Castleton and Saul, 1981). It is therefore possible that the chloroplast transcription and translation system could allow the expression of prokaryotic DNA in plants. To expore this possibility it would be necessary to develop a transformation system to incorporate foreign DNA sequences into the chloroplast.

Transformation systems used successfully in *S. cerevisiae*, *N. crassa* and *P. anserina* could be relevant. The *S. cerevisiae* 2-micron DNA, a high copy number (50 copies per haploid genome) autonomously replicating circular plasmid, is the basis for many 'shuttle' vectors which can replicate in both *S. cerevisiae* and *E. coli* since they carry replication origins and gene-encoded functions which can be selected in either organism. The 2-micron DNA has an inverted repeat which recombines at specific sequences (a function in which a 2-micron DNA encoded product is involved), and shuttle vectors carrying one inverted repeat and transformed into *S. cerevisiae* are often observed to recombine with the 2-micron DNA plasmid (Broach, 1982). This interaction is much more frequent than that attributable to normal homologous recombination. The interconversion of chloroplast DNA mediated by inverted repeats (reported by Palmer, 1983) could be caused by a similar mechanism, and it would facilitate the integration of foreign genes on vectors carrying one repeat into the chloroplast genome. An *N. crassa*, *E. coli* shuttle vector carrying part of an *N. crassa* mitochondrial plasmid which could replicate autonomously in both the mitochondria and cytoplasm of the fungus was observed to integrate into the mitochondrial DNA (Stohl and Lambowiz, 1983). Similarly, a hybrid plasmid carrying a mitochondrial replicon from *Podospora anserina* could replicate autonomously in *P. anserina* (Stahl *et al.*, 1982).

If mitochondrial replication origins support replication of hybrid plasmids in the cytoplasm or mitochondria of fungi, it is feasible that organelle DNA origins could support replication of hybrid plasmids in plant cells. Since transformation systems for fungi (described above) and dicotyledonous plants using T-DNA (*see* next section) are now available, the feasibility of directed integration into organelle DNA or autonomous replicating vectors located within organelles could be tested. However, it is necessary to recall the constraints on *nif* expression: even if the promotors could be transcribed and the RNA translated, the enzymes would not function unless the physiological requirements were met. One problem with chloroplasts is that during photosynthesis, oxygen is evolved which would inactivate the nitrogenase, so no N fixation would be possible in the light, even if an O_2-protection system were provided (such as the *Azotobacter* protein described previously). However, during darkness the conditions would be more suitable.

Symbiotic N_2 fixation (*rhizobium* and legumes)

The most important N_2-fixing symbiosis is that of *Rhizobium* and leguminous plants, in which the bacteria fix nitrogen in root nodules. The association is rather specific: only certain *Rhizobium* strains can nodulate a particular host plant, and this has led to the definition of *Rhizobium* 'species' according to the host plants which they can nodulate. For example, *R. trifolii* nodulates *Trifolium* sp.; *R. phaseoli* nodulates *Phaseolus* sp.; *R. meliloti* nodulates *Melilotus* sp.; *R. leguminosarum* nodulates several legume spp. including *Pisum*, *Lathyrus*, *Lens* and *Vicia*, and *R. japonicum* nodulates *Glycine*.

The rhizobia include two distinct subgroups which can be classified according to their behaviour in culture as 'fast' or 'slow' growers. The slow-growing rhizobia have been shown to be able to fix N_2 *in vitro* under micro-aerobic conditions with trace amounts of fixed N and a carbon compound such as succinate for an energy source (Kurz and LaRue, 1975; McComb, Elliot and Dilworth, 1975; Pagan *et al.*, 1975). However, partly because of their growth characteristics, very little genetic analysis has been possible with the slow growers. In contrast, great progress has been made with genetic analysis in the fast growers:

however, no conditions under which they can fix N_2 outside the host have been found, although recently a relatively fast-growing strain of *Rhizobium* that nodulates *Sesbania* has been shown to fix N_2 in pure culture (Dreyfus, Elmerich and Dommergues, 1983).

Although nitrogenase is O_2 sensitive, rhizobia are obligate aerobes and require some O_2 to fix N_2. Nodules containing N_2-fixing bacteroids contain large amounts of haemoglobin which, like animal haemoglobins, is involved in oxygen transport. The globin moiety is encoded by plant genes (Dilworth, 1969; Sullivan *et al.*, 1981) and synthesis appears to be regulated by the rhizobia, since it is induced only when rhizobia are present (Verma and Long, 1983). The haem moiety is probably synthesized by the rhizobia and exported into the cytoplasm (Cutting and Schulman, 1969; Dilworth, 1969).

The formation of the N_2-fixing symbiosis is complex and many genes from both plant and bacterium must be involved. Only a few plant genes involved in the establishment of the symbiosis have been identified and the functions of these are not clear (*see* Verma and Long, 1983).

RHIZOBIUM GENETICS

Auxotrophic and antibiotic-resistant mutants were isolated in *R. leguminosarum* and wide host-range antibiotic-resistant plasmids of the IncP1 group were found to be able to mobilize genes between different strains (Beringer and Hopwood, 1974) and species of *Rhizobium* (Johnston and Beringer, 1977). Subsequently similar circular linkage maps of chromosomal genes in *R. meliloti* (Kondorosi *et al.*, 1977; Meade and Signer, 1977) and *R. leguminosarum* (Beringer, Hoggan and Johnston, 1978) have been constructed. Chromosomal genes in *R. leguminosarum*, *R. trifolii* and *R. phaseoli* were found to be interchangeable (Johnston and Beringer, 1977) implying that nodulation specificity might be extrachromosomal, although mutations resulting in no nodulation (Nod^-) or no N_2 fixation (Fix^-) have been mapped to the chromosome (Beringer, Brewin and Johnston, 1980; Forrai *et al.*, 1983).

The host specificity of nodulation is now known to be plasmid-determined in *R. leguminosarum*, *R. phaseoli* and *R. trifolii.* A conjugative plasmid was identified in *R. leguminosarum* (Hirsch, 1979) and subsequently shown to carry genes specifying both nodulation of *Pisum*, and nitrogen fixation (Johnston *et al.*, 1978). This plasmid, pRL1JI, is about 200 kb in size (Hirsch *et al.*, 1980). A kanamycin-resistant derivative, pJB5JI, has been made by introducing the transposon Tn*5* into pRL1JI (Johnston *et al.*, 1978). It has facilitated transfer to many different *Rhizobium* sp., converting Fix^- *R. leguminosarum* mutants to Fix^+ and enabling *R. trifolii* and *R. phaseoli* to nodulate *Pisum* (Johnston *et al.*, 1978). A conjugative plasmid carrying *Trifolium* nodulation and N_2-fixation genes has been found in an *R. trifolii* strain which on transfer to *Agrobacterium tumefaciens* conferred the ability to nodulate clovers, although no N_2 fixation was detected (Hooykaas *et al.*, 1981), and an *A. tumefaciens* strain carrying pJB5JI has been shown to nodulate *Vicia hirsuta* (Van Brussel *et al.*, 1982). These experiments imply that many of the genes involved in nodulation and N_2 fixation are plasmid-encoded. Indirect evidence for the location of symbiotic genes on plasmids has been obtained from Nod$^-$ mutants found to have plasmid deletions or to have lost plasmids (e.g. Beynon, Beringer and Johnston, 1980; Hirsch *et al.*, 1980; Zurkowski, 1982).

The homology observed between nitrogenase genes from different species has been used to look for *nif* genes in *Rhizobium*. They have been reported to be plasmid-borne in *R. leguminosarum*, *R. phaseoli* and *R. trifolii* (Nuti *et al.*, 1979; Hombrecher, Brewin and Johnston, 1981; Prakash, Schilperoort and Nuti, 1981), *R. meliloti* (Banfalvi *et al.*, 1981) and some fast-growing rhizobia which nodulate soybeans (Masterson, Russell and Atherly, 1982). In some cases the plasmids were very large, greater than 500 kb (Banfalvi *et al.*, 1981; Rosenberg *et al.*, 1982), and could be detected physically only by lysing bacteria *in situ* on agarose gels, separating plasmid DNA from the cellular debris and linear fragmented chromosomal DNA by electrophoresis. When transferred from the gel to a nitrocellulose filter, by 'Southern blotting', homologous DNA sequences on the plasmid can be detected by hybridization to a radioactive DNA probe. It is possible that in cases where *nif* homology could be detected in total DNA but not in the plasmid fractions, e.g. some *R. japonicum* strains described by Masterson, Russell and Atherly (1982), a very large plasmid was involved, or that these genes are carried in the chromosome of some species and strains.

Identification and mapping of genes involved in both nodulation and N_2 fixation has been facilitated by the technique of transposon mutagenesis using the transposon Tn*5*, which specifies kanamycin resistance, a convenient selectable marker for most *Rhizobium* strains. An unstable 'suicide' vector, pJB4JI, has been constructed to transfer Tn*5* to *Rhizobium* from *E. coli* (Beringer *et al.*, 1978): it consists of an IncP1 plasmid, Tn5, and the bacteriophage Mu which renders IncP1 plasmid replication unstable in many *Rhizobium* strains. *Rhizobium* exconjugants selected for kanamycin resistance are stable when Tn*5* has transposed from the unstable vector to chromosomal or plasmid DNA, and since Tn*5* appears to transpose randomly to *Rhizobium* DNA a great number of different mutants can be obtained. The Tn*5* mutations can be mapped genetically, by transduction, conjugation (Beringer *et al.*, 1978) and physically, using Tn*5*-DNA probes (Dhaese *et al.*, 1979). Any Tn*5* mutations can be isolated in *E. coli* by digesting DNA with a restriction endonuclease that does not cut internal sites in Tn*5*, cloning into a small plasmid, and selecting *E. coli* transformants. Such clones will contain the regions of the target gene which border the Tn*5* insertion. These can be mapped for restriction sites, and also used to detect *E. coli* transformants carrying the intact gene cloned from unmutated *Rhizobium* (usually from a 'library' cloned into cosmids, which are vectors that select for the insertion of large DNA fragments). The cloned gene can then be transferred back to *Rhizobium* mutants to check phenotypic complementation (Scott *et al.*, 1982). Other methods have utilized the homology of *K. pneumoniae* and *Rhizobium nif* genes to select for transformants carrying cloned *Rhizobium nif* DNA. This method has facilitated a molecular analysis of the *nif* genes of *R. japonicum* which was not possible using 'classical' genetics. Using defined Tn*5* insertions selected in *E. coli*, mutated genes could be transferred back to *Rhizobium* and recombined into the homologous DNA, causing 'reverse mutagenesis' (Ruvkun and Ausubel, 1981).

Using such methods, the organization of the plasmid regions carrying Nod and Fix genes in *R. meliloti* (Ruvkun, Sundaresan and Ausubel, 1982) and *R. leguminosarum* (Ma *et al.*, 1982; Downie *et al.*, 1983) have been shown to be similar. Two regions in which Fix^- mutations map are separated by a relatively long section, within which Nod^- mutations map. In *R. leguminosarum*, a region with *nifHD* homology (in which Fix^- mutants map) is separated by 20 kb from the region in which Nod^- mutants map; a further 6 kb from this is the second Fix region, which has homology with *K. pneumoniae nifA* DNA (Downie *et al.*, 1983). This was similar to the arrangement in *R. meliloti* (Long, Buikema and Ausubel, 1982). Site-directed mutagenesis in the *nifHDK* region of *R. meliloti* has shown that the direction of transcription is the same as that in *K. pneumoniae* and that the three genes are transcribed from one promotor (Ruvkun, Sundaresan and Ausubel, 1982). This implies that some aspects of the organization of the two *nif* regions are conserved (although the *nifK* gene in *Rhizobium* is sufficiently different from its *K. pneumoniae* counterpart not to show DNA sequence homology). The *nifHDK* genes from a slow-growing *R. japonicum* strain have been cloned (Hennecke, 1981) and shown to be arranged and transcribed in the same direction as a single unit, as in *K. pneumoniae* (Fuhrmann and Hennecke, 1982).

The report that the *nifHDK* promotor from *R. meliloti* is activated in *K. pneumoniae* by the *nifA* gene product (Sundaresan *et al.*, 1983) is especially interesting, since little is understood about the regulation of *nif* in *Rhizobium*. It does not seem to be inhibited in bacteroids exposed to 7·5 mM NH_4^+ (Scott, Hennecke and Lim, 1979), a level which would repress *K. pneumoniae nif* (due to activation of the repressor, the *nifL* gene product). This may imply that although a *nifA*-like product is involved in activation of transcription, it is itself not controlled by a *nifL*-like gene.

The methodologies of molecular biology have greatly increased our knowledge of the number and location of some of the bacterial genes involved in the *Rhizobium*–legume symbiosis, but there are still many unknown factors. The recognition between compatible bacteria and host plant, the steps leading to infection and bacteroid formation, and the induction of nitrogenase synthesis in the bacteroid and leghaemoglobin in the host, are not understood. Bacterial exopolysaccharides may be important in the first step (Chakravorty *et al.*, 1982), and hormone synthesis by rhizobia in the infection process and induction of meristematic activity in the root cortex (*see* Beringer *et al.*, 1979). Since nitrogenase synthesis does not appear to be repressed by fixed N_2, understanding the mechanisms by which it is induced is important.

The potential for transferring the ability to be nodulated from legumes to other plant groups is extremely limited, because we understand very little of the plant factors involved. Some improvements may be made to the rhizobia, but this may be limited to recognizing strains which are particularly good at fixing N_2 in a certain host, and inoculating the seed before planting. Provided that conditions are suitable for infection, and the inoculant strain can outcompete native rhizobia in the soil, great improvements in crop yield and seed quality are possible. It is possible to manipulate the rhizobia genetically, although what improvements could be made are not clear. The only factor so far which appears to vary among strains is the presence of, or absence of, an uptake hydrogenase (Evans *et al.*, 1981, 1982) and genes for this have been transferred to strains which lack it (Brewin *et al.*, 1980) but it is not clear whether the efficiency (i.e. amount of fixed carbon consumed per molecule of N_2 fixed) is improved.

Future research with *Rhizobium* will reveal more information about the processes involved, and will certainly present new opportunities for genetic manipulation of strains, once the techniques for cloning and transfer of genes between strains have been fully developed.

Conclusions

Our knowledge of biological nitrogen fixation has advanced rapidly over the last few years, partly because of the amount of time and money invested in research, but largely because of the rapid advances in genetic methodology that have occurred in this time. For example, our understanding of the number, function and regulation of the *nif* genes of *Klebsiella pneumoniae*, which is described in this review, has depended largely upon the use of procedures to clone DNA, to move it into genetically different hosts and to study it and gene expression *in vitro*. That so much remains to be learnt is not so much a criticism of the research being done as a reflection of the great complexity involved in the production, functioning and especially the regulation of nitrogenase.

We have discussed briefly the concept of moving *nif* genes into new hosts to widen the range of nitrogen-fixing species in agriculture. That it will be possible to move these genes into plants which are susceptible to *Agrobacterium* is not in doubt; indeed it will probably be done during 1984. However, as stated previously, for expression to occur, promotor and other regulatory regions for these genes will need to be replaced with appropriate eukaryotic DNA sequences; in addition, some methods will be needed to regulate the *nif* genes to prevent the hosts being damaged or killed by excess ammonia and to ensure that energy is used efficiently. Whether sites can be exploited in which oxygen damage can be avoided remains to be demonstrated. While we feel that efficient nitrogen-fixing plants are unlikely to be produced in the near future, it is a tribute to modern genetics that we can consider their production as a rational goal in research on biological nitrogen fixation.

However, in striving to widen the range of species that are not dependent on fixed nitrogen, we should recognize that animals are major consumers of protein feeds, which in many cases could be used for human nutrition. Most domestic animals, particularly ruminants, have digestive systems which are often anaerobic and contain extremely large numbers of micro-organisms. There would appear to be enormous potential in producing nitrogen-fixing organisms to colonize these animals so that poorer quality feed containing much less protein (e.g. straw) could be utilized effectively. Undoubtedly there would be serious problems in producing micro-organisms which could colonize animal digestive systems and real problems in introducing *nif* genes into such micro-organisms. However, would these problems be as great as those which we expect to surmount in introducing *nif* genes into plants?

References

ANDERSEN, K. AND SHANMUGAM, K.T. (1977). Energetics of biological nitrogen fixation: determination of the ratio of formation of H_2 to NH_4^+ catalysed by nitrogenase of *Klebsiella pneumoniae in vivo. Journal of General Microbiology* **103,** 107–122.

ANONYMOUS (1979). *Fertiliser Recommendations.* Her Majesty's Stationery Office, London.

AUSUBEL, F.M. AND CANNON, F.C. (1980). Molecular genetic analysis of *Klebsiella pneumoniae* nitrogen fixation (*nif*) genes. *Cold Spring Harbour Symposia on Quantitative Biology Volume XLV: Movable Genetic Elements,* pp. 487–492.

BANFALVI, Z., SAKANYAN, V., KONCZ, C., KISS, A., DUSHA, I. AND KONDOROSI, A. (1981). Location of nodulation and nitrogen fixation genes on a high molecular weight plasmid of *Rhizobium meliloti. Molecular and General Genetics* **184,** 318–325.

BERINGER, J.E. (1983). The significance of nitrogen fixation in plant production. *CRC Reviews in Plant Sciences,* in press.

BERINGER, J.E. AND HOPWOOD, D.A. (1976). Chromosomal recombination and mapping in *Rhizobium leguminosarum. Nature* **264,** 291–293.

BERINGER, J.E., BREWIN, N.J. AND JOHNSTON, A.W.B. (1980). The genetic analysis of *Rhizobium* in relation to symbiotic nitrogen fixation. *Heredity* **45,** 161–186.

BERINGER, J.E., HOGGAN, S.A. AND JOHNSTON, A.W.B. (1978). Linkage mapping in *Rhizobium leguminosarum* by means of R plasmid-mediated recombination. *Journal of General Microbiology* **104,** 201–207.

BERINGER, J.E., BEYNON, J.L., BUCHANAN-WOLLASTON, A.V. AND JOHNSTON, A.W.B. (1978). Transfer of the drug resistance transposon Tn5 to *Rhizobium. Nature* **276,** 633–634.

BERINGER, J.E., BREWIN, N.J., JOHNSTON, A.W.B., SCHULMAN, H.M. AND HOPWOOD, D.A. (1979). The *Rhizobium*–legume symbiosis. *Proceedings of the Royal Society* **B204,** 219–233.

BEVAN, M.W., FLAVELL, R.B. AND CHILTON, M.D. (1983). A chimaeric antibiotic resistance gene as a selectable marker for plant cell transformation. *Nature* **304,** 184–187.

BEYNON, J.L., BERINGER, J.E. AND JOHNSTON, A.W.B. (1980). Plasmids and host range in *Rhizobium leguminosarum* and *Rhizobium phaseoli. Journal of General Microbiology* **120,** 421–429.

BEYNON, J., CANNON, M., BUCHANAN-WOLLASTON, V. AND CANNON, F. (1983). The *nif* promotors of *Klebsiella pneumoniae* have a characteristic primary structure. *Cell,* **34,** 665–671.

BREWIN, N.J., DEJONG, T.M., PHILLIPS, D.A. AND JOHNSTON, A.W.B. (1980). Co-transfer of determinants for hydrogenase activity and nodulation ability in *Rhizobium leguminosarum. Nature* **288,** 77–79.

BROACH, J.R. (1982). The yeast plasmid 2μ circle. *Cell* **28,** 203–204.

BUCHANAN-WOLLASTON, V., CANNON, M.C., BEYNON, J.L. AND CANNON, F.C. (1981). Role of the *nifA* gene product in the regulation of *nif* expression in *Klebsiella pneumoniae. Nature* **294,** 776–778.

CANNON, F.C. AND POSTGATE, J.R. (1976). Expression of *Klebsiella* nitrogen fixation genes (*nif*) in *Azotobacter. Nature* **260,** 271–272.

CANNON, F.C., DIXON, R.A. AND POSTGATE, J.R. (1976). Derivation and properties of F-prime factors in *Escherichia coli* carrying nitrogen fixation genes from *Klebsiella pneumoniae. Journal of General Microbiology* **93,** 111–125.

CANNON, F.C., RIEDEL, G.E. AND AUSUBEL, F. (1979). Overlapping sequences of *Klebsiella pneumoniae nif* DNA cloned and sequenced. *Molecular and General Genetics* **174,** 59–66.

CHAKRAVORTY, A.K., ZURKOWSKI, W., SHINE, J. AND ROLFE, B.G. (1982). Symbiotic nitrogen fixation: molecular cloning of *Rhizobium* genes involved in exopoly-saccharides synthesis and effective nodulation. *Journal of Molecular and Applied Genetics* **1,** 585–596.

CHILTON, M.-D. (1983). A vector for introducing new genes into plants. *Scientific American* **248,** 36–45.

CLARKE, L. AND CARBON, J. (1980). Isolation of a yeast centromere and construction of functional small circular chromosomes. *Nature* **287,** 504–509.

CLARKE, L.E., PEARSON, L. AND HEPBURN, A.G. (1983). Expression of T-DNA functions. *John Innes Institute Annual Report 1981–2* pp. 63–65.

COCKING, E.C., DAVEY, M.R., PENTAL, D. AND POWER, J.B. (1981). Aspects of plant genetic manipulation. *Nature* **293,** 265–270.

COHEN, J.D., ECCLESHALL, T.R., NEEDLEMAN, R.B., FEDEROFF, H., BUCHFERER, B.A. AND MARMUR, J. (1980). Functional expression in yeast of the *Escherichia coli* plasmid gene coding for chloramphenicol acetyltransferase. *Proceedings of the National Academy of Sciences of the United States of America* **77,** 1078–1082.

CUTTING, J.A. AND SCHULMAN, H.M. (1969). The site of heme synthesis in soybean root nodules. *Biochimica et biophysica acta* **192,** 486–493.

DAVEY, M.R., COCKING, E.C., FREEMAN, J., PEARCE, N. AND TUDOR, I. (1980). Transformation of *Petunia* protoplasts by isolated *Agrobacterium* plasmids. *Plant Science Letters* **18,** 307–313.

DHAESE, P., DEGREVE, H., DECRAEMER, H., SCHELL, J. AND VAN MONTAGU, M. (1979). Rapid mapping of transposon insertion and deletion mutations in the large Ti plasmids of *Agrobacterium tumefaciens. Nucleic Acids Research* **7,** 1837–1849.

DILWORTH, M.J. (1969). The plant as the genetic determinant of leghaemoglobin production in the legume root nodule. *Biochimica et biophysica acta* **184**, 432–441.

DIXON, R.A. AND POSTGATE, J.R. (1972). Genetic transfer of nitrogen fixation from *Klebsiella pneumoniae* to *Escherichia coli. Nature* **237,** 102–103.

DIXON, R., CANNON, F. AND KONDOROSI, A. (1976). Construction of a P plasmid carrying nitrogen fixation genes from *Klebsiella pneumoniae. Nature* **260,** 268–271.

DIXON, R., EADY, R., ESPIN, G., HILL, S., IACCARINO, M., KAHN, D. AND MERRICK, M. (1980). Analysis of regulation of the *Klebsiella pneumoniae* nitrogen fixation (*nif*) gene cluster with gene fusions. *Nature* **286,** 128–132.

DIXON, R.A., KENNEDY, C., KONDOROSI, A., KRISHNAPILLAI, V. AND MERRICK, M. (1977). Complementation analysis of *Klebsiella pneumoniae* mutants deficient in nitrogen fixation. *Molecular and General Genetics* **157,** 189–198.

DOBEREINER, J. AND BODDEY, R.M. (1981). Nitrogen fixation in association with Graminae. In *Current Perspectives in Nitrogen Fixation* (A.H. Gibson and W.E. Newton, Eds), pp. 305–312. Australian Academy of Science, Canberra.

DOWNIE, J.A., MA, Q.-S., KNIGHT, C.D., HOMBRECHER, G. AND JOHNSTON, A.W.B. (1983). Cloning the symbiotic region of *Rhizobium leguminosarum*: the nodulation genes are between the nitrogenase genes and a *nifA*-like gene. *EMBO Journal* **2,** 947–952.

DREYFUS, B., ELMERICH, C. AND DOMMERGUES, Y. (1983). Free-living *Rhizobium* strain able to grow on N_2 as sole nitrogen source. *Applied and Environmental Microbiology* **45,** 911–913.

DRUMMOND, M., CLEMENTS, J., MERRICK, M. AND DIXON, R. (1983). Positive control and autogenous regulation of the *nifLA* promotor in *Klebsiella pneumoniae. Nature* **301,** 302–307.

EADY, R.R., ISSACK, R., KENNEDY, C., POSTGATE, J.R. AND RATCLIFFE, H.D. (1978). Nitrogenase synthesis in *Klebsiella pneumoniae*: Comparison of ammonium and oxygen regulation. *Journal of General Microbiology* **104,** 277–285.

ELMERICH, C., SIBOLD, L., GUERINEAU, M., TANDEAU DE MARSAC, N., CHOCAT, P., GERBAUD, C. AND AUBERT, J.-P. (1981). The *nif* genes of *Klebsiella pneumoniae*: characterisation of a *nif* specific constitutive mutant and cloning in yeast. In *Current Perspectives in Nitrogen Fixation* (A.H. Gibson and W.E. Newton, Eds), pp. 157–160. Australian Academy of Science, Canberra.

EMERICH, D.W. AND BURRIS, R.H. (1978). Complementary functioning of the component proteins of nitrogenase from several bacteria. *Journal of Bacteriology* **134,** 936–943.

EVANS, H.J., EISBRENNER, G., CANTRELL, M.A., RUSSELL, S.A. AND HANUS, F.J. (1982). The present status of hydrogen recycling in legumes. *Israel Journal of Botany* **31,** 72–88.

EVANS, H.J., PUROHIT, K., CANTRELL, M.A., EISBRENNER, G., RUSSELL, S.A., HANUS, F.J. AND LEPO, J.E. (1981). Hydrogen losses and hydrogenases in nitrogen-fixing micro-organisms. In *Current Perspectives in Nitrogen Fixation* (A.H. Gibson and W.E. Newton, Eds), pp. 84–96. Australian Academy of Science, Canberra.

FORRAI, T., VINCZE, E., BANFALVI, Z., KISS, G.B., RANDHAWA, G.S. AND KONDOROSI, A. (1983). Localisation of symbiotic mutations in *Rhizobium meliloti. Journal of Bacteriology* **153,** 635–643.

FRALEY, R. AND PAPAHADJOPOULOUS, D. (1982). Liposomes: The development of a new carrier system for introducing nucleic acids into plant and animal cells. *Current Topics in Microbiology and Immunology* **96,** 171–191.

FUHRMANN, M. AND HENNECKE, H. (1982). Coding properties of cloned nitrogenase structural genes from *Rhizobium japonicum. Molecular and General Genetics* **187,** 419–425.

GATENBY, A.A., CASTLETON, J.A. AND SAUL, M.W. (1981). Expression in *E. coli* of maize and wheat chloroplast genes for large subunit of ribulose biphosphate carboxylase. *Nature* **291,** 117–121.

GERMERAAD, S. (1976). Genetic transformation in *Drosophila* by microinjection of DNA. *Nature* **262,** 229–231.

HENNECKE, H. (1981). Recombinant plasmids carrying nitrogen fixation genes from *Rhizobium japonicum. Nature* **291,** 354–355.

HEPBURN, A.G., BLUNDY, K.S., CLARKE, L.E. AND WHITE, J. (1983). T-DNA inserts in a flax genome. *John Innes Institute Annual Report 1981–2* pp. 59–63.

HERRERA-ESTRELLA, L., DEPICKER, A., VAN MONTAGU, M. AND SCHELL, J. (1983). Expression of chimaeric genes transferred into plant cells using a Ti plasmid-derived vector. *Nature* **303,** 209–213.

HILL, S. (1976). The apparent ATP requirement for nitrogen fixation in growing *Klebsiella pneumoniae. Journal of General Microbiology* **95,** 297–312.

HILL, S. AND KAVANAGH, E.P. (1980). Roles of *nifF* and *nifJ* gene products in electron transport to nitrogenase in *Klebsiella pneumoniae. Journal of Bacteriology* **141,** 470–475.

HILL, S., KENNEDY, C., KAVANAGH, E., GOLDBERG, R.B. AND HANAU, R. (1981). Nitrogen fixation gene (*nifL*) involved in oxygen regulation of nitrogenase synthesis in *K. pneumoniae. Nature* **290,** 424–426.

HIRSCH, P.R. (1979). Plasmid-determined bacteriocin production by *Rhizobium leguminosarum. Journal of General Microbiology* **113,** 219–228.

HIRSCH, P.R., VAN MONTAGU, M., JOHNSTON, A.W.B., BREWIN, N.J. AND SCHELL, J. (1980). Physical identification of bacteriocinogenic, nodulation and other plasmids in strains of *Rhizobium leguminosarum. Journal of General Microbiology* **120,** 403–412.

HOHN, T., RICHARDS, K. AND LEBEURIER, G. (1982). Cauliflower Mosaic virus on its way to becoming a useful plant vector. *Current Topics in Microbiology and Immunology* **96,** 193–236.

HOLLENBERG, C.P. (1982). Cloning with 2-μm DNA vectors in *Saccharomyces cerevisiae. Current Topics in Microbiology and Immunology* **96,** 118–144.

HOMBRECHER, G., BREWIN, N.J. AND JOHNSTON, A.W.B. (1981). Linkage of genes for nitrogenase and nodulation ability on plasmids in *Rhizobium leguminosarum* and *R. phaseoli. Molecular and General Genetics* **182,** 133–136.

HOOYKAAS, P.J.J., VAN BRUSSEL, A.A.N., DEN DULK-RAS, H., VAN SLOGTEREN, G.M.S. AND SCHILPEROORT, R.A. (1981). Sym plasmid of *Rhizobium trifolii* expressed in different rhizobial species and *Agrobacterium tumefaciens. Nature* **291,** 351–353.

HUGHES, K., CASE, M.E., GEEVER, R., VAPNEK, D. AND GILES, N.H. (1983). Chimeric plasmid that replicates autonomously in both *Escherichia coli* and *Neurospora crassa. Proceedings of the National Academy of Sciences of the United States of America* **80,** 1053–1057.

JIMINEZ, A. AND DAVIES, J. (1980). Expression of a transposable antibiotic resistance element in *Saccharomyces. Nature* **287,** 869–871.

JOHNSTON, A.W.B. AND BERINGER, J.E. (1977). Chromosomal recombination between *Rhizobium* species. *Nature* **267,** 611–613.

JOHNSTON, A. W.B., BEYNON, J.L., BUCHANAN-WOLLASTON, A.V., SETCHELL, S.M., HIRSCH,

P.R. AND BERINGER, J.E. (1978). High frequency transfer of nodulating ability between strains and species of *Rhizobium*. *Nature* **276,** 634–636.

KENNEDY, C. (1977). Linkage map of the nitrogen fixation (*nif*) genes in *Klebsiella pneumoniae*. *Molecular and General Genetics* **157,** 199–204.

KONDOROSI, A., KISS, G.B., FORRAI, T., VINCZE, E. AND BANFALVI, Z. (1977). Circular linkage map of *Rhizobium meliloti* chromosome. *Nature* **268,** 525–527.

KOZAK, M. (1983). Comparison of initiation of protein synthesis in procaryotes, eucaryotes and organelles. *Microbiological Reviews* **47,** 1–45.

KRENS, F., MOLENDIJK, L., WULLEMS, G.J. AND SCHILPEROORT, R.A. (1982). *In vitro* transformation of plant protoplasts with Ti plasmid DNA. *Nature* **296,** 72–74.

KURZ, W.G.W. AND LARUE, T.A. (1975). Nitrogenase activity in rhizobia in absence of plant host. *Nature* **256,** 407–408.

LEVINGS, C.S. (1983). The plant mitochondrial genome and its mutants. *Cell* **32,** 659–661.

LJONES, T. AND BURRIS, R.H. (1978). Evidence for one-electron transfer by the Fe protein of nitrogenase. *Biochemistry and Biophysics Research Communications* **80,** 22–25.

LONG, S.R., BUIKEMA, W.J. AND AUSUBEL, F.M. (1982). Cloning of *Rhizobium meliloti* nodulation genes by direct complementation of Nod^- mutants. *Nature* **298,** 485–488.

MA, Q.-S., JOHNSTON, A.W.B., HOMBRECHER, G. AND DOWNIE, J.A. (1982). Molecular genetics of mutants of *Rhizobium leguminosarum* which fail to fix nitrogen. *Molecular and General Genetics* **187,** 166–171.

MCCOMB, J.A., ELLIOT, J. AND DILWORTH, M.J. (1975). Actylene reduction by *Rhizobium* in pure culture. *Nature* **256,** 409–410.

MACNEIL, D., ZHU, J. AND BRILL, W.J. (1981). Regulation of nitrogen fixation in *Klebsiella pneumoniae*; Isolation and characterisation of strains with *nif-lac* fusions. *Journal of Bacteriology* **145,** 348–357.

MACNEIL, T., MACNEIL, D., ROBERTS, G.P., SUPIANO, M. AND BRILL, W.J. (1978). Fine-structure mapping and complementation analysis of *nif* (nitrogen fixation) genes in *Klebsiella pneumoniae*. *Journal of Bacteriology* **136,** 253–266.

MASTERSON, R.V., RUSSELL, P.R. AND ATHERLY, A.G. (1982). Nitrogen fixation (*nif*) genes and large plasmids of *Rhizobium japonicum*. *Journal of Bacteriology* **152,** 928–931.

MAZUR, B.J., RICE, D. AND HASELKORN, R. (1980). Identification of blue-green algal nitrogen fixation genes by using heterologous DNA hybridisation probes. *Proceedings of the National Academy of Sciences of the United States of America* **77,** 186–190.

MEADE, H.M. AND SIGNER, E.R. (1977). Genetic mapping of *Rhizobium meliloti*. *Proceedings of the National Academy of Sciences of the United States of America* **74,** 2076–2078.

MERRICK, M.J. (1982). A new model for nitrogen control. *Nature* **297,** 326–363.

MERRICK, M., FILSER, M., DIXON, R., ELMERICH, C., SIBOLD, L. AND HOUMARD, J. (1980). The use of translocatable genetic elements to construct a fine-structure map of *Klebsiella pneumoniae* nitrogen fixation (*nif*) gene cluster. *Journal of General Microbiology* **117,** 509–520.

MERRICK, M., HILL, S., HENNECKE, H., HAHN, M., DIXON, R. AND KENNEDY, C. (1982). Repressor properties of the *nifL* gene product in *Klebsiella pneumoniae*. *Molecular and General Genetics* **185,** 75–81.

NUTI, M.P., LEPIDI, A.A., PRAKASH, R.K., SCHILPEROORT, R.A. AND CANNON, F.C. (1979). Evidence of nitrogen fixation (*nif*) genes on indigenous *Rhizobium* plasmids. *Nature* **282,** 533–535.

O'GARA, F. AND SHANMUGAM, K.T. (1976). Regulation of nitrogen fixation by rhizobia. Export of fixed N_2 as NH_4^+. *Biochimica et biophysica acta* **437,** 313–321.

OW, D.W. AND AUSUBEL, F.M. (1983). Regulation of nitrogen metabolism genes by *nifA* gene product in *Klebsiella pneumoniae*. *Nature* **301,** 307–313.

PAGAN, J.D., CHILD, J.J., SCOWCROFT, W.R. AND GIBSON, A.H. (1975). Nitrogen fixation by *Rhizobium* cultured on defined medium. *Nature* **256,** 406–407.

PALMER, J.D. (1983). Chloroplast DNA exists in two orientations. *Nature* **301,** 92–93.

PALMER, J.D., SINGH, G.P. AND PILLAY, D.T.N. (1983). Structure and sequence evolution of three legume chloroplast DNAs. *Molecular and General Genetics* **190,** 13–19.

PETERS, G.A., CALVERT, H.E., KAPLAW, D., ITO, O. AND TOIA, R.E. (1982). The *Azolla–Anobaera* symbiosis: morphology, physiology and use. *Israel Journal of Botany* **31,** 305–323.

POSTGATE, J.R. AND KRISHNAPILLAI, V. (1977). Expression of *Klebsiella nif* and *his* genes in *Salmonella typhimurium. Journal of General Microbiology* **98,** 379–385.

PRAKASH, R.K., SCHILPEROORT, R.A. AND NUTI, M.P. (1981). Large plasmids of fast-growing rhizobia: homology studies and location of structural nitrogen fixation (*nif*) genes. *Journal of Bacteriology* **145,** 1129–1136.

PÜHLER, A. AND KLIPP, W. (1981). Fine-structure analysis of the gene region for N_2-fixation (*nif*) of *Klebsiella pneumoniae.* In *Biology of Inorganic Nitrogen and Sulphur* (H. Boethe and A. Trebst, Eds), pp. 276–286. Springer-Verlag, Berlin.

RIEDEL, G.E., AUSUBEL, F.M. AND CANNON, F.C. (1979). Physical map of chromosomal nitrogen fixation (*nif*) genes of *Klebsiella pneumoniae. Proceedings of the National Academy of Sciences of the United States of America* **76,** 2866–2870.

ROBERTS, G.P. AND BRILL, W.J. (1980). Gene product relationships of the *nif* regulon of *Klebsiella pneumoniae. Journal of Bacteriology* **144,** 210–216.

ROBERTS, G.P., MACNEIL, T., MACNEIL, D. AND BRILL, W.J. (1978). Regulation and characterisation of protein products coded by the *nif* (nitrogen fixation) genes of *Klebsiella pneumoniae. Journal of Bacteriology* **136,** 267–279.

ROBSON, R.L. (1979). Characterisation of an oxygen-stable nitrogenase complex isolated from *Azotobacter chroococcum. Biochemical Journal* **181,** 569–575.

ROSENBERG, C., CASSE-DELBART, F., DUSHA, I., DAVID, M. AND BOUCHER, C. (1982). Megaplasmids in the plant-associated bacteria *Rhizobium meliloti* and *Pseudomonas solanacearum. Journal of Bacteriology* **150,** 402–406.

RUVKUN, G.B. AND AUSUBEL, F.M. (1980). Interspecies homology of nitrogenase genes. *Proceedings of the National Academy of Sciences of the United States of America* **77,** 191–195.

RUVKUN, G.B. AND AUSUBEL, F.M. (1981). A general method for site directed mutagenesis in prokaryotes: construction of mutations in symbiotic nitrogen fixation genes of *Rhizobium meliloti. Nature* **289,** 85–88.

RUVKUN, G.B., SUNDARESAN, V. AND AUSUBEL, F.M. (1982). Directed transposon Tn*5* mutagenesis and complementation analysis of *Rhizobium meliloti* symbiotic nitrogen fixation genes. *Cell* **29,** 551–559.

SCHELL, J. AND VAN MONTAGU, M. (1983). The Ti plasmids as natural and as practical gene vectors for plants. *Biotechnology* **1,** 175–180.

SCHUBERT, K.R. AND WOLK, C.P. (EDS) (1982). *The Energetics of Biological Nitrogen Fixation.* American Society of Plant Physiologists, Rockville. 30 pp.

SCOTT, D.B., HENNECKE, H. AND LIM, S.T. (1979). The biosynthesis of nitrogenase MoFe protein polypeptides in free-living cultures of *Rhizobium japonicum. Biochimica et biophysica acta* **565,** 365–378.

SCOTT, K.F., HUGHES, J.E., GRESSHOFF, P.M., BERINGER, J.E., ROLFE, B.G. AND SHINE, J. (1982). Molecular cloning of *Rhizobium trifolii* genes involved in symbiotic nitrogen fixation. *Journal of Molecular and Applied Genetics* **1,** 315–326.

SHETHNA, Y.I., DERVARTANIAN, D.V. AND BEINERT, H. (1968). Non-haeme (iron–sulphur) proteins of *Azotobacter vinelandii. Biochemical and Biophysical Research Communications* **31,** 862–868.

SPIELMAN, A., ORITZ, W. AND STUTZ, E. (1983). The soybean chloroplast genome. *Molecular and General Genetics* **190,** 5–12.

STAHL, U., TUDZYNSKI, P., KUCK, U. AND ESSER, K. (1982). Replication and expression of a bacterial–mitochondrial hybrid plasmid in the fungus *Podospora anserina. Proceedings of the National Academy of Sciences of the United States of America* **79,** 3641–3645.

STOHL, L.L. AND LAMBOWITZ, A.M. (1983). Construction of a shuttle vector for the filamentous fungus *Neurospora crassa. Proceedings of the National Academy of Sciences of the United States of America* **80,** 1058–1062.

SULLIVAN, D., BRISSON, N., GOODCHILD, B., VERMA, D.P.S. AND THOMAS, D.Y. (1981).

Molecular cloning and organisation of two leghaemoglobin genomic sequences of soybean. *Nature* **289,** 516–518.

SUNDARESAN, V., JONES, J.D.G., OW, D.W. AND AUSUBEL, F.M. (1983). *Klebsiella pneumoniae nifA* product activates the *Rhizobium meliloti* nitrogenase promotor. *Nature* **301,** 728–732.

VAN BRUSSEL, A.A.N., TAK, T., WETSELAAR, A., PEES, E. AND WIJFFELMAN, C.A. (1982). Small leguminosae as test plants for nodulation of *Rhizobium leguminosarum* and other rhizobia and Agrobacteria harbouring a *leguminosarum* sym-plasmid. *Plant Science Letters* **27,** 317–325.

VERMA, D.P.S. AND LONG, S. (1983). The molecular biology of *Rhizobium*–legume symbiosis. *International Review of Cytology*, Supplement 14, pp. 211–245.

VON BÜLOW, J.F.W. AND DOBEREINER, J. (1975). Potential for nitrogen fixation in maize genotypes in Brazil. *Proceedings of the National Academy of Sciences of the United States of America* **72,** 2389–2393.

WARD, B.L., ANDERSON, R.S. AND BENDICH, A.J. (1981). The mitochondrial genome is large and variable in a family of plants (curcurbitaceae). *Cell* **25,** 793–803.

ZAMIR, A., MAINA, C.V., FINK, G.R. AND SZALY, A.A. (1981). Stable chromosomal integration of the entire nitrogen fixation gene cluster from *Klebsiella pneumoniae* in yeast. *Proceedings of the National Academy of Sciences of the United States of America* **78,** 3496–3500.

ZURKOWSKI, W. (1982). Molecular mechanism for loss of nodulation properties of *Rhizobium trifolii*. *Journal of Bacteriology* **150,** 999–1007.

4
Biosensors and Biofuel Cells

W. J. ASTON AND A. P. F. TURNER

Biotechnology Centre, Cranfield Institute of Technology, Cranfield, Beds., MK43 0AL, UK

Introduction

Revival of interest in the art of coupling biological systems with electrochemical techniques has resulted from a subtle blend of scientific advances and commercial requirement. The science of bioelectrochemistry has a distinguished academic lineage stemming from the classic work of Galvani on frog muscle, in the 18th century, and encompassing fundamental work on the redox potentials of biological materials fostered by Ehrlich's investigations into dye reduction by animal tissue, published in 1885. In retrospect, perhaps one of the most premonitory papers published on the subject of this review was a short work by Cohen (1931), in which he described a bacterial cell acting as an electrical half cell, using mediators such as ferricyanide and benzoquinone to shuttle electrons from the biological catalyst to an electrode. This principle forms the basis of some of the most commercially attractive bioelectrochemical systems demonstrated to date. Effective coupling of redox proteins to electronic systems by direct electron transfer has potential applications ranging from biological memories for computers to electrically driven biocatalysts for chemical synthesis, but probably the best short-term practical proposition is the development of biosensors for industrial and clinical monitoring. The technology required to produce efficient biocatalytic electrodes may also be exploited in biofuel cells, but as in the case of bio-organic synthesis, the advantages are more subtle and success may eventually depend on the vagaries of political and economic pressure.

The combination of a large number of studies revealing the elegance of biological catalysts, together with developments in the production and purification of enzymes, has led to their increasing use for analysis, therapy and industrial catalysis. Analytical applications of enzymes represent by far the largest market, the principal usage being in diagnostics and food analysis. It is difficult to estimate future trends in commercial biochemical analysis and authoritative opinions on the matter are generally coveted secrets. It is very obvious, however, that enzymatic methods of analysis have found increasing acceptance and are likely to continue to rise in popularity (*see also* Chapter 1). *Appendix A* to this chapter shows a survey of enzymes used routinely in either clinical or quality

Biotechnology and Genetic Engineering Reviews—Vol. 1, February 1984
0264-8725/84/01/89-32$10.00 + $0.00 © Intercept Ltd

control laboratories; all the enzymes listed are widely available and marketed either for automated procedures or as test kits. Although relatively pure enzymes with high specific activity may now be obtained at reasonable prices, significant economic advantages are offered by systems using immobilized enzymes and it is likely that these will occupy an increasing share of the market (*see* Chapter 5).

The ubiquity of the spectrophotometer in conventional assays (*Appendix A*) is meeting mounting opposition from electrochemical alternatives. Electrochemistry has made notable inroads into analytical instrumentation over the past two decades, matching a rapidly expanding range of applications with improved reproducibility and stability. Elegant, yet simple devices such as the ion-selective potentiometric electrode, the polarographic oxygen membrane electrode and coulometric electrochemical detectors, provide a tempting alternative to the complex instrumentation and/or methods traditionally associated with enzyme assays. The proposal, originally attributable to Clark and Lyons (1962), to combine the specificity of immobilized enzymes with such electrochemical devices to produce the enzyme electrode, has blossomed into a new multidisciplinary field. With a range of over 2000 enzymes now available, effecting dozens of readily measurable physical or chemical changes, the breadth of research activity proposed or under way is hardly surprising. While there is undoubtedly a shortage of novel sensors compatible with microprocessor control of a variety of processes and environments, it may be argued that the greatest immediate demand is for more efficient versions of the assays represented in *Appendix A*. A large number of sensors based on enzymes and whole organisms have already been proposed as alternatives to these routine assays, these generally relying on the electrochemical detection of enzyme products or substrates. It is worth reflecting, however, on a common feature of the assays shown in *Appendix A*: that all but the formino-glutamate assay involve an oxidoreductase. This key group of enzymes is widely distributed with commercially available members representing only a fraction of the redox catalysts available from nature. One premise of this contribution is that these electron transfer proteins may be more effectively coupled to a transducer by direct electrochemistry than by the widely propounded indirect methods.

In this short review it is not intended to give a complete guide to biosensors and biofuel cells, as many excellent works have already been written on the numerous possible configurations of these devices. It is hoped, however, to present a commercially aware picture of some of the most important principles and to draw attention to a few of the exciting recent developments in this rapidly expanding field (*see also* Chapter 7).

Indirect and direct systems

Bioelectrochemistry has principally involved the study of either indirect or direct interactions of biological material with an inert electrode. The practical advantages conferred on an electrochemical system by the biological element are the ability to operate at ambient temperatures and under mild chemical conditions, coupled with extraordinary catalytic specificity and high substrate affinity. While whole organisms, tissue slices and organelles have been exploited for their stable wide-ranging activity, the ultimate goal must be to harness the specific (sometimes even stereospecific) nature of enzymes, membrane recognition sites and the manipulable properties of the antibody.

Conventional enzymatic analysis involves the determination of the concentration of either substrates or products of the catalysed reaction or of some coupled reaction (*Appendix A*). Two principal techniques have been used, depending either on dynamic or end-point measurements. In the former method, a physical or chemical change is followed continuously for a short period, giving a rapid result. The initial velocity of the reaction is dependent on enzyme concentration, activity, substrate affinity and the concentration of non-saturating substrates. In the second method relatively large amounts of enzyme are used and the reaction is allowed to reach equilibrium, rendering the techniques relatively insensitive to physical and chemical conditions affecting enzyme activity. Whereas the latter technique is favoured for substrate determination, the former may be applied to inhibitor and activator assays, although care must be taken to control other conditions that may affect the rate of reactions, especially pH and temperature. Immobilization of the biocatalyst introduces a further dimension; most importantly, such heterogeneous systems offer barriers to the free passage of molecules both to and from the enzyme, either by diffusion limitation or by some partitioning effect. Many of the advantages (and disadvantages) of the end-point assay may thus be conferred on the immobilized enzyme system by making substrate access limiting and measuring the rate of reaction. Enzyme electrodes generally rely on achieving an appropriate steady state for the required operational characteristics, by balancing diffusion of substrates and products against conversion rates (*Figure 1*). Some of the parameters that may be

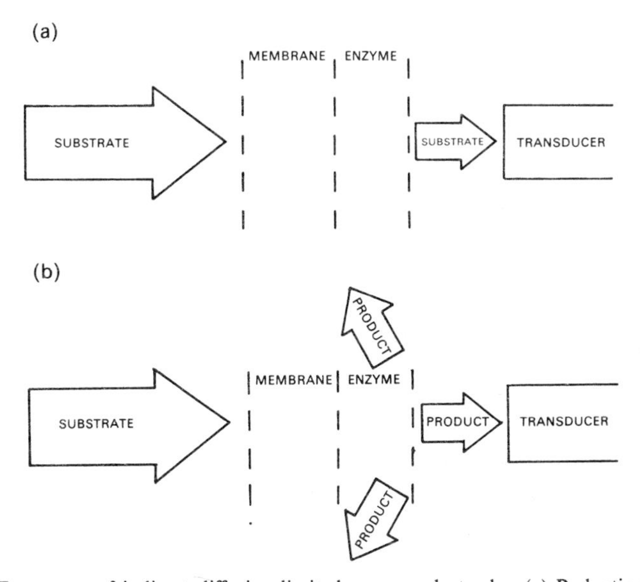

Figure 1. Two types of indirect diffusion-limited enzyme electrodes. (a) Reduction in substrate concentration determined, e.g. oxygen consumption, catalysed by an oxidase, monitored polarographically. (b) Product formation determined, e.g. ammonia liberation, catalysed by urease, monitored with an ion-selective electrode.

manipulated are discussed more fully by Carr and Bowers (1980); the principal advantage of diffusion-limited enzyme electrodes is stability, but this is achieved at the expense of increased response times.

The enzyme electrodes shown in *Figure 1* represent the vast majority of previously proposed systems, consisting of an enzyme immobilized on a membrane and placed in the vicinity of a transducer. These sensors may be considered to be indirect systems because the enzyme-catalysed reaction and the electrochemical sensor involve two quite discrete reactions. A wide variety of secondary transducers have been proposed for use in biosensors (*Table 1*)

Table 1. Some transducers used in indirect enzyme electrodes.

Transducer	Species detected
Amperometric electrodes	O_2, H_2O_2, NADH, I_2
Ion-selective electrodes	H^+, NH_4^+, NH_3, CO_2, I^-, CN^-
Field-effect transistors	H^+, H_2, NH_3
Photomultiplier (in conjunction with fibre optics)	Light emission or chemiluminescence
Photodiode (in conjunction with a light-emitting diode)	Light absorption
Thermistor	Heat of reaction
Piezoelectric crystal	Mass adsorbed

based on both physical and chemical measurements. The end-product of these devices is an electric current and yet in many cases the enzyme-catalysed reaction of interest involves electron transfer (*Appendix A*) and could theoretically be coupled directly to an electrode or semi-conductor device. In order to achieve such direct electron transfer from a redox protein to an electrode, the electron acceptor must replace one of the natural redox partners in the biological system. A variety of possibilities exist for achieving direct interactions (*Figure 2*), the most elegant and desirable solution being an electrode surface that resembles the natural partner. Practical electrodes based on this premise have been difficult to achieve. The demonstration of rapid and reversible electron transfer between cytochrome *c* and a gold electrode modified with 4,4′-bipyridyl (Albery *et al.*, 1981), however, provided the basis for a variety of bioelectrochemical systems, based on cytochrome *c* as a redox partner for oxidoreductases. Other mechanisms shown in *Figure 2* illustrate some practical alternatives to the ideal solution. The principle of using a solution mediator of low molecular weight to shuttle electrons from biological systems to electrodes has been known for many years, but has recently been shown to provide the basis for highly sensitive assays for methanol (Plotkin, Higgins and Hill, 1981; Aston *et al.*, 1983) and microbial activity (Turner, Ramsay and Higgins, 1983).

The main problems associated with the use of mediators are retention of these small molecules at the electrode surface and their potential reactivity giving rise to a source of interference. Immobilization of the mediator, either on the electrode surface or on the enzyme, solves retention problems (*Figure 2*). Reactivity of the mediator, especially with oxygen, remains a more serious problem, however, necessitating anaerobic conditions for efficient electron transfer from most mediators to electrodes. Exceptions to this rule are various derivatives of the organometallic compound, ferrocene, which form the basis of a recently

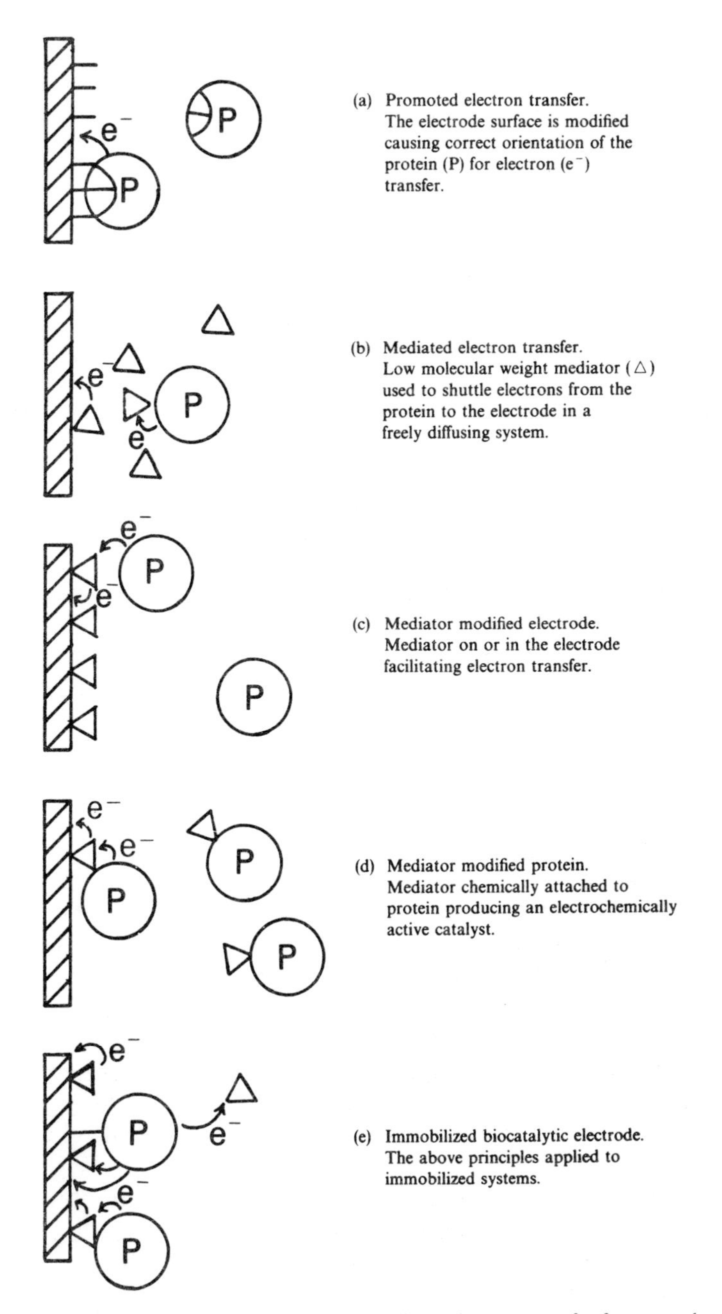

Figure 2. Some possible configurations facilitating direct electron transfer from a redox protein to an electrode.

patented series of oxygen-insensitive devices (Higgins, Hill and Plotkin, 1983; Aston *et al.*, 1983).

The advantages of the tightly coupled direct enzyme-based sensor lie in its simplicity, cheapness, reliability and accuracy. Interferences are restricted to those electrochemically active substances that reach the electrode surface and can be minimized by the use of membranes and low working potentials. The production of an effective biocatalytic electrode, however, finds wider application than sensors: biofuel cells may also be considered in the two classes of indirect and direct. The most highly developed indirect biofuel cell reported to date relies on the microbial production of hydrogen and its subsequent oxidation in a conventional hydrogen/oxygen fuel cell (Suzuki, Karube and Matsunaga, 1979). Cells based on direct electrochemistry potentially offer advantages in simplicity and cost, but limitations imposed by the need to achieve high current densities may necessitate the use of soluble mediators (Davis *et al.*, 1983). The limitation on current is imposed by the large size of protein molecules (and even larger size of micro-organisms) which have to interact with the surface of the electrode. Similar considerations are intrinsic to the successful application of direct electrochemistry for chemical synthesis. Biotransformations of particular relevance to the future of the chemical industry are a variety of alkane, aromatic and steroid hydroxylations catalysed by external mono-oxygenases (Higgins, Best and Hammond, 1980). A major problem in the exploitation of these enzymes, however, is their requirement for reduced cofactors, usually NADH. One proposed solution involves the direct electrochemical reduction of the enzyme prosthetic group (Higgins and Hill, 1978; 1979). For both chemical and energy production, however, the overriding factor will be the relative economics of the proposed processes, with the balance undoubtedly proving more subtle than the clear case for biosensors.

Indirect biofuel cells

The present state of development of biofuel cells has been excellently discussed by Wingard, Shaw and Castner (1982). Indirect biofuel cells typically involve the microbial production of an electroactive species such as hydrogen by anaerobic fermentation and the subsequent oxidation of the gaseous fuel at the anode of a conventional hydrogen/oxygen fuel cell (Suzuki *et al.*, 1980a, b):

at the anode

$$H_2 \rightarrow 2H^+ + 2e^- \qquad E^0 = 0\,V$$

at the cathode

$$2H^+ + 2e^- + 1/2O_2 \rightarrow H_2O \qquad E^0 = 1{\cdot}23\,V$$

overall

$$H_2 + 1/2O_2 \rightarrow H_2O \qquad E^0 = 1{\cdot}23\,V$$

Whereas the electrochemical conversion may be highly efficient, there are inevitable losses associated with the microbial transformation of the fuel to hydrogen. The microbial component of the system, however, effectively broadens the range of fuels available to the fuel cell and is particularly suited to the utilization of dilute aqueous fuels or wastes. Despite recent efforts to achieve a commercially viable indirect biofuel cell, however, continuous net power production from such a device remains to be reported. The principal hurdle to be overcome is the high energy input in the form of stirring, gassing and temperature control required for the continuous efficient microbial production of hydrogen. Biochemical processes may also be adapted to sustain the cathodic reaction; photosynthetic systems from plants and algae may be used to produce oxygen for a conventional fuel cell (Suzuki *et al.*, 1980a). Alternatively, chloroplasts may be coupled directly to an electrode via a mediator to produce electricity (Pan, Bhardwaj and Gross, 1983).

The concept of a 'hydrogen economy' has been widely debated (McAuliffe, 1980; McGown and Bockris, 1980; Beghi, 1981). Such an approach is dependent upon the economic availability of large quantities of the gas and a suitable means of handling and transporting it. Hydrogen gas, even when compressed, contains relatively little energy per unit volume; the ideal storage medium would be a liquid which would readily release the gas when required. Two compounds meeting these requirements are methanol and formate, both of which provide substrates for microbial production of hydrogen and can be produced cheaply either from fossil fuels or from vegetable matter (Egorov, Recharshky and Berezin, 1981). Hydrogen may be burnt, providing pollution-free power to generate electricity, but the efficiency of this process is limited according to the Second Law of Thermodynamics. The advantage of direct energy conversion was recognized as early as 1894 (Oswald), with the proposal that fuel oxidation could be performed at an electrode, effecting the direct conversion of chemical energy to electric current without heat losses. The hydrogen/oxygen fuel cell is now relatively well developed, being used in certain situations for domestic power generation (Daggitt, 1982). Despite intense research effort, however, suitable inorganic catalytic electrodes for the oxidation of other fuels remain elusive. One obvious goal is the efficient direct electrochemical oxidation of methanol, avoiding hydrogen as an intermediate.

Experiments performed in the early part of this century first showed that microbial cultures could develop electrode potentials directly (Potter, 1911). Two half cells containing glucose were connected by a salt bridge. Addition of yeast to one side resulted in an observable current. Yudkin (1935) studied the redox potentials of washed suspensions of a facultative anaerobic, a strict anaerobic and a strict aerobic bacterium. Although these studies did not involve electron transfer to an external circuit to derive useful energy, they did show a good correlation between redox potentials measured at an electrode and those determined using redox dyes. The potentials developed, however, were attributed to electromotively active molecules capable of passing through a collodion sac and not to any direct interaction between the bacteria and the electrode. Interest in the subject of biofuel cells was revised in the 1960s by NASA in the United States. Work was primarily concerned with microbial cells and fermentation broths within the anodic compartment (Austin, 1967; Cenek, 1968), but performance was poor when compared with that of inorganic fuel cells (Rourback,

Scott and Canfield, 1962). A number of patent applications were granted for biofuel cells both in England and America during the 1960s, including alcoholic fermentation by yeast (Hunger and Perry, 1966), hydrocarbon oxidation by bacteria (Young, 1965), hydrocarbon oxidation by actinomycetes (Davis and Yarborough, 1967) and methane oxidation by *Pseudomonas methanica* (Van Hees, 1965). The latter type of cell was capable of developing a maximum power density of 2·8 W/cm^2 at 0·34 volts.

Explanations of the mechanism by which current is generated in this group of bioelectrochemical fuel cells fall into two categories: one possibility is that the bioelectrode reaction is achieved through the discharge of organic substrates irrespective of the presence of bacteria; the other requires active participation of the biological system in the bioelectrode reaction. It has been shown (Disalvo, Videla and Arvia, 1979) that the kinetics of the bioelectrode system are due to two consecutive reactions which occur at the electrode surface. The early work of Yudkin (1935) has been related to these more recent systems by Videla and Arvia (1975) who showed that current was still produced in these biofuel cells when the biologically active material was isolated from the electrode, although the current was reduced due to ohmic and diffusional effects. While it is not clear exactly what electrochemical reactions were responsible for the currents observed, it is likely that the majority of previously reported glucose-powered (Wingard, Shaw and Castner, 1982) and hydrocarbon-powered (Higgins *et al.*, 1980) biofuel cells did not operate by direct electron transfer from the catalyst to the electrode. It is more likely that changes in dissolved oxygen concentration (where appropriate) and formation of electroactive products such as hydrogen and ammonia were responsible for the currents observed.

Direct biofuel cells

An alternative to electrochemical oxidation of microbial products as a basis for a biofuel cell is the development of promoted or mediated systems. The attraction of fuel cells lies in their thermodynamic efficiency with their energy output being dependent merely on the difference in Gibbs free energy (ΔG) between the reactants and the products, with small losses due to entropy effects. The net release of Gibbs free energy is related to the reversible potential difference (E) between the two half cells and the number of electrons transferred per mole of fuel (n) according to:

$$-\Delta G = n\mathrm{FE}$$

where F = the Faraday Constant.

The use of either mediators or redox proteins as intermediates in the electron transfer from substrate to electrode effectively reduces the potential difference between the two half cells, since the redox potential seen by the electrode is that of the last component of the chain. The efficiency of electron transfer (coulomic efficiency) using intermediates, however, can be high (Davis *et al.*, 1983; Roller *et al.*, 1983) and they facilitate reasonable current densities otherwise hampered by the large size of biological catalysts (especially whole organisms). The theoretical output of a fuel cell is also proportional to the number of electrons (n) transferred per mole of substrate. Complete biological oxidation of complex substrates, such as dairy wastes or sewage, involves numerous enzymatic steps and is therefore more efficiently carried out by whole organisms containing an effective package of the correct enzymes. The complete oxidation of methanol, however, is catalysed by just two enzymes, methanol dehydrogenase (MDH) and formate dehydrogenase (FDH):

$$\text{Methanol} \xrightarrow{\text{MDH}} \underset{+2H^+ + 2e^-}{\text{Formaldehyde}} \xrightarrow[H_2O]{\text{MDH}} \underset{+2H^+ + 2e^-}{\text{Formate}} \xrightarrow{\text{FDH}} \underset{2H^+ + 2e^-}{CO_2}$$

and is probably the most commercially interesting enzyme-based fuel cell investigated to date.

The first clear example of a direct biofuel cell may be attributed to Cohen (1931) who demonstrated that a current of 2 mA at a potential of 35 V could be produced by connecting several bioelectrochemical cells in series. Interest during the 1960s in the United States was focused on the biofuel cell as a method of efficient energy production for silent military generators and vehicle power supplies (Huff and Orth, 1960) and as an implantable power source for cardiac pacemakers (Drake, 1968). Methylene blue was shown to act as a mediator for glucose oxidase in a glucose-powered biofuel cell (Davis and Yarbrough, 1962; Scott and Cohn, 1962) and hydrogenase activity was coupled to an anode using either methylene blue or methyl viologen (Mizuguchi, Suzuki and Takahashi, 1966). More recently high coulomic efficiencies in the region of 90% have been reported using dichloroindophenol with glucose oxidase (Weibel and Dodge, 1975). The methanol biofuel cell has been developed and studied in our laboratory (Plotkin, Higgins and Hill, 1981; Turner, Higgins and Hill, 1982; Aston *et al.*, 1983; Davis *et al.*, 1983) using the quinoprotein methanol dehydrogenase (MDH) principally with phenazine etho-sulphate (PES) or *N, N, N', N'*-tetramethyl-*p*-phenylenediamine (TMPD) as mediator. While PES gave the largest currents, TMPD-based cells were more stable, yielding a steady current output decreasing by less than 10% over a 24-hour period of continuous operation. The basic scheme of the cell is shown in *Figure 3*. The maximum current density achieved at a platinum gauze anode was 0·2 A/m^2 geometric area; power density was 20 mW/m^2 representing 14 kW/mol catalyst. Various designs have been tested and cells requiring no power input in the form of gassing or stirring have been constructed (Turner, Higgins and Hill, 1982; Aston, unpublished work). Roller *et al.* (1983) reported a biofuel cell based on whole organisms for utilization of lactose wastes, using thionine to mediate electron transfer from *Escherichia coli* to the anode and the reduction of ferricyanide at the cathode. The 20 ml cell yielded approximately 0·4 mW for over two weeks with continuous additions of lactate. While an equivalent-sized MDH fuel cell might be expected to yield 10 mW, whole organisms offer potential advantages of wide substrate range and stability. A major disadvantage shared by all these mediated fuel cells is the reactivity of the mediators with oxygen. Substantial power losses occur if the anode compartment is not kept strictly anaerobic,

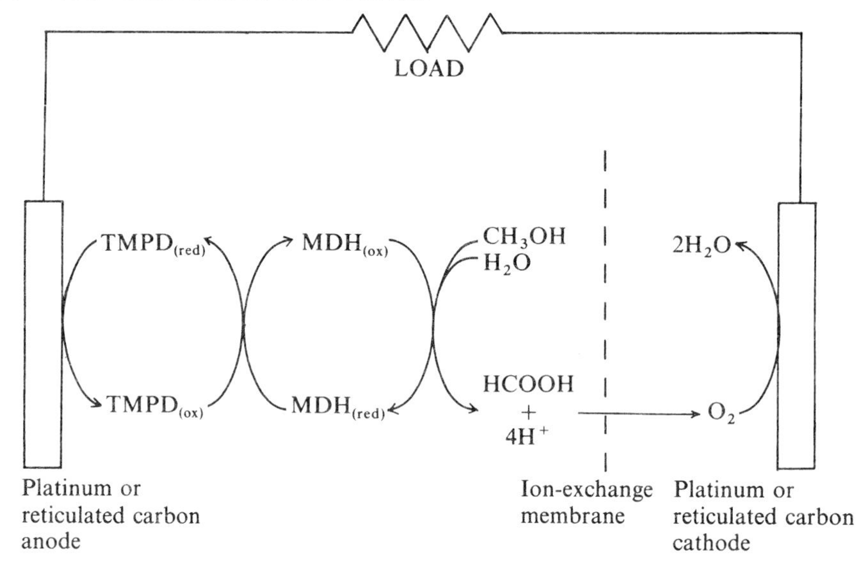

Figure 3. Methanol dehydrogenase-based biofuel cell.

because electrons are passed to oxygen as well as to the electrode (Turner, Higgins and Hill, 1982). Realistic configurations for power generation without these losses are difficult to conceive, especially as the cathodic reaction usually proposed is the reduction of oxygen to water, and proton-permeable membranes required to separate the anodic and cathodic compartments also allow the diffusion of oxygen (Turner, Higgins and Hill, 1982).

Although ferrocene and its derivatives previously were largely ignored, presumably because of their apparent insolubility and unexciting optical properties, they provide an alternative to conventional electron acceptors (Higgins, Hill and Plotkin, 1983; A. E. G. Cass *et al.*, unpublished work). Preliminary experiments have shown that these compounds will rapidly accept electrons from several oxidases and NAD-independent dehydrogenases and undergo reoxidation at carbon electrodes. Moreover, they remain stable in the reduced form allowing enzyme half cells to be constructed which show no variation in current output on changing from oxygen-free nitrogen to pure oxygen saturation. Ferrocene and its insoluble derivatives such as dimethyl ferrocene may be used to produce modified electrodes for use in immobilized systems, whereas more soluble ferrocenes, e.g. carboxyferrocene, are of interest in homogeneous configurations.

The cathodic reaction of the fuel cell has also received some attention. Reduction of molecular oxygen is catalysed by cytochrome oxidase at a gold/bipyridyl electrode (Hill, Walton and Higgins, 1981):

$2e^-$ — Cytochrome *c* — Oxidase red — $1/2O_2 + 2H^+$

Cytochrome *c* red — Oxidase — H_2O

Laccase also catalysed the reduction of oxygen at an electrode, an effect that was enhanced by hydroquinone (Mizuguchi, Suzuki and Takahashi, 1966). The replacement of laccase with an analogous non-biological system, however, such as ammonium chloride and copper sulphate, was found to be as effective as the enzyme.

Although the power densities of biofuel cells demonstrated to date are extremely low compared with their inorganic counterparts, some practical applications may be envisaged. The potential of biofuel cells to utilize waste products such as urine, carbon dioxide and faecal material led to the proposal that they may be valuable for space programmes, producing electricity, oxygen and food while removing waste materials. Systems were devised on the basis of previous work (Sisler, 1962; McNeil, 1969) using both bacteria and algae. Bioelectrochemical fuel cells may be of significance for electricity generation in the Third World where various plant and animal wastes could be converted directly to small quantities of electricity. Specialized military needs may be met by, for example, a noiseless battery recharger operating at ambient temperatures using readily available diesel or methanol/water anti-freeze mixtures as fuels. The direct conversion of industrial wastes to electricity as part of a detoxification process may also find some application, one particularly interesting example being the carbon monoxide biofuel cell (Turner, Higgins and Hill, 1982). It is unlikely that biological fuel cells will offer realistic alternatives for general power transduction, but they can operate under peculiarly mild and dilute chemical conditions.

Indirect biosensors

The obvious appeal of straightforward and inexpensive measurement of industrial and clinical biochemicals has led to a rapid expansion of the basic principle of combining immobilized biological material with a secondary detector (*Figure 4*). Surprisingly, the bacterial electrode was suggested as late as 1975 (Diviès), but its development has since been intensely pursued, particularly in Japan (Suzuki, Satoh and Karube, 1982). Equally, the range and form of secondary detectors has expanded, aided particularly by concurrent interest in ion-selective electrodes and semi-conductor devices. Recent reviews on biosensors (Carr and Bowers, 1980; Kobos, 1980; Wingard, Katzir and Goldstein, 1981; Guilbault, 1980, 1982; Suzuki, Satoh and Karube, 1982; Lowe, Goldfinch and Lias, 1983; Mosbach, 1983) detail the numerous configurations of biological catalysts coupled with either potentiometric, amperometric, calorimetric or photometric transducers that have been reported in the literature. *Appendix B* of this chapter shows some examples of these.

Clark and Lyons (1962) coined the term 'enzyme electrode' when they proposed that glucose oxidase could be held between two cuprophane membranes at a polarographic oxygen electrode and glucose concentration determined by measuring the oxygen consumption. Updike and Hicks (1967) reported a more practical system where glucose oxidase was immobilized in a polyacrylamide gel over an oxygen electrode and a second electrode containing heat-inactivated enzymes was introduced to correct for fluctuations in oxygen tension

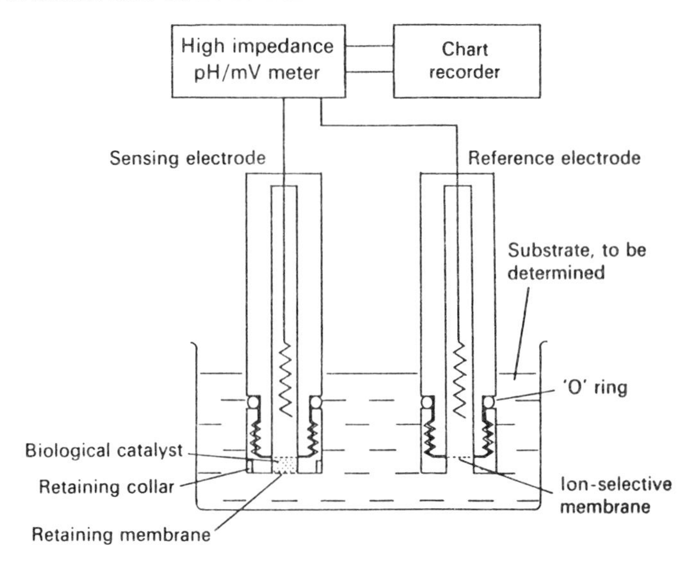

Figure 4. Schematic diagram of a typical ion-selective electrode biosensor.

and interfering substances. When the enzyme was placed in contact with a biological solution or tissue, glucose and oxygen diffused into the gel layer of immobilized enzyme. Instability of the Clark electrode, however, was caused by hydrostatic pressure variations, through loosening of the membranes (Severinghaus, 1968) and fluctuations in oxygen tension, although the latter may be overcome by supplying sufficient oxygen. (Rossette, Froment and Thomas, 1979). Oxygen-based sensors have been used industrially to monitor a variety of oxidase substrates (Karube *et al.*, 1977; Okada, Karube and Suzuki, 1981; Karube, Okada and Suzuki, 1982).

Hydrogen peroxide, a product of the enzymatic oxidation of glucose, is itself electrochemically active and may be measured in as low as picomolar concentrations at an electrode (Sittampalam and Wilson, 1982). Enzyme electrodes based on this system have been developed for both *in vitro* (Chua and Tan, 1978) and *in vivo* blood glucose analysis (Shichiri *et al.*, 1982). Glucose oxidase from *Aspergillus niger* has been shown to contain a glycoprotein structure (Pazur, Kleppe and Cepure, 1965) conferring resistance to inactivation by sodium dodecyl sulphate, urea and heating (Nakamura, Hayashi and Koya, 1976), but the enzyme is rapidly inactivated by hydrogen peroxide (Greenfield, Krittrell and Laurence, 1975) which must be removed from an enzyme electrode. The intrinsic dependence of the reaction on oxygen also presents a problem for process control and *in vivo* applications, where oxygen tensions may fluctuate. The production of oxygen by electrolysis has been used to produce an oxygen-stabilized enzyme electrode for glucose analysis (Enfors, 1981). Electrochemical interference from substances such as ascorbic acid and drugs can occur (Farrance and Aldons, 1981; Lidh *et al.*, 1982), but much of this may be eliminated by electrode modification such as the incorporation of ascorbate oxidase (Nagy, Rice and Adams, 1982) used for *in vivo* electrochemical analysis of catecholamines, or the use of membranes (Newman, 1976; Lobel and Rishpon, 1981).

Acetyl cellulose membranes, for example, enhance peroxide diffusion under alkaline conditions and may effectively be used between the enzyme and the electrode to reduce ascorbate interference. Alternatively, ascorbate may be assayed concurrently using an enzyme-based ascorbate sensor (Schenic, Miller and Adams, 1982). Hydrogen peroxide production may be linked to iodide oxidation (Nanjo, von Strop and Guilbault, 1973), catalysed by molybdate ions (Malmstadt and Pardue, 1961; Pardue, 1963) or a peroxidase enzyme, allowing the use of an iodine-selective electrode (Llenado and Rechnitz, 1973; Nagy, Rice and Adams, 1982).

The use of ion-selective electrodes has increased enormously over the last few years (Fricke, 1980; Freiser, 1980); their mode of operation and basic principles are well documented (Koryta, 1975; Moody and Thomas, 1975; Gammann, 1977). They have been applied successfully to physiological monitoring of blood sodium, potassium, calcium and lithium (Berman, 1974; Meir *et al.*, 1980; Zhukov *et al.*, 1981). Modification of ion-selective membrane materials (Rechnitz, 1967) has led to the use of liquid membrane electrodes for *in vivo* analysis of chlorpheniramine (Fukamachi and Ishibashi, 1978), novocaine (Negoiu, Ionescu and Cosofret, 1981), sulpha drugs (Ionescu *et al.*, 1981), naproxinate (Hogue and Landgraf, 1981), codeine and morphine (Goiha, Hobai and Rozenberg, 1978). Miniaturization of ion-selective electrodes has enabled research to be directed towards microscale enzyme electrodes (Brown and Flaming, 1974) which will provide a valuable tool in medicine, biology and physiological research (Gammann, 1977). With the exception of the Corning liquid membrane acetylcholine electrode (Baum and Ward, 1971), all biosensors based on ion-selective electrodes function by the indirect determination of ion formation or utilization. An example (*see Appendix B* for further examples) is the urea sensor incorporating urease (Katz and Rechnitz, 1963; Guilbault and Montalvo, 1970; Guilbault and Nagy, 1973a; Herman and Rechnitz, 1975). Two products are formed, both of which may be assayed using an ion-selective electrode.

$$CO(NH_2)_2 + 2H_2O \rightarrow 2NH_4^+ + CO_3^{2-}$$

Ammonium ions may be measured using an ammonium-selective liquid membrane electrode (Montalvo and Guilbault, 1969) and carbonate ions determined using a carbonate liquid membrane electrode. These two methods have the disadvantage that they are susceptible to interference caused by other ions present in biological samples (Katz and Cowans, 1965). This interference may be overcome by the use of gas-sensitive electrodes and air gap electrodes, which have been the subject of excellent reviews serving to describe the study, design and use of gas-sensing electrodes (Bailey and Riley, 1975; Fricke, 1980). Air gap electrodes introduced in the urease assay were later refined for measurements in both serum and whole blood (Hansen and Růzuička, 1974), measuring the ammonia evolved upon exposure to strong alkali. A more rapid and reliable one-step assay was developed (Guilbault and Tarp, 1974) using immobilized urease. The other product under acidic conditions, carbon dioxide, may also be measured using a gas-sensitive electrode (Guilbault and Shu, 1972). These methods are not affected by ions present in the measuring solution and exhibit superb selectivity, although hydrogen ions may affect the equilibrium (Ross, Riseman and Krueger, 1974).

The enzyme electrode generally offers superior specificity to that given by sensors based on whole organisms and tissue. Workers in Japan, however, have shown that a variety of practical devices may be based on immobilized whole organisms at an oxygen electrode (Suzuki, Satoh and Karube, 1982). Judicious selection of organisms and membrane configurations has yielded devices that are sufficiently selective for use in a variety of industrial processes, particularly for fermentation control. Whole organisms also provide the opportunity to exploit enzymes that become unstable on purification, for example methane mono-oxygenase (Higgins, Best and Hammond, 1980).

While the amperometric oxygen electrode and potentiometric ion-selective electrode have attracted the most attention as secondary transducers for biosensors, considerable effort has been expended on other systems. The technique of microcalorimetry has been studied in relation to automated analysis for clinical and process control situations resulting in the development of the enzyme thermistor (Mosbach, 1983). The ion-selective field effect transistor (ISFET) has been used in place of traditional probes to detect product formation in enzyme-based sensors (Janata and Huber, 1980; Danielsson *et al.*, 1983) and miniaturized optical systems have been devised using light-emitting diodes combined with photodiodes to detect colorimetric reactions in transparent films (Lowe, Goldfinch and Lias, 1983). These developments are indicative of a trend towards the fusion of biological systems and electronics in an attempt to achieve low-cost miniaturized sensors. It is questionable, however, whether these relatively complex indirect systems will provide the key to computer-compatible micro-biosensors.

Direct biosensors

The advantages in simplicity, cost and accuracy of achieving a direct interaction between biological recognition sites and electronic systems has focused attention on this area. As in the case of earlier sensors, effort has polarized into potentiometric and amperometric approaches, the former involving charge and capacitance effects with the latter exploiting direct electron transfer.

Yamamoto and Hiroshi (1981) have described immunosensors based on immobilized antibodies at a platinum electrode. Small potential changes were observed when the antigen was bound. This type of interaction may be more sensitively monitored using chemically sensitive field effect transistors (CHEMFET) with a polarized interface to measure the interfacial charge density resulting from the immunochemical reaction (Janata and Huber, 1980). Immunochemically sensitive field effect transistors (IMFET) consist of either antibodies or antigens covalently attached to an inert hydrophobic layer at the gate of an FET. The insulated gate FET measures the change in charge resulting from the formation of the antibody/antigen complex. Considerable effort is being expended to achieve reliable commercial devices based on this principle, with some attention also being paid to measuring charge differences associated with the formation of enzyme/substrate complexes.

While direct amperometric sensors may be made immunosensitive by protein binding interfering either with an electrochemical or a coupled enzymatic reaction, the most obvious application of this technique is for the assay of whole organisms, oxidoreductases and their substrates. The output of direct biofuel cells, both enzyme-based (Turner, Higgins and Hill, 1982) and those based on whole organisms (Turner, Ramsay and Higgins, 1983), is proportional to the concentration of the biological catalyst and this can be used as an assay technique. In place of the fuel cell configuration, a potential may be applied to the biological half cell by means of a defined power source or a potentiostat. Using a standard reference such as a calomel or silver/silver chloride electrode and a mediator to facilitate electron transfer to the working electrode, a sensitive bioactivity monitor may be constructed (Turner, Ramsay and Higgins, 1983). Such a technique may be suitable for assessing microbial populations, especially when rapid methods are required and the sample is coloured or opaque, for example, fermentation monitoring and microbial contamination of milk or cutting oils. Of wider application, however, is the use of bioelectrochemical systems to detect either substrates, inhibitors or activators of the catalyst. Recent work on the use of the MDH fuel cell to detect methanol contamination in drinking water (Aston *et al.*, 1983) confirmed the remarkable sensitivity of this procedure. Using very simple equipment final concentrations of less than 10^{-3} ppm methanol were detectable, well below the limits of routine gas–liquid chromatography and mass spectrometry.

The performance of the direct enzyme-based sensor may be attributed to the tight coupling of the redox reaction allowing expression of the enzymes' high substrate affinity. Current density is far less critical in this application since it may be calculated that even a monolayer of enzyme may be expected to yield tens of microamperes per square centimetre of electrode, which is well within the sensitivity of very simple electronic metering. In addition, there is plenty of scope for miniaturization of such electrodes. More important to the commercial success of a biosensor is the development of a stable immobilized configuration, which is free from interference. Previous enzyme sensors using redox compounds of low molecular weight, such as ferricyanide, have suffered from mediator leakage and oxygen sensitivity, the latter caused by the reaction of the mediator with molecular oxygen.

There are many suitable substrates for an oxidoreductase-based sensor, some of which are detailed in *Table 1*, and others include formaldehyde, explosives, carbon monoxide, methane, NADH and NADPH. Probably the most widely beneficial and consequently highly marketable enzyme sensor, however, is for glucose. Diabetes is one of the commonest chronic diseases and affects about 6% of the adult population of the Western World (WHO, 1980). Associated complications include heart disease, strokes, amputations and blindness; these should be reduced by improved control (Albisser and Spencer, 1982). Several devices may be envisaged that would improve general diabetes management: cheaper and more reliable instruments for clinical analysis; convenient subcutaneous sensors activating a hypoglycaemia alarm and giving a continuous read-out; ultimately, a continuous sensor activating a portable insulin pump as required. Each of these applications possesses specialized problems: longevity is paramount for *in vivo* sensors, but not essential for disposable personalized testers, whereas a fast response time is critical for sample analysis, yet less important in continuous sensors. Despite considerable research effort in the area there is, as yet, no glucose sensor which has proved to be suitable for *in vivo* use. It is hoped, however, that the use of dimethyl ferrocene modified carbon electrodes will rectify this situation (Aston *et al.*, 1983; Higgins, Hill and Plotkin, 1983). These electrodes accept electrons from immobilized glucose oxidase or NAD-independent glucose dehydrogenase, but do not react with oxygen. The electrode may be operated at low potentials (100 mV–160 mV) thus reducing electrochemical interference from blood components such as ascorbate. Prototype electrodes based on plastic strips with a coat of carbon produce a linear range of 0·1–35 mM and have a response time of less than 30 s (to 95% of maximum current).

Exploitation of the range of oxidoreductases that may be coupled to modified electrodes offers an important industrial opportunity. Not only may the substrates for these enzymes be detected, but inhibiting and activating reactions may also be monitored. The demands of micro-miniaturization are stimulating investigation of direct electron transfer from enzymes and whole organisms to materials compatible with silicon-based devices; such enzyme-effected transistors are likely to play a major part in future monitoring and control systems.

Acknowledgements

The authors are indebted to Professor I. J. Higgins (Cranfield), Dr H. A. O. Hill (Oxford University), Mr G. Davis (Oxford University) and our colleagues at Cranfield for numerous invaluable discussions and ideas. The generous sponsorship of the joint Cranfield/Oxford sensor programme by Genetics International Inc., Boston, is greatly appreciated. A. P. F. Turner is a Senior Research Fellow of the British Diabetic Association and W. J. Aston's studentship was provided by the SERC.

Appendix A. Enzymes commonly used in analysis.

Recommended name of enzyme [Systematic name in square brackets]	Used to determine	Reaction(s)	Principle
1. Acetyl-CoA synthetase [Acetate: CoA ligase (AMP-forming)] EC 6.2.1.1	Acetic acid in foodstuffs.	Acetate + CoASH + ATP $\xrightarrow{1}$ acetyl-CoA + AMP + PP_i. Malate + $NAD^+ \overset{31}{\rightleftarrows}$ oxaloacetate + NADH + H^+ Acetyl-CoA + oxaloacetate + $H_2O \xrightarrow{10}$ citrate + CoASH + H^+	Malate dehydrogenase reaction[31] used as an indicator. Formation of NADH determined spectrophotometrically.
2. Alcohol dehydrogenase [Alcohol: NAD^+ oxidoreductase] EC 1.1.1.1	Ethyl alcohol in foodstuffs, blood, serum or plasma.	Ethanol + $NAD^+ \xrightarrow{2}$ acetaldehyde + NADH + H^+	Formation of NADH determined spectrophotometrically.
3. Alkaline phosphatase [Orthophosphoric-monoester phosphohydrolase (alkaline optimum)] EC 3.1.3.1	Lecithin in foodstuffs	Lecithin + $H_2O \xrightarrow{36}$ 1,2-diglyceride + phosphorylcholine Phosphorylcholine + $H_2O \xrightarrow{3}$ choline + P_i Choline + ATP $\xrightarrow{8}$ phosphorylcholine + ADP ADP + phospho(enol) pyruvate $\xrightarrow{37}$ pyruvate + ATP Pyruvate + NADH + $H^+ \xrightarrow{29}$ Lactate + NAD^+ + H_2O	Lecithin converted to a kinase substrate (36 & 3). ADP produced by choline kinase (8) measured using pyruvate kinase and lactate dehydrogenase (37 & 29). Oxidation of NADH determined spectrophotometrically.
4. Amyloglucosidase (Exo-1,4-α-D-glucosidase) [1,4-α-D-Glucan glucohydrolase] EC 3.2.1.3	Starch in foodstuffs	Starch + $n\,H_2O \xrightarrow{4} n$ glucose Glucose + ATP $\xrightarrow{27}$ glucose-6-phosphate + ADP Glucose-6-phosphate + NADP $\xrightarrow{20}$ 6-phosphogluconate + NADPH	Starch converted to a dehydrogenase substrate (4 & 27). NADPH formed by glucose-6-phosphate dehydrogenase (20) determined spectrophotometrically.
5. Ascorbate oxidase [L-Ascorbate:oxygen oxidoreductase] EC 1.10.3.3	L-Ascorbic acid in foodstuffs.	Ascorbate + $\frac{1}{2}O_2$ + chromogen $\xrightarrow{5}$ dehydro-ascorbic acid + reduced chromogen	Reduced dye determined spectrophotometrically.

Appendix A (continued)

Recommended name of enzyme [Systematic name in square brackets]	Used to determine	Reaction(s)	Principle
6. Cholesterol esterase [Sterol-ester acylhydrolase] EC 3.1.1.13	Total cholesterol in serum or plasma	Cholesterol esters $\xrightarrow{6}$ cholesterol + fatty acids Cholesterol + $O_2 \xrightarrow{7}$ cholest-4-en-3-one + H_2O_2 $2H_2O_2$ + 4-amino antipyrine + phenol $\overset{32}{\rightleftharpoons}$ quinoeimine dye + $4H_2O$	Cholesterol esters converted to cholesterol (6) Peroxide produced by cholesterol oxidase (7) reacted to produce dye.
7. Cholesterol oxidase [Cholesterol: oxygen oxidoreductase] EC 1.1.3.6	Cholesterol in foodstuffs, serum or plasma	As above (6)	As above (6).
8. Choline kinase [ATP: choline phosphotransferase] EC 2.7.1.32	Lecithin in foodstuffs	As above (3)	As above (3)
9. Citrate (*pro*-3*S*)-lyase [Citrate oxaloacetate-lyase (*pro*-3*S*-$CH_2COO \rightarrow$ acetate)] EC 4.1.3.6	Citric acid in foodstuffs	Citrate $\xrightarrow{9}$ acetate + oxaloacetate Oxaloacetate + NADH + $H^+ \overset{31}{\rightleftharpoons}$ malate + NAD^+ (Oxaloacetate $\xrightarrow{\text{decarboxylase?}}$ pyruvate + CO_2) Pyruvate + $NADH^+ \overset{29}{\rightleftharpoons}$ lactate + NAD^+	Citrate converted to dehydrogenase substrates (9). NADH oxidation (29) determined spectrophotometrically.
10. Citrate (*si*)-synthase [Citrate oxaloacetate-lyase (*pro*-3*S*-CH_2COO- $\rightarrow$ acetyl-CoA)] EC 4.1.3.7	Acetic acid in foodstuffs	As above (1)	As above (1)
11. Diaphorase	L-Glutamic acid in foodstuffs and	Glutamate + NAD^+ + $H_2O \overset{22}{\rightleftharpoons}$	NADH produced by a dehydrogenase

		Lactate + $NAD^+ \xrightarrow{LDH}$ pyruvate + NADH + H^+ *THEN* NADH + H^+ + chromogen $\xrightarrow{11}$ NAD^+ + reduced chromogen	
12. 2,3-Diphosphoglyceric acid phosphatase (phosphoglyceromutase) [2,3-Bisphospho-D-glycerate: 2-phospho-D-glycerate phosphotransferase] EC 2.7.5.3	2,3-Diphosphoglyceric acid in blood	2,3-Diphosphoglycerate $\xrightarrow{12}$ 3-phosphoglycerate + P_i 3-phosphoglycerate + ATP $\xrightarrow{33}$ 1,3-diphosphoglycerate + ADP 1,3-diphosphoglycerate + NADH $\xrightarrow{25}$ glyceraldehyde-3-phosphate + NAD^+	2,3-Diphosphoglycerate converted to a dehydrogenase substrate (12 & 33). NADH oxidation (25) determined spectrophotometrically.
13. Formimino-L-glutamic acid transferase (glutamate formiminotransferase) [5-Formiminotetrahydrofolate: L-glutamate *N*-forminotransferase] EC 2.1.2.5.	Formimino-L-glutamic acid in urine	Formimino-L-glutamate + tetrahydrofolic acid $\xrightarrow{13}$ formimino-tetrahydrofolic acid + glutamate Formimino-tetrahydrofolic acid $\xrightarrow{14}$ 5,10-methenyl-tetrahydrofolic acid 4 + NH_4	Formation of 5,10-methenyl-tetrahydrofolic acid determined spectrophotometrically.
14. Formiminotetrahydrofolate cyclodeaminase [5-Formiminotetrahydrofolate ammonia-lyase (cyclizing)] EC 4.3.1.4	As above (13)	As above (13)	As above (13)
15. β-D-Fructofuranosidase [β-D-Fructofuranoside fructohydrolase] EC 3.2.1.26	Sucrose and glucose in foodstuffs	Sucrose + $H_2O \xrightarrow{15}$ β-D-fructose + D-glucose D-glucose + ATP $\xrightarrow{27}$ glucose-6-phosphate + ADP Glucose-6-phosphate + $NADP^+ \xrightarrow{20}$ gluconate-6-phosphate + NADPH + H^+	Sucrose converted to dehydrogenase substrate (15 & 27). NADP reduction (20) determined spectrophotometrically.

Appendix A (continued)

Recommended name of enzyme [Systematic name in square brackets]	Used to determine	Reaction(s)	Principle
16. α-D-Galactosidase [α-D-Galactosidase galactohydrolase] EC 3.2.1.22	Raffinose in foodstuffs	Raffinose + $H_2O \xrightarrow{16}$ Sucrose + D-galactose D-galactose + $NAD^+ \xrightarrow{18}$ galactono-γ-lactone + NADH + H^+	Raffinose converted to a dehydrogenase substrate (16). NAD reduction (18) determined spectrophotometrically.
17. β-D-Galactosidase [β-D-Galactosidase galactohydrolase] EC 3.2.1.23	Lactose and galactose in foodstuffs	Lactose + $H_2O \xrightarrow{17}$ galactose + glucose Galactose + $NAD^+ \xrightarrow{18}$ galactono-γ-lactone + NADH + H^+	Lactose converted to a dehydrogenase substrate (17). NAD reduction (18) determined spectrophotometrically.
18. Galactose dehydrogenase [D-Galactose: NAD^+ 1-oxidoreductase] EC 1.1.1.48	Raffinose (16) or lactose and galactose (17) in foodstuffs	As above (16 + 17)	Formation of NADH determined spectrophotometrically
19. Gluconokinase [ATP: D-gluconate 6-phosphotransferase] EC 2.7.1.12	D-Gluconic acid and D-glucono-δ-lactone in foodstuffs	D-Gluconate + ATP $\xrightarrow{19}$ D-gluconate-6-phosphate + ADP D-Gluconate-6-phosphate + $NADP^+$ $\xrightarrow{34}$ D-ribulose-5-phosphate + NADPH + CO_2 + H^+	Gluconate converted to a dehydrogenase substrate (18). NADP reduction (34) determined spectrophotometrically.
20. Glucose-6-phosphate dehydrogenase [D-Glucose-6-phosphate: $NADP^+$ 1-oxidoreductase] EC 1.1.1.49	Glucose in serum or plasma and glucose, starch (4), sucrose (15) or fructose (35) in foodstuffs. Kinase enzymes.	Glucose + ATP $\xrightarrow{27}$ glucose-6-phosphate + ADP Glucose-6-phosphate + $NADP^+ \xrightarrow{20}$ gluconate-6-phosphate + NADPH + H^+	Glucose or ATP determined by conversion to a dehydrogenase substrate. NADPH determined spectrophotometrically.
21. Glucose oxidase [β-D-Glucose: oxygen 1-oxidoreductase] EC 1.1.3.4	Glucose in blood, serum, plasma or urine	Glucose + $O_2 \xrightarrow{21}$ δ-gluconolactone + H_2O_2 H_2O_2 + chromogen $\xrightarrow{32}$ H_2O + oxidized chromogen	Oxidase used to produce peroxide (21) which reacts to give coloured product (32) determined spectrophotometrically.

22. Glutamate dehydrogenase ($NAD(P)^+$) [L-Glutamate: $NAD(P)^+$ oxido-reductase (deaminating)] EC 1.4.1.3	Urea and ammonia in serum, plasma or urine. Ammonia in foodstuffs and urine. Glutamate in foodstuffs (11).	Urea + $H_2O \overset{38}{\rightarrow} 2NH_4 + CO_2$ NH_4 + α-Ketoglutarate + NADH $\overset{22}{\rightarrow}$ glutamate + NAD^+	NH_4 required for reductive deamination of α-ketoglutarate. NADH oxidation determined spectrophotometrically.
23. Glutamate–oxaloacetate transaminase (aspartate aminotransferase) [L-Aspartate: 2-oxoglutarate aminotransferase] EC 2.6.1.1	L-Malic acid in foodstuffs	Malate + $NAD^+ \overset{31}{\rightleftharpoons}$ oxaloacetate + NADH + H^+ Oxaloacetate + glutamate $\overset{23}{\rightarrow}$ α-oxoglutarate + aspartate	Formation of NADH determined spectrophotometrically
24. Glutamate-pyruvate transaminase (alanine aminotransferase) [L-Alanine: 2-oxoglutarate aminotransferase] EC 2.6.1.2	Lactic acid in foodstuffs	L-Lactate + $NAD^+ \overset{29}{\rightleftharpoons}$ pyruvate + NADH + H^+ Pyruvate + glutamate $\overset{24}{\rightarrow}$ α-oxoglutarate + alanine	Formation of NADH determined spectrophotometrically.
25. Glyceraldehyde-phosphate dehydrogenase [D-Glyceraldehyde-3-phosphate: NAD^+ oxidoreductase (phosphorylating)] EC 1.2.1.12	2,3-Diphosphoglyceric acid in blood	As above (12)	NADH oxidation determined spectrophotometrically
26. Glycerol kinase [ATP: glycerol 3-phosphotransferase] EC 2.7.1.30	Triglycerides in serum or blood and glycerol in foodstuffs	Triglycerides $\overset{30}{\rightarrow}$ glycerol + fatty acids Glycerol + ATP $\overset{26}{\rightarrow}$ glycerol-1-phosphate + ADP ADP + phospho(enol) pyruvate $\overset{37}{\rightarrow}$ pyruvate + ATP Pyruvate + NADH + $H^+ \overset{29}{\rightarrow}$ lactate + NAD^+ + H_2O	Triglycerides converted to a dehydrogenase substrate (30, 26 & 37). NADH oxidation determined spectrophotometrically.
27. Hexokinase [ATP: D-hexose-6-phosphotransferase] EC 2.7.1.1	Glucose in serum or plasma and glucose, starch (4), sucrose (15) or fructose (35) in foodstuffs. Creatine phosphokinase and kinase in serum or plasma.	ATP + glucose $\overset{27}{\rightarrow}$ ADP + glucose-6-phosphate Glucose-6-phosphate + $NADP^+ \overset{20}{\rightarrow}$ gluconate-6-phosphate + NADPH + H^+	ATP or glucose determined by conversion to a dehydrogenase substrate. NADPH determined spectrophotometrically.

Appendix A (continued)

Recommended name of enzyme [Systematic name in square brackets]	Used to determine	Reaction(s)	Principle
28. Isocitrate dehydrogenase ($NADP^+$) [*threo*-D_S-Isocitrate : $NADP^+$ oxidoreductase (decarboxylating)] EC 1.1.1.42	D-Isocitric acid in foodstuffs	Isocitrate + $NADP^+ \overset{28}{\rightleftarrows}$ 2-ketoglutarate + CO_2 + NADPH + H^+	NADPH determined spectrophotometrically
29. Lactate dehydrogenase [L-Lactate : NAD^+ oxidoreductase] EC 1.1.1.27	Glycerol (26) or L-lactate in foodstuffs. Triglycerides (26) or lactate in blood, serum or plasma	L-Lactate + $NAD^+ \overset{29}{\rightleftarrows}$ pyruvate + NADH + H^+	NADH determined spectrophotometrically
30. Lipase (triacylglycerol lipase) [Triacylglycerol acylhydrolase] EC 3.1.1.3	Triglycerides in serum or plasma (26)	As above (26)	As above (26)
31. Malate dehydrogenase [L-Malate : NAD^+ oxidoreductase] EC 1.1.1.37	Malate, citrate (9) or acetate (1) in foodstuffs and aspartate aminotransferase in serum or plasma	Malate + $NAD^+ \overset{31}{\rightleftarrows}$ oxaloacetate + NADH + H^+	Reaction used as an indicator. NAD(H) followed spectrophotometrically
32. Peroxidase [Donor : hydrogen-peroxide oxidoreductase] EC 1.11.1.7	Cholesterol (6) or glucose (21) in food, serum or plasma	As above (6, 21)	Dye-linked determination of peroxide
33. Phosphoglycerate kinase [ATP : 3-phospho-D-glycerate 1-phosphotransferase] EC 2.7.2.3	2,3-Diphosphoglyceric acid in blood (12)	As above (12)	As above (12)
34. Phosphogluconate dehydrogenase (decarboxylating) [6-Phospho-D-gluconate : $NADP^+$ 2-oxidoreductase (decarboxylating)] EC 1.1.1.44	D-Gluconic acid and D-glucono-δ-lactone in foodstuffs (19)	As above (19)	As above (19)

35. Phosphoglucose isomerase (glucosephosphate isomerase) [D-Glucose-6-phosphate ketol-isomerase] EC 5.3.1.9	Fructose and glucose in foodstuffs (27)	Fructose + ATP $\xrightarrow{27}$ fructose-6-phosphate + ADP Fructose-6-phosphate $\xrightarrow{35}$ glucose-6-phosphate Glucose-6-phosphate + $NADP^+$ $\xrightarrow{20}$ gluconate-6-phosphate + NADPH + H^+	Fructose converted to a dehydrogenase substrate (27 & 35). NADPH determined spectrophotometrically.
36. Phospholipase C [Phosphatidylcholine cholinephosphohydrolase] EC 3.1.4.3	Lecithin in foodstuffs (3)	As above (3)	As above (3)
37. Pyruvate kinase [ATP: pyruvate 2-*O*-phosphotransferase] EC 2.7.1.40	Glycerol (26) or lecithin (3) in foodstuffs and triglycerides (26) in serum or blood	As above (3 & 26)	Substrate (pyruvate) produced for a dehydrogenase
38. Urease [Urea amidohydrolase] EC 3.5.1.5	Urea and ammonia in foodstuffs and urine (22)	As above (22)	As above (22)
39. Uricase (urate oxidase) [Urate: oxygen oxidoreductase] EC 1.7.3.3	Uric acid in serum or urine	Urate + O_2 + $2H_2O$ $\xrightarrow{39}$ allantoin + H_2O_2 + CO_2	Allantoin determined spectrophotometrically

Appendix B. Some typical indirect biosensors.

a) Enzyme-based sensors

Type	Enzyme	Sensor	Immobilization	Stability	Response	Linear range	Reference
Alcohol	Alcohol oxidase	O_2	Glutaraldehyde	>2 weeks	1–2 min	1–25 mg/ℓ	Verduyn, Van Dijken and Scheffers (1983)
Acetyl choline	Acetyl cholinesterase	pH	Gelatin	40 days	Several weeks	1×10^{-5}–1×10^{-4} M	Durland, David and Thomas (1978)
Ascorbic acid	Ascorbate oxidase	O_2	Collagen paste, glutaraldehyde	3 weeks	1–5 min	5×10^{-5}–5×10^{-4} M	Matsumoto, Yamata and Osajima (1981)
L-Amino acids	L-Amino acid oxidase	O_2	Glutaraldehyde	4 months	2 min	$1 \times 10^{-6} - 6 \times 10^{-5}$ M	Nanjo and Guilbault (1974)
L-Asparagine	Asparaginase	NH_3	Polyacrylamide	1 month		10^{-4}–10^{-3} M	Guilbault and Hrabenkova (1971)
L-Amino acids	L-Amino acid oxidase	NH_3	Polyacrylamide	1 month	–	10^{-4}–10^{-3} M	Guilbault and Hrabenkova (1970)
Choline	Choline oxidase	H_2O_2	Glutaraldehyde	>20 days	7 seconds	0–1×10^{-4} M	Matsumoto *et al.* (1980)
β-Glucose	β-Glucose oxidase	O_2	Chemical	3 weeks	–	0·5–4% weight	Chotani and Constantinides (1982)
Glutamate	Glutamate decarboxylase	CO_2	Glutaraldehyde	>7 days	10 min	$4{\cdot}4 \times 10^{-4}$–$4{\cdot}7 \times 10^{-3}$ M	Kuriyama and Rechnitz (1981)
Glutamic acid	Glutamine decarboxylase	CO_2	Glutaraldehyde	–	18 min	0·4–>10 mg%	Ahn, Wolfson and Yao (1975)
L-Lysine	L-Lysine carboxylase	CO_2	Glutaraldehyde	500 assays	3 min	$1 \times 10^{-2} - 1{\cdot}0$ g/ℓ	Tran, Romette and Thomas (1983)
Oxalate	Oxalate decarboxylase	CO_2	Glutaraldehyde	1 month	8–10 min	2×10^{-4}–1×10^{-2} M	Kobos and Ramsay (1980)
Penicillin	Penicillinase	pH	Polyacrylamide	>2 weeks	15–30 seconds	10^{-4}–10^{-2} M	Papariello, Mukherji and Shearer (1973)
L-Phenyl alanine	L-Phenyl alanine oxidase and peroxidase	I^-	Chemical	>3 weeks	60–180 sec steady state 30 seconds slope	3×10^{-5}–1×10^{-4} M	Guilbault and Nagy (1973b)
Proteins	L-Amino acid oxidase	NH_3	Glutaraldehyde	>30 days	7 mins	0·1–100 μg/ml	Mascini and Giardini (1980)

Sucrose	Invertase, glucose oxidase and mutarotase	O_2	Glutaraldehyde	10 days	3–7 min	0–10 mM	Satoh, Karube and Suzuki (1976)
Sulphate	Arylsulphatase	4-Nitrocatechol	Chemical	25–50 min	5 min	10^{-6}–10^{-4} M	Cserfalvi and Guilbault (1976)
Urea	Urease	NH_4^+	Polyacrylamide	> 19 days	20–40 seconds	5×10^{-4}–1×10^{-1} M	Montalvo and Guilbault (1969)

b) Whole-organism-based sensors

Type	Organism	Sensor	Immobilization	Stability	Response time	Range	Reference
Ammonia	*Nitrosomonas europea*	O_2	Physically entrapped	> 2 weeks	8 min	$0{\cdot}5 \times 10^{-2} - 1{\cdot}3$ mg/ℓ	Hikuma *et al.* (1980)
Arginine	*Streptococcus faecium*	NH_3	Physically entrapped	> 20 days	20 min	5×10^{-5}–1×10^{3} M	Rechnitz *et al.* (1977)
Aspartate	*Bacterium cadaveris*	NH_3	Physically entrapped	2 days	5 min	3×10^{-4}–7×10^{-3} M	Kobos and Rechnitz (1977)
Cholesterol	*Nocardia erythropolis*	O_2	Physically entrapped	4 weeks	35–70 sec	15×10^{-6}–130×10^{-6} M	Wollenberg, Scheller and Atrat (1980)
Cysteine	*Proteus morganii*	H_2S gas-sensing electrode	Physical entrapment	6 days	5–8 min	$5 \times 10^{-5} - 9 \times 10^{-4}$ M	Jensen and Rechnitz (1978)
Formic acid	*Clostridium butyricum*	Fuel cell using H_2 produced	Adsorbed	> 20 days	20 min	10–1000 mg/ℓ	Matsunaga, Karube and Suzuki (1980)
Glutamine	*Sarcina flava*	NH_3	Physically entrapped	2 weeks	5 min	10^{-4}–10^{-2} M	Kobos and Rechnitz (1977)
Methane	*Methylomonas flagellata*	O_2	Physical entrapment	720 days	1 min	$1{\cdot}3 \times 10^{-6}$–$6{\cdot}6 \times 10^{-3}$ M	Okada, Karube and Suzuki (1981)
Nitrate	*Azotobacter vinelandii*	NH_3	Physical entrapment	2 weeks	7–8 min	1×10^{-5}–8×10^{-4} M	Kobos, Rice and Flourney (1979)
Nitroacetic acid	*Pseudomonas* sp.	NH_3	Physical	1 month	5 min	1×10^{-4}–1×10^{-3} M	Kobos and Pyon (1981)
L-Serine	*Clostridium acidiurici*	NH_3	Physical	> 3 days	3–5 min	$1{\cdot}6 \times 10^{-2}$–$1{\cdot}8 \times 10^{-4}$ M	Di Paolanionio, Arnold and Rechnitz (1981)

c) Organelle-based sensors

Substrate	Tissue or organelle	Sensor	Immobilization	Stability	Response time	Linear range	Reference
Adenosine	Mouse small intestine containing adenosine deaminase	NH_3	Glutaraldehyde	–	11 min	$3{\cdot}1 \times 10^{-4}$–1×10^{-3} M	Arnold and Rechnitz (1981)
Sulphite	Microsomes	O_2	Cellulose acetate	2 days	10 min	–	Karube *et al.* (1983)

(d) Hybrid sensors

Substrate	Biological components	Sensor	Immobilization	Stability	Response time	Linear range	Reference
Nicotinamide adenine dinucleotide	NADase and *Escherichia coli*	NH_3	Dialysis	1 week	5–10 min	$2{\cdot}5 \times 10^{-1}$–$2{\cdot}5 \times 10^{-3}$ M	Riechel and Rechnitz (1978)
Phenylalanine	Lactate oxidase and *Leuconostoc mesenteroides*	O_2	–	–	5 min (total test time 6 h)	$0{\cdot}75$–6×10^{-7} g/ml	Karube *et al.* (1980)

References

AHN, B.K., WOLFSON, S.K. AND YAO, S.J. (1975). An enzyme electrode for the determination of L-glutamic acid. *Bioelectrochemistry and Bioenergetics* **2,** 142–153.

ALBERY, W.J., EDDOWES, M.J., HILL, H.A.O. AND HILLMAN, A.R. (1981). Mechanism of the reduction and oxidation of cytochrome *c* at a modified gold electrode. *Journal of the American Chemical Society* **103,** 3904–3910.

ALBISSER, A.M. AND SPENCER, W.J. (1982). Electronics and the diabetic. *IEEE Transactions on Biomedical Engineering* **29,** 239–247.

ARNOLD, M.A. AND RECHNITZ, G.A. (1981). Selectivity enhancement of a tissue-based adenosine sensing membrane electrode. *Analytical Chemistry* **53,** 575–578.

ASTON, W.J., SCOTT, L.D., HIGGINS, I.J., TURNER, A.P.F. AND HILL, H.A.O. (1983). Enzyme based methanol sensor. In *Proceedings, Bioelectrochemistry and Bioenergetics Symposium, Nottingham, 4–9 September, 1983* (in press).

AUSTIN, L.G. (1967). *Fuel Cells, a Review of Government-Sponsored Research, 1950–1964.* NASA SP 120.

BAILEY, D.L. AND RILEY, M. (1975). Performance characteristics of gas-sensing membrane probes. *Analyst* **100,** 145–156.

BAUM, G. AND WARD, F.B. (1971). General enzyme studies with a substrate selective electrode. *Analytical Biochemistry* **42,** 487–493.

BEGHI, G., ED. (1981). *Hydrogen: Energy Vector of the Future.* Graham and Trotman Ltd, London.

BERMAN, H.J. (1974). Ion selective microelectrodes: Their potential in the study of living matter. In *Ion Selective Electrodes* (H.J. Berman and N.C. Herbert, Eds), pp. 3–11. Plenum Press, New York.

BROWN, K.T. AND FLAMING, D.G. (1974). Beveling of fine micropipette electrodes by rapid precision method. *Science* **185,** 693–695.

CARR, P.W. AND BOWERS, L.D. (1980). *Immobilised Enzymes in Analytical and Clinical Chemistry.* John Wiley, New York.

CENEK, M. (1968). Biochemical fuel cells. *US Air Force AD* 694072. *August* 1969: (translated from *Chemiche Listy* **62,** 927–974).

CHOTANI, G. AND CONSTANTINIDES, A. (1982). On line glucose analyser for fermentation studies. *Biotechnology and Bioengineering* **24,** 2743–2745.

CHUA, K.S. AND TAN, I.K. (1978). Plasma glucose measurements with the Yellow Springs Glucose Analyser. *Clinical Chemistry* **24,** 150–152.

CLARK, L.C. AND LYONS, C. (1962). Electrode systems for continuous monitoring in cardiovascular surgery. *Annals of the New York Academy of Sciences* **102,** 29–45.

COHEN, B. (1931). The bacterial cell as an electrical half cell. *Journal of Bacteriology* **21,** 18–19.

CSERFALVI, T. AND GUILBAULT, G.G. (1976). An enzyme electrode based on immobilised arylsulfatase for the selective assay of sulphate ion. *Analytica chimica acta* **84,** 259–270.

DAGGITT, G.E.G. (1982). *Fuel Cells.* Science and Engineering Research Council, Didcot, Oxon.

DANIELSSON, B., WINQUIST, F., MOSBACH, K. AND LUNDSTROM, I. (1983). Enzyme transistors. In *Biotech 83: Proceedings of the International Conference on the Commercial Applications and Implications of Biotechnology*, pp. 629–688. Online Publications, Northwood, London.

DAVIS, G., HILL, H.A.O., ASTON, W.J., TURNER, A.P.F. AND HIGGINS, I.J. (1983). Bioelectrochemical fuel cell and sensor based on a quinoprotein. *Enzyme and Microbial Technology* **5,** 383–388.

DAVIS, J.B. AND YARBOROUGH, H.F. (1962). Preliminary experiments on a microbial fuel cell. *Science* **137,** 615–616.

DAVIS, J.B. AND YARBOROUGH, H.F. (1967). US Patent 3331, 848.

DI PAOLANIONO, C.L., ARNOLD, M.A. AND RECHNITZ, G.A. (1981). Serine-selective membrane probe based on immobilised anaerobic bacteria and a potentiometric ammonia gas sensor. *Analytica chimica acta* **128,** 121–127.

DISALVO, E.A., VIDELA, H.A. AND ARVIA, A.J. (1979). Kinetic model of a depolarisation-type bioelectrochemical fuel cell. *Bioelectrochemistry and Bioenergetics* **6,** 493–508.

DIVIÈS, C. (1975). Remarques sur l'oxidation de l'ethanol par une 'electrode microbienne' de *Acinetobacter xylinium. Annals of Microbiology* **126A,** 175–186.

DRAKE, R.F. (1968). *Implantable Fuel Cell for an Artificial Heart.* US Government PB 177695.

DURLAND, P., DAVID, A. AND THOMAS, D. (1978). An enzyme electrode for acetylcholine. *Biochimica et biophysica acta* **527,** 277–281.

EGOROV, A.M., RECHARSHKY, M.V. AND BEREZIN, I.V. (1981) Coal-derived liquid fuels provide hydrogen for fuel cell power vehicle propulsion. *Chemical Institute* **5,** 20–26.

ENFORS, S.O. (1981). Oxygen stabilised enzyme electrode for D-glucose analysis in fermentation broths. *Enzyme and Microbial Technology* **3,** 29–32.

FARRANCE, I. AND ALDONS, J. (1981). Paracetamol interference with YS1 glucose analyser. *Clinical Chemistry* **27,** 782–783.

FREISER, H., ED (1980). *Ion Selective Electrodes in Analytical Chemistry*, volumes 1 and 2. Plenum Press, New York.

FRICKE, G.M. (1980). Ion selective electrodes. *Analytical Chemistry* **52,** 259R–275R.

FUKAMACHI, K. AND ISHIBASHI, N. (1978). Chlorpheniramine sensitive electrode. *Bunseki Kagaku* **27,** 152–155.

GAMMAN, K. (1977). Biosensors based on ion-selective electrodes. *Analytical Chemistry* **287,** 1–9.

GOIHA, T., HOBAI, S. AND ROZENBERG, L. (1978). Liquid membrane ion selective electrodes for some alkaloids. *Farmaco (Bucharest)* **26,** 141–147.

GREENFIELD, P.F., KRITTRELL, J.R. AND LAURENCE, R.L. (1975). Inactivation of immobilised glucose oxidase by hydrogen peroxide. *Analytical Chemistry* **1,** 109–124.

GUILBAULT, G.G. (1980). Enzyme electrode probes. *Enzyme and Microbial Technology* **2,** 258–264.

GUILBAULT, G.G. (1982). Ion selective electrodes applied to enzyme systems. *Ion Selective Electrode Reviews* **4,** 187–231.

GUILBAULT, G.G. AND HRABENKOVA, E. (1970). An electrode for the determination of amino acids. *Analytical Chemistry* **42,** 1779–1783.

GUILBAULT, G.G. AND HRABENKOVA, E. (1971). New enzyme electrode probes for D-amino acids and asparagine. *Analytica chimica acta* **56,** 285–290.

GUILBAULT, G.G. AND MONTALVO, J.G. (1970). An enzyme electrode of the substrate urea. *Journal of the American Chemical Society* **92,** 2533–2538.

GUILBAULT, G.G. AND NAGY, G. (1973a). Improved urea electrode. *Analytical Chemistry* **45,** 417–419.

GUILBAULT, G.G. AND NAGY, G. (1973b). Enzyme electrodes for the determination of L-phenylalanine. *Analytical Letters* **6,** 301–312.

GUILBAULT, G.G. AND SHU, F.R. (1972). Enzyme electrodes based on the use of a carbon dioxide sensor, urea and L-tyrosine electrodes. *Analytical Chemistry* **44,** 2161–2166.

GUILBAULT, G.G. AND TARP, M. (1974). A specific enzyme electrode for urea. *Analytica chimica acta* **73,** 355–365.

HANSEN, E.H. AND RŮZUIČHA, J. (1974). Enzymatic analysis by means of the air gap electrode, determination of urea in blood. *Analytica chimica acta* **72,** 353–364.

HERMAN, H.B. AND RECHNITZ, G.A. (1975). Preparation and properties of a carbonate ion selective membrane electrode. *Analytica chimica acta* **76,** 155–164.

HIGGINS, I.J. AND HILL, H.A.O. (1978). British Patent 33388/78.

HIGGINS, I.J. AND HILL, H.A.O. (1979). Microbial generation and interconversions of energy sources. In *Proceedings, Society for General Microbiology Symposium on Microbial Technology* (A.T. Bull, D.C. Ellwood and C. Ratledge, Eds), volume 29, pp. 359–377. Society for General Microbiology, UK.

HIGGINS, I.J., BEST, D.J. AND HAMMOND R.C. (1980). New findings in methane-utilising bacteria highlighting their importance in the biosphere and commercial potential. *Nature* **286,** 561–564.

HIGGINS, I.J., HILL, H.A.O. AND PLOTKIN, E.V. (1983). Sensor for components of a liquid mixture. European Patent Application 0078636.

HIGGINS, I.J., HAMMOND, R.C., PLOTKIN, E.V., HILL, H.A.O., UOSAKI, K., EDDOWES, M.J. AND CASS, A.E.G. (1980). Electroenzymology and biofuel cells. In *Hydrocarbons in Biotechnology* (D.E.F. Harrison, I.J. Higgins and R. Watkinson, Eds), pp. 181–193. Heyden, London.

HIKUMA, M., KUBO, T., YASUDA, T., KARUBE, I. AND SUZUKI, S. (1980). Ammonia electrode with immobilised nitrifying bacteria. *Analytical Chemistry* **52,** 1020–1024.

HILL, H.A.O., WALTON, W.J. AND HIGGINS, I.J. (1981). Electrochemical reduction of dioxygen using a terminal oxidase. *FEBS Letters* **126,** 282–284.

HOGUE, E.R. AND LANDGRAF, W.C. (1981). Naproxinate ion specific electrode: development and use. *Analytical Letters* **14,** 1757–1766.

HUFF, J.R. AND ORTH, J.C. (1960). The USACOM-MERC fuel cell. Electric power generation Program. *Advances in Chemistry Series*, volume 90, pp. 315–327.

HUNGER, H.F. AND PERRY, J. (1966). US Patent 3,284,239.

IONESCU, M., CILIANU, S., BUNACIN, A.A. AND COSSOFRET, V.V. (1981). Determination of sulpha drugs with ion selective membrane electrodes. *Talanta* **28,** 383–387.

JANATA, J. AND HUBER, R.J. (1980). Chemically sensitive field effect transistors: In *Ion Selective Electrodes in Analytical Chemistry* (H. Freiser, Ed.), pp. 107–174. Plenum, New York.

JENSEN, M.A. AND RECHNITZ, G.A. (1978). Bacterial membrane electrode for L-cysteine. *Analytica chimica acta* **101,** 125–130.

KARUBE, I., OKADA, T. AND SUZUKI, S. (1982). A methane gas sensor based on methane oxidising bacteria. *Analytica chimica acta,* **135,** 61–67.

KARUBE, I., MATSUNAGA, T., MITSUDA, S. AND SUZUKI, S. (1977). Microbial Electrode BOD Sensor. *Biotechnology and Bioengineering* **19**, 1535–1547.

KARUBE, I., MATSUNAGA, T., TERAOKA, N. AND SUZUKI, S. (1980). Microbioassay of phenylalanine in blood sera with a lactate electrode. *Analytica chimica acta* **119,** 271–276.

KARUBE, I., SOGABE, S., MATSUNAGA, T. AND SUZUKI, S. (1983). Sulfite ion sensor with use of immobilised organelle. *European Journal of Applied Biotechnology* **17,** 216–220.

KATZ, S.A. AND COWANS, J.A. (1965). Direct potentiometric study of the urea, urease system. *Biochimica et biophysica acta* **107,** 605–608.

KATZ, S.A. AND RECHNITZ, G.A. (1963). Direct potentiometric determination of urea after urease hydrolysis. *Analytische Chemie* **196,** 248–257.

KOBOS, R.K. (1980). Potentiometric enzyme methods. In *Ion-Selective Electrodes in Analytical Chemistry* (H. Freiser, Ed.), pp. 1–84. Plenum, New York.

KOBOS, R.K. AND PYON, H.Y. (1981). Application of microbial cells as multistep catalysts in potentiometric biosensing electrodes. *Biotechnology and Bioengineering* **23,** 627–633.

KOBOS, R.K. AND RAMSAY, T.A. (1980). Enzyme electrode system for oxalate determination utilising oxalate decarboxylase immobilised on a carbon dioxide sensor. *Analytica chimica acta* **121,** 111–118.

KOBOS, R.K. AND RECHNITZ, G.A. (1977). Regenerable bacterial membrane electrode for L-aspartate. *Analytical Letters* **10,** 757–758.

KOBOS, R.K., RICE, D.J. AND FLOURNEY, D.S. (1979). Bacterial membrane electrode for the determination of nitrate. *Analytical Chemistry* **57,** 1122–1125.

KORYTA, J. (1975). *Ion Selective Electrodes.* Cambridge University Press.

KURIYAMA, S. AND RECHNITZ, G.A. (1981). Plant tissue based bioselective membrane electrode for glutamate. *Analytica chimica acta* **131,** 91–96.

LIDH, M., LINDGREN, K., CARLSTROM, A. AND MOUSSON, P. (1982). Electrochemical interferences with the YSI glucose analyser. *Clinical Chemistry* **28,** 726.

LLENADO, R.A. AND RECHNITZ, G.A. (1973). Ion electrode based autoanalysis system for enzymes. *Analytical Chemistry* **45,** 2165.

LOBEL, E. AND RISHPON, J. (1981). Enzyme electrode for the determination of glucose. *Analytical Chemistry* **53,** 51–53.

LOWE, C.R., GOLDFINCH, M.J. AND LIAS, R.T. (1983). Some novel biomedical biosensors. In *Biotech 83: Proceedings of the International Conference on the Commercial Applications of Biotechnology,* pp. 633–641. Online Publications, Northwood, London.

MCAULIFFE, C.A. (1980). *Hydrogen and Energy.* Macmillan Press Ltd, London.

MCGOWN, L.B. AND BOCKRIS, J.O.M. (1980). *How to Obtain Abundant Clean Energy.* Plenum Press, New York.

MCNEIL, R.J. (1969). The development of a bioelectrochemical fuel cell for utilisation in space flight. *Proceedings of South Dakota Academy of Science* **48,** 174–187.

MALMSTADT, H.V. AND PARDUE, H.L. (1961). Quantitative analysis by an automatic potentiometric reaction rate method. Specific enzymatic determination of glucose. *Analytical Chemistry* **33,** 1040–1047.

MASCINI, M. AND GIARDINI, R. (1980). Potentiometric determination of proteins with an ammonia sensor. *Analytica chimica acta* **114,** 329–334.

MATSUMOTO, K., YAMATA, K. AND OSAJIMA, Y. (1981). Ascorbate electrode for determination of L-ascorbic acid in food. *Analytical Chemistry* **53,** 1974–1979.

MATSUMOTO, K., SEIJO, H., KARUBE, I. AND SUZUKI, S. (1980). Amperometric determination of choline with use of immobilised choline oxidase. *Biotechnology and Bioengineering* **22,** 1071–1086.

MATSUNAGA, T., KARUBE, I. AND SUZUKI, S. (1980). A specific microbial sensor for formic acid. *European Journal of Applied Microbiology and Biotechnology* **10,** 235–243.

MEIR, P.L., AMMANN, D., MORT, W.E. AND SIMON, W. (1980). Liquid membrane ion selective electrodes and their biomedical applications. In *Medical and Biological Applications of Electrochemical Devices* (J. Koryter, Ed.), pp. 13–91. John Wiley and Sons, New York.

MIZUGUCHI, J., SUZUKI, S. AND TAKAHASHI, F. (1966). Biochemical reaction cell: *Bulletin of the Tokyo Institute of Technology* **78,** 27–33.

MONTALVO, J.G. AND GUILBAULT, G.G. (1969). Sensitised carbon selective electrode. *Analytical Chemistry* **41,** 1897–1899.

MOODY, J. AND THOMAS, J.D.R. (1975). The analytical role of ion selective and gas sensing electrodes in enzymology. *Analyst* **100,** 609–619.

MOSBACH, K. (1983). New biosensor devices. In *Biotech 83: Proceedings of the International Conference on the Commercial Applications and Implications of Biotechnology,* pp. 665–678. Online Publications. Northwood, London.

NAGY, G., RICE, M.E. AND ADAMS, R.N. (1982). A new type of enzyme electrode: the ascorbic acid eliminator electrode. *Life Sciences* **31,** 2611–2616.

NAKAMURA, S., HAYASHI, S. AND KOYA, K. (1976). Effect of periodic oxidation on the structure and properties of glucose oxidase. *Biochimica et biophysica acta* **445,** 294–308.

NANJO, M. AND GUILBAULT, G.G. (1974). Enzyme electrode for L-amino acids and glucose. *Analytica chimica acta* **73,** 367–373.

NANJO, M., VON STROP, L.M. AND GUILBAULT, G.G. (1973). Enzyme electrode for glucose, based on an iodine membrane sensor. *Analytica chimica acta* **66,** 443–455.

NEGOIU, D., IONESCU, M.S. AND COSOFRET, V.V. (1981). Determination of novocaine in pharmaceutical preparation with novocaine selective membrane electrodes. *Talanta* **28,** 377–381.

NEWMAN, D.P. (1976). US Patent, No. 3979274.

OKADA, T., KARUBE, I. AND SUZUKI, S. (1981). Microbial sensor system which uses *Methylomonas* sp. for the determination of methane. *European Journal of Applied Microbiology and Biotechnology* **12,** 102–106.

OSWALD, W. (1984). Die Wissenschaftliche Elektrochemie der Gergenwart und die Technische der Zukunft, 2. *Elektrochemie* **1,** 122–125.

PAN, R.L., BHARDWAJ, R. AND GROSS, E.L. (1983). Photochemical energy conversion by a thiazine photosynthetic-photoelectrochemical cell. *Journal of Chemical Technology and Biotechnology* **33A,** 39–48.

PAPARIELLO, G.J., MUKHERJI, A.K. AND SHEARER, C.M. (1973). A penicillin selective enzyme electrode. *Analytical Chemistry* **45,** 790–792.

PARDUE, H.L. (1963). Automatic amperometric measurement of reaction rates: Enzymatic determination of glucose in serum, plasma and whole blood. *Analytical Chemistry* **35,** 1240–1243.

PAZUR, J.H., KLEPPE, K. AND CEPURE, A. (1965). A glycoprotein structure for glucose oxidase from *Aspergillus niger. Archives of Biochemistry and Biophysics* **11,** 351–357.

PLOTKIN, E.V., HIGGINS, I.J. AND HILL, H.A.O. (1981). Methanol dehydrogenase bioelectrochemical cell and alcohol detector. *Biotechnology Letters* **3,** 187–192.

POTTER, M.C. (1911). Electrochemical effects accompanying the decomposition of organic compounds. *Proceedings of the University of Durham Philosophical Society* **4,** 260–274.

RECHNITZ, G.A. (1967). Ion Selective Electrodes. *Chemical and Engineering News* June 12th, pp. 146–158.

RECHNITZ, G.A., KOBOS, R.K., RIECHEL, S.J. AND GEBAUER, C.R. (1977). A bioselective membrane electrode prepared with living bacterial cells. *Analytica chimica acta* **945,** 357–365.

RIECHEL, T.L. AND RECHNITZ, G.A. (1978). Hybrid bacterial and enzyme sensor with NAD response. *Journal of Membrane Science* **4,** 243–250.

ROLLER, S.D., BENNETTO, H.P., DELANEY, G.M., MASON, J.R., STIRLING, J.L., THURSTON, C.F. AND WHITE, D.R., JR (1983). A biofuel cell for utilisation of lactose wastes. In *Biotech 83: Proceedings of the International Conference on the Commercial Applications and Implications of Biotechnology*, pp. 655–663. Online Publications, Northwood, London.

ROSS, J.W., RISEMAN, J.M. AND KRUEGER, J.A. (1974). Potentiometric gas sensing electrodes. *Pure and Applied Chemistry (IUPAC)* **36,** 473–487.

ROSSETTE, J.L., FROMENT, B. AND THOMAS, D. (1979). Glucose oxidase electrode measurements of glucose in samples exhibiting high variability in oxygen content. *Clinica chimica acta* **95,** 249–253.

ROURBACK, G.M., SCOTT, W.R. AND CANFIELD, J.H. (1962). Biochemical fuel cells. In *Proceedings of the American Power Sources Conference, Fort Monmouth, New Jersey, May 1962*, volume 16, pp. 18–21. Committee of the American Power Sources Conference, Fort Monmouth, New Jersey.

SATOH, I., KARUBE, I. AND SUZUKI, S. (1976). Enzyme electrode for sucrose. *Biotechnology and Bioengineering* **18,** 269–272.

SCHENIC, J.O., MILLER, E. AND ADAMS, R.N. (1982). Electrochemical assay for brain ascorbate with ascorbate oxidase. *Analytical Chemistry* **54,** 1452–1454.

SCOTT, W.R. AND COHN, E.M. (1962). *Proceedings of the Biochemical Fuel Cell Session, Santa Monica Pic-Bat 209/5, Power Information Center, Philadelphia*, p. 9.

SEVERINGHAUS, J.W. (1968). Measurement of blood glucose, O_2 and CO_2. *Annals of the New York Academy of Sciences* **148,** 115–132.

SHICHIRI, M., KAWAMORI, R., YAMASKI, Y., HAKAI, N. AND ABE, H. (1982). Wearable artificial endocrine pancreas with needle type glucose sensor. *Lancet* **ii,** 1129–1131.

SISLER, F.D. (1962). Electrical energy from microbiological processes. *Journal of the Washington Academy of Sciences* **52,** 181–187.

SITTAMPALAM, G. AND WILSON, G.S. (1982). Amperometric determination of glucose at parts per million levels with immobilised glucose oxidase. *Journal of Chemical Education* **59,** 70–73.

SUZUKI, S., KARUBE, I. AND MATSUNAGA, T. (1979). Application of a biochemical fuel cell to wastewaters. In *Biotechnology in Energy Production and Conservation. Biotechnology and Bioengineering Symposium*, volume 8, pp. 501–511. John Wiley and Sons, Inc., New York.

SUZUKI, S., SATOH, I. AND KARUBE, I. (1982). Recent trends of biosensors in Japan. *Applied Biochemistry and Biotechnology* **7,** 147–155.

SUZUKI, S., KARUBE, I., MATSUNAGA, T. AND KAYANO, H. (1980a). Biochemical energy conversion system. *Enzyme Engineering* **5,** 143–145.

SUZUKI, S., KARUBE, I., MATSUNAGA, T., KURIYAMA, S., SUZUKI, N., SHIROGAMI, T. AND TAKAMURA, Y., (1980b). Biochemical energy conversion using immobilised whole cells of *Clostridium butyricum*. *Biochimie* **62,** 353–358.

TRAN, N.D., ROMETTE, J.L. AND THOMAS, D. (1983). An enzyme electrode for specific determination of L-lysine: Real time control sensor. *Biotechnology and Bioengineering* **25,** 329–340.

TURNER, A.P.F., HIGGINS, I.J. AND HILL, H.A.O. (1982). Biological fuel cells. In *Fuel Cells* (G.E.G. Daggit, Ed.), pp. 107–122. Science and Engineering Council, Didcot, Oxon.

TURNER, A.P.F., RAMSAY, G. AND HIGGINS, I.J. (1983). Applications of electron transfer between biological systems and electrodes. *Biochemical Society Transactions* **11,** 381–384.

UPDIKE, J.W. and HICKS, J.P. (1967). The enzyme electrode. *Nature* **214,** 986–988.

VAN HEES, W. (1965). A bacterial fuel cell. *Journal of the Electrochemical Society* **112,** 258–262.

VERDUYN, C., VAN DIJKEN, J.P. AND SCHEFFERS, W.A. (1983). A simple, sensitive and accurate alcohol electrode. *Biotechnology and Bioengineering* **21,** 1049–1055.

VIDELA, H.A. AND ARVIA, A.J. (1975). The response of a bioelectrochemical cell with *Saccharomyces cerevisiae* metabolising glucose under fermentation conditions. *Biotechnology and Bioengineering* **17**, 1529–1543.

WEIBEL, M.K. AND DODGE, C. (1975). Biochemical fuel cells: demonstration of an obligatory pathway involving an external circuit for the enzymatically catalysed aerobic oxidation of glucose. *Archives of Biochemistry and Biophysics* **169,** 146–151.

WHO (1980). *WHO Expert Committee Report on Diabetes Mellitus*. World Health Organization, Geneva.

WINGARD, L.B., KATZIR, E.K. AND GOLDSTEIN, L., EDS (1981). *Analytical Applications of Immobilised Enzymes and Cells. Applied Biochemistry and Bioengineering*, volume 3. Academic Press, New York.

WINGARD, L.B., SHAW, C.H. AND CASTNER, J.F. (1982). Bioelectrochemical fuel cells. *Enzyme and Microbial Technology* **4,** 137–147.

WOLLENBERGER, U., SCHELLER, F. AND ATRAT, P. (1980). Microbial membrane electrode for the determination of cholesterol. *Analytical Letters* **13,** 825–836.

YAMAMOTO, N. AND HIROSHI, T. (1981). Immunoelectrodes. *Kagaku* **36,** 538–542.

YOUNG, T.G. (1965). British Patent 981,803.

YUDKIN, J. (1935). The reduction potentials of bacterial suspensions. *Biochemical Journal* **229,** 1130–1135.

ZHUKOV, A.F., ERNIE, D., AMMANN, D., GUGGI, M., PRETSCH, E. AND SIMON, W. (1981). Improved lithium ion selective electrodes based on a lipophilic diamide as neutral carrier. *Analytica chimica acta* **131,** 117–122.

5
Current Applications of Immobilized Enzymes for Manufacturing Purposes

POUL BØRGE POULSEN

Novo Industri A/S, Novo Allė, DK-2880 Bagsværd, Denmark

Introduction

Ever since the industrial use of enzymes began almost 100 years ago, people have discussed whether and how it would be possible to reuse them. In fact, what was being discussed was whether and how immobilized enzymes could be made industrially important.

According to the Working Party on Immobilized Biocatalysts within the European Federation of Biotechnology (1983), immobilized biocatalysts are defined as enzymes, cells or organelles (or combinations of these) which are in a state that permits their reuse. That means that immobilized enzymes can be, for example, insoluble enzymes used in a fixed-bed reactor, or soluble enzymes used in a semipermeable membrane reactor. The scope of this chapter is to describe the enzymes (and *not* organelles or whole cells) which today are used for manufacturing purposes in industry. That means that this chapter does not intend to cover, for example, immobilized enzymes for analytical purposes, or those for medical use. The term 'whole cells' is not well defined. Some of the applications which will be dealt with in this article involve immobilized whole cells; however, these cells contain only one enzyme which is utilized in the process. Furthermore, such cells are often partly broken down before or during the immobilization step.

The author also interprets the term 'manufacturing purposes' in such a way that immobilized enzymes being studied under pilot plant conditions will not be covered.

The final limitation is that this chapter is based on published data, which are extremely sparse, even for the longest-established example of an immobilized enzyme—immobilized penicillin G acylase. This is because such products mainly have been developed and manufactured by the self-same company which uses each particular immobilized enzyme, or by an industrial consortium. Most data for these processes are still regarded as confidential 'know-how' by the companies involved.

This leaves those enzymes which are reused today in industrial plants for manufacturing purposes, and with the additional limitation that most data can

Biotechnology and Genetic Engineering Reviews—Vol. 1, February 1984

be given only for those enzymes which are marketed (i.e. produced by one company and used by another).

Immobilized enzymes in general

THE SCOPE OF ENZYME TECHNOLOGY

This is indicated by the list of the types of reactions catalysed by enzymes (Dixon and Webb, 1979; *see also* Chapter 1):

—Oxidation
—Reduction
—Inter- and intramolecular transfer of groups
—Hydrolysis
—Cleavage of covalent bonds by elimination
—Addition of groups to double bonds
—Isomerization

Hence virtually all organic and many inorganic reactions can be catalysed by an enzyme or several enzymes acting in sequence. Of course most of these reactions can also be catalysed by non-biological, chemical catalysts. One may therefore ask: When should enzymes be considered as catalysts for a certain reaction and when should they be avoided and non-biological, chemical catalysts chosen?

Enzymes have the advantage of being able to differentiate between chemicals of closely related structures, e.g. between stereoisomers, and of being effective at low temperatures (0–110°C) and at 'neutral' pH values (pH 2–12). This usually gives fewer by-products, a simpler purification process, and improved quality of the final product. Furthermore, enzymes are non-toxic and readily degradable (non-polluting), and they can be produced in unlimited quantities.

Enzymes can often replace strong acids and bases. This means that the use of specially resistant materials in reaction vessels is not necessary, thus saving money. It also means that the presence of large amounts of salts (which otherwise would have to be removed after the reaction) can be avoided. This can contribute both to financial savings and to a reduction in environmental pollution.

Enzymes have their limitations in the relatively fragile nature of the amino acid building blocks and the even more fragile tertiary structure of the protein molecule. These factors make enzymes sensitive to high temperature, extreme pH values, aggressive chemicals, and in some cases, to organic solvents.

It will not be possible to apply enzymes in reactions involving chemicals that denature proteins, i.e. destroy their tertiary structure. In addition, the presence of chemicals that inhibit the catalytic properties of the enzymes must be avoided. Furthermore, there are some processes, particularly in the petrochemical industry, which are so highly developed that one cannot expect further improvement by use of enzymes. Finally, the cost of enzymes will preclude their application in process stages that function well with inexpensive chemicals.

The important factor, costs, encouraged trials for reusing enzymes at a very early stage. Enzymes adsorbed to charcoal were described by Nelson at the beginning of this century but the system was very unstable. In the 1950s, Georg Manecke was the first really to succeed in making relatively stable systems: however, he could not convince industry of the importance of further development of his systems. It was the group of chemists working with Ephraim Katchalski-Katzir in Israel in the 1960s who really opened the eyes of industry to the world of immobilized enzymes. The first immobilized enzyme products to be scaled up to pilot plant level and industrial manufacture (in 1969) were immobilized amino acid acylases (i.e. Chibata and colleagues at Tanabe Seiyaku Company in Japan), penicillin G acylase (M. D. Lilly, University College, London, and Beecham Pharmaceuticals, England) and glucose isomerase (Clinton Division of Standard Brands, now Nabisco Brands, USA, from the 'know-how' of Takasaki and his colleagues at the Fermentation Research Institute, Chiba City, Japan) (Klibanov, 1983).

IMMOBILIZATION TECHNIQUES

During the past ten years, more than a hundred immobilization techniques have been worked out (Zaborsky, 1973; Mosbach, 1976; Trevan, 1980). They can be divided into the following five groups (Klibanov, 1983):

1. Covalent attachment of enzymes to solid supports

In the laboratory a variety of supports have been used, e.g. porous glass and ceramics, stainless steel, sand, charcoal, cellulose, synthetic polymers, and metallic oxides. Enzymes are usually immobilized through their amino or carboxyl groups. In most instances, the immobilization procedure consists of at least two steps: activation of the support and enzyme attachment *per se*.

2. Adsorption of enzymes on to solid supports

Ion exchangers readily adsorb most proteins and have therefore been used for enzyme immobilization. This immobilization procedure is simple: an enzyme solution is added to the support, mixed, and surplus enzyme is then removed by washing.

3. Entrapment of enzymes in polymeric gels

In this approach an enzyme is added to a solution of monomers before the gel is formed. Gel formation is then initiated either by altering the temperature or by adding a gel-inducing chemical. As a result, the enzyme becomes trapped in the gel.

4. Cross-linking of enzymes with bifunctional reagents

Among the most popular cross-linkers are glutaraldehyde, dimethyl adipimidate, and aliphatic diamines. The first two directly cross-link enzymes through their amino groups. Diamines cross-link enzymes through carboxyl groups following activation of these groups with carbodiimides. Cross-linking may be both intermolecular and intramolecular.

5. Encapsulation of enzymes/soluble enzymes in semipermeable membrane reactors

In this approach, enzymes are enveloped within various forms of membranes (e.g. between sheets or within hollow capsules or fibres consisting of semipermeable membranes) that are impermeable to enzymes and other macromolecules but permeable to substrates and products of low molecular weight.

Which of these immobilization techniques has become most popular in industry? Unfortunately there is no clear answer, for the immobilized enzyme products which will be discussed later cover all five groups.

From the above, it can be deduced that there are several optimal immobilization methods. Every enzyme product and every enzyme process each needs its own optimization. The following are some examples of critical optimization factors:

1. Necessary degree of purification of the enzyme;
2. Apparent immobilization yield;
3. Pressure/shear stability of the insoluble enzyme;
4. Degree of leakage (enzyme, monomers, etc.);
5. (Apparently) changed kinetics (e.g. increased by-product formation);
6. System stability during application;
7. Variable costs, necessary investments;
8. Acceptable storage stability;
9. Can the immobilized enzyme be transported to the user?
10. Does the user like the immobilized enzyme product/ process (easy to use, inexpensive, etc.)?

Why are enzymes sometimes immobilized? The primary reason concerns running costs: in some cases it is much more economical to reuse the enzymes instead of using them once only. The second reason is that sometimes this leads to easier purification of the product. The third reason is that sometimes this entails lower investment costs.

On the negative side, immobilized enzymes cost more than those enzymes used once only. Usually, more sophisticated process equipment is needed. In the production process other problems arise, such as increased risk of contamination, need for extra control of temperature and pH, and substrate purity. Diffusion resistance may cause increased levels of by-products and may also reduce the apparent activity.

The conclusion is that the result of a cost–benefit analysis must be the factor to decide whether immobilization of an enzyme is needed.

Immobilized enzymes currently in industrial use

The immobilized enzymes used today in industry are listed in *Table 1*, together with names of producers. This appears to be an impressive list—eight different

Table 1. Immobilized enzymes currently used in industry.

Industrially used immobilized enzymes	Producers (listed alphabetically)
Aminoacylase (EC 3.5.1.14)	Amano, Japan; Tanabe, Japan
Amyloglucosidase/glucoamylase (exo-1,4-α-D-glucosidase, EC 3.2.1.3)	Tate & Lyle, UK
Glucose isomerase (xylose isomerase, EC 5.3.1.5)	CPC, US; Finn Sugar, SF; Gist Brocades, NL; Godo Shusei, Japan; Miles, US; MKC, West Germany; Nagase, Japan; Novo Industri, DK; UOP, US
Hydantoinase (dihydropyrimidinase, EC 3.5.2.2)	Amano, Japan (?); Snam Progetti, It. (?)
Lactase (β-D-galactosidase, EC 3.2.1.23)	Corning Glass, US, Snam Progetti, Italy; Valio, SF
Nitrilase (EC 3.5.5.1)	Nitto, Japan
Penicillin G acylase (EC 3.5.1.11)	Bayer, West Germany; Beecham, UK; Gist Brocades, NL; Snam Progetti, Italy; Pfizer, US; Toyo Jozo, Japan
Penicillin V acylase (EC 3.5.1.11)	Biochemie, Austria; Novo Industri, DK

enzymes and possibly 20 different suppliers. However, a more precise description of the present situation is that only one immobilized enzyme is used in large tonnage, and several in considerably smaller amounts. The same picture holds good for the producers, where less than five companies cover more than 90–95% of the market (*Table 2*).

Immobilized glucose isomerase

Glucose isomerase is an intracellular enzyme which is found in several micro-organisms and used in the production of fructose from glucose. Fructose syrup is competing with sucrose on the industrial market. Ideally, glucose isomerase will produce an equilibrium mixture of glucose and fructose, although in practice (for economic reasons) the mixture will contain about 42% fructose, 52% glucose and 6% dextrins. This mixture is sweeter, weight for weight, than glucose and about as sweet as sucrose or invert sugar (a 50/50 mix of fructose and glucose).

From the outset, in the 1950s and 1960s, it was obvious that the enzyme would have to be reused in order to prove economical. The price of raw sugar in the sixties was US \$0.05/lb (0·45 kg) and fructose syrups could not be produced more cheaply. The enzymatic productivity was only 50–100 kg syrup d.s. (dry solids) per kg of the expensive cells containing the enzyme. Furthermore, cobalt salts had to be added to the reaction mixture and quantitatively removed after the isomerization. The cells autolysed during the reaction, thus making it difficult to produce a pure syrup.

Table 2. Immobilized enzymes—annual product consumption (1982—worldwide).

Immobilized enzyme product (listed alphabetically)	Size
Aminoacylase	Presumably less than 250 t L-amino acid produced per year Estimated enzyme amount: less than 5 t/year Largest producer: Amano, Japan (?)
Amyloglucosidase/glucoamylase	Presumably less than 5000 t syrup d.s. produced per year Estimated enzyme amount: less than 1 t/year Largest producer: Tate & Lyle, UK
Glucose isomerase	Approx. 2150000 t 42% HFCS and 1450000 t 55% HFCS produced per year (after Poulsen 1981/82) Estimated enzyme amount: 1500–1750 t/year Largest producers: Novo Industri, DK and Gist Brocades, NL
Hydantoinase	Presumably less than 50 t D-phenylglycine produced per year Estimated enzyme amount: less than 1 t/year Largest producer: ?
Lactase	Presumably less than 10 000 tons d.s. lactose hydrolysates produced per year Estimated enzyme amount: less than 5 t/year Largest producer: Valio, SF
Nitrilase	Presumably less than 5 t acrylamide produced per year Estimated enzyme amount: less than 0·1 t/year Largest producer: Nitto, Japan
Penicillin G acylase	Approx. 4000 t 6-APA produced per year Estimated enzyme amount: 3–4 t/year (after Godfrey & Reichelt) Largest producers: Gist Brocades, NL, Beecham, UK, Toyo Jozo, Japan
Penicillin V acylase	Approx. 500 t 6-APA produced per year Estimated enzyme amount: approx. 1 t/year Largest producers: Biochemie, Austria and Novo Industri, DK

For comparison, the largest enzyme for once-only use is amyloglucosidase, of which 15000 tons are produced annually (substrate approx. 20×10^6 t dry matter) and used for syrup, ethanol, and beer production

The Clinton Division of the American Standard Brands Co. (now Nabisco Brands Co.) was the first to produce an immobilized glucose isomerase. Their product was based on technical expertise developed by Takasaki and his co-workers at the Fermentation Research Institute in Chiba City, Japan. However, 1969 was not the best time for marketing this product: the disadvantages were obvious; it was a new process for the industry with a complicated system of filter presses, and it necessitated equipment to remove the cobalt which was necessary for the reaction.

In November 1974, the spot prices of raw sugar reached US $0.6/lb (0·45 kg) and this made immobilized glucose isomerase commercially viable. At the same time, a group at Novo Industri A/S succeeded in producing an immobilized glucose isomerase product (Zittan, Poulsen and Hemmingsen, 1975; Hemmingsen, 1979) which could compete economically, could be used without the addition of cobalt, and could withstand the pressure in the large industrial fixed-bed reactors where heights of 7 m are not unusual.

Because nearly all the disadvantages had been eliminated the sales boomed. In 1976, 750 tonnes (t) of immobilized glucose isomerase were used to produce around 800 000 t of 42% fructose in the US alone. Thus, by the time that the sugar price fell again to US $0·15 in the last months of 1976, the new process and product were established. The 42% high fructose corn syrup (HFCS) was sold at a lower price than sucrose, but in 1978 the next development took place: a 55% HFCS became available at a price only 15–25% higher than that of the 42% syrup. This product was the result of a new fractionation technology, involving chromatographic separation of fructose from glucose by means of ion-exchange resins. In test panels, this 55% fructose syrup was found to be equivalent to invert syrup; in comparison to 42% syrup, the glucose and polysaccharide contents were lowered.

Table 3. Actual sales of HFCS and sucrose—USA†

Product	1977	1978	1979	1980	1981	1982
Industrial sucrose, 1000 t	6110	5990	5820	5500	5120	4790
HFCS						
42%	965	1100	1225	1320	1450	1450
55%	0	75	365	590	950	1450
Total, 1000 t	7075	7165	7410	7410	7520	7690
HFCS penetration, %	14	16	21	26	32	35

† Poulsen and Christensen (1982)

It is estimated that, in 1982, 1200–1500 t of immobilized glucose isomerase were used to produce 1450 000 t (dry weight) of 42% HFCS and 1450 000 t of 55% HFCS (*Table 3*).

Table 4 lists the major markets for HFCS in the USA. Japan is the second largest producer of high fructose syrups with an annual production of about $0{\cdot}5 \times 10^6$ t HFS. Other countries with high fructose production are Belgium, France, Italy, Spain, Yugoslavia, Hungary, Greece, Pakistan, S. Korea, Canada and Argentina, but in all these countries the scale of production is much less than that in the US and in Japan.

Table 4. Major markets for HFCS, USA†

Use	Sugar (1000 t)	1981 HFCS (DS) (1000 t)	HFCS penetration (%)	Estimated long-term penetration (%)
Baking	1160	385	25	25
Confections	873	11	1	5
Dairy products	408	181	31	35
Beverage	1659	1483	47	90
Processed foods	204	102	33	40
Canning	226	308	58	75
Other food and non-food uses	644	0	—	—
Total	5174	2470	32	52

† Poulsen and Christensen (1982).
Source: McKeany-Flavell Co. Inc., Sweeteners' Brokers San Francisco, June 7, 1982 and United States Department of Agriculture.

Table 5. Glucose isomerase products commercially available on a tonne scale (listed alphabetically)

Company	Immobilization method
Finn Sugar, SF	—
Gist Brocades, NL	*Actinoplanes missouriensis* cells mixed with gelatin and cross-linked with glutaraldehyde
Godo Shusei, Japan	Enzyme source: *Streptomyces phaechromogenes* (?)
Miles Labs, US Miles Kali Chemie, D	*Streptomyces olivaceus* cells cross-linked with glutaraldehyde or purified enzyme from *Streptomyces rubiginosus* immobilized on SiO_2 beads
Nagase, Japan	*Streptomyces* sp. (?)
Novo Industri, DK	*Bacillus coagulans* cell homogenate cross-linked with glutaraldehyde
UOP, US	*Streptomyces olivochromogenes* purified and immobilized with polyethyleneimine and glutaraldehyde on ceramic carrier Al_2O_3

Table 6. Glucose isomerase—typical application data.

Temperature	55–60°C
pH	7·5–8·0
Dextrose concentration	40–50% (w/w)
Additive	0–1 g $MgSO_4$, $7H_2O/\ell$
Inhibitor	Ca^{2+}
Application time†	2–4 months
Productivity	2000–22000 kg/kg

†Time for which the enzyme product may be employed—typically down to 25% residual activity.

We learn from this that the use of immobilized glucose isomerase seems to be a success only in those nations which are net sugar importers. There must be substantial consumption of liquid sugar through a well-developed processed food and beverage industry. An abundant starch supply must be available. A labour force educated in the operation of sophisticated processing equipment must be available. Finally, a very good market incentive must exist during the start-up of the new process. The general layout of the HFCS production process is shown in *Figure 1*. All commercially available immobilized glucose isomerase products (*Table 5*) are used under more or less the same application conditions (*Table 6*). The development of an immobilized glucose isomerase process has entailed considerable effort by the companies involved (Hemmingsen, 1979; van Tilburg, 1983). The first step was to find a suitable micro-organism which produced an acceptable enzyme cheaply and in large quantities. The discovery by R. O. Marshall and E. R. Kooi in 1957 that xylose isomerase from *Pseudomonas hydrophila* could isomerize D-glucose to D-fructose was quite a novelty. For the organisms investigated, the conversion of glucose to fructose required the previous isomerization of the phosphorylated sugar by the enzyme glucosephosphate isomerase (EC 5.3.1.9). Marshall and Kooi further pointed out, however, that the presence of arsenate in the reaction mixture was essential in order to obtain a proper conversion. This fact, and the fact that the formation of xylose isomerase was absolutely dependent on the presence of xylose in the

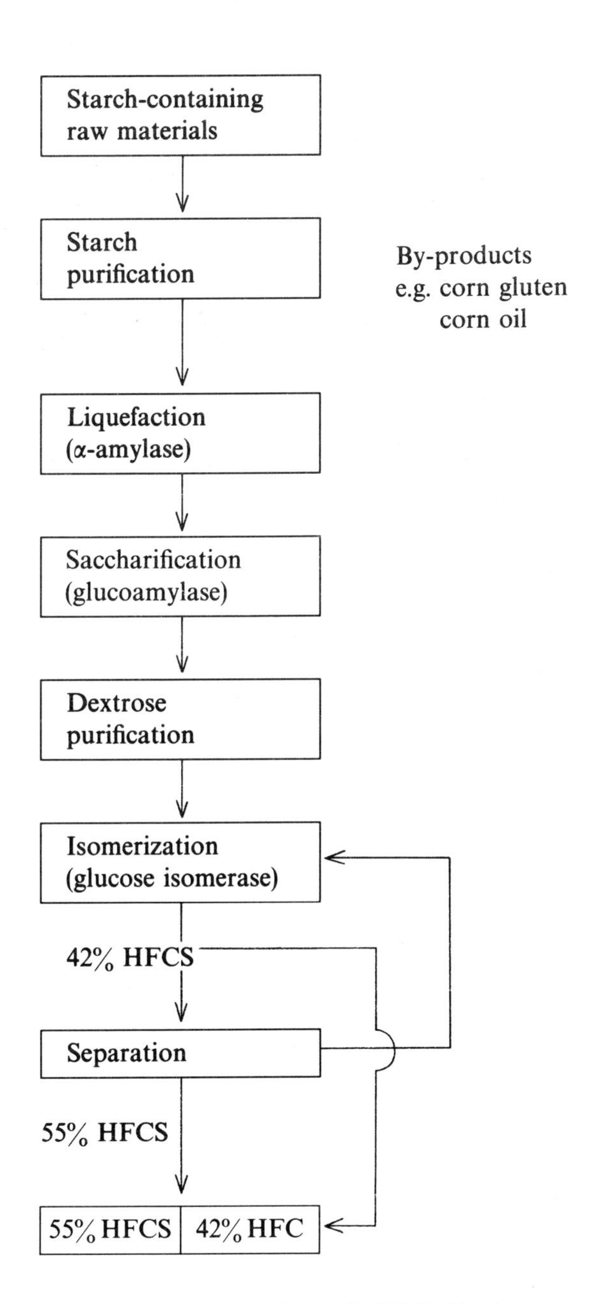

Figure 1. General process layout for HFCS (high fructose corn syrup).

growth medium, meant that Marshall and Kooi's enzyme could not be exploited commercially.

The important step forward came in the mid-1960s when enzyme-producing micro-organisms of the genera *Streptomyces*, *Actinomyces* and *Bacillus* began to be investigated. The glucose isomerases obtained here were fairly thermostable (i.e. they could withstand 10 minutes at 80–90°C). The only cofactors necessary were Mg^{2+} and Co^{2+}. However, as Co^{2+} is undesirable in industrial processes, it was important that the enzyme also worked reasonably well without this. A problem which initially caused considerable concern was the fact that the pH optimum was undesirably high (pH 9·3–9·5): under alkaline conditions, apart from the formation of coloured by-products, a non-metabolizable sugar (D-psicose) is produced in hot glucose and fructose solution. New mutants were soon found, with pH optima about 7–8 and a concomitant satisfactory reduction in psicose formation. This development has been described in more detail by van Tilburg (1983).

The next step was to find an acceptable immobilization procedure. Why immobilize glucose isomerase? The main reason is that this is necessary in order to keep the holding time within satisfactory limits. A holding time longer than 3 hours will easily cause by-product formation. The same activity per volume with soluble once-only enzyme would be considerably more expensive. As the holding time is minimized in a packed-bed or fluidized-bed reactor, the research soon concentrated on finding an immobilization procedure which would yield a product that could withstand such conditions. The immobilization procedure also had to be cheap (i.e. involve food-grade chemicals), to result in high yield, and to avoid the involvement of too many unit operations. The first product marketed by Clinton Corn in the US was simply glucose isomerase fixed in whole cells by heat treatment: however, as this material is very soft, the product had to be used in a very expensive filterpress system. The technology soon improved to granulated flocculated cross-linked whole cells. The final winners in the mid-1970s were products such as granulated glutaraldehyde cross-linked cell homogenate, and cells entrapped in gelatine cross-linked with glutaraldehyde. These particles of the immobilized enzyme were so strong that they could withstand the pressure in columns 7 m high with a holding time of less than 1 hour.

The final step comprised process optimization, i.e. finding the temperature, pH, concentration and necessary degree of substrate purification, which resulted in the highest productivity and therefore the most economical process.

The development of immobilized glucose isomerase has been documented fully by Zittan, Poulsen and Hemmingsen (1975), Hupkes (1978), Hemmingsen (1979), Norsker, Gibson and Zittan (1979), Roels and van Tilburg (1979a), van Tilburg and Roels (1982) and van Tilburg (1983); *see also* Chapter 2.

Immobilized penicillin G/V acylases

The only penicillins which can be produced in high yields by fermentation are penicillin G (phenyl acetyl-6-APA) and penicillin V (phenoxy acetyl-6-APA). However, micro-organisms resistant to these two penicillins have always existed

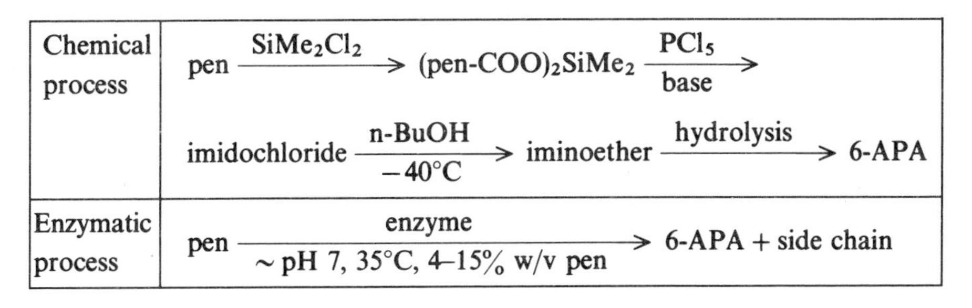

Figure 2. Conversion of penicillin G or V to 6-APA.

and more and more resistant mutants have appeared. In order to enlarge the antibiotic spectrum and improve the pharmacokinetic properties of these antibiotics towards the resistant strains, the G and V side chains have been removed and new side chains have been coupled to 6-APA (6-aminopenicillanic acid). Several semisynthetic penicillins, in particular ampicillin (D-phenyl glycyl-6-APA) and amoxycillin (*p*-OH-D-phenyl glycyl-6-APA) have proved to be effective against strains resistant to the parent penicillins V and G.

Two types of process are possible for the deacylation step—a chemical process and an enzymatic process—(*Figure 2*). The chemical route has been used for longer than the enzymatic route: it proceeds at −40°C, involves organic solvents and must be absolutely free from water—factors which all make the process difficult. The enzymatic process was developed in the 1960s and soon became an accepted alternative to the chemical route as the process conditions were very undemanding; however, the weak point was the enzyme costs.

The penicillin acylases (both the V and the G acylases) are intracellular enzymes produced by several micro-organisms. Typically, G acylases are found in bacteria such as *E. coli* and V acylases in fungi such as *Bovista plumbea.* The first major problem to overcome was elimination of β-lactamases from the acylase preparations: β-lactamase (EC 3.5.2.6) is unwanted because it opens the lactame ring in 6-APA, thereby destroying it. This problem was solved by mutation and chemical treatments.

Initially, the cell suspensions containing acylase were used only once, as in the case of glucose isomerase. Productivity was very low, only about 0·5–1 kg 6-APA per kg *E. coli* suspension. In 1969, Beechams and Malcolm Lilly discovered a method of immobilizing the *E. coli* cells and were able to increase productivity thereby. Few technical data have been published but the immobilization method was probably the DEAE cellulose method. In the 1970s this method was abandoned: Bayer switched to a polymethacryl glutaraldehyde method and Beechams to a cyanogen-bromide-activated dextran/sephadex method; productivity increased to approximately 50 kg/kg (early 1970s).

In principle, these methods are still used today. The productivities obtained are presumably now in the range 100–2000 kg/kg. The current annual 6-APA production is also thought to be about 8–10 × 10^3 t, more than half of which is produced by the enzymatic route. However, these figures are tentative as little is published by the companies involved.

Beechams is probably the largest producer of 6-APA from penicillin G by the enzymatic route. Their process data are presumably those given by Godfrey and Reichelt (1983) and shown in *Table 7*. Toyo Jozo, Japan has also developed an immobilized penicillin G acylase; their product information is shown in *Table* 8. Pfizer also claims to have had a penicillin G acylase system operating for several years.

Table 7. Typical process data for immobilized penicillin G acylase†

Enzyme product	Rigid granules or dextran/sephadex
Reactor	Column
Substrate	Penicillin G or certain semisynthetic cephalosporins (4–15% w/v)
Temperature	35–40°C
Inlet pH	7–8
Operating life	2000–4000 hours
Productivity	1000–2000 kg/kg enzyme

‡ According to Godfrey and Reichelt (1983)

Table 8. Process data for immobilized penicillin G acylase†

Enzyme product	Pen G acylase (*B. megatherium*) covalent bound to glutaraldehyde-activated porous polyacrylonitrile fibres
Reactor	Parallel column system
Substrate	Penicillin G (10% w/v)
Temperature	30–36°C
Inlet pH	8·4 ± 0·1
Operating life	Approx. 1200 hours
Productivity	500–700 kg/kg enzyme

† According to Toyo Jozo, Japan.

Table 9. Process data from immobilized penicillin V acylase†

Enzyme product	Penicillin V acylase cell homogenates covalently bound with glutaraldehyde
Reactor	Stirred batch or column
Substrate	Penicillin V (4–12% w/v)
Temperature	35°C
Inlet pH	7·5–8·0
Operating life	1000–2000 hours
Productivity	200–600 kg/kg enzyme

† Data from Novo Industri A/S, Denmark.

Penicillin V acylases are manufactured by Biochemie and Novo. However, the only published data are of the Novo process (*Table 9*).

Immobilized penicillin G and V acylases have become a success because the current very low enzyme consumption has made this process the most economical one. The process involves an aqueous system (without organic solvents) and proceeds at room temperature (not at energy-demanding temperatures of −20°C and lower). Interesting future developments are expected as the use of semisynthetic cephalosporins rapidly becomes more and more widespread.

Immobilized lactase

To produce 1 kg of cheese, about 10 ℓ of milk are needed, and 9 ℓ of whey are produced as a by-product (Harju and Kreula, 1980). Whey is a yellowish liquid containing 6% dry matter (4·7% lactose, 0·7% protein and 0·5% salts). In a few countries whey is now being used either as pig food or, in a converted form, for human consumption; however, in many countries whey is still discarded as waste. Lactose, the main component of whey, is causing some problems because of properties such as low solubility and low sweetness. If hydrolysis of lactose is combined with ultrafiltration, demineralization, etc., several potentially useful lactose hydrolysates can be produced (*see* Chapter 12).

Lactose in whey can be hydrolysed to glucose and galactose either enzymatically or with acid. Hydrolysis of lactose in milk is a more difficult problem: in this case the enzymatic routes are the only feasible ones. The lactases available can be divided into two groups: neutral lactases (yeasts and bacteria, pH optimum 6–8) and acid lactases (moulds, pH optimum 3–5). Whey is an excellent substrate for microbes and usually the microbial growth is the decisive factor when a process is to be set up.

As previously mentioned, lactose hydrolysis can be accomplished with acid. Because of its high temperature and very low pH, acid-catalysed hydrolysis is suitable only for fairly pure lactose solutions, because otherwise the dark colour and unpleasant taste arising from this process will necessitate costly purification processes. When mineral acids (1–3 M HCl, 0·5 M H_2SO_4) are used, the temperature must be fairly low (50–60°C) to limit side reactions, and up to 48 h may be needed to reach a high degree of hydrolysis. With strong cationic resins the process with deproteinized whey can be improved because no acid is added and a high temperature (90–100°C) can be used. The rate of reaction is greatly accelerated and, in addition, a continuous process may be developed. In a column operation the residence time needed is 0·5–2·0 h (Harju and Kreula, 1980).

An alternative to this acid process is the enzyme process. First, a simple batch process utilizing soluble (once-only) lactase was proposed for both milk and whey. However, in order to limit microbial contamination, a low temperature (10°C) was chosen. This again demanded a high level of lactase in order to keep the hydrolysis time down to 24 h. With such an enzyme level, the cost of the hydrolysis was relatively high.

Because of the drawbacks in batch hydrolysis, several methods have been developed for immobilization of lactase. When lactase is immobilized, the cost of hydrolysis may be reduced if the operational life of the system is long enough. The microbial growth in the reactor, however, needs special attention. The use of sanitizers is difficult because they may easily inactivate the enzyme. If an acid-stable mould lactase is immobilized, the pH of whey can be lowered to 3·5 where the growth of most of the microbes is inhibited.

Many factors are to be considered in optimization of an immobilized lactase system. Besides the temperature, pH and substrate concentration, the activity and operational life of the preparation, the microbial growth and the reactor

Table 10. Large-scale production of immobilized lactase

Manufacturer	Enzyme	Carrier
Corning Glass Works	*A. niger* lactase	Silica beads
Snam Progetti	Yeast lactase	Cellulose acetate fibres
Valio Finnish Co-op Dairies Association	*A. niger* lactase	Adsorption resin

design may all affect the economics of the operation. At least three immobilized lactase systems have been developed on a large scale (*Table 10*).

The first industrial concern commercially to hydrolyse lactose in milk by means of immobilized lactase was Centrale del Latte of Milan, Italy, utilizing Snam Progetti hollow-fibre technology (Marconi and Morisi, 1979). A plant with a capacity of 10000 ℓ/day was opened in 1975, using a low-temperature batch process. In 1978, Valio started their first industrial operation in Koria, Finland (Harju, 1982). Two 300 ℓ columns were able to treat 20000 ℓ of whey per day. Today, the capacity has been increased to two 600 ℓ columns with a whey capacity of 80000 ℓ/day. The Valio process can operate with ordinary whey as well as demineralized whey and the ultrafiltration permeate.

As described, the immobilized lactase can now be designated as used on an industrial scale. However, it is remarkable that, after nearly 10 years, fewer than 10 plants are utilizing this technology. Reasons for this slow development could be:

1. It is cheaper to produce sweeteners from starch than from whey;
2. The need for milk improvement is apparently not great enough when all factors are considered;
3. If the whey causes effluent disposal problems, there are several other ways of using it (e.g. production of fodder or of ethanol);
4. The immobilized enzyme process still has too many weak or undesirable points, such as the risk of contamination, slow rate, and unacceptable cost.

Other immobilized enzymes

IMMOBILIZED AMINOACYLASE

When amino acids are produced by chemical synthetic methods, optically inactive racemic mixtures of L- and D-isomers are obtained. To obtain the L-amino acid (which is the form existing in nature) from the chemically synthesized D,L forms, optical resolution is necessary. The enzymatic method for optical resolution using aminoacylase is one of the most advantageous procedures for industrial production of optically pure L-amino acids (*Figure 3*). *Aspergillus oryzae* aminoacylase has versatile substrate specificity and can be used for the production of many L-amino acids. Since 1954, the enzymatic resolution method has been used by Tanabe Seiyaku Co., Osaka, Japan for the industrial production of several L-amino acids (Chibata *et al.*, 1976).

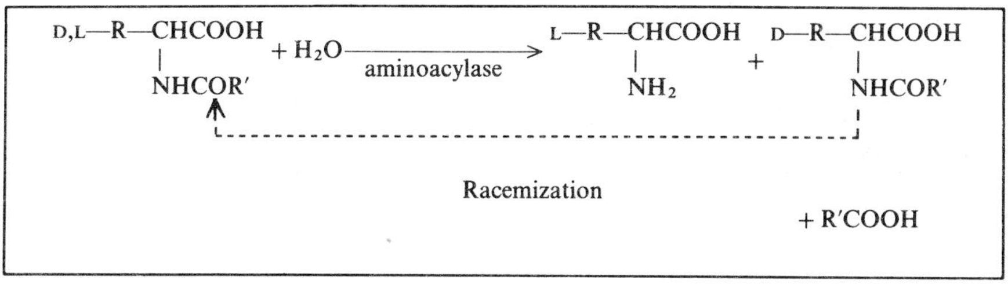

$$\text{D,L—R—}\underset{\text{NHCOR}'}{\underset{|}{\text{CHCOOH}}} + H_2O \xrightarrow[\text{aminoacylase}]{} \text{L—R—}\underset{\text{NH}_2}{\underset{|}{\text{CHCOOH}}} + \text{D—R—}\underset{\text{NHCOR}'}{\underset{|}{\text{CHCOOH}}}$$

Racemization

+ R′COOH

Figure 3. Optical resolution using aminoacylase.

Until 1969, the enzyme reaction was carried out in a batch process by incubating a mixture containing substrate and soluble enzyme, a procedure which had some disadvantages for industrial use. To overcome these disadvantages, the immobilization of aminoacylase and the continuous optical resolution of D,L-amino acids using a column packed with immobilized enzyme was started in 1969. The immobilization method selected was adsorption of aminoacylase to DEAE-Sephadex. When this product was used at 50°C, it lost 50% of its activity after 65 days, but could be reactivated, thus increasing its useful life. Chibata *et al.* (1976) claim that, when using the immobilized enzyme, purification becomes simpler and the yield higher than when native enzyme is used; labour costs are also significantly reduced. The overall operating cost of the immobilized enzyme process has been described as 60% of the conventional batch process using the native enzyme.

Immobilized aminoacylase has been used for production of L-alanine (214 kg after 24 h in 1000 ℓ reactor), L-methionine (715 kg), L-phenylalanine (594 kg), L-tryptophan (441 kg) and L-valine (505 kg). In 1982, Degussa in West Germany reported that they had started a plant with annual output 60 t. They use the soluble (native) aminoacylase from Amano, Nagoya, Japan in a membrane reactor, which has resulted in lower enzyme activity losses, reduced contamination, and easier sterilization than if an insolubilization technique was used. Degussa are currently producing 240 t/year.

IMMOBILIZED AMYLOGLUCOSIDASE/GLUCOAMYLASE

Soluble glucoamylase has been used industrially for more than 20 years. It is an extracellular enzyme produced by fungi such as *Aspergillus niger* and *Rhizopus* spp. The enzyme is used in the saccharification of dextrins to glucose, primarily in the starch industry, but also in such industries as distilling and brewing.

The object of immobilizing glucoamylase was to improve the economics of the process, to reduce the size of the saccharification tanks, and at the same time to maintain the quality of the product.

Perhaps the most crucial factor is the percentage of glucose obtainable. The residence time with soluble glucoamylase is normally in the range of 48–110 h in the dextrin saccharification process. This means that the tank volume required becomes very large, about 7–15 m^3/t d.s. (dry solids) daily production capacity. In this process, glucose percentages of about 95–96 (at a total solids concentration of approximately 30 w/w%) are routinely obtained in industry. However, the highest figures reported in the literature with immobilized glucoamylase products have been 94% which is considered to be significantly lower than 95–96% because this depression has a marked effect on the yield of crystalline dextrose or on the yield of HFCS. The reason for the lower figure is thought to be diffusion resistance. The product glucose is resynthesized, for example to isomaltose (Roels and van Tilburg, 1979b).

This situation might be acceptable if the economics of the process were significantly improved, but this is not yet the case. The important factor as far as the economics are concerned is productivity at an acceptable temperature. A low reaction temperature requires larger tank volumes and special equipment to reduce (or eliminate) the risk of contamination. On the other hand, a high reaction temperature gives a faster enzyme decay. Rugh, Nielsen and Poulsen (1979) report that 55°C is the optimal temperature: however, the immobilized glucoamylase had to be used for 700 h (and there is only 25% residual activity after 700 h at 55°C) before the process became economically superior to that using soluble glucoamylase. Taking the lower product quality into consideration, the whole immobilized glucoamylase concept becomes non-viable (for the dextrin-to-glucose process under discussion).

The concept of immobilized glucoamylase seems to be viable for one process only, i.e. the saccharification of the so-called raffinate. Raffinate (which has a typical composition of 7% fructose, 84% glucose, 4% maltose and maltulose, 2% isomaltose, and 1% each of other disaccharides, DP3 and higher saccharides) is the side stream (or mother liquors) which appears when 42% fructose syrups are upgraded to 55%. The reason why this stream is easier to saccharify is primarily that the dry substance level is 20% w/w or lower. Tate and Lyle are marketing an immobilized glucoamylase for that purpose (*A. niger* adsorbed to granulated charcoal).

IMMOBILIZED HYDANTOINASE

Immobilized hydantoinase is used in the production of optically active amino acids, starting from the racemic amino acid hydantoins which are hydrolysed stereospecifically by the hydantoinase to the corresponding carbamoyl derivatives and finally transformed into the optically active amino acids.

The principle of the preparation of D-phenylglycine, an important intermediate for the synthesis of semisynthetic penicillin and cephalosporins, is illustrated in *Figure 4*. Several hydantoinases apparently exist—some are strictly specific for D-hydantoins, others hydrolyse only the L-form.

According to newspaper reports, the Kanegafuchi Company of Singapore use immobilized hydantoinase(s) for the production of the semisynthetic penicillin and cephalosporin side chains, D-phenylglycine and *p*-OH-phenylglycine.

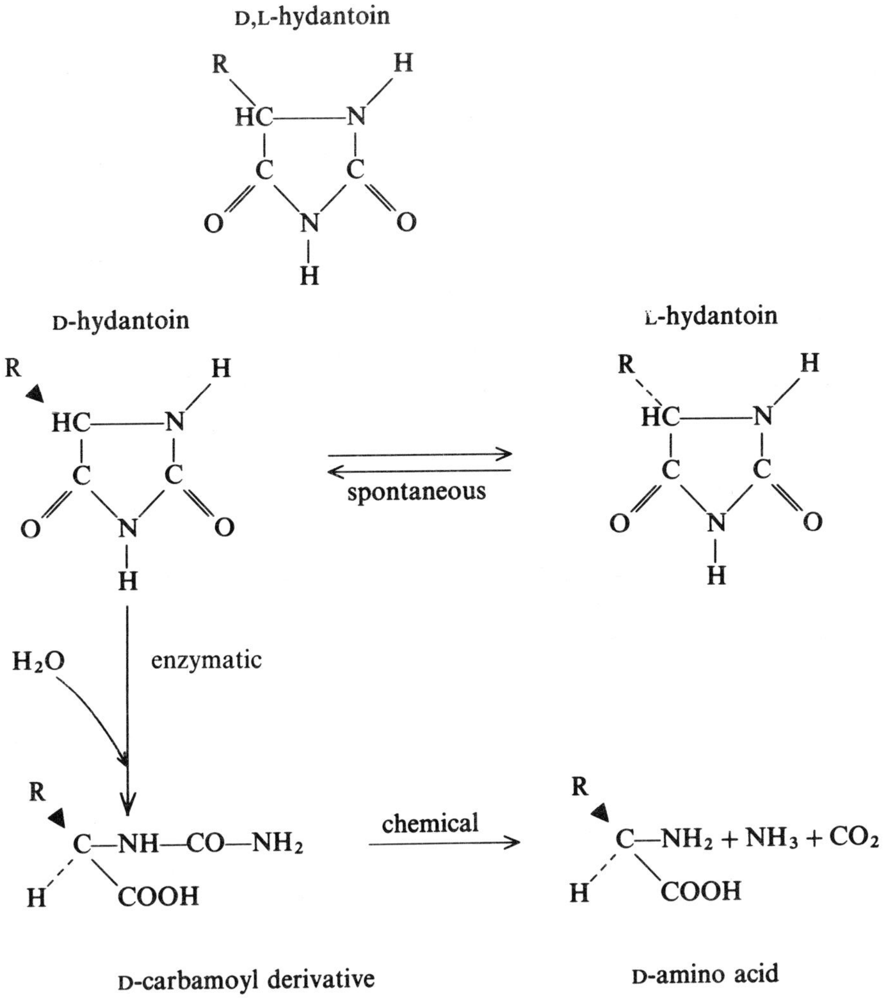

D,L-hydantoin

D-hydantoin ⇌ (spontaneous) L-hydantoin

H_2O — enzymatic

D-carbamoyl derivative: $R(H)C(COOH)\text{—NH—CO—NH}_2$ —chemical→ $R(H)C(COOH)\text{—NH}_2 + NH_3 + CO_2$ D-amino acid

Figure 4. Preparation of an optically active amino acid (D-phenylglycine) using immobilized hydantoinase.

IMMOBILIZED NITRILASE

A new and interesting use of immobilized enzymes is just about to be introduced on a commercial scale, namely the use of immobilized nitrilase for the production of acrylamide from acrylonitrile. Acrylamide has been widely used as a starting material for the production of various polymers for such uses as flocculants, stock additives and polymers of petroleum recovery.

The 'chemical process' for acrylamide involves a copper catalyst. This catalyst is difficult to regenerate and the separation and purification of acrylamide is complex. Furthermore, it would be preferable to produce acrylamide under moderate reaction conditions because compounds like acrylamide contain double bonds and therefore are easily polymerized. Nitto Chemical Industry is about to introduce an enzymatic process on a commercial scale, but only their patents have been published so far. According to these patents (Watanabe, 1982; Watanabe, Sakashita and Ogawa, 1981; Watanabe, Satoh and Takano, 1981), nitrilase from *Corynebacterium* is immobilized in a cationic acrylamide-based polymer gel. One hundred per cent conversion is claimed when a concentration of 5% (w/w) acrylonitrile or less is used in aqueous solution at 0–10°C and pH 7–8·5. Further news of this promising development is awaited with interest.

Conclusions—the future

The first conclusion is that, among the numerous projects which have been described in the literature, there have been very few immobilized enzyme systems which can be described as an industrial success. The only two (or perhaps three) are immobilized glucose isomerase and the immobilized penicillin G and V acylases, and even these have been produced for a long time. All the established processes pertain to carbohydrate processing, to penicillin/cephalosporin processing, and to separation of amino acids. It seems that many projects have been started before the real need and the characteristics of the market have been determined. To obtain a technical success is very seldom sufficient and many other criteria must be fulfilled before industrial success is assured. The immobilized glucose isomerase did not become a success until the market was ready and the technical problems were solved. The same could be said for the immobilized penicillin G and V acylases. It should be borne in mind that the immobilized enzyme systems will have to be very cheap, very heat-stable, and will have to work in reliable optimized process systems. In addition, if there is a cheap soluble enzyme on the market or if other processes are well established, then it is very doubtful whether a newly developed immobilized enzyme system can survive.

No single immobilization method seems to be generally applicable if an optimized immobilized enzyme system is required. One has to be flexible and let the process conditions and the nature of the enzyme determine the optimal method of immobilization. The choice of immobilization method therefore appears to be only a question of engineering.

What will the future bring with regard to new immobilized enzyme systems? That is, of course, a very difficult question. However, the newest immobilized enzyme system, the immobilized nitrilase, could become an opening to a new area. In the past, immobilized enzymes opened routes to new products or alternative routes to chemical routes. Besides continuation of this development, the future could also show more examples of substitution of inorganic catalysts within organic synthesis. Organic synthesis could mean synthesis of lipids, fats and petrochemicals. This would further mean that, whereas all existing immobilized enzyme systems work in an aqueous phase, more research and development will be needed on topics such as enzyme behaviour in two-phase systems, organic solvents, etc. If organic chemists could be convinced that the use of immobilized enzyme systems would also offer them a choice in their process optimization, then the immobilized enzyme field could gain a new, strong impetus.

To sum up, there are interesting perspectives within the field of immobilized enzyme systems, but we will have to develop better process techniques with complicated systems, we will have to be very selective with respect to the processes we try to scale up, and we will have to improve our analysis of the target market.

Acknowledgements

I should like to express my deep thanks to the following companies who answered my request for data to form the basis for this article: CPC, Corning Glass Works, Gist Brocades, Nitto Chemical Industry, Novo Industri, Pfizer, Snam Progetti/Assoreni, Toyo Jozo, Tanabe Seiyaku, Valio.

References

CHIBATA, I., TOSA, T., SATO, T. AND MORI, T. (1976). Production of L-amino acids by aminoacylase adsorbed on DEAE-Sephadex. *Methods in Enzymology* **44,** 746.

DIXON, M AND WEBB, E.C. (1979). *Enzymes*, 3rd edn, chapter 5. Academic Press, New York.

EUROPEAN FEDERATION OF BIOTECHNOLOGY (1983). Report of EFB Working Party on Immobilized Biocatalysts. *Enzyme and Microbial Technology*, pp. 293–307. Butterworths, London.

GODFREY, T. AND REICHELT, J. (1983). *Industrial Enzymology: the Application of Enzymes in Industry.* Macmillan, London.

HARJU, M. (1982). Meieriteknikk. *Norsk Meieriteknisk Forening*, 43–46.

HARJU, M. AND KREULA, M. (1980). *Carbohydrate Sweeteners in Foods and Nutrition*, pp. 233–242. Academic Press, New York.

HEMMINGSEN, S.H. (1979). Development of an immobilized glucose isomerase for industrial application. *Applied Biochemistry and Bioengineering* **2,** 157–183. Academic Press, New York.

HUPKES, J.V. (1978). Practical process conditions for the use of immobilized glucose isomerase. *Stärke* **30,** 24–28.

KLIBANOV, A.M. (1983). Immobilized enzymes and cells as practical catalysts. *Science* **219,** 722–727.

MARCONI, W. AND MORISI, F. (1979). Industrial applications of fiber-entrapped enzymes. *Applied Biochemistry and Bioengineering* **2,** 219–258. Academic Press, New York.

MARSHAL, R.O. AND KOOI, E.R. (1957). Enzymatic conversion of D-glucose to D-fructose. *Science* **125,** 648–649.

MOSBACH, K., ED. (1976). *Methods in Enzymology*, volume 44 (S.P. Colowick and N.O. Kaplan, Eds), Academic Press, New York.

NORSKER, O., GIBSON, K. AND ZITTAN, L. (1979). Experience with empirical methods for evaluating pressure drop properties of immobilized glucose isomerase. *Stärke* **31,** 13–16.

POULSEN, P.B. (1981). European and American trends in the industrial application of immobilized biocatalysts. *Enzyme and Microbial Technology* pp. 271–273. Butterworths, London.

POULSEN, P.B. AND CHRISTENSEN, W. (1982). Current commercial developments in the field of biotechnology. In *Proceedings, ECMRA Conference* 1982, *Oslo*, pp. 239–264. EVAF, Morley House, 320 Regent Street, London, W1R 5AB.

ROELS, J.A. AND VAN TILBURG, R. (1979a). Temperature dependence of stability and activity of an immobilized glucose isomerase in a packed bed. *Stärke* **31,** 17–24.

ROELS, J.A. AND VAN TILBURG, R. (1979b). Kinetics of reactions with amyloglucosidase and their relevance to industrial applications. *Stärke* **31,** 338–345.

RUGH, S., NIELSEN, T. AND POULSEN, P.B. (1979). Application possibilities of a novel immobilized glucoamylase. *Stärke* **31,** 333–337.

TREVAN, M.D. (1980). *Immobilized Enzymes: Introduction and Application in Biotechnology.* Wiley, New York.

VAN TILBURG, R. (1983). *Engineering Aspects of Biocatalysts in Industrial Starch Conversion Technology.* Doctoral thesis, Technische Hogesschool, Delft, The Netherlands.

VAN TILBURG, R. AND ROELS, J.A. (1982). An improved productivity and stability test for immobilized enzymes with reference to glucose isomerase. *Stärke* **34,** 134–140.

WATANABE, I. (1982). *Process for the Production of Acrylamide using Microorganisms.* (Nitto Chemical Industries), US Patent 4,343,900.

WATANABE, I., SAKASHITA, K. AND OGAWA, Y. (1981). *Process for the Production of Acrylamide using Immobilized Cells.* (Nitto Chemical Industries), UK Patent Application 2,086,376.

WATANABE, I., SATOH, Y. AND TAKANO, T. (1981). *Process for Producing Acrylamide or Methacrylamide Utilizing Microorganisms.* Nitto US Patent 4,248,968.

ZABORSKY, O.R. (1973). *Immobilized Enzymes.* CRC Press, Cleveland, Ohio.

ZITTAN, L., POULSEN, P.B. AND HEMMINGSEN, S.H. (1975). Sweetzyme—a new immobilized glucose isomerase. *Stärke* **27,** 236.

Recommended key books

CRC Immobilized Enzymes (1973), Oskar Zaborsky.

Methods in Enzymology (1976), Klaus Mosbach, Ed. volume 44. Immobilized Enzymes (S.P. Colowick and N.O. Kaplan, Eds). Academic Press, New York.

Enzyme Engineering (1972, 74, 76, 78, 80, 82), volumes 1–6.

Applied Biochemistry and Bioengineering (1979), ff. volume 2.

Dechema Monograph No. 84. (1979). Characterization of Immobilized Biocatalysts.

Industrial Enzymology: the Application of Enzymes in Industry (1983) (T. Godfrey and J. Reichelt, Eds). Macmillan, London.

6

Safety in Microbiology: a Review

C. H. COLLINS

Public Health Laboratory, Dulwich Hospital, East Dulwich Grove, London SE22 8QF, UK

Introduction

Several generations of scientists have been aware of the health risks involved in work with certain microbes. Attention was first drawn to the problems and hazards of laboratory-associated infections by German workers long before the Second World War (Paneth, 1915; Kisskalt, 1929). Stories of microbiological martyrdom, ranging from the emotive (de Kruif, 1926, 1933) to the sober (Hunter, 1936) aroused more public interest, however, than the factual accounts of laboratory-acquired infections, over 400 of which are cited by Collins (1983). The number of individuals affected is difficult to determine because many cases were probably missed and others, for various reasons, were not reported. However, a world-wide survey started in 1950 by Sulkin and Pike (1951) and continued until 1978 (Pike, 1978, 1979), revealed by literature research and questionnaires that there had been (at least) 4079 infections with 168 deaths. These numbers included not only single cases, but institutional or common source outbreaks each involving more than 20 people.

Since these figures were published they have been augmented by further accounts, e.g. of Banerjee, Gupta and Goverdhan (1979) who reported 87 cases of Kyasanur Forest disease (caused by KF virus), all in, or associated with, one laboratory; and of over 30 cases of laboratory-acquired typhoid fever in different parts of the United States, and all connected with the distribution of proficiency-testing material (Anonymous, 1979). There have also been many more single or single-figure incidents.

Surveys in the United Kingdom have been concerned more with morbidity among laboratory workers than with the identification of laboratory-acquired infections. This is understandable in view of the difficulties in identifying these infections. Not only is there the possibility of under-reporting, mentioned above, but there are the problems of connecting an infection, especially one with a long or uncertain incubation period, with an incident which has been forgotten or never recorded. These problems have been considered by Phillips (1961) and Collins (1983). Thus Reid (1957) investigated 153 cases of tuberculosis among laboratory workers and concluded that the various categories of such workers

Biotechnology and Genetic Engineering Reviews—Vol. 1, February 1984
0264–8725/84/01/141–25$10.00 + $0.00 © Intercept Ltd

were from two to nine times more likely to contract that disease than matched controls. In a single year (1971) in England and Wales, Harrington and Shannon (1976) found 21 cases of tuberculosis, 38 of hepatitis and 45 of shigellosis among about 22000 laboratory workers who replied to a questionnaire. In a similar survey for the year 1979, Grist (1981) identified five cases of tuberculosis, three of salmonellosis, one each of malaria, shigellosis and hepatitis and four of chicken pox. For 1980–81 (Grist, 1983) he recorded nine cases of hepatitis, 19 of tuberculosis, 13 bowel infections, and seven 'others'.

Between the surveys of Harrington and Shannon and those of Grist there were two important events that stimulated interest, concern and positive action in the United Kingdom. The first of these was an increased incidence of hepatitis B in medical laboratory workers: 17 cases in 1970–72 (Grist, 1975). The second was the laboratory origin of smallpox cases in London in 1973. While the first of these was largely a local problem, the smallpox incident had world-wide repercussions, especially when it was followed, in 1978, by another incident in which a Birmingham laboratory was involved.

At about the same time, and on an international scale, public concern was aroused about two other potential microbiological hazards. There were fears that some of the 'new' diseases, like Lassa and other haemorrhagic fevers, which have a high mortality rate and for which there was no vaccine nor specific treatment, might be spread rapidly by air travellers. The other, which appeared to frighten some scientists as much as the less well-informed public, was the possibility that the newly developed recombinant DNA research might produce 'doom bugs' which would threaten the whole human race. Although both of these fears are now known to be largely conjectural, they contributed to the pressures which compelled governments and other bodies to act, even if only to set up committees.

Events in the United Kingdom

We are concerned here with official actions that were designed to regulate the activities of microbiologists, not only those who work with viruses and bacteria known to be highly pathogenic, but also those who use comparatively harmless organisms or microbes which have never been known to cause human disease. It is an unfortunate fact of life that restrictions which might reasonably be placed at one end of the scale are frequently applied to the other, because the officials who enforce them are unfamiliar with the principles involved and do not trust the scientists to regulate their own activities. Nor is this lack of trust entirely unmerited. Although most scientists are indeed responsible citizens, most of us who have been engaged in the study of microbiological hazards have encountered the biochemist with no microbiological training, who engages, or proposes to engage, in 'bucket bacteriology' with pathogenic micro-organisms and no precautions. Two incidents come to my mind: we were asked to recommend a medium for growing *Brucella melitensis* in winchester bottles in a college incubator so that the organisms could then be separated in an open continuous flow centrifuge; and to advise on the proposal to work with *Salmonella typhi* in a room with sole access through a staff canteen.

Although smallpox is officially extinct, and all known world stocks of virus have been deposited in specified laboratories, a senior WHO scientist expressed to me his fears that 'there may still be an ampoule of unlabelled virus left in the bottom of a deep freeze by some guy who has forgotten all about it; someone else finds it and then what?'.

The smallpox incident in London in March 1973, which claimed the lives of two people who had no connection with the smallpox laboratory, was the subject of a detailed and expensive inquiry (Report, 1974). In November 1973, the Secretary of State for Social Services set up a Working Party on the Laboratory Use of Dangerous Pathogens under the chairmanship of Sir George Godber, then Chief Medical Officer. This Working Party reported 18 months later (Report, 1975a) and made certain recommendations. It also appended a Code of Practice for laboratories working with the most dangerous organisms, designated Category A pathogens. One result was the formation of the Dangerous Pathogens Advisory Group (DPAG) which then published its constitution and a revised Code of Practice (Department of Health and Social Security (DHSS), 1976). The activities of this group were confined to advising on the suitability of particular laboratories for work with viruses such as those of smallpox and haemorrhagic fevers, and on the precautions that should be taken.

The (Godber) Working Party also recommended that a Code of Practice should be established for the less dangerous pathogens, designated Category B, which are encountered mainly in clinical and diagnostic laboratories. In 1975 a Working Party to formulate a Code of Practice for the Prevention of Infection in Clinical Laboratories was convened with Sir James Howie, formerly Director of the Public Health Laboratory Service, as Chairman. This Working Party presented a Report, which was never published, making several recommendations, none of which has been effectively taken up, and a Code of Practice (DHSS, 1978) which was circulated to all laboratories and to individual laboratory workers in the National Health Service. The Code of Practice, now known as the Howie Code, was well received among laboratory workers (*sensu stricto*) in clinical laboratories but many objections to its provisions and requirements were made by administrators and others who were not exposed to laboratory infections.

The (Howie) Working Party recognized the temporary nature of parts of their Code and that progress and new developments would soon make some of its provisions out of date and in need of revision. They recommended the setting up of a 'small permanent advisory group to keep this Code of Practice up-to-date, to ensure that it remains relevant to the needs of safety in clinical laboratories ... and to give advice and such other help that may be requested'. Unfortunately, this very sensible recommendation was not accepted. There have been attempts to tinker with the Howie Code but the printed words remain, regrettably, unchanged.

The Health Departments in the UK laid down a time scale for National Health Service laboratories to meet the requirements of the Howie Code. Coincidentally it became possible and practicable for the Health and Safety Executive (HSE) to apply the Health and Safety at Work Act to clinical and other laboratories. The Howie Code was a convenient, although not mandatory, instrument for the Inspectorate to use. Thus the Inspectors were not only applying the Act, they were fulfilling the Health Departments' ambitions for them. These moves aroused some controversy and two non-governmental bodies were formed to consider the effects of the Code and its application.

The first of these was the Joint Working Party on the Prevention of Infection in Clinical Laboratories. Its members are drawn from societies and organizations concerned with clinical pathology, i.e. who are directly affected by the Howie Code. This committee is uniquely placed for advising the Health Departments, the Health and Safety Executive and clinical laboratory workers in general, but so far it has not published any alternative code of practice.

The other group, formerly known as the Joint Coordinating Committee for the Implementation of Safe Practices in Microbiology, is now known as the Microbiological Consultative Committee (MCC). Its members represent the learned societies that are engaged in non-clinical microbiology, for which the Howie Code was not designed and cannot be sensibly applied. This committee provides a channel of communication between members of the societies and official bodies such as the Health and Safety Commission, the Department of Health and Social Security, the Genetic Manipulation Advisory Group (GMAG: *see below*) and the Research Councils. Its aim is 'to promote the highest standard of cost-effective safe practice with the minimum of bumbledom'. It maintains close links with the Joint Working Party on the Prevention of Infection in Clinical Laboratories and with the Royal Society safety groups. A major task of the MCC has been to offer practical advice to local safety committees who were charged under the Safety Representatives and Safety Committee Regulations (Statutory Instrument 1977, No. 500) with the responsibility of supervising safety procedures in many establishments which had previously been exempt from statutory regulations. Although each local safety committee is expected to formulate its own code of safe practice, the MCC published its guidelines for microbiological safety in 1980, and because of its favourable reception, not only among laboratory workers, but also by HSE, GMAG and the Association of Scientific, Technical and Managerial Staffs, a second, revised edition was produced two years later (Microbiological Consultative Committee, 1980, 1982).

In 1978 there was another death from smallpox associated with a laboratory in Birmingham, the Director of which was a member of DPAG. The subsequent inquiry (Report, 1980) called into question the duties of that group and, as the result of recommendations made by a committee of permanent officials of DHSS and HSE, a new and larger body was created. This is the Advisory Committee on Dangerous Pathogens (ACDP) whose remit covers not only the truly dangerous pathogens, formerly the province of DPAG, and the less dangerous pathogens considered in the Howie Code, but in spite of its name, organisms which are not pathogenic at all! The ACDP has already published and invited comments on its first report (Report, 1983). Apart from venturing into the minefield of classifying micro-organisms into groups on the basis of hazard (considered later in this review), it would seem that the ACDP proposes to construct a new code of practice which will embrace all fields of microbiology.

Personally, I do not think that it is possible for any one group of individuals (unless it is a very large group indeed, which creates its own problems) to formulate a code which will satisfy the needs of workers in such diverse laboratory activities as work with very dangerous viruses, clinical microbiology, control of microbial spoilage in food, and the industrial use of organisms of any kind. It would be more realistic to have existing codes such as those mentioned above brought up to date by very small groups of individuals whose experience in those particular fields cannot be challenged. This would have the added advantage of avoiding vested and political interests. The codes might then be linked in one publication, suitably cross-referenced and containing an abstract of common ground.

Events outside the UK

The real or conjectural hazards of the rapid movement, e.g. by air transport, of dangerous pathogens and infected individuals, as well as recombinant DNA material, across national boundaries and into areas where they might cause epidemics, was the concern of some consultations held by the World Health Organization (WHO) in Geneva in 1976. The possibility of aircraft accidents, releasing pathogens of medical or economic importance into the environment, was also considered. The outcome of these discussions was the creation by WHO of the Special Programme on Safety Measures in Microbiology. Small international groups of advisors and consultants with experience in specific hazards in microbiology were convened. One group has produced guidelines for the management of accidents involving micro-organisms (WHO, 1979a, 1980) and another, minimum standards for laboratory safety in microbiology (WHO, 1979b). Other material relating to microbiological safety problems has been circulated but not yet formally published and a *Laboratory Biosafety Manual* has been printed (WHO, 1983).

Although activities in the field of microbiological safety have been patchy in Europe there have been strenuous efforts particularly in relation to biotechnology laboratories in the Federal Republic of Germany (FDR), in the Netherlands, and by the European Federation of Biotechnology, which, of course, includes the United Kingdom. The committee on Microbiological Safety of the Dutch Society for Microbiology has circulated a set of laboratory safety guidelines. Laboratory safety in the FDR is covered by an impressive but confusing set of regulations. Nevertheless, the Committee on Biotechnology of the Deutsche Gesellschaft für Chemisches Apparatwesen (DECHEMA) has produced a study document on safe biotechnology. Neither of these has yet been published formally but are regarded as very useful documents by the Safety in Biotechnology Group of the European Federation of Biotechnology. Members of this group are using the DECHEMA document as the basis for a set of guidelines.

American microbiologists, who have had their share of laboratory accidents and laboratory-acquired infections, have not had the trauma of committee decisions. They have worked within the framework of rules and recommendations made by professional microbiologists employed by the Centers for Disease

Control and the National Institutes of Health (United States Public Health Service (USPHS), 1974a, b, c).

Action on genetic manipulation experiments

The technique of transferring genetic material from one organism to another and allowing the latter to reproduce so that it will 'express' characteristics of the former is variously known as genetic manipulation, genetic engineering, bioengineering, gene engineering and recombinant DNA research. Fears that these experiments could produce new and highly pathogenic micro-organisms that might escape into the community were voiced by a group of scientists (Berg *et al.*, 1974) who, in a letter to *Science*, called for a moratorium on all such activities. The Berg letter had national and international consequences.

In America an Azilomar Conference considered that certain experiments should not be done (*see* Berg *et al.*, 1975) but that other work could continue with certain specified precautions. The National Institutes of Health Recombinant DNA Advisory Committee produced its guidelines a year later (USPHS, 1976, 1980a) and then a supplement, which included a great deal of additional information on microbiological safety in 1978 (USPHS, 1978). Thereafter, pending decisions by Congress about legislation on genetic engineering, the whole problem seemed to become a political football.

In the UK a Working Party on the Experimental Manipulation of Microorganisms with Lord Ashby as Chairman concluded (Report, 1975b) that research should be continued, but with rigorous safeguards. A Working Party on the Practice of Genetic Manipulation, Chairman Sir Robert Williams, was then convened. It described categories of risk, produced a code of practice, with conditions for containment and recommended the formation of a Genetic Manipulations Advisory Committee (GMAG) to which intending workers in this field would be required to submit their experimental protocols (Report, 1976). GMAG was indeed constituted, in 1976, and has since then visited and advised scientists on genetic manipulations. In 1978 effective control passed to the Health and Safety Executive (HSE) under the Health and Safety (Genetic Manipulation) Regulations and scientists were required to give notice of their intentions and activities to HSE as well as to GMAG.

GMAG published a new categorization scheme in 1978, based on risk assessment, and a new Code of Practice in January 1981 (slightly amended later the same year). At present the future of GMAG is under review because, among other considerations, time and experiments have shown that the hazards of this kind of work, always conjectural, are now regarded as overstated.

In addition, there is now more emphasis on large-scale industrial applications of the research. HSE inspectors, who have experience of production methods in many fields, are, perhaps, better qualified than academic scientists to monitor the industries.

Activities in other countries have followed on much the same lines, with varying amounts of political concern. The Economic and Social Committee of the European Economic Community considered the subject in 1981 without arriving at any particularly new conclusions.

The Assessment of Risk

> 'If reasonable precautions are to be taken against laboratory-acquired infections it is necessary to assess realistically the hazards that might be imposed on the laboratory worker and the community during and as a result of work with any particular micro-organisms. It is a waste of time and resources to take elaborate precautions when the risks are negligible, but foolish to take none if they are considerable. The precautions should be appropriate to the organisms being investigated and the techniques used.' (Collins, 1983).

To assess the risks in order to determine the precautions, the following information is needed:

1. The history of the micro-organisms in so far as their ability to infect laboratory workers is concerned;
2. The routes of infection or the ways in which micro-organisms can enter the body of the laboratory worker (as distinct from the ways in which infection is spread in the community);
3. The dose required to initiate infection;
4. The availability of vaccines for preventing or attenuating infections and of specific therapy in case infections do occur;
5. The ways in which micro-organisms can 'escape' from their containers.

This information will enable:

1. A hierarchial classification of micro-organisms to be made according to the risks they offer to the laboratory worker, and, through him and his work, to the community;
2. Suitable barriers to be placed around the workers to protect them and around the organisms to 'contain' them. The nature and extent of these barriers would be determined by the organisms handled and the nature of the work.

In all this, one must not lose sight of host susceptibility, which varies with age, health and medication, and it must be assumed that the potential hosts, i.e. the laboratory workers, are healthy adults of working age, who are not receiving steroids or immunosuppressive drugs.

HISTORY OF LABORATORY-ACQUIRED INFECTIONS

It has long been appreciated that some micro-organisms (the term here includes viruses) are much more likely than others to infect those who work with them. *Table 1* shows the 'top ten' in Pike's (1978) survey, with *Brucella* in the lead. At the other end of the scale, in the same period *Vibrio parahaemolyticus* was incriminated only once. Nevertheless, it has become fashionable, among non-microbiologists and financial overlords, to discount as anecdotal the surveys and accounts of laboratory-acquired infection by Sulkin and Pike (1951), Sulkin (1961), Pike, Sulkin and Schulze (1965); Pike (1976, 1978, 1979) as well as the more restricted reports of Reid (1957), Harrington and Shannon (1976) and the

Table 1. The 'top ten' laboratory acquired infections.

Infection	*No. of cases*
Brucellosis	426
Q fever	280
Hepatitis	268
Typhoid fever	258
Tularaemia	225
Tuberculosis	194
Dermatomycosis	162
Venezuelan equine encephalitis	146
Psittacosis	116
Coccidioidomycosis	93
TOTAL	2168

Reproduced by permission of the author and publishers from Pike, R. M. (1978). *Archives of Pathology and Laboratory Medicine* **102,** (July), 333–336; Copyright 1978, American Medical Association.

series by Grist (1975–83). Indeed, Grist's figures, which show a dramatic decline in the incidence of hepatitis B among laboratory workers (especially biochemists and haematologists, who are most at risk), have been used as 'evidence' that all other laboratory-acquired infections have declined to the point that microbiological safety, as required by the Howie Code, is not cost effective (Cohen, 1982). One must agree, up to a point, with Williams (1981) that safety has become a bit of a 'bonanza', but it is very unwise to ignore these earlier accounts (summarized by Collins (1983)), although they have no factor in common and have not been subjected to statistical examination, unlike those of Harrington and Shannon and of Grist. There can be no argument whatever for not accepting the evidence of the more recent surveys.

ROUTES OF INFECTION

The ways in which micro-organisms can enter the human body and perhaps initiate infections are well known to practitioners in community medicine. In the laboratory, however, the route may be different. An example of this is brucellosis. Among the general public (veterinarians are an exception) this disease is usually acquired by drinking unpasteurized milk from infected animals. In the laboratory it is almost invariably acquired by the inhalation of the organisms released in aerosols during certain manipulations.

The eye is seldom a portal of entry outside the laboratory but there is adequate evidence and an impressive list of infections acquired in this way by laboratory workers (Papp, 1959; Collins, 1983).

The percutaneous route also is almost peculiar to laboratory workers. Accidents in which individuals stab or prick themselves with infected hypodermic needles or broken, contaminated glassware are not uncommon (Phillips, 1969; Pike, 1976). Apart from overt accidents, organisms may enter through cuts and scratches or microscopic abrasions in the skin as a result of contact with infected droplets or splashes, which might well pass unnoticed.

Infection by the oral route is almost invariably associated with aspiration during mouth pipetting and recorded incidents go back to the beginning of this century (Pike, 1978; Collins, 1983). On the other hand, although eating, drinking and smoking in laboratories are rightly regarded as unhygienic and undesirable activities, there is little but anecdotal evidence of infections acquired in this way in the laboratory (Lubarsch, 1931; Pike, 1979; Collins, 1983).

INFECTIOUS DOSES

Although little is known about the numbers of organisms required to initiate naturally acquired infection, some information, derived from experiments on volunteers, has been accumulated at the National Institutes of Health (USPHS, 1978). The results show, for example, that 10 particles of *Coxiella burnetii* or *Francisella tularensis* are sufficient to initiate infection, but that more than 1300 anthrax organisms are necessary. At least 32 particles of adenovirus 24 caused infection by the ocular route. By the percutaneous route (subcutaneous or intradermal) one particle of Venezuelan equine encephalitis virus, three of *Rickettsia tsutsugamushi* and 57 of *Treponema pallidum* were required. Infection by the oral route required 10^5 organisms for typhoid fever; 10^8 for cholera and 10^9 for shigellosis.

The smaller of these doses demonstrates the extreme vulnerability of laboratory workers who handle these organisms. The larger doses are easily achieved in small droplets of culture material. In this context it should be noted that pathogens such as *Salmonella* serotypes, certain staphylococci and streptococci are frequently present in sub-infective doses in foodstuffs, animal by-products etc. and are easily concentrated into infectious doses by the ordinary cultural methods employed in quality control, and that their colonies may not be recognized.

PROPHYLAXIS AND THERAPY

Although many vaccines are available and are of undoubted value in the prevention of naturally acquired infection, it is by no means certain that they will prevent infection in laboratory workers who may be exposed to very large doses of micro-organisms, or if those organisms enter the body through an unusual route. Nor does vaccination automatically confer immunity. Even when vaccines exist, it may not be possible to give them to some individuals, e.g. those who are sensitive to egg products. Nor is it possible or practicable to vaccinate all laboratory workers against all the organisms they are likely to encounter. There remains, furthermore, a whole group of viruses for which no vaccine is available, and even some bacterial vaccines are known to be relatively ineffective.

In spite of remarkable advances in antibiotics and chemotherapy, and the high hopes expressed for interferon compounds, there are still a number of diseases to which laboratory workers are exposed where no useful therapeutic measures are available.

'ESCAPE' OF MICRO-ORGANISMS

Accidents in which organisms were injected or ingested, spilled or dropped on to the skin, and the bites and scratches sustained from arthropods and mammals during laboratory manipulations, are well recorded, but account for only about 20% of the total number of infections (3921) in Pike's survey (Pike, 1976); Pike considered that the causes of the other 80% were a matter for speculation. However, between 1947 and 1962, a period which Kubica, in his lectures on biosafety at the Centers for Disease Control refers to as 'the frightening fifteen years', an impressive number of papers appeared which described how micro-organisms, in aerosol clouds or in dried material, are dispersed into the laboratory air during many ordinary laboratory manipulations. These technique- and equipment-related hazards include work with bacteriological loops, pipettes, centrifuges, homogenizers and hypodermic syringes, harvesting of eggs, and opening culture dishes, tubes and screw-capped bottles. These investigations did not stop at Kubica's 1962 deadline but continued for the next two decades. It is now clearly understood that the smaller of these aerosol droplets dry rapidly, leaving their nuclei of bacteria or virus particles freely floating in the air, to be moved around rooms and buildings on quite small air currents. If such nuclei are less than 0·5 μm in diameter and are inhaled, they pass into the lungs and may initiate an infection. Larger droplets and particles sediment rapidly and contaminate hands and work surfaces, from which they may be transferred to the eyes or the mouth.

Classification of micro-organisms on the basis of hazard

Armed with all this information it should be possible to place micro-organisms in groups according to the risks they offer to those who work with them and to the community if they 'escape' from the laboratory. If we accept four possible groups we can define them as follows:

1. Those that are not known to cause human disease or which do so only under special or unusual circumstances.
2. Commonplace pathogens that are usually present in the community but which may be handled quite safely by competent workers using good techniques and provided with adequate equipment. Prophylaxis and specific therapy are available.
3. Pathogens that can cause serious disease in laboratory workers, are comparatively rare in the community but offer little threat to it if they escape. These include organisms most likely to cause infection by the air-borne route (inhalation) as well as those capable of initiating infection by very small numbers and any route. Prophylaxis and specific therapy are available.
4. Pathogens that can cause very serious, often fatal disease in laboratory workers and which might give rise to epidemics. Neither vaccines nor specific therapy are available.

Several reasonably successful attempts have been made to classify micro-organisms on this basis and three of these are shown in *Table 2*.

The US system of classification evolved gradually between 1969 and 1974 and recognized four classes, 1—4 in ascending order of risk. In the UK

Table 2. Summary of systems for classifying micro-organisms on the basis of hazards to laboratory workers and the community, and the type of laboratory required.

	Low		*Hazard*	→ *High*
USPHS (1974)	Class 1 none or minimal	Class 2 ordinary potential	Class 3 special, to individual	Class 4 high, to individual and community
UK (DHSS 1978)		Category C no special potential	Category B special, to individual	Category A high, to individual and community
WHO (1979)	Risk Group I low individual low community	Risk Group II moderate individual, limited community	Risk Group III high individual low community	Risk Group IV high, to individual and community
Appropriate laboratory	Basic	Basic	Containment	Maximum containment

From Collins (1983) by courtesy of Butterworths Ltd.

classification there are three categories, A, B, C in descending order of risk. Category A, the 'dangerous pathogens' were defined by the Godber Working Party (Report, 1975a) and the DPAG (DHSS, 1976). The less hazardous pathogens, in Categories B and C, were defined by the Howie Working Party (DHSS, 1978). There is no fourth category in this classification because the terms of reference of the Howie Working Party restricted it to 'infections' and a 'Category D' would have contained non-pathogens.

Although I was involved in formulating the UK system, I now consider that of the WHO (1979a, b) to be the best. Like the US classification, this has four classes—Risk Groups I–IV in ascending order of risk. It is the simplest system and has a wider and more international application than the others. Individual states, however, are still free to go their own way. The Dutch have their own scheme and two new US schemes have been proposed since that of 1974 (USPHS, 1981, 1983). Both are complicated and neither has found much favour among American microbiologists, but one seems to have been adopted at least in part by the ACDP (Report, 1983).

LISTS OF MICRO-ORGANISMS

Having participated in the construction of three sets of lists, I repeat my observation (above) that this area is a 'minefield'. It is not possible to create universal lists because of the uneven geographical distribution of the pathogens themselves, their vectors and reservoirs, and of variable standards of hygiene. An organism which merits a place in a high risk group in one country may deserve only a low category in another. It was for these reasons that the WHO workers felt unable to present lists, and recommended that each member state should create its own. Various other problems arise, however, in compiling 'local' lists.

There is a tendency, on the part of some committees that make lists, to elevate into higher risk groups organisms of which they have no personal experience and which have been used industrially without incident for decades. In addition, organisms are sometimes allocated to risk groups or categories at the generic level. This is too broad. One should not argue, for example, that because one organism is a pathogen, all other species in the same genus should be in the same risk group. List-makers should also guard against too many 'exceptions' or these may well outnumber the species or strains in the main list, thus bringing it into disrepute.

It should also be possible for an organism to be listed in two different risk groups, depending on the amount used and the technique employed. Such differences occur between, for example, diagnostic and quality control laboratories on the one hand, and industrial research establishments on the other. Such a system is used in the newer American classifications (USPHS, 1981, 1983) but it makes them unwieldy and complicated. It would be better, as suggested above, to have separate codes for the various applications of microbiological work, each with its own but strictly relevant lists. In addition, at the diagnostic and 'pure' research levels in parasitology it should be recognized that certain stages in a life cycle may be more hazardous than others to the laboratory worker and this should be reflected in the categorization.

There are three possible ways of listing organisms within the risk groups or categories. One is to name, species by species and strain by strain, those that are known to be pathogenic and to allocate them to the three (or two) highest groups; and then to say that all others are in the lowest group. This approach was used in the US (USPHS, 1974, 1980b) and the UK (Report, 1975a; DHSS, 1978). An alternative would be to attempt to name all the organisms in the lowest-risk group as well. Although this idea has been put forward, hopefully, as a desirable long-term solution, it would be a monumental task and would require several standing committees. It appeals only as an exercise in bureaucracy.

The third way would permit lists of organisms in the lowest group but only in a well-defined context, e.g. in a particular field or industry. Review and revision of the lists would then be a comparatively simple task, easily done by the small groups of experts referred to above who would keep the appropriate code of practice up to date. Such a list, within a defined context, is presented in *Table 3*. This contains only the organisms discussed in a particular book (Collins and Lyne, 1984). Within such a context it is possible to use the expression 'other species'.

Problems also arise with 'difficult' and 'new' organisms. An example of the first is the hepatitis B virus, which is 'difficult' only because of the high incidence of laboratory- and hospital-acquired infection in 1969–71, which created so much emotion that the Howie Working Party were compelled by the threat of a minority report to place this virus (and material that might contain it) in a higher category than it deserved, thus getting out of step with other countries. Further attempts to place this virus in its appropriate category have failed, because no committee has yet grasped the nettle and the subject has become a political chess-piece. Members of the 'high-category' lobby still demand the kind of protection known to be ineffective (microbiological safety cabinets;

Table 3. Risk Groups of some genera and species discussed in one particular book (Collins and Lyne, 1984). Where only the generic name is given all the species mentioned in the text may be treated as if they belong to the Risk Group specified.
(Courtesy of Butterworths Ltd).

Bacteria

Organism	Risk Group
Acetobacter	I
Achromobacter	I
Acinetobacter	I
Actinobacillus	I
Actinomyces bovis	II
eriksonii	II
israelii	II
naeslundii	II
other spp.	I
Aerococcus	I
Aeromonas	I
Alcalescens	I
Alkaligenes	I
Arizona	I
Bacillus anthracis	II
other spp.	I
Bacteroides	I
Bifidobacterium	I
Bordetella	II
Borrelia	II
Branhamella	I
Brevibacterium	I
Brochothrix	I
Brucella	III
Campylobacter	I
Chromobacterium	I
Citrobacterium	I
Clostridium botulinum	III
difficile	II
fallax	II
novyi	II
perfringens	II
septicum	II
sordelli	II
tetani	II
other spp.	I
Corynebacterium diphtheriae	II
equi	I
pyogenes	I
renale	I
ulcerans	II
other spp.	I
Edwardsiella	I
Eikenella	I
Enterobacter	I
Erwinia	I
Erysipelothrix	II
Escherichia	I
Flavobacterium meningosepticum	I
other spp.	I
Francisella	III
Fusobacterium	I
Gardnerella	I
Gemella	I
Gluconobacter	I
Klebsiella	I
Lactobacillus	I
Legionella	II
Leptospira	II
Leuconostoc	I
Listeria	II
Microbacterium	I
Micrococcus	I
Moraxella	I
Mycobacterium africanum	III
avium	III
bovis	III
chelonei	II
fortuitum	II
intracellulare	III
kansasii	III
leprae	II
marinum	II
malmoense	III
scrofulaceum	III
simiae	III
szulgai	III
tuberculosis	III
ulcerans	II
xenopi	III
other spp.	I
Neisseria gonorrhoea	II
meningitidis	II
other spp.	I
Nocardia	II
Pasteurella	II
Pediococcus	I
Photobacterium	I
Plesiomonas	I
Propionibacterium	I
Proteus	I
Providence	I
Pseudomonas mallei	III
pseudomallei	III
other spp.	I
Salmonella paratyphi A	III*
typhi	III*
other serotypes	II
Serratia	I**
Shigella	II
Staphylococcus aureus	II
other spp.	I
Streptobacillus	II
Streptococcus human and animal pathogens	II
Food and milk spp.	I

(*continued overleaf*)

Table 3 (continued)

Organism	Risk Group	Organism	Risk Group
Streptomyces madurae	II	*Vibrio cholerae*	II
pelleteri	II	*parahaemolyticus*	I
somaliensis	II	other spp.	I
Treponema	II	*Yersinia pestis*	III
Veillonella	I	other spp.	I

* Require a Containment laboratory but not a safety cabinet; airborne infection unlikely
** Use a safety cabinet for experiments with aerosols.

Fungi and Yeasts

Organism	Risk Group	Organism	Risk Group
Acremonium	I	*Madurella*	II
Alternaria	I	*Microsporon*	II
Aspergillus	I♯	*Neurospora*	I
Blastomyces	II	*Paecilomyces*	I
Botrytis	I	*Paracoccidioides braziliensis*	II
Candida	I	*Penicillium*	I
Cladosporium	I	*Phialophora*	I
Coccidioides immitis	III	*Pichia*	I
Cryptococcus neoformans	II	*Pullularia*	I
Debaromyces	I	*Rhodotorula*	I
Endomyces	I	*Saccharomyces*	I
Epidermophyton	II	*Scopulariopsis*	I
Fonseceae	I	*Sporobolomyces*	I
Geotrichum	I	*Sporothrix*	II
Gliocladium	I	*Torulopsis*	I
Hansenula	I	*Trichoderma*	II
Helminthosporium	I	*Trichophyton*	II
Histoplasma capsulatum	III		
Kloeckera	I		

♯ Use a safety cabinet for experiments which generate spores. These may be allergenic.

controlled ventilation), but reject the simpler and more effective methods (good technique).

Two examples of 'new' organisms are *Legionella* (and its associates) and the conjectural agent of acquired immune deficiency syndrome (AIDS). Others might be naturally occurring organisms suddenly elevated to prominence for industrial reasons, or even newly created organisms (although these should be adequately covered by genetic manipulation regulations). It is not unreasonable to place any such organisms in a high category or risk group until it is shown that it is safe to handle them without extra precautions. The trouble is that once an organism has been so placed it is very difficult, as in the case of hepatitis B virus, to bring it down to its proper level.

Pathogenicity tests

It may not be possible to determine the history of potential pathogenicity of an organism that has been found suitable for some laboratory or industrial process. The question then arises: Can pathogenicity tests be done? This is another 'minefield'.

Laboratory animals have been used widely in medical microbiology to assess virulence or toxin production, but it is well known that some animals are immune to organisms which cause diseases in others, or in humans. If animal pathogenicity tests are to be used, it would become necessary to employ a fairly large number of different test species. A wide range of doses would also be needed, with and without steroids. Even then, it might be difficult to distinguish between specific pathogenicity, and intoxication or shock due to the injection of large amounts of foreign protein. It would also be necessary to use the various routes, oral and inhalation, as well as injection, to challenge different defence mechanisms. At the end of the experiment it would be difficult, in any case, to argue from some test animals, susceptible or not, that the organism is or is not pathogenic to man.

Some workers have turned to tissue culture tests but, as all virologists know only too well, some commensals quite harmless to man are cytotoxic. Whether an organism which fails to affect, say, human fibroblasts in tissue culture, is also incapable of causing human disease, is also open to conjecture. (I have grown tubercle bacilli in HeLa cells without apparently disturbing them.) This area of risk assessment is still very much in need of exploration.

Assessment of risk in genetic manipulation

Much time and labour has been devoted to this problem but, at present, any other than a very brief review should best appear in a paper dealing with the history of the subject. Changes may be expected.

Unlike risk assessment in conventional microbiology, it is the kind of experiments, not the micro-organisms, that are the centre of a classification scheme. Consideration is given to the source and purity of the nucleic acid used, the host/vector system, and the techniques employed. Certain factors are given probability values and the product of these values indicates the category of risk. The 'Access' factor measures the chances that the organisms will enter a human body and be able to initiate infection; the 'Expression' factor relates to the efficiency with which the foreign DNA is translated into protein; the 'Damage' factor the chance that a genetic element will cause damage, e.g. produce a toxin. This system gives four categories of risk (GMAG, 1978, 1981).

Assessment of risks in large-scale biotechnology

The production of amounts of 20 litres or of much greater volumes of natural or genetically manipulated organisms or their products can offer hazards to the health of the workers very similar to those from experiments conducted in a laboratory. Fortunately there are a number of reasons why larger-scale production is more likely to be safer than research.

Any hazards associated with the organisms, their possible mutants and their products, will already have been assessed in the research and development laboratory. Alternatively, the organisms will have been used in the industry for many years with no untoward incidents.

An 'intermediate technology' between the laboratory and the larger production plant has demonstrated that quite large volumes of even hazardous micro-organisms can be grown safely (Harris-Smith and Evans, 1968). Equipment such as they describe can be scaled up.

Many years of experience, accumulated by mechanical, hydraulic and electrical engineers in fermentation and similar industries, have given the products a high degree of protection from the environment. Those forms of protection can be adapted to protect the worker from the product. Examples of safety measures in downstream processing were presented at a recent meeting of the Society for Chemical Industry and the European Federation of Biotechnology (Lawrence and Barry, 1982; Turner, 1982; Walker and Foster, 1982; van Hemert, 1982).

Protection of the worker from infection (or allergy) is not the only consideration, however, in the assessment of risk. The possible effects on plant and animal life of effluents discharged into rivers or the sea, and of bacterial masses dumped on land, must also be considered, not only because the presence of living organisms may adversely affect the environment but because other economic or ecological 'undesirables' may flourish on their (living or dead) cells and their products.

Lastly, attention to hazards is the major task of health and safety authorities, and many processes involving micro-organisms are closely monitored by trade unions and conservationists.

Classification of laboratories for microbiology

Each of the three systems for classifying micro-organisms on the basis of hazards specifies laboratory facilities and precautions suitable for work within each risk group. These include physical containment in terms of building construction, ventilation, waste disposal and equipment; techniques used; personal precautions; and limitations on access. These all increase with increasing risk. In the American, but not the other systems, degrees of competence of staff are also specified. Although there are four risk groups (three in the UK), only three grades of laboratory are considered necessary. In the WHO (1979a, b) and USPHS (1983) classifications these are Basic, Containment and Maximum Containment, and correspond to the UK (DHSS, 1978) Categories C, B and A laboratories in order of containment. The Maximum Containment or Category A laboratories are sometimes referred to as Special Pathogens Units. The problem of fitting four Risk Groups of micro-organisms into three classes of laboratory was solved by the Americans (USPHS, 1983) by having four Biosafety Levels, which specify equipment and techniques. Two of these, Levels 1 and 2, apply in Basic laboratories according to whether the laboratory is used mainly for Risk Group I or for Risk Group II micro-organisms.

It would be tedious to reproduce here the precise requirements for each kind of laboratory and each Biosafety Level. They are therefore summarized below. For further information the references cited above and the appropriate national regulatory or advisory body should be consulted.

It is, of course, essential that all these laboratories are designed to meet the requirements of the microbiologists who will be expected to work in them. It follows that these professionals should have a major voice in the design, not only because inept planning may directly influence the spread of airborne infection (Phillips, 1961; Collins, 1983), but to avoid any expensive alterations that might be required by the regulating authorities after the building is finished.

BASIC LABORATORIES (BIOSAFETY LEVELS 1 AND 2)

These are intended for work with organisms in Risk Groups I and II. Examples are college and quality control laboratories dealing with Risk Group I organisms (Biosafety Level 1); and clinical or diagnostic laboratories where many of the organisms encountered are in Risk Group II (Biosafety Level 2). No special building or engineering facilities are necessary at either level, but there should be hand-washing basins and also access to an autoclave so that all waste materials can be made safe. The principal hazards to be controlled, especially in clinical laboratories, are ingestion (e.g. by mouth pipetting) and injection. These call for an adequate level of technical competence. Similarly, the standards of hygiene, such as hand washing and wearing protective clothing, and also of access and vaccination, need to be maintained.

CONTAINMENT LABORATORIES (BIOSAFETY LEVEL 3)

These laboratories are necessary for work with organisms in Risk Group III which are most likely to infect laboratory workers by the airborne route or with relatively small doses by other routes. Special ventilation arrangements are therefore required to prevent the dispersal of infectious airborne particles into the room during manipulations and their transfer to other parts of the building. This is achieved by using microbiological safety cabinets (Classes I or II; *see below*) and maintaining pressure gradients so that air flows from the relatively 'clean' areas, e.g. Basic laboratories or corridors, into the Containment laboratories, and thence to atmosphere, filtered or otherwise according to local circumstances. This is not as difficult as some people believe and there are descriptions of simple systems (Clark, 1983; Collins, 1983). Access should be controlled so that members of the public and other unauthorized people cannot casually or inadvertently enter a Containment laboratory. The international Biohazard sign, with a cautionary notice for those who do not understand it, should be displayed on the doors of these laboratories. All equipment should be designed to minimize aerosol production and dispersal. An autoclave should be available.

A high standard of technical competence, again aimed at controlling aerosols, is essential, and the standards of hygiene and medical supervision should be higher than those specified for work in Basic laboratories. Appropriate vaccinations should be mandatory.

MAXIMUM CONTAINMENT LABORATORIES (BIOSAFETY LEVEL 4)

These are essential for work with Risk Group IV pathogens which offer serious 'life-threatening' hazards to the individual worker and the community. In most countries there are laws or regulations about the construction and operation of these laboratories.

Sophisticated engineering facilities are essential to control air flows and to ensure negative pressure gradients in the rooms, the filtration of effluent air and the decontamination, by steam or chemicals, of all liquid effluents and sewage. Physical separation of the laboratory rooms from the same or nearby buildings is required, with access and egress through airlocks and showers. Access is strictly controlled. Infectious material is taken in via a separate airlock and is processed in Class III cabinets or negative-pressure flexible-film isolators which contain centrifuges, incubators and all other equipment. Alternatively, the work is done on the open bench with the operators enclosed in positive-pressure flexible suits with an external air supply from flexible 'umbilical' tubes. All discarded material leaves through a double-doored autoclave with safety locks so that the outer door can be opened only after a sterilization cycle is complete. Very high standards of technical competence, personal protection and medical supervision are essential, as is specific vaccination if available. When work is in progress at least two persons should be present.

Classification of laboratories for genetic manipulation

The principles are the same as those outlined above for conventional microbiology but there are four grades of laboratories, corresponding to four categories of risk. In the UK these are designated Categories I-IV in ascending order of containment (GMAG, 1981), and in the US P1–P4 where P means Physical Containment (USPHS, 1978). The two systems are roughly comparable. Genetic manipulation laboratories are subject to inspection by regulatory bodies and in the UK at present Categories III and IV laboratories are visited by the Genetic Manipulation Advisory Group.

CATEGORY I

This is a Basic Laboratory which must have a hand-basin, and a Class I safety cabinet if any of the work is likely to generate aerosols. An autoclave must be available and all waste material rendered safe before disposal. The laboratory should not, as far as possible, be used for other purposes.

CATEGORY II

The requirements for this kind of laboratory are almost the same as those for a Containment Laboratory for conventional microbiology: designation for a specific purpose; air lock; restricted access; continuous airflow from 'clean' to 'dirty' areas and thence to atmosphere through a HEPA filter; a Class I microbiological safety cabinet and an autoclave in the laboratory or nearby.

CATEGORY III

This is somewhere between the Containment and a Maximum Containment laboratory used for conventional microbiology. The laboratory must be physically isolated, with controlled access and not near to such hazardous areas as solvent stores; it should not be liable to flooding. Entry is through an airlock. A negative pressure gradient is necessary and exhaust air must be passed through a HEPA filter before dispersal to atmosphere. An autoclave must be provided, either in an adjoining room and used solely for Category III work, or double-ended with appropriate safeguards. Effluents must be made safe before disposal. One or more Class III microbiological safety cabinets are required. There are also requirements for protective clothing, respirators, lockable refrigerators and telephones.

CATEGORY IV

This is a full Maximum Containment laboratory and, in addition to all the requirements of a Category III laboratory, there must be showers between the clean side of the airlock and the laboratory, and an emergency electric power generator that automatically operates if the main power source fails.

The containment requirements for genetic manipulations are therefore slightly different from those for conventional microbiology. It is unfortunate that these two systems are not in step. This can cause irritation, and even problems and confusion in establishments (and, at the lower level, even rooms) where the two activities coexist. Now that we have a new broom in the form of ACDP, and GMAG may well become reconstituted, it should be possible to equate the two laboratory classifications and containment requirements.

Microbiological safety cabinets

These very important pieces of containment equipment are mentioned above and merit a note here. For more detailed information about design, siting, installation, testing and use, the current British Standard (BS 5726: 1979) and the books by Clark (1983) and Collins (1983) should be consulted.

There are three Classes, I, II and III, but these numbers bear little relation to the degree of protection afforded. Classes I and III belong to the same genus; Class II works on an entirely different principle.

CLASS I

This is shown diagrammatically in *Figure 1a.* The operator works with bare or gloved hands in the centre or rear of the cabinet while his face is protected by the glass screen. Room air is pulled in through the working face, past the operator's arms at a velocity (0·75–1·0 m/s at the face) calculated to entrain any aerosols and prevent their escape into the room. Aerosols and airborne particles are removed by a coarse filter and a High Efficiency Particulate Air (HEPA) filter; the clean effluent is exhausted outside the building.

Class I cabinets are suitable for work with most micro-organisms up to and including those in Risk Group III, and are intended to protect the operator.

CLASS II

One of these is shown in *Figure 1b*. The operator works and observes as in a Class I cabinet. Air is re-circulated round the cabinet by an integral fan, passing through a HEPA filter before descending through the working space at 0·4–0·5 m/s, and returning through grilles in the floor and rear of the working area. The vertical curtain of air at the working face acts as a barrier to prevent contaminants entering the cabinet from the room. About 30% of the recirculated air is exhausted to atmosphere and is replaced by room air which enters through a grille in the aerofoil at the front of the cabinet floor. This additional curtain prevents the escape of organisms into the room.

Class II cabinets are intended to protect both the worker and the work and are thus particularly useful for handling tissue cultures with organisms up to and including those in Risk Group III. There is an official prejudice against them in UK clinical laboratories because the Howie Working Party (DHSS, 1978) considered that some of the cabinets available at that time did not provide enough protection to the worker. Since then, however, cabinets made to the British Standard (1979) and, in America, to the National Sanitation Foundation Standard (1976) have become available and these do offer adequate operator protection. Class II cabinets are extensively used in the US, where Class I cabinets are unpopular, and there is no evidence that they have failed to protect laboratory workers from infection. They are more expensive, however, to purchase, install and maintain than are Class I cabinets. It is of interest that the amended GMAG Code of Practice (GMAG 1981) permits Class II cabinets to be used for certain procedures with tissue cultures 'which may produce only small amounts of aerosol'.

CLASS III

A Class III cabinet is shown in *Figure 1c*. It resembles a Class I cabinet but the front is closed and sealed. The cabinet is gas-tight. The operator works with gloves, sealed into glove ports. Air enters through one HEPA filter and is exhausted to atmosphere through at least one other. Materials and equipment may be loaded, before work is started, by unclamping the front, or loaded and removed through side ports with air locks or dunk tanks. These also allow the cabinets to be connected in series.

FLEXIBLE-FILM ISOLATOR

This is a modification for laboratory use of the patient isolator developed by Trexler (van der Groen, Trexler and Pattyn, 1980; *see also* Collins and Yates, 1982). A heavy-duty plastic envelope, fitted with glove ports, is mounted on a trolley and maintained at a negative pressure relative to the room by an exhaust fan and valves. Air enters through one HEPA filter and is ducted to atmosphere

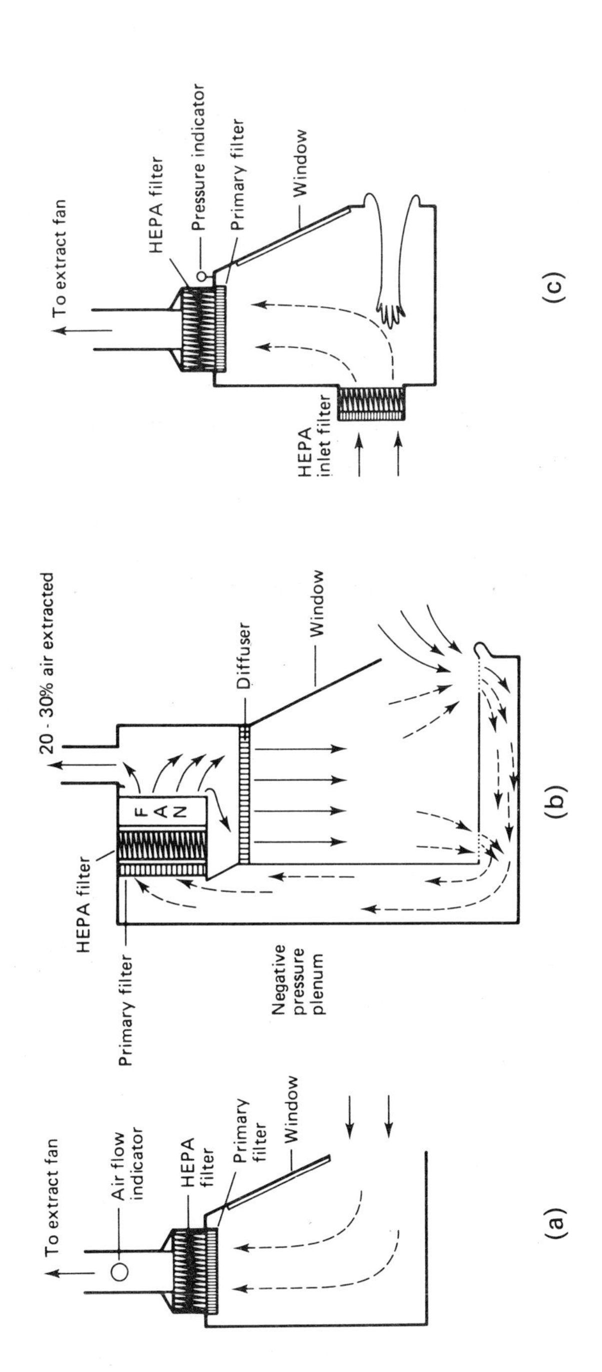

Figure 1. Microbiological safety cabinets showing air flows and general features. (a) Class I. (b) Class II. (c) Class III (from Collins, 1983, by courtesy of Butterworths Ltd). ____ Clean air ---- Contaminated air.

through another. The isolator is mobile, with ports which enable materials to be 'bagged' in and out in plastic bags. It can be docked with other isolators including those accommodating patients. There is space for quite large equipment, with air-tight service connections. As it is double-sided at least four people can work at the same time.

Conclusions

It is quite impossible for anyone who has studied the history of laboratory-associated infections to conclude that no hazards exist in work with micro-organisms. Only those who are unaware of the facts, and are not themselves at risk, will dismiss these hazards as acceptable or non-existent.

At the same time, such studies will confirm that all, or very nearly all, of these hazards have been identified and that means exist to neutralize or overcome them. The paradox is that although little or no interest or action was taken in the well-documented risks in conventional microbiology (until the smallpox incidents described on p. 143, 144), a great deal of concern has been directed at the conjectural hazards of genetic manipulation. But for the smallpox episodes, codes of practice and regulations for genetic engineering would have antedated those for microbiology, probably by many years. Nevertheless, although the means to deal with the hazards are mostly within the power and competence of the scientists themselves, the laboratory workers can cope properly only if they have adequate and suitable accommodation, proper equipment and good training facilities.

In the interests of public confidence, laboratory workers also need codes of practice and advisory and regulatory bodies, as much to protect themselves as to protect the general public, and to ensure that resources are directed where they are most needed. This, of course, means committees, and in this age of decision-taking by committees, all vested (and some unvested) interests will fight to have representation. Some of the representatives on these committees may not be well informed and may lack practical experience, and these may even outnumber the practising scientists. There is also the danger that these committees may become self-perpetuating bureaucracies serving no purpose but their own. There seems to be a belief that scientists cannot be trusted, but there is no evidence that decisions made by lay persons about, say, containing a newly discovered pathogen such as *Legionella*, or a postulated one such as the agent of AIDS, are likely to be more correct than those of scientists. Scientists are, generally, responsible citizens, taxpayers and (usually) trade unionists.

References

ANONYMOUS (1979). Laboratory-associated typhoid fever. *Morbidity and Mortality Weekly Reports* **28,** 44.

BANERJEE, K., GUPTA, N.P. AND GOVERDHAN, M.K. (1979). Viral infections in laboratory personnel. *Indian Journal of Medical Research* **69,** 363–373.

BERG, P., BALTIMORE, D., BOYER, H.W., COHEN, S.N., DAVIS, R.W., HOGNESS, D.S., NATHANS, D., ROBLIN, R.O., WATSON, J.D., WEISSMAN, S. AND ZINDER, N.D. (1974). Potential biohazards of recombinant DNA molecules. *Science* **185,** 303.

BERG, P., BALTIMORE, D., BRENNER, S.O., ROBLIN, R.O. AND SINGER, M.F. (1975), Summary statement of the Azilomar Conference on recombinant DNA molecules. *Science* **188,** 991, and *Nature* **225,** 442.

BRITISH STANDARD (1979). *Specification for Microbiological Safety Cabinets.* BS 5726. British Standards Institution, London.

CLARK, R.P. (1983). *The Performance, Installation, Testing and Limitation of Microbiological Safety Cabinets.* Occupational Hygiene Monograph No. 9. Science Reviews, Northwood.

COHEN, D.R. (1982). The Howie Code: is the price of safety too high? *Journal of Clinical Pathology* **35,** 1018–1023.

COLLINS, C.H. (1983). *Laboratory-Acquired Infections: History, Incidence, Causes and Prevention.* Butterworths, London.

COLLINS, C.H. AND LYNE, P.M. (1984). *Microbiological Methods,* 5th edn, chapter 1. Butterworths, London.

COLLINS, C.H. AND YATES, M.D. (1982). The use of a flexible film isolator in a diagnostic laboratory. *Laboratory Practice,* September 1982, 892–894.

DE KRUIF, P. (1926). *The Microbe Hunters.* Jonathan Cape, London.

DE KRUIF, P. (1933). *Men Against Death.* Jonathan Cape, London.

DEPARTMENT OF HEALTH AND SOCIAL SECURITY (1976). *Control of Laboratory Use of Pathogens very Dangerous to Humans.* Department of Health & Social Security, London.

DEPARTMENT OF HEALTH AND SOCIAL SECURITY (1978). *Code of Practice for the Prevention of Infection in Clinical Laboratories and Post-mortem Rooms.* HMSO, London.

GENETIC MANIPULATION ADVISORY GROUP (1978). Genetic manipulations: new guidelines for UK. *Nature* **276,** 104–108.

GENETIC MANIPULATION ADVISORY GROUP (1981). *Code of Practice for Genetic Manipulation: Containment Facilities.* Medical Research Council, London.

GRIST, N.R. (1975). Hepatitis in clinical laboratories; a three year survey. *Journal of Clinical Pathology* **28,** 255–259.

GRIST, N.R. (1976). Hepatitis in clinical laboratories 1973–1974. *Journal of Clinical Pathology* **29,** 480–483.

GRIST, N.R. (1978). Hepatitis in clinical laboratories 1975–1976. *Journal of Clinical Pathology* **31,** 415–417.

GRIST, N.R. (1980). Hepatitis in clinical laboratories 1977–1978. *Journal of Clinical Pathology* **33,** 471–473.

GRIST, N.R. (1981). Hepatitis and other infections in clinical laboratory staff. *Journal of Clinical Pathology* **34,** 655–658.

GRIST, N. (1983). Infections in British clinical laboratories 1980–81. *Journal of Clinical Pathology* **36,** 121–126.

HARRINGTON, J.M. AND SHANNON, H.S. (1976). Incidence of tuberculosis, hepatitis, brucellosis and shigellosis in British medical laboratory workers. *British Medical Journal* **1,** 759–762.

HARRIS-SMITH, R. AND EVANS, C.G.T. (1968). The Porton mobile enclosed chemostat (POMEC) In *Continuous Cultivation of Micro-organisms.* Proceedings of 4th Symposium. Academia, Prague.

HUNTER, D. (1936). Saints and martyrs. *Lancet* **ii,** 1131–1134.

KISSKALT, K. (1929). Laboratoriumsinfectionen mit Typhuzbazillen und anderen Bakterium. *Archiv für Hygiene und Bakteriologie* **101,** 137–160.

LAWRENCE, A. AND BARRY, A. (1982). Potential hazards associated with the large-scale manufacture of bacterial vaccines. *Chemistry and Industry* 20 November, 880–884.

LUBARSCH, O. (1931). *Ein bewegtes Gelehrtenleben,* p. 68. Springer, Berlin.

MICROBIOLOGICAL CONSULTATIVE COMMITTEE (1980, 1982). *Guidelines for Microbiological Safety.* Society for General Microbiology, Reading.

NATIONAL SANITATION FOUNDATION (1976). *Standard No.* 49. *Class II (Laminar Flow) Biohazard Cabinetry.* Ann Arbor, Michigan.

Paneth, L. (1915). The prevention of laboratory infections. *Medizinische Klinik* **11,** 1398–1399.

Papp, K. (1959). The eye as a portal of entry of infections. *Bulletin of Hygiene* **34,** 969.

Phillips, G.B. (1961). *Microbiological Safety in US and Foreign Laboratories. Technical Report No.* 35. *US Army Chemical Corps.* US Army, Washington.

Phillips, G.B. (1969). Control of microbiological hazards in the laboratory. *American Industrial Hygiene Association Journal,* **30,** 170–176.

Pike, R.M. (1976). Laboratory associated infections. Summary and analysis of 3921 cases. *Health Laboratory Science* **13,** 105–113.

Pike, R.M. (1978). Past and present hazards of working with infectious materials. *Archives of Pathology and Laboratory Medicine* **102,** 333–336.

Pike, R.M. (1979). Laboratory associated infections; incidents, fatalities, causes and prevention. *Annual Review of Microbiology* **33,** 41–66.

Pike, R.M., Sulkin S.E. and Schulze, M.L. (1965). Continuing importance of laboratory acquired infections. *American Journal of Public Health* **55,** 190–199.

Reid D.D. (1957). The incidence of tuberculosis among workers in medical laboratories. *British Medical Journal* **ii,** 10–14.

Report (1974). *Report of the Committee of Inquiry into the Smallpox Outbreak in London, March–April,* 1973 (*Chairman P.J. Cox*). Cmnd. 5626. HMSO, London.

Report (1975a). *Report of the Working Party on the Laboratory Use of Dangerous Pathogens* (*Chairman Sir George Godber*). Cmnd. 6054. HMSO, London.

Report (1975b). *Report of the Working Party on the Experimental Manipulation of the Genetic Composition of Micro-organisms* (*Chairman Lord Ashby*). Cmnd. 5880. HMSO, London.

Report (1976). *Report of the Working Party on the Practice of Genetic Manipulation* (*Chairman Sir Robert Williams*). Cmnd. 5880. HMSO, London.

Report (1980). *Report of the Inquiry into the Causes of the* 1978 *Birmingham Smallpox Occurrence* (*Chairman R.A. Shooter*). House of Commons Paper 79–80. HMSO, London.

Report (1983). *Report No.* 1 *of the Advisory Committee on Dangerous Pathogens: Categorisation of Pathogens According to Risk and Categories of Containment* (*Chairman D.A.J. Tyrrell*). Department of Health and Social Security and the Health and Safety Executive, London.

Sulkin, S.E. (1961). Laboratory acquired infections. *Bacteriological Reviews* **25,** 203–209.

Sulkin, S.E. and Pike, R.M. (1951). Laboratory acquired infections. *Journal of the American Medical Association* **147,** 1740–1745.

Turner, M. (1982). Downstream processing: hygiene or containment? *Chemistry and Industry,* 20 November, p. 876.

United States Public Health Service (1974a). *Classification of Etiologic Agents on the Basis of Hazard.* Centers for Disease Control. Government Printing Office, Washington.

United States Public Health Service (1974b). *National Institutes of Health Biosafety Guide.* Government Printing Office, Washington.

United States Public Health Service (1974c). *Lab Safety at the Center for Disease Control.* Government Printing Office, Washington.

United States Public Health Service (1976). National Institutes of Health. Recombinant DNA Research: Guidelines. *Federal Register* **41,** No. 131, Part II (July).

United States Public Health Service (1978). *National Institutes of Health Laboratory Safety Monograph. Supplement to the NIH Guidelines on Recombinant DNA Research.* Government Printing Office, Washington.

United States Public Health Service (1980a). National Institutes of Health: Guidelines for Research Involving Recombinant DNA Molecules. *Federal Register* **45,** No. 20, Parts V and VI (January).

United States Public Health Service (1980b). *Proposed Biosafety Guidelines for Biomedical and Microbiological Laboratories.* Centers for Disease Control, Atlanta.

UNITED STATES PUBLIC HEALTH SERVICE (1981). Classification of Etiologic Agents on the Basis of Hazard. *Federal Register* **46,** 59379–59380.

UNITED STATES PUBLIC HEALTH SERVICE (1983). *Biosafety in Microbiological and Biomedical Laboratories.* Centers for Disease Control, Atlanta; National Institutes of Health, Bethesda.

VAN DER GROEN, G., TREXLER, P.C. AND PATTYN, S.R. (1980). Negative pressure flexible film isolator for work with Class IV viruses in a maximum security laboratory. *Journal of Infection* **2,** 165–170.

VAN HEMERT, P. (1982). Biosafety aspects of a closed system Westfalia type continuous flow centrifuge. *Chemistry and Industry* 20 November, 889–891.

WALKER, P.D. AND FOSTER, W.H. (1982). Containment in vaccine production. *Chemistry and Industry*, 20 November, 884–887.

WILLIAMS, R.E.O. (1981). In pursuit of safety. *Journal of Clinical Pathology* **34,** 232–239.

WORLD HEALTH ORGANIZATION (1979a). Safety measures in microbiology; the development of emergency services. *Weekly Epidemiological Record* No. 20, 154–156.

WORLD HEALTH ORGANIZATION (1979b). Safety measures in microbiology; minimum standards of laboratory safety. *Weekly Epidemiological Record* No. 44, 340–342.

WORLD HEALTH ORGANIZATION (1980). Guidelines for the management of accidents involving micro-organisms. A WHO Memorandum. *Bulletin of the World Health Organization* **58,** 245–256.

WORLD HEALTH ORGANIZATION (1983) *Laboratory Biosafety Manual.* WHO, Geneva.

7
Biosensors for Environmental Control

HALINA Y. NEUJAHR

Department of Biochemistry and Biotechnology, The Royal Institute of Technology, S-100 44 Stockholm, Sweden

Introduction

Biosensors for environmental control do not differ in principle from biosensors aimed at other applications. The range of substances of interest and the enzymes or micro-organisms required may differ. However, many of the biosensors that have been developed for fermentation or food industries, or for clinical use, may also be utilized for selected environmental purposes. The spark for the development of biosensors came from the major breakthrough in enzyme immobilization in the mid-1960s, as surveyed by Silman and Katchalski (1966) (*see* Chapter 5). An immobilized enzyme system employing a colorimetric method for measuring urea in body fluids was described by Riesel and Katchalski (1964).

The earliest description of a biosensor constructed as an enzyme electrode was given by Clark and Lyons (1962) for the measurement of glucose. The term 'enzyme electrode' was, however, coined by Updike and Hicks (1967) in their presentation of the electrode for glucose based on glucose oxidase entrapped in polyacrylamide gel.

The measurement of glucose using biosensors has subsequently been the subject of a large number of papers and has been carried through the stage of several commercial instruments. The subject is reviewed by Keyes, Semersky and Gray (1979) and is also dealt with in Chapter 4 of this volume. Biosensors for glucose and urea have been favourite model systems for the development and study of enzyme electrodes and are invariably described in the numerous review papers on enzyme electrodes that have been published during recent years (Barker and Somers, 1978; Guilbault, 1980; Vadgama, 1981; Suzuki and Karube, 1981; Suzuki, Satoh and Karube, 1982). Only one group of workers explicitly mentions a glucose electrode as being aimed at environmental applications (Sternberg, Apoteker and Thevenot, 1980). An early review by Guilbault and Rohm (1975) describes enzyme electrodes in 'environmental and clinical studies'.

What are biosensors?

Biosensors are analytical applications of biologically derived catalysts. Such biocatalysts can consist of isolated enzymes, immunosystems, tissues, organelles

Biotechnology and Genetic Engineering Reviews—Vol. 1, February 1984
0264–8725/84/01/167–20$10.00 + $0.00 © Intercept Ltd

or whole cells. In a biosensor, the biocatalyst is usually immobilized in conjunction with a physico-chemical device which monitors its activity in chemical transformation of the substances under analysis, the substrates. The chemical change is translated by the biosensor into a physical response, e.g. an electrical signal. Various means of transduction between the chemical reaction and the electrical signal have been employed, namely redox electrodes, ion-selective electrodes, thermistors, photon counters, fluorimetry.

Electrochemical monitoring devices using immobilized biocatalysts offer many advantages, such as: (1) the specificity of biocatalysis; (2) rapid response; (3) applicability to coloured and/or turbid samples; (4) repeated use of the biocatalyst.

Although the biocatalyst need not necessarily be derived from microbial sources, it is often so, in practice. This is because of the great versatility of microorganisms in synthesizing adequate biocatalysts and the relative ease of growing them in desired quantities. One should, however, also keep in mind spectacular biosensor devices such as those using insects as detectors during gas chromatography of insect pheromones or the more recently described rabbit tissue-based membrane electrode for adenosine 5′-monophosphate (Arnold and Rechnitz, 1981). In most cases, however, the biocatalyst consists of an immobilized enzyme. The topic of enzyme immobilization is thus intimately related to the field of biosensors. A review on enzyme immobilization in connection with the construction of biosensors is given by Bowers and Carr (1980). The extensive general topic of enzyme immobilization has been covered in many monographs and reviews (e.g. Mosbach, 1976). It is also dealt with in other chapters of this book, particularly in Chapters 4 and 5.

BIOPROBES AND BIOREACTORS

There are two main designs of biosensors, the bioprobes and the bioreactors. In a bioprobe, the biocatalyst is directly affixed to a part of the sensing system. Typical examples of this design are enzyme electrodes or bioprobe electrodes. The latter employ whole cells instead of isolated enzymes. In an immobilized bioreactor, a flowing stream of buffer and reactants moves through the biocatalyst, sensing devices being placed before and/or after the reactor. The biocatalyst can be located in a packed bed or it can be shaped as an open tubular reactor.

The main advantage of bioreactors compared with bioprobes is the possibility of using almost any detection technique. However, the use of photometric methods, for example, would eliminate one of the foremost advantages of biosensors in general, namely that the samples need not be clear and colourless.

THE SENSING DEVICE

Electrochemical sensors can be classed as potentiometric or amperometric electrodes, according to the principle of their operation. In a potentiometric electrode, the response (mV) is a logarithmic function of the concentration according to the Nernst equation:

$$E = E^\circ - \frac{RT}{nF} \ln Q$$

where E is the potential measured, E° is the standard potential, R is the gas constant, T is the absolute temperature, n is the number of electrons transferred in the electrode reaction, F is the Faraday constant and Q is the ratio of activities ($\cong$ concentrations) of the reactant(s) and product(s). In an amperometric electrode, the electrode response (μA) is a linear function of the concentrations. The current measured arises from the diffusion towards the electrode of the species (substance) which has been removed from its vicinity by the electrode reaction (the so-called 'diffusion current'). Diffusion is a first-order process with respect to the concentration of the solute which diffuses.

The potentiometric principle is used in ion-selective electrodes for H^+ or NH_4^+ and also in gas-sensing electrodes for CO_2 or NH_3. The amperometric principle is used in oxygen electrodes, which are polarographic or galvanic electrodes covered by a gas-permeable membrane or uncovered platinum electrodes. Oxygen can be analysed with a platinum electrode at $-0{\cdot}8$ V versus a standard calomel electrode (SCE) as reference. Similarly, hydrogen peroxide can be analysed at $+0{\cdot}6$ V versus an SCE. A more convenient arrangement is that offered by the Clark-type combination electrodes, at least for the measurement of oxygen. I find the YSI Model 4004 (Yellow Springs, Ohio, USA), consisting of a platinum–silver combination electrode covered with a teflon membrane, particularly useful. This is because of its small dimensions (2 mm across the platinum cathode) and efficient protection of the enzyme layer from heavy metals derived from the silver reference.

Another class of sensing devices employs microcalorimetry. Such enzyme sensors are called enzyme thermistors (Mosbach and Danielsson, 1974; Mosbach *et al.*, 1975; Danielsson and Mosbach, 1976; Weaver *et al.*, 1976; Weaver *et al.*, 1977). A sensing device employing solid surface fluorescence methods has been developed by Guilbault and his co-workers and successfully applied to several enzyme systems (Guilbault, 1976).

KINETIC OR EQUILIBRIUM METHOD OF ANALYSIS?

There are two fundamentally different methods of enzymatic analysis—the kinetic method, and the 'end-point' or equilibrium method. Only the latter is applicable to bioreactors, where the signal of the sensing device necessarily corresponds to the total amount of substrate transformed during its passage through the reactor. Both the kinetic and the equilibrium method can be employed with bioprobes. In this context, the equilibrium method is often (probably erronously) called the 'steady state' method.

The kinetic method eliminates possible instances of end product (product-caused) inhibition. It may also offer quicker response and greater sensitivity, especially when the derivative of the signal is recorded, as has been done, for example, using enzyme electrodes for phenol or catechol (Neujahr and Kjellén, 1979; Kjellén and Neujahr, 1980; Neujahr, 1980, 1982). There is a definite advantage of the kinetic method when oxygen is one of the reactants. This is not only because the comparatively low solubility of O_2 in water (250 μM at 25°C) may severely limit the range of linear response when using the equilibrium method, but also because of possible interference from oxygen leaking in. On the other hand, the response using the kinetic method is dependent on the thickness of the enzyme layer immobilized on the surface of the electrode. Recalibration of the probe is therefore required, whenever this thickness might have been affected, for example by accidental external pressure or change of the layer.

ANALYSIS WITH ENZYMES IMMOBILIZED ON ELECTRODE SURFACES

In homogeneous aqueous solutions, even the fastest enzyme-catalysed reactions do not appear to be diffusion controlled. The situation becomes quite different in immobilized enzyme systems. If an excess of enzyme is present, diffusion of the substrate to the immobilized enzyme will now become the rate-limiting step. Mass transfer by diffusion is a first-order reaction with respect to substrate concentration. Imposing a diffusion barrier by immobilizing the enzyme thus has the effect of extending the linear range of initial reaction velocity beyond the K_m value of the enzyme for a substrate. Because of the linear relationship, however, the observed rate of reaction is lower than it would have been in a kinetically controlled enzyme reaction conforming to the rectangular hyperbola of Michaelis–Menten kinetics. This is illustrated in *Figure 1*.

The performance of an enzyme electrode carrying on its surface a layer of immobilized enzyme, however thin, will therefore be greatly influenced by diffusional factors. A diffusion barrier is here beneficial because it can make the transport of the substrate under analysis rate limiting for the enzyme reaction and can thus extend the range of the linear response. By making the rate of diffusion very slow, the linearity can be extended over a large range of concentrations, but the response will then also become very slow. High concentrations of enzyme in the layer immobilized on the electrode will ensure, however, that the enzymatic reaction is fast enough to give a reasonable overall response time while keeping the system under diffusional control. An overcapacity of enzyme activity will also prolong the 'life time' of the probe, because the response will remain stable for as long as the overall reaction rate will remain diffusion controlled.

Diffusional considerations become more complicated in relation to enzyme systems where more than one substrate participates. This is the case, for example, with cofactor-linked enzymes, when the low-molecular substrate usually has a higher diffusivity than the cofactor. In such cases, the response of the bioprobe may become controlled by the concentration of the cofactor rather than by that of the substrate to be measured. An example of a three-substrate system (phenol, NADPH, O_2), further complicated by the side reaction of NADPH-oxidase, is illustrated by the enzyme electrode for phenol employing immobilized phenol hydroxylase (Kjellén and Neujahr, 1980). The discussion in that paper further develops the diffusional aspects of enzyme electrodes with multi-substrate enzymes when the size and hence the diffusion constants of the substrates differ significantly from each other.

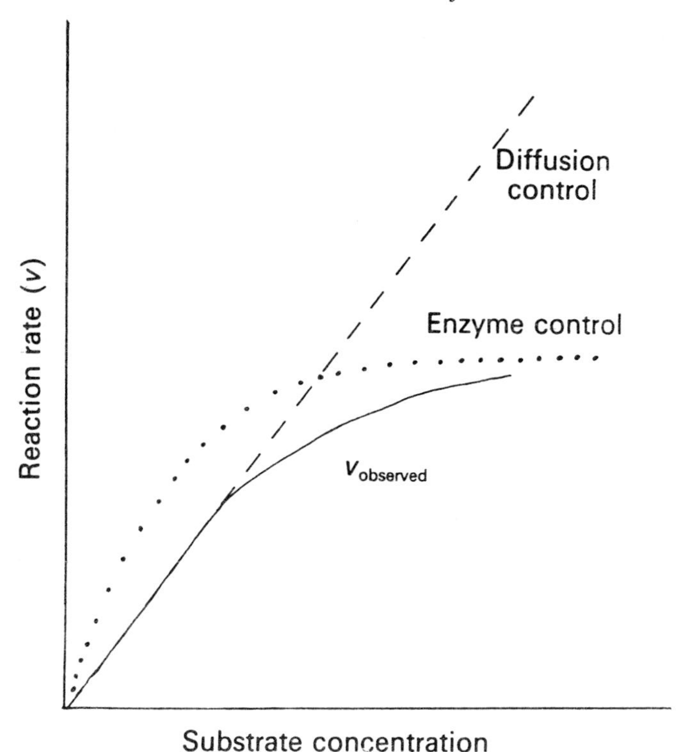

Figure 1. Reaction kinetics in an immobilized enzyme system.

The field of enzyme electrodes has developed along purely empirical lines. Basic theories and concepts in this area are yet to be defined. From a theoretical point of view, the systems are very complex, encompassing elements of diffusional mass transfer, heterogenous catalysis, enzyme kinetics and electrochemistry. Ambitious attempts at theoretical analysis have been published (Engasser and Horvath, 1976; Brady and Carr, 1980; Morf, 1980; Gough and Leypoldt, 1981).

BIOPROBES EMPLOYING WHOLE MICROBIAL CELLS

The importance of a diffusional barrier for the linearity of a bioprobe response may lie behind the success obtained with whole microbial cells attached to the surface of an electrode (Rechnitz *et al.*, 1977, 1978; Neujahr and Kjellén, 1979; Hikuma *et al.*, 1979a, b, 1980). There are several rather obvious advantages, such as the elimination of the laborious and time-consuming work of enzyme isolation and its immobilization, and elimination of the need for cofactors. On the negative side of such bioprobes one could quote a possibly decreased specificity and a lower sensitivity because the enzyme in question may be 'diluted' by other proteins of the cell. However, these circumstances are by no means general. In cases when the enzyme concerned, or an enzyme sequence, acts at the start of a metabolic route, has a high turnover number and/or possibly has been induced in the cells by the type of substrate which has to be measured by the bioprobe, sensitivities comparable to those obtained with isolated enzymes can easily be obtained. Such a case is illustrated by the bioprobe electrode for phenol (Neujahr and Kjellén, 1980).

The bioprobes employing whole microbial cells can be considered as an outgrowth of the erstwhile 'analytical microbiology'. This methodology flourished during the 1950s and 1960s, often as the only feasible method available for the measurement of antibiotics, vitamins, amino acids and other growth factors in natural materials (Kavanagh, 1963). During that era, however, the only parameter which could be measured was growth or the formation of growth-related products, such as acid production in batch cultures or growth inhibition zones on agar plates. This was a time-consuming process requiring incubation at least overnight, and in many cases for several days. With the variety and refinement of electrochemical sensors now available, and with the development of microcalorimetry and photochemical devices, new and much more varied possibilities have opened up for monitoring of microbial activities. Small changes in substrates, metabolic intermediates or products can now be detected by a method unrelated to growth. This reduces to a minimum the number of the cells required and the time of assay.

Biosensors for substances of environmental interest

Such substances can be of two different kinds. One group consists of synthetic organic compounds derived from industrial wastes, biocides and detergents. They may contain in their structures organic phosphates or carbamates or the inert aromatic rings, which have to be prepared for cleavage and degradation by hydroxylation to phenols. Many phenols also occur naturally, arising mostly from the lignins of decaying wood. The highly toxic effects of phenol(s) on humans and their environment are well documented. The second group consists of compounds, which normally occur in nature and, indeed, between certain concentrations may be essential for the maintenance of 'normal' conditions in the biosphere and lithosphere. This group includes nitrite, nitrate, sulphate, phosphate, and even some heavy metals. When such compounds accumulate in abnormal quantities, they upset the established ecological equilibria and become an environmental risk. To this group one could also add naturally occurring organic poisons such as antibiotics and enzyme inhibitors. However, the production of such substances has mainly been studied under 'artificial' laboratory conditions. Their actual occurrence in nature is very difficult to measure. They will not be dealt with in this review.

PESTICIDES

Few biosensors have been constructed with the specific aim of environmental control. The most advanced such biosensor system is that described by Goodson and Jacobs (1974, 1976) for monitoring of air and water. The sensor employs immobilized cholinesterase of the non-specific type (EC 3.1.1.8) as a detector which, after exposure to the inhibitors in air or water, is then analysed for its residual activity. Any decrease in enzyme activity can be related to the quantity of inhibitor to which the enzyme was exposed. The system possesses the advantage of 'biological self-amplification'. This is inherent in the large turnover number of cholinesterase, which is of the order of 80 000/min. Inhibition of one active site can thus reduce the formation of products by 80 000 molecules in tests lasting one minute. Other advantages are specificity of inhibition, ease of automation, good correlation with animal toxicities, and the possibility of concentrating pollutants by absorbing them from a large volume of air into a small volume of water. Goodson and Jacobs claim that, with such a concentrator, they were able to detect quantities of sarin of less than a nanogram per litre of air in 9 min.

For detection of the inhibitors, butyrylthiocholine iodide is chosen as substrate. This thioester is stable in the presence of platinum electrodes under a constant electric current. After hydrolysis by cholinesterase, however, the resulting thiol is readily oxidized to a disulphide, in an anodic reaction. This generates a potential, which is highest with the uninhibited enzyme and decreases with the degree of enzyme inhibition.

$$C_3H_7COS{-}R \xrightarrow{\text{cholinesterase}} C_3H_7COOH + R{-}SH$$

$$2R{-}SH \xrightarrow{-2e^-} RS - SR + 2H$$

$$(R = {-}CH_2CH_2N^+(CH_3)_3I^-)$$

Good contact between the substrate or inhibitor solutions and the biocatalyst is ensured by immobilizing the enzyme on an open-pore polyurethane foam pad. Such a pad is placed in an electrochemical cell between the electrodes. It can easily be replaced by a fresh one when the enzyme activity decreases below a

Table 1. Biosensors for environmental poisons (pesticides), operating on the enzyme inhibition principle.

Type/substance	Enzyme/sensor†	Detectable level (ppm)	Reference‡
Organic phosphates			
Azodrin	A	20·0	1
DDVP	A	1·0	1
Diazinon	A	1·2	2
Dursban	A	4·5	2
Malathion	A	16·5	1
Paraoxon	A	0·1	2
Parathion	A	5·0	1
Carbamates			
Dimetilan	A	10·0	1
Sevin	A	20·0	2
Temik	A	0·5	2
NaF as an inhibitor model	B	4·2	3

† A = cholinesterase/potentiometric cell (platinum electrodes). The cholinesterase is immobilized in partially hydrolysed potato starch ('electrophoresis grade') and open-pore polyurethane foam.
B = urease/CO_2 gas electrode. The urease is chemically bound with glutaraldehyde to a silicon membrane.
‡ Reference 1: Goodson and Jacobs (1974).
Reference 2: Goodson and Jacobs (1976).
Reference 3: Tran-Minh and Beaux (1979).

desired level. Using the above components, Goodson and Jacobs (1976) constructed a continuous aqueous monitor (CAM-1) for an unattended operation. The system works on the basis of a computerized detection cycle. The sensitivity of the monitor to several pesticides in water is listed in *Table 1*.

Table 1 also quotes another enzyme–inhibitor system, namely urease with sodium fluoride as an inhibitor model. This study defines various parameters of an enzyme electrode in relation to a given inhibitor without carrying the system through practical application (Tran-Minh and Beaux, 1979).

PHENOL(S)

Two types of enzymes have been employed to construct enzyme electrodes for phenol (*Table 2*). One employs members of those copper-containing proteins designated by the vague term 'phenol oxidases'; they are obtained from fungi (mushrooms), potatoes and several other sources. The other type is the well-defined flavoprotein phenol 2-monoxygenase (2-hydroxylating), EC 1.14.13.7, isolated from the soil yeast *Trichosporon cutaneum* (Neujahr and Gaal, 1973).

The phenol oxidases are listed under several numbers by the Enzyme Commission (International Union of Biochemistry, 1979). Catechol oxidase (EC 1.10.3.1) refers to enzymes that act on a variety of substituted catechols as well as on monophenols. The synonymous names are diphenol oxidase, *o*-diphenolase and tyrosinase. Laccase (EC 1.10.3.2) comprises enzymes that will act on both *o*- and *p*-quinones and often also on aminophenols and phenylenediamine. The synonymous names are phenolase, polyphenol oxidase and urishiol oxidase. Finally, the enzyme called monophenol monooxidase (EC 1.14.18.1) will also act on 1,2-benzene diols when no monophenols are available. The synonymous names are tyrosinase, phenolase, cresolase.

An enzyme electrode with immobilized 'polyphenol oxidase' (EC 1.14.18.1) has been described by Macholan and Schanel (1977). In this, glutaraldehyde is used to immobilize a mixture of potato or fungal polyphenol oxidase and bovine serum albumin on the surface of a polyamide netting. This is then stretched over the hydrophobic membrane of a Clark oxygen cell. The electrode response (measured amperometrically) is quoted as good for phenol, *p*-cresol, pyrocatechol and pyrogallol in waste water, but inadequate for tyrosine, chlorogenic acid and DL-dihydroxyphenylalanine; there is no response to hydroquinone and phloroglucinol. Calibration curves for the above phenols, when assayed individually, are linear in the range 6·6–66 μM. The relative activities towards these phenols are different. No attempts were made to quantitate individual phenols when assayed in mixtures. However, linear and stable response, corresponding to 'total phenols' was obtained with graded amounts of coking water.

A different principle is utilized in an enzyme electrode employing fungal 'tyrosinase' (EC 1.14.18.1), described by Schiller, Chen and Liu (1978). In this device, the enzyme is immobilized by entrapment in polyacrylamide gel cast around a thin platinum grid. Phenol is oxidized by the immobilized enzyme in the presence of saturating levels of oxygen. The oxidation product, ortho-

Table 2. Biosensors for phenol(s).

Substance	Biocatalyst	Sensor	Range ($M \times 10^6$)	Reference
Phenol(s)	Polyphenol oxidase, EC 1.14.18.1	Pt–Ag Clark-type	6·6–66	Macholan and Schanel, 1977
Phenol(s)	Tyrosinase (from mushrooms)	Pt–SCE	0·38–100	Schiller *et al.*, 1978
Phenol(s)	Phenol hydroxylase, EC 1.14.13.7	Pt–Ag Clark-type YSI4004	0·5–50†	Kjellén and Neujahr, 1979, 1980; Neujahr, 1982
Phenol(s)	Whole cells of phenol-induced *Trichosporon cutaneum*	Pt–Ag Clark-type YSI4004	20–150†	Neujahr and Kjellén, 1979; Neujahr, 1982
Catechol(s)	Catechol-1,2-dioxygenase, EC 1.13.11.1	Pt–Ag Clark-type YSI4004	10–500†	Neujahr, 1980, 1982
Nitrocatechol-sulphate	Aryl sulphatase, EC 3.1.6.1	Pt–SCE	1–100	Cserfalvi and Guilbault, 1976

† Refers to unsubstituted phenol (or catechol); sensitivity towards simply substituted phenols is much lower, linear range shorter.

benzoquinone, is then chemically reduced in the presence of an excess of ferrocyanide ions. The coupled oxidation of ferrocyanide ions to ferricyanide ions gives a measurable potential difference in the electrochemical system. The resulting zero current potentials in these equilibrium potentiometric measurements are directly proportional to the logarithm of phenol concentration in the range 0·38–100 μM.

Individual calibration curves for four substituted phenols show that the response of the enzyme electrode to these substrates correlates fairly well with the corresponding activities of the enzyme in solution. An exception is catechol, which is a less efficient substrate with the enzyme electrode than it is with the soluble enzyme. The authors state that the sensing system was developed with the specific objective of producing a portable and self-sustaining field unit for the detection of phenol and related compounds. However, they do not describe any such unit in practical operation. Instead, they test the response of the enzyme electrode when graded levels of phenol are measured in the presence of compounds likely to occur together with phenol in aqueous washings from a commercial polymerization process. The sensor output with phenol is shown to be unchanged in the presence of saturating concentrations of either chloro- or nitrobenzene, as well as in the presence of 10^{-7}–10^{-3} M butylamine. Both aniline and benzoic acid (10^{-4} M) decrease the sensor output with phenol. The detection level for phenol increases to 10 μM in the presence of aniline and to 1 μM in the presence of benzoic acid. However, the slope of the calibration curve for phenol is not affected by these compounds.

An enzyme electrode for phenol using the inducible FAD-containing protein phenol *o*-hydroxylase from the soil yeast *Trichosporon cutaneum* is described by Kjellén and Neujahr (1979, 1980). This enzyme (EC 1.14.13.7) has a strict requirement for NADPH as an external electron donor. The function of NADPH is to reduce the FAD-prosthetic group in order to activate it towards oxygen. Phenol hydroxylase reacts also with several monosubstituted phenols, carrying halogen, hydroxy- or methyl-, but not carboxy- groups (Neujahr and Gaal, 1973; Neujahr and Kjellén, 1978). The activity towards most of them is considerably lower than that towards unsubstituted phenol.

The enzyme was immobilized by several different methods (Kjellén and Neujahr, 1979), employing entrapment, adsorption or covalent coupling. These preparations were tested directly on a Clark Oxygen Electrode (YSI Model 4004) employing the kinetic assay method and recording the derivative of the signal. Of the immobilized preparations tested, those with enzyme entrapped in polyacrylamide gels gave probes which were very slow responding, whereas those with enzyme adsorbed to various DEAE (diethyl-aminoethyl) carriers or covalently attached to AH-Sepharose 4B or to nylon nets gave probes of high sensitivity which were very fast responding. It is noteworthy that successful covalent coupling could only be achieved employing —COOH groups on the enzyme, whereas the most widespread methods of covalent immobilization such as coupling to CNBr-Sepharose or crosslinking with glutaraldehyde, gave completely inactive preparations. This could later be related to the presence of a reactive lysyl residue in the catalytic site of the enzyme (Neujahr and Kjellén, 1980).

The limit of detection, using the most active enzyme electrodes prepared with phenol hydroxylase, is 0·5 μM. Deviations from mean values of repeated determinations are less than 5%. The linear range, with enzyme adsorbed to DEAE-carriers or covalently coupled to AH-Sepharose 4B, is at least up to 50 μM. With other types of immobilized enzyme, the linear range is shorter. Heavy metals, at the highest concentrations encountered in municipal sewage (0·5–10 mg/ℓ, depending on which metal) slightly reduce the response of the electrode. However, this can be counteracted completely by, for example, 1 mM dithiothreitol in the assay medium, an addition which does not affect the performance of the electrode in any other way. Like soluble phenol hydroxylase, the immobilized enzyme is active towards several monosubstituted phenols. Because of diffusional differences in the immobilized enzyme system, the relative activities differ from those with soluble enzyme. An equimolar mixture of phenol and a monosubstituted phenol usually gives a response that is higher than that with phenol alone, but much lower than the sum of the responses to the individual phenols. Thus, this type of electrode cannot be used for accurate measurement of 'total phenols', although it gives reproducible calibration curves for phenol in the presence of comparatively high levels of defined mixtures or individual phenol analogues. Because the method is rapid, it seems well suited for use as a 'phenol indicator': for example, it may be placed before the inlet to the biological step in sewage purification plants.

The enzyme electrode was tested in different media supplemented with known amounts of phenol. These media included deionized water, chlorinated tap water, lake water, spent culture medium of *Trichosporon cutaneum*, with or without suspended cells, and certain fractions of municipal sewage. There was good stability and accuracy of the probe with no interference from substances present in these media. However, the sewage samples had to be pH-adjusted and aerated immediately before the assay to ensure an adequate supply of dissolved oxygen.

A disadvantage of phenol hydroxylase as a biocatalyst in an enzyme electrode for phenol is the requirement for NADPH. The cost of this requirement may amount to as much as 80–90% of the total cost of the assay. Because of side reactions of phenol hydroxylase (NADPH-oxidase) and because of the diffusional phenomena discussed above (*see also* Kjellén and Neujahr, 1980), incorporation of a NADPH-regenerating system into the enzyme layer does not appear to be feasible. However, the cofactor requirement as well as the rather cumbersome work of enzyme purification and immobilization can, in this case, be circumvented by employing whole cells of phenol-induced *T. cutaneum* instead of isolated and immobilized phenol hydroxylase (Neujahr and Kjellén, 1979). Such a 'bioprobe electrode' enables one to perform rapid quantitative determination of phenol in the range of 20–150 μM. The assay can be carried out in 2–10 ml and is complete 15 s after adding the sample. Phenol-induced cells can come from shake cultures or agar plates. They can be freeze-stored for several months before they are mounted on the electrode. Once mounted, the ready bioprobe is stable for at least a week at room temperature and for at least several weeks in a refrigerator. It lasts for at least 100 assays, the stability improving when in frequent use. The problem of stability, however, is of minor importance, because mounting of the bioprobe is extremely simple and can be carried out within a minute.

With whole cells mounted on an oxygen electrode, possible interference from oxidizable non-phenolic substrates, present together with phenol, must be considered. With phenol-induced *T. cutaneum*, this problem is of minor importance (*see Figure 2*), because the two (serial) enzymes initiating phenol metabolism, i.e. phenol hydroxylase and the ring-cleaving enzyme, both utilize oxygen, one equivalent O_2 each. Phenol hydroxylase alone constitutes nearly 2% of all soluble cell proteins (Neujahr and Gaal, 1973). Oxygen consumption

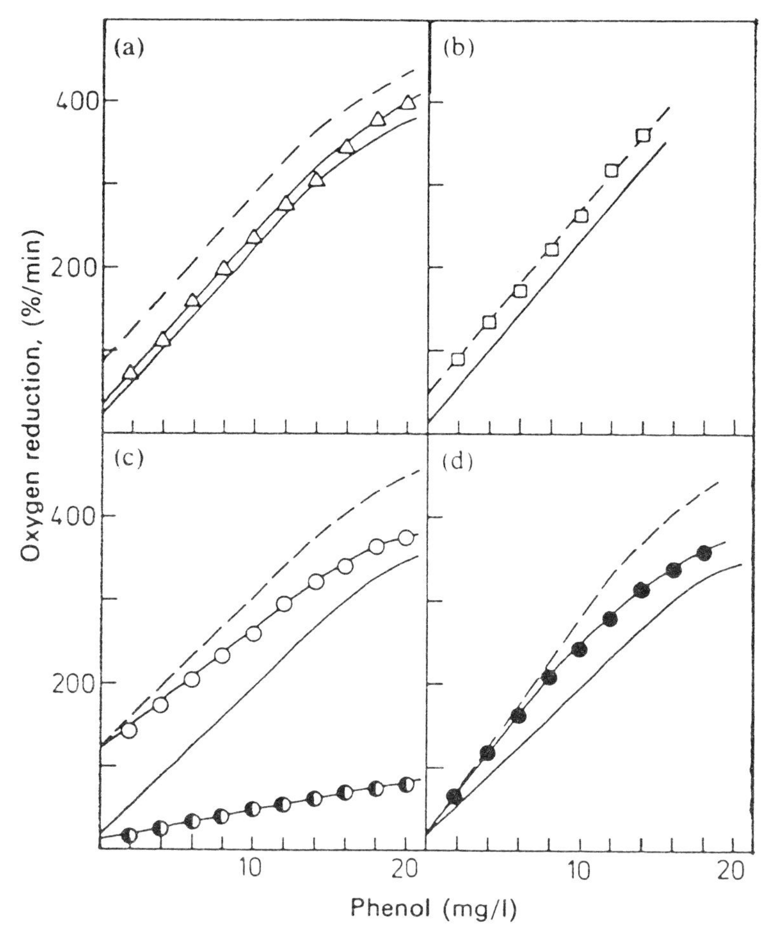

Figure 2. The effect of phenolic and non-phenolic oxygen-consuming compounds on the response of a bioprobe for phenol. Phenol response of a bioprobe (Clark oxygen electrode, coated with cell paste of phenol-induced *T. cutaneum*: (——) phenol alone; (– – –) calculated curve if response to phenol is added to that of other sample components. (a) (△) Phenol in presence of 200 mg glucose/litre. (b) (□) Phenol in presence of 456 mg/litre (anhydrous) sodium acetate. (c) Mixture of eight phenol derivatives including cresols, chlorophenols, resorcinol and catechol in equimolar amounts (◐); phenol in the presence of 36 mg/litre of the same mixture (○). (d) Mixture of the eight phenol derivatives together with an equimolar amount of phenol (●). (Reproduced, with permission, from *Biotechnology and Bioengineering* (1979) **21**, 671.)

by dosed phenol in the presence of comparatively high glucose concentrations (1 mM) is linear up to at least 150 μM, i.e. in the same range as in the absence of glucose. The two calibration curves have the same slope; that obtained in the presence of glucose is shifted upwards and thus does not pass through the origin (*Figure 2a*). The intercept with the *y*-axis permits estimation of the 'non-phenolic' oxygen consumption, which is far from additive to that caused by phenol(s). With acetate, another oxidizable non-phenolic substrate, the response is additive. However, here also, the calibration curve for phenol in the presence of as much as 5·5 mM acetate is parallel to that for phenol alone and shifted upwards only slightly (*Figure 2b*).

A different type of response to phenol is obtained in the presence of oxidizable phenolic substrates. Here, both the *y*-intercept and the slope of the calibration curve is affected (*Figure 2c, d*). Estimates of the content of unsubstituted phenol can be made as described (Neujahr and Kjellén, 1979). With equimolar concentrations of phenol (20–100 μM) and a mixture of eight simple phenol derivatives, the increase in the rate of oxygen consumption, compared with that in the absence of phenol analogues, was in the range 5–17%, the higher values referring to the lowest phenol concentrations. Thus, when unsubstituted phenol is a major component, the measurement error caused by phenol analogues seems negligible. The bioprobe is routinely used in our laboratory to monitor phenol concentrations in large-scale cultures of *Trichosporon cutaneum*, directly in cell suspensions taken from the fermenter. The bioprobe also proved applicable to measurements of phenol in sewage, after the samples have been pH-buffered and aerated.

Enzyme electrodes for catechol are prepared using catechol 1,2-oxygenase (EC 1.14.18.1) coupled to CNBr-activated Sepharose or cross-linked with glutaraldehyde (Neujahr, 1980). The enzyme is obtained from the soil yeast *Trichosporon cutaneum* (Varga and Neujahr, 1970). Similar and related enzymes are present in many other organisms, notably in the soil pseudomonads. The enzyme from yeast specifically cleaves catechol in the *o*-fashion; it does not react with monohydric phenols at all. Enzyme electrodes with immobilized catechol 1,2-oxygenase give specific responses, linearly dependent on catechol concentrations over the range 0–200 μM with a lower limit of sensitivity at about 5 μM. The response of the electrodes is practically unaffected by the presence of high amounts of biological materials, e.g. urine (50%), blood serum (25%) or liver extract (corresponding to 75 g fresh weight/ℓ). The presence of methyl catechols affects the slope and/or the *y*-intercept in a characteristic manner, permitting estimates of the approximate proportions of these compounds.

PHENOXYACIDS, NITRILOTRIACETATE, HEAVY METALS

Phenoxyacids especially the hormone herbicides, CPA (4-chlorophenoxyacetate), 2,4-D (2,4-dichlorophenoxyacetate) and MCPA (4-chloro-2-methyl-phenoxyacetate) as well as their derivatives, are of world-wide importance because of their efficiency in crop protection. The degree of their toxicity to the environment is still under discussion, provoking recurring passionate actions from 'ecologist' groups around the world. It is now generally agreed that these herbicides are ultimately (although slowly) degraded by soil microflora. Several microbial species, isolated from enrichment cultures, have been shown to attack these compounds (Kearney and Kaufman, 1969). Metabolic routes have been established for some of them, showing the participation of inducible enzyme sequences similar to those acting in phenol degradation (Evans *et al.*, 1971a, b; Gaunt and Evans, 1971). These metabolic routes involve monohydric phenol derivatives, which are *o*-hydroxylated and then ring-cleaved. There is no general consensus, however, as to the stage at which the halogen is removed, before or after ring cleavage. By analogy with the bioprobe for phenol (Neujahr and Kjellén, 1979), those microbial species which are inducible for the degradation of phenoxyacids might be considered to be attractive biocatalysts for use in biosensors for determination of these compounds. However, no attempt to employ them in this manner has been described.

Nitrilotriacetic acid (NTA) is often employed as a chelating and sequestering agent as a component of detergent compositions. Its effects on the environment have been the subject of considerable concern. The metabolism of this compound by soil bacteria is now fully established (Firestone and Tiedje, 1978 and references therein). A potentiometric bioprobe for NTA employing whole cells of a *Pseudomonas* sp. and an electrode sensing ammonia gas is described by Kobos and Pyon (1981). The bacterial cells carry out a four-step reaction sequence to produce the measured compound, ammonia, from NTA. The response of the biosensor to NTA is useful in the range 100–700 μM with a slope of 35–40 mV/decade*. The specificity of the biosensor is not impressive: it responds also to glycine and serine, two metabolites of NTA, and, in addition, to urea and, of course, to ammonia. For practical measurement of NTA (or glycine, serine?) any urea and ammonia present in the sample have to be removed before determination. The authors do not state how they intend to cope with the interference from serine and/or glycine.

Heavy metals can, in principle, be detected and roughly estimated as a group, using any enzyme that carries catalytically essential SH groups. This method has evidently not inspired investigators to develop biosensors for heavy metals. An interesting approach to the specific determination of Cu^{2+} is represented by a tyrosinase apoenzyme electrode described by Mattiasson, Nilsson and Olsson (1979). The device was originally aimed at demonstrating the usefulness of apoenzyme electrodes for specific determination of cofactors: as the specific cofactor of tyrosinase (EC 1.14.18.1) is Cu^{2+}, the biosensor can also be employed to measure Cu^{2+}. The enzyme is immobilized on nylon netting and stripped of Cu^{2+} by washing with 0·1 M NaCN. The resulting immobilized apoenzyme is mounted on an oxygen electrode. This device has no tyrosinase activity, unless Cu^{2+} is supplied to reconstitute the holoenzyme. For repeated use, the enzyme has again to be stripped of Cu^{2+} and then brought in contact with Cu^{2+} solution, followed by the aromatic substrate. The system is operated in a semicontinuous (cyclic) fashion to permit recurrent regeneration of the apoenzyme in a flow

* Decade = order of (concentration) magnitude.

system. The useful response to Cu^{2+} is up to 50 μM Cu^{2+}, the linear range being about half that level.

NATURALLY OCCURRING SUBSTANCES OF POTENTIAL ENVIRONMENTAL HAZARD WHEN ACCUMULATED

Table 3 lists biosensors developed for some such substances, namely nitrite, nitrate, phosphate, sulphate, urea. These biosensors are not specifically aimed at environmental control, but can be used for such a purpose if necessary. They will not be reviewed here in detail: the interested reader is referred to the published work cited in the Table.

COMMON CHARACTERISTICS OF DESCRIBED BIOSENSORS

The linear response of most enzyme electrodes is at least in the range 10^{-2}–10^{-4} M, while some electrodes will respond to concentrations as low as 5×10^{-7} M and/or as high as 10^{-1} M. The stability of an enzyme electrode is difficult to define in an 'all-or-none' manner. The immobilized enzyme may gradually lose activity, which results in a gradual downward shift of the calibration curve. Daily recalibration of an enzyme electrode is necessary, but this is also a requirement with electrochemical sensors in general, whether carrying biocatalysts or not. As long as the remaining activity of the enzyme ensures diffusional control and thus linear response in a reasonable range, the enzyme electrode may be considered to be 'stable'. With most enzyme electrodes described, this length of life is at least 3–4 weeks; with some it is several months.

The corresponding life of bioprobes based on whole microbial cells has been quoted as anything between 5 and 30 days. However, the problem of stability becomes of minor importance with the microbial probes, because of their extreme ease of preparation. It is interesting to note that, whereas the stability of isolated (and immobilized) enzymes may be adversely affected by increasing numbers of assays, the reverse seems to be true with regard to whole microbial cells. Other factors affecting the stability of the probes are, of course, the conditions of storage and the risk of microbial contamination.

Concluding remarks

The list of biosensors in *Table 3* is by no means exhaustive with respect to urea, which, as mentioned in the introduction, has been one of the two favourite model substrates when developing enzyme electrodes. In fact, about 40% of all published papers on enzyme electrodes (which total over 500) deal with glucose oxidase (EC 1.1.3.4) or urease (EC 3.5.1.5). It is hard to assess whether this is because these two enzymes are commercially available and inexpensive, or whether there are other reasons. One is involuntarily left with the impression that, in spite of the optimistic appraisals of enzyme electrodes in almost every published paper, their application to industrial use has been rather less than enthusiastic.

The largest potential market for enzyme electrodes is undoubtedly the medical

Table 3. Biosensors for substances of environmental interest, operating on an enzyme–substrate† principle

Substance	Enzyme	Sensor	Detection limit (μM)	Reference
Nitrite	Nitrite reductase	gas (NH_3)	50	Kiang, Kuan and Guilbault, 1975
Nitrate	Nitrate reductase/nitrite reductase	NH_4	50	Kiang, Kuan and Guilbault, 1978a, b
	Whole cells of *Azotobacter vinelandii*	gas (NH_3)	10	Kobos, Rice and Flournay, 1979
Sulphate†	Aryl sulphatase	Pt	100	Cserfalvi and Guilbault, 1976
Phosphate	Phosphatase/glucose oxidase	Pt (O_2)	100	Guilbault and Nanjo, 1975
				Guilbault and Cserfalvi, 1976
Urea	Urease	NH_4^+	10	Fritz *et al.*, 1976
				Guilbault and Nagy, 1973
				Guilbault, Smith and Montalvo, 1969
			100	Montalvo, 1969
				Montalvo and Guilbault, 1969
		gas (NH_3)	100	Anfält, Granell and Jagner, 1973
				Johanson and Ögren, 1976
				Mascini and Guilbault, 1977
		gas (CO_2)		Guilbault and Shu, 1972
		pH	100	Nilsson, Åkerlund and Mosbach, 1973

† The enzyme electrode for sulphate is, instead, based on inhibition by SO_4^{2-} of the reaction: 4-nitrocatechol sulphate + $H_2O \rightarrow$ 4-nitrocatechol + SO_4^{2-}, catalysed by aryl sulphatase (EC 3.1.6.1). The equilibrium current arises from oxidation of 4-nitrocatechol, measured at 0·8 V vs. SCE. Linear calibration curves also for PO_4^{3-} (competitive inhibition) and for F^- (activation).

field. However, most published works dwell in the domain of the analytical chemist dealing with basic chemistry in simple aqueous solutions. Few papers contain accounts of *in vitro* measurements in biological solutions, let alone *in vivo* measurements. The reasons may include problems of selectivity, sensitivity or speed of response. Such problems become less important in the field of environmental control: here, the goal is often not so much the analysis of one single substance, but rather of a group of related compounds; not so much accurate measurements of their concentrations, but rather of safe threshold values. A further advantage may be the large inductive capacity of soil microbes to attack many chemicals. Once the required enzymes are induced, they can be isolated and/or stabilized and used for assays of their unusual substrates. A prerequisite of a more extensive use of biosensors in environmental control thus seems to be the availability of a larger array of suitable organisms and enzymes. So far, investigators have mostly used the few readily available commercial enzymes instead of undertaking purpose-orientated screening programmes to find the most suitable ones. The outcome of such programmes could open the way to a truly widespread use of biosensors in environmental control. After all, what can more adequately measure biologically hazardous substances than the biological systems themselves?

References

ANFÄLT, T., GRANELL, A. AND JAGNER, D. (1973). A urea electrode based on the ammonia probe. *Analytical Letters* **6,** 969–975.

ARNOLD, M.A. AND RECHNITZ, G.A. (1981). Tissue-based membrane electrode with high biocatalytic activity for measurement of adenosine 5′-monophosphate. *Analytical Biochemistry* **53,** 1837–1842.

BARKER, A.S. AND SOMERS, P.J. (1978). Enzyme electrodes and enzyme based sensors. In *Topics in Enzyme and Fermentation Technology* (A. Wiseman, Ed.), volume 2, pp. 120–151. John Wiley, St Louis.

BOWERS, L.D. AND CARR, P.W. (1980). Immobilized enzymes in analytical biochemistry. In *Advances in Biochemical Engineering* (A. Fiechter, Ed.), volume 15, pp. 89–129. Springer Verlag, Berlin.

BRADY, J.E. AND CARR, P.W. (1980). Theoretical evaluation of the steady-state response of potentiometric enzyme electrodes. *Analytical Chemistry* **52,** 977–980.

CLARK, L.C. JR AND LYONS, C. (1962). Electrode systems for continuous monitoring in cardiovascular surgery. *Annals of the New York Academy of Sciences* **102,** 29–45.

CSERFALVI, T. AND GUILBAULT, G.G. (1976). An enzyme electrode based on immobilized arylsulfatase for the selective assay of sulfate ion. *Analytica chimica acta* **84,** 259–270.

DANIELSSON, B.D. AND MOSBACH, K. (1976). In *Methods in Enzymology* (S.P. Colowick and N.O. Kaplan, Eds), volume 44, pp. 667–676. Academic Press, New York, San Francisco, London.

ENGASSER, J.M. AND HORVATH, C. (1976). Diffusion and kinetics with immobilized enzymes. In *Immobilized Enzyme Principles*, pp. 127–220. Academic Press, New York.

EVANS, W.C., SMITH, B.S.W., MOSS, P. AND FERNLEY, H.N. (1971a). Bacterial metabolism of 4-chlorophenoxyacetate. *Biochemical Journal* **122,** 509–517.

EVANS, W.C., SMITH, B.S.W., FERNLEY, H.N. AND DAVIES, J.J. (1971b). Bacterial metabolism of 2,4-dichlorophenoxyacetate. *Biochemical Journal* **122,** 543–551.

FIRESTONE, M.K. AND TIEDJE, J.M. (1978). Pathway of degradation of nitrilotriacetate by a *Pseudomonas* species. *Applied Environmental Microbiology* **35,** 955–961.

FRITZ, J., NAGY, G., FODOR, L. AND PUNGOR, E. (1976). *In vitro* studies on the dissolution rate of industrial retarded urea feedingstuffs by use of a selective electrode. Application of the potentiometric urea enzyme electrode in measurements of dissolution rate. *Analyst* **101,** 439–444.

Gaunt, J.K. and Evans, W.C. (1971). Metabolism of 4-chloro-2-methylphenoxyacetate by a soil pseudomonad. *Biochemical Journal* **122,** 519–526.

Goodson, L.H. and Jacobs, W.B. (1974). Application of immobilized enzymes to detection and monitoring. In *Enzyme Engineering* (K.E. Pye and L.B. Wingard, Jr., Eds), volume 2, pp. 393–400. Plenum Press, New York and London.

Goodson, L.H. and Jacobs, W.B. (1976). Monitoring of air and water for enzyme inhibitors. In *Methods in Enzymology* (S.P. Colowick and N.O. Kaplan, Eds), volume 44, pp. 647–658. Academic Press, New York, San Francisco, London.

Gough, D.A. and Leypoldt, J.K. (1981). Theoretical aspects of enzyme electrode design. *Applied Biochemistry and Bioengineering* **3,** 175–200.

Guilbault, G.G. (1976). Enzyme electrodes and solid surface fluorescence methods. In *Methods in Enzymology* (S.P. Colowick and N.O. Kaplan, Eds), volume 44, pp. 579–633. Academic Press, New York, San Francisco, London.

Guilbault, G.G. (1980). Enzyme electrode probes. *Enzyme and Microbial Technology* **2,** 258–264.

Guilbault, G.G. and Cserfalvi, T. (1976). Ion-selective electrodes for phosphate using enzyme systems. *Analytical Letters* **9,** 277–289.

Guilbault, G.G. and Nagy, G. (1973). Improved urea electrode. *Analytical Chemistry* **45,** 417–419.

Guilbault, G.G. and Nanjo, M. (1975). A phosphate-selective electrode based on immobilized alkaline phosphatase and glucose oxidase. *Analytica chimica acta* **78,** 69–80.

Guilbault, G.G. and Rohm, T.J. (1975). Ion-selective electrodes and enzyme electrodes in environmental and clinical studies. *International Journal of Environmental and Analytical Chemistry* **4,** 51–61.

Guilbault, G.G. and Shu, F. (1972). Enzyme electrodes based on the use of a carbon dioxide sensor. Urea and L-tyrosine electrodes. *Analytical Chemistry* **44,** 2161–2166.

Guilbault, G.G., Smith, R.M. and Montalvo, J., Jr (1969). Use of ion-selective electrodes in enzymic analysis. Cation electrodes for deaminase enzyme system. *Analytical Chemistry* **41,** 600–605.

Hikuma, M., Kubo, T., Yasuda, T., Karube, I. and Suzuki, S. (1979a). Amperometric determination of acetic acid with immobilized *Trichosporon brassicae*. *Analytica chimica acta* **109,** 33–38.

Hikuma, M., Kubo, T., Yasuda, T., Karube, I. and Suzuki, S. (1979b). Microbial electrode sensor for alcohols. *Biotechnology and Bioengineering* **21,** 1845–1853.

Hikuma, M., Obana, H., Yasuda, T., Karube, J. and Suzuki, S. (1980). A potentiometric microbial sensor based on immobilized *Escherichia coli* for glutamic acid. *Analytica chimica acta* **116,** 61–67.

International Union of Biochemistry (1979). *Enzyme Nomenclature. Recommendations (1978) of the Nomenclature Committee of the International Union of Biochemistry.* Academic Press, New York and London.

Johansson, G. and Ögren, L. (1976). An enzyme reactor electrode for urea determinations. *Analytica chimica acta* **84,** 23–29.

Kavanagh, F., ed. (1963). *Analytical Microbiology*. Academic Press, New York and London.

Keyes, M.H., Semersky, F.E. and Gray, D.N. (1979). Glucose analysis utilizing immobilized enzymes. *Enzyme and Microbial Technology* **1,** 91–94.

Kearney, P.C. and Kaufman, D.D. (1969). *Degradation of Herbicides*, pp. 1–49. Marcel Dekker Inc., New York.

Kiang, C., Kuan, S. and Guilbault, G.G. (1975). A novel enzyme electrode method for the determination of nitrite based on nitrite reductase. *Analytica chimica acta* **80,** 209–214.

Kiang, C.H., Kuan, S.S. and Guilbault, G.G. (1978a). Enzymatic determination of nitrate: electrochemical detection after reduction with nitrate reductase and nitrite reductase. *Analytical Chemistry* **50,** 1319–1322.

KIANG, C.H., KUAN, S.S. AND GUILBAULT, G.G. (1978b). Enzymatic determination of nitrate: fluorometric detection after reduction with nitrate reductase. *Analytical Chemistry* **50,** 1323–1325.

KJELLÉN, K.G. AND NEUJAHR, H.Y. (1979). Immobilization of phenol hydroxylase. *Biotechnology and Bioengineering* **21,** 715–719.

KJELLÉN, K.G. AND NEUJAHR, H.Y. (1980). Enzyme electrode for phenol. *Biotechnology and Bioengineering* **22,** 299–310.

KOBOS, R.K. AND PYON, H.Y. (1981). Application of microbial cells as multistep catalysts in potentiometric biosensing electrodes. *Biotechnology and Bioengineering* **23,** 627–633.

KOBOS, R.K., RICE, D.J. AND FLOURNAY, D.S. (1979).Bacterial membrane electrode for the determination of nitrate. *Analytical Chemistry* **51,** 1122–1125.

MACHOLAN, L. AND SCHANEL, L. (1977). Enzyme electrode with immobilized polyphenol oxidase for determination of phenolic substrates. *Collection of Czechoslovak Chemical Communications* **42,** 3667–3675.

MASCINI, M. AND GUILBAULT, G. (1977). Urease coupled ammonia electrode for urea determination in blood serum. *Analytical Chemistry* **49,** 795–798.

MATTIASSON, B., NILSSON, H. AND OLSSON, B. (1979). An apoenzyme electrode. *Journal of Applied Biochemistry* **1,** 377–384.

MONTALVO, J., JR (1969). Electrode for measuring urease enzyme activity. *Analytical Chemistry* **41,** 2093–2094.

MONTALVO, J., JR AND GUILBAULT, G.G. (1969). Sensitized cation selective electrode. *Analytical Chemistry* **41,** 1897–1899.

MORF, W.E. (1980). Theoretical evaluation of the performance of enzyme electrodes and of enzyme reactors. *Microchimica Acta (Vienna)* **2,** 317–332.

MOSBACH, K., ED. (1976). Immobilized Enzymes. Volume 44 of *Methods in Enzymology* (S.P. Colowick and N.O. Kaplan, Eds). Academic Press, New York, San Francisco, London.

MOSBACH, K. AND DANIELSSON, B. (1974). An enzyme thermistor. *Biochimica et biophysica acta* **364,** 140–145.

MOSBACH, K., DANIELSSON, B., BORGERUD, A. AND SCOTT, M. (1975). Determination of heat changes in the proximity of immobilized enzymes with an enzyme thermistor and its use for the assay of metabolites. *Biochimica et biophysica acta* **403,** 256–265.

NEUJAHR, H.Y. (1980). An enzyme probe for catechol. *Biotechnology and Bioengineering* **22,** 913–918.

NEUJAHR, H.Y. (1982). Determination of phenol and catechol concentrations with oxygen probes coated with immobilized enzymes or immobilized cells. *Applied Biochemistry and Biotechnology* **7,** 107–111.

NEUJAHR, H.Y. AND GAAL, A. (1973). Phenol hydroxylase from yeast. Purification and properties of the enzyme from *Trichosporon cutaneum. European Journal of Biochemistry* **35,** 386–400.

NEUJAHR, H.Y. AND KJELLÉN, K.G. (1978). Phenol hydroxylase from yeast. Reaction with phenol derivatives. *Journal of Biological Chemistry* **253,** 8835–8841.

NEUJAHR, H.Y. AND KJELLÉN, K.G. (1979). Bioprobe electrode for phenol. *Biotechnology and Bioengineering* **21,** 671–678.

NEUJAHR, H.Y. AND KJELLÉN, K.G. (1980). Phenol hydroxylase from yeast: a lysyl residue essential for binding of reduced nicotinamide adenine dinucleotide phosphate. *Biochemistry* **19,** 4967–4972.

NILSSON, H., ÅKERLUND, A. AND MOSBACH, K. (1973). Determination of glucose, urea and penicillin using enzyme–pH electrodes. *Biochimica et biophysica acta* **320,** 529–534.

RECHNITZ, G.A., KOBOS, R.K., RIECHEL, S.J. AND GEBAUER, C.R. (1977). A bio-selective membrane electrode prepared with living bacterial cells. *Analytica chimica acta* **94,** 357–365.

RECHNITZ, G.A., RIECHEL, T.L., KOBOS, R.K. AND MEYERHOFF, M.E. (1978). Glutamine-selective membrane electrode that uses living bacterial cells. *Science* **199,** 440–441.

RIESEL, E. AND KATCHALSKI, E. (1964). Preparation and properties of water-insoluble derivatives of urease. *Journal of Biological Chemistry* **239,** 1521–1524.

SCHILLER, J.G., CHEN, A.K. AND LIU, C.C. (1978). Determination of phenol concentrations by an electrochemical system with immobilized tyrosinase. *Analytical Biochemistry* **85,** 25–33.

SILMAN, I.H. AND KATCHALSKI, E. (1966). Water-insoluble derivatives of enzymes, antigens, and antibodies. In *Annual Reviews of Biochemistry* (P. Boyer, Ed.), volume 35, Part 2, pp. 873–908.

STERNBERG, R., APOTEKER, A. AND THEVENOT, D.R. (1980). Trace glucose electrode for clinical, food and environmental determinations. *Analytical Chemistry Symposia Series* **2,** 461–473.

SUZUKI, S. AND KARUBE, I. (1981). Bioelectrochemical sensors based on immobilized enzymes, whole cells and proteins. *Applied Biochemistry and Bioengineering* **3,** 145–174.

SUZUKI, S., SATOH, J. AND KARUBE, J. (1982). Recent trends of biosensors in Japan. *Applied Biochemistry and Biotechnology* **7,** 147–155.

TRAN-MINH, C. AND BEAUX, J. (1979). Enzyme electrode for inhibitors determination: Urease–fluoride system. *Analytical Chemistry* **51,** 91–95.

UPDIKE, S.J. AND HICKS, G.P. (1967). The enzyme electrode. *Nature* **214,** 986–988.

VADGAMA, P. (1981). Enzyme electrodes as practical biosensors. *Journal of Medical Engineering and Technology* **5,** 293–298.

VARGA, J.M. AND NEUJAHR, H.Y. (1970). Purification and properties of catechol 1,2-oxygenase from *Trichosporon cutaneum. European Journal of Biochemistry* **12,** 427–434.

WEAVER, J.C., COONEY, C.L., FULTON, S.P., SCHULER, P. AND TANNENBAUM, S.R. (1976). Experiments and calculations concerning a thermal enzyme probe. *Biochimica et biophysica acta* **452,** 285–291.

WEAVER, J.C., COONEY, C.L., TANNENBAUM, S.R. AND FULTON, S.P. (1977). Possible medical applications of the thermal enzyme probe. In *Biomedical Applications of Immobilized Enzymes and Proteins* (T.M.S. Chang, Ed.), pp. 191–205. Plenum Press, New York.

8
Microbiological Methods for the Enhancement of Oil Recovery

DEREK G. SPRINGHAM

School of Biological Sciences, Queen Mary College, University of London, Mile End Road, London E1 4NS, UK

Introduction

Conventional processes for extraction of oil from reservoir rocks are surprisingly inefficient. The proportion of the original oil in place which is extracted can vary from about 5% to about 90% but the average recovery is not much more than 30%.

Oil is a vital ingredient in the economy of most nations, and its availability and price have profound and widespread economic and political impact. The total quantity available in the earth's crust, although not exactly known, is finite, and certainly insufficient to meet the future needs and aspirations of either consuming or producing nations in the next 50–100 years.

Some idea of the potential significance of methods aimed at the extraction of extra oil (Enhanced Oil Recovery, EOR) can be gained from figures presented by Geffen (1976). The known oil deposits in the United States, at the beginning of 1975 totalled 440×10^9 barrels. Of this total, some 106×10^9 barrels (24%) had already been produced. The reserves which could be produced by conventional methods amounted to 39×10^9 barrels (9%), equivalent to about seven times the US consumption in 1982. Forty per cent (176×10^9 barrels) were considered to be irrecoverable by any present or envisaged technologies, leaving 120×10^9 barrels (27%) as the target for enhanced oil recovery techniques. On a world-wide basis the target for EOR is correspondingly greater, and it has been calculated that the world's oil fields contain roughly $1{\cdot}2 \times 10^{12}$ barrels of oil with a current value of about US $\$3{\cdot}5 \times 10^{13}$ which is not recoverable by conventional techniques (V. Moses, personal communication). Engineers have been working for some years on chemical and physical methods of EOR. So far, applications have been limited to specially favourable circumstances and results have often been disappointing in terms of the extra oil recovered and of the economics of the process. An alternative approach is to make use of micro-organisms, deliberately injected into the oil-bearing rock, to release oil. Such processes, known as Microbial Enhancement of Oil Recovery (MEOR), are less well established than the chemical and physical methods, but may have

Biotechnology and Genetic Engineering Reviews—Vol. 1, February 1984
0264–8725/84/01/187–35$10.00 + $0.00 © Intercept Ltd

special advantages such as low operating costs and, perhaps, improved efficiency, which would make them especially attractive in the long run (Moses and Springham, 1982). They form the subject matter of this review.

Conventional procedures for oil recovery

To understand the potential uses of micro-organisms an outline knowledge of the conventional processes for oil recovery is necessary. A simplified account, aimed at the microbiologist, has been presented by Moses and Springham (1982) and a broader account of this and other aspects of the oil industry is that by Stockil (1977).

Crude oil is an extremely complex mixture consisting mainly of alkanes, cycloalkanes and aromatics. Unsaturated compounds are rare. Sulphur compounds are present, usually in small amounts and there are traces of oxygen, nitrogen and metal-containing compounds. The specific gravities of crude oils range from 0·78 to 1·0. Viscosities are usually higher than that of water, values ranging over seven orders of magnitude. Gases, mainly methane, are found in solution. Oil is believed to have been formed from plant or animal remains deposited in marine sediments. It is found in porous sandstone or carbonate rocks having been forced upwards by displacement with water. It is found in commercially useful quantities only where a layer of impermeable clay or shale, the cap rock, prevents further upward movement. Water is always present in the oil-bearing layer (connate water). It often has a high salt content and is referred to as brine. Frequently a layer of brine underlies the oil.

The reservoir rock can be characterized in terms of two very important properties: porosity and permeability. The porosity is the proportion of the rock volume not occupied by rock particles; it defines the maximum quantity of oil which a given volume of rock can contain. Permeability, measured in darcies ($d = 1\ \mu m^2$), is a measure of the ease with which a fluid can be forced through the rock. Permeabilities of reservoir rocks vary from fractions of a millidarcey up to several darcies. In the former case oil will flow only with great difficulty.

If a well is drilled into the correct region of a reservoir, oil will be forced into the well bore by one or more so-called 'primary mechanisms'. The gas in the cap (usually under considerable pressure) will expand, gas may come out of solution from the oil, and water may flow in from surrounding rocks. The pressure may be sufficient to force oil to the surface as a 'gusher' or alternatively a pump may be used to raise it. Sooner or later the flow of oil due to primary mechanisms will decrease and secondary recovery will usually be undertaken. This involves drilling one or more injection wells and pumping in fluid to maintain pressure or to displace oil directly. Waterflooding is the most common method of secondary recovery. The extra wells (which may be in a five-spot pattern usually with four injection wells surrounding one production well, or may form a more extensive array in a larger field) are drilled down into the brine layer beneath the oil. Water quality is critical for injection. If the salinity is unsuitable, clay particles in the reservoir rock may swell and reduce the permeability. Careful filtration is usually practised to remove organic and inorganic materials which might block ('plug') the formation, together with de-aeration to prevent corrosion, and the addition of scale inhibitors, corrosion inhibitors, and biocides as considered appropriate.

As primary and secondary production proceeds, the oil produced will be mixed with increasing proportions of water. Production will normally cease, not because there is no more oil, but because the proportion of oil is too low to justify continued operation. We have seen that at this point some two-thirds of the oil, on average, remains in place: it is now appropriate to consider why this is so.

In particular fields some oil will be cut off from the wells by impermeable layers; this can be released only by engineering procedures such as drilling more wells, and will not be considered further. Some oil will not be displaced because of the nature of flow patterns of fluids between injection wells and production wells. *Figure 1a* shows a portion of a reservoir in plan view with injection wells and production wells. To achieve 100% recovery of the oil, the area enclosed by the rectangle must be swept by water from the injection well. *Figure 1b* shows a more realistic flooding pattern. Water is almost always less viscous

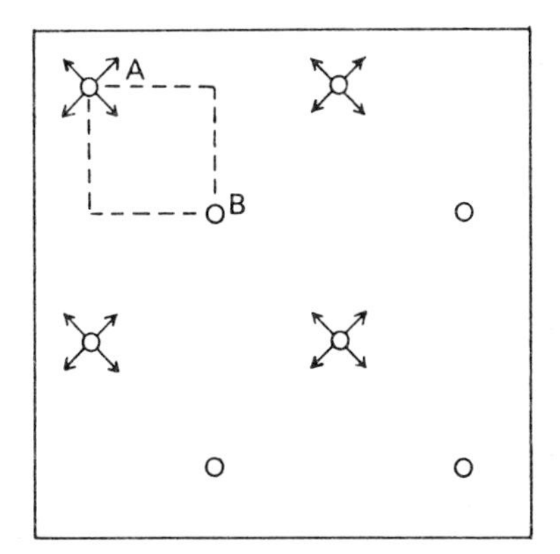

Figure 1(a). Part of a reservoir in plan view showing injection wells (✕) and production wells (○). The dotted rectangle shows the area which, ideally, should be swept by flow of water from A to B.

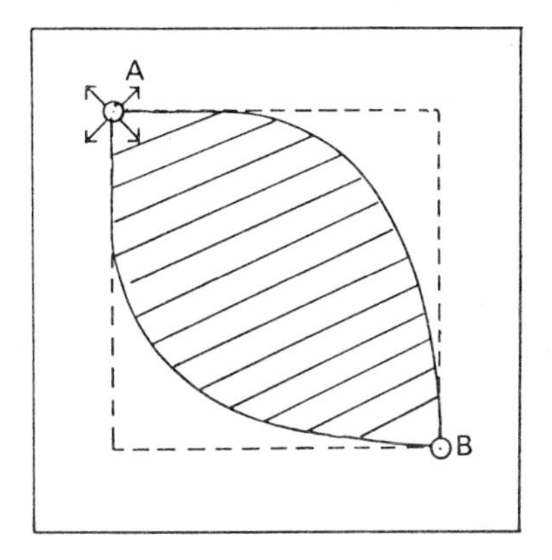

Figure 1(b). Enlarged view of part of the same area. The shaded area shows a more likely flow pattern of water between A and B. Waterflooding will not displace oil from outside this zone.

than the oil it is displacing and takes the easiest path between the two routes so the swept area is only a proportion of the whole. In thick oil-bearing layers, water tends to slide under the oil instead of displacing it evenly, thus reducing the sweep efficiency still further. A consideration of *Figure 1* suggests that it should be possible to improve the sweep efficiency by drilling more wells; this is indeed the case. More oil could be obtained by drilling wells at shorter intervals but the high cost of drilling largely determines the drilling pattern in the first instance.

Another departure from the optimum flow pattern is attributable to a process known as fingering. When a fluid such as water displaces a more viscous fluid such as oil in a porous matrix, the displacement fronts are highly uneven, with irregular fingers of water extending into the oil region. Once water flows in a particular zone, the resistance to flow is decreased because of its lower viscosity and the uneven flow is enhanced. In this way water can reach the production well at a comparatively early stage. As fingering becomes more severe the oil/water ratio in the produced fluid will decrease. If the structure of the reservoir rock is uneven (as it usually is), fingering will be promoted by preferential flow in regions of high permeability. Such regions, which may be due to high permeability strata such as the Brent Sand layers in many of the North Sea reservoirs or to rock fractures, are known as thief zones. In extreme cases preferential flow along thief zones may be so severe that only a small percentage of the oil in place is recoverable.

A further mechanism by which oil is cut off occurs on a microscale and involves the stranding of globules (ganglia) of oil in capillaries by the waterflood. *Figure 1c* shows an oil ganglion which has been cut off by water. The pressure differential across the ganglion tends to move it in the direction of the waterflood but to pass through the narrow throat of the capillary work must be done to distort the shape, because the surface area has to increase. With the range of normal parameters applying during a waterflood, a substantial proportion of the oil will be permanently cut off in this fashion.

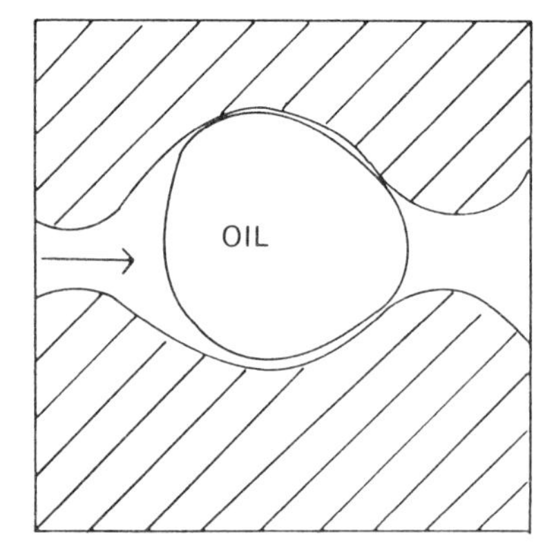

Figure 1(c). Oil globule trapped in a rock pore. The arrow shows the direction of the waterflood tending to displace the oil globule. In order to escape from the pore the globule must be deformed. This would increase the oil:water interfacial area and would require work to be done.

Chemical and physical methods for enhanced oil recovery

Only a brief consideration of non-biological methods is appropriate here. Detailed accounts are available (Groszek, 1977; Stewart, 1977; Stockil, 1977; Brown, 1979; Van Poollen, 1980; Shah, 1981). The methods fall into four classes: thermal methods, miscible displacement, chemical flooding and selective plugging. Thermal methods raise the temperature of regions of the reservoir by injection of hot water, steam or other gas or by conducting a combustion *in situ* of oil or gas. The principal aim is to reduce oil viscosity.

Miscible displacement methods involve the injection of fluids which are fully miscible with the oil, thus avoiding many of the difficulties of waterflooding. Carbon dioxide is a useful displacing fluid. As it dissolves in oil, the oil swells and its viscosity is reduced; at high pressures the fluids are fully miscible. Carbon dioxide flooding is an attractive method but large quantities of carbon dioxide are required: estimates range from 112 to 280 m^3 per barrel of oil (Meyer, 1977; Shah and Wittmeyer, 1978; Doscher and Kuuskraa, 1979).

In a chemical flood, chemicals are injected with the waterflood to improve the displacement efficiency. In polymer flooding, a viscous solution of polyacrylamide or polysaccharide is introduced as a discrete slug in the waterflood. By raising the viscosity of the displacing fluid to equal or exceed that of the oil, the mobility ratio and efficiency of displacement can be greatly improved. Another way of making displacement more efficient is to lower the interfacial tension between the oil and the displacing fluid by the use of surfactant solutions. The preferred method is to produce a microemulsion from a mixture of surfactants (often sulphonated petroleum fractions) together with other components such as alcohols, brine and sometimes chelating agents. The aim is to push an oil-miscible micellar slug into the oil-bearing region to flush out residual oil. The displaced oil droplets coalesce to form a continuous layer or oil 'bank' which is itself very effective at dislodging more oil. The micellar slug follows behind the oil bank preventing oil ganglia being left behind. Usually a polymer will be used in conjunction, to optimize the sweep pattern.

Micellar flooding and polymer flooding work well on the laboratory scale but tend to perform less well in the reservoir. Success undoubtedly depends on choosing the right reservoir and treating it in the optimum manner: the formulation of the surfactant mix is especially critical. Several difficulties are peculiar to large-scale operations. Because of the high cost of the chemicals used they are not injected continuously but as a discrete slug. Correct function depends on the integrity of the slug so that piston-like displacement is achieved and optimum concentration of surfactant is maintained. The slug, however, tends to disintegrate as it proceeds through the reservoir, due to adsorption, degradation, precipitation or backmixing of components.

The last method, selective plugging, aims to inject viscous polymers or time-setting gels into thief zones. Since these zones accept the bulk of the injected aqueous flow the expectation is that they will be effectively plugged: floodwater will then be forced into the less permeable regions which will have received only minor amounts of polymer.

Enhanced oil recovery procedures are undoubtedly becoming more effective as they become better understood but they are expensive to operate and their cost-effectiveness can be assessed from the large subsidies paid by the US Department of Energy to companies using them. Dafter (1980) has reviewed the prospects for the chemical and physical methods. It should be noted that in some cases EOR methods are best applied at an early stage.

Microbial enhancement of oil recovery (MEOR)

EARLY EXPERIMENTS ON MEOR

The first suggestion that bacteria might have a useful role in enhanced oil recovery came from Beckman in 1926. Claude ZoBell followed up the suggestion with an extensive series of laboratory experiments (*see*, for example, ZoBell, 1946, 1947a, b, 1953). He used cultures of sulphate-reducing bacteria designated as *Desulfovibrio hydrocarbonoclasticus* and *D. halohydrocarbonoclasticus*, claiming that treatment of oil-bearing formations with bacteria could potentially release oil by a number of mechanisms. Crude oil hydrocarbons were utilized by bacteria, leading to the production of carbon dioxide and acids. The acids reacted with rock carbonates, increasing permeability and producing more carbon dioxide.

Carbon dioxide, and other gases which might be produced, increased reservoir pressure and dissolved in the oil, forcing it to swell and reducing its viscosity, thus aiding displacement. It was also reported that the growth of bacteria on the surface of oil-bearing rocks in contact with nutrient media led to films of oil forming on the (aqueous) liquid surface, and it was suggested that displacement of oil from surfaces by bacterial growth was occurring. Some degree of oil utilization was reported and it was suggested that oil viscosity fell as a result of a reduction in the average molecular weight of oil components. The bacteria were also said to produce surface-active agents, reducing surface tensions from about 70 to about 50 mN/m, thus potentially aiding oil release. MacKenzie (1952) supported many of ZoBell's claims.

Beck (1947), using similar bacteria, found that utilization of oil was slow, and was unable to demonstrate enhanced oil release attributable to bacteria in model systems. Updegraff and Wren (1954), working in consultation with petroleum engineers, were unable to demonstrate consistent enhancement of oil release by sulphate-reducing bacteria alone under conditions relevant to reservoir operations. La Rivière (1955) expressed some reservations about ZoBell's conclusions but was able to demonstrate the reduction of surface tension by a number of bacteria. He pointed out that peptone, used by ZoBell, would itself reduce surface tension. On the other hand, Updegraff and Wren (1953) and Updegraff (1957) patented techniques in which a variety of other organisms including *Escherichia coli*, *E. freundii*, *Aerobacter aerogenes* and members of the genera *Clostridium* and *Bacillus* or sulphate-reducing bacteria in conjunction with other types, were claimed to produce significant enhancements of oil release when supplied with carbohydrates and mineral nutrients.

Possible mechanisms of oil release

The early publications of ZoBell proposed a series of possible mechanisms by which bacteria might release oil: several more have been added since. Not all are equally plausible.

Modification of reservoir permeability by acids. Many bacteria certainly produce acids under anaerobic conditions. Many reservoir rocks consist partly or entirely of carbonate. Bacterial action may result in dissolution of reservoir rock carbonate (*see* for example Yarbrough and Coty, 1983) and changes visible to the naked eye can be produced in calcite and dolomite cores by bacterial action (Bubela, 1983a), but it is not obvious that effects on permeability will necessarily prove favourable. There is a danger that fine particles, released by acid, might be swept into pore throats resulting in plugging. Extensive dissolution of a pure carbonate rock would require massive amounts of acid and hence of substrate. Acid might well preferentially attack the largest channels and there appear to be signs of this in the photographs presented by Bubela (1983a). This would increase permeability but would result in bypassing most of the oil. On the other hand, production of acid resulting in pore enlargement might well be favourable where the vital region immediately round the injection well has become partly plugged, and Johnson (1979) appears to have exploited this possibility.

2. *Repressurization by production of carbon dioxide, hydrogen, methane, nitrogen.* This seems a plausible mechanism and many field trials report pressure increases on bacterial treatment, sometimes of substantial magnitude and duration. Yarbrough and Coty (1983) reported that *Clostridium acetobutylicum* could generate pressures up to 8·8 MPa in laboratory culture in the absence of a gas phase, mainly due to production of hydrogen which has a very low solubility in water. Carbon dioxide generated from rock carbonate could contribute to this mechanism.

3. *Solution of gases in oil resulting in swelling and decreased viscosity.* Gases, particularly carbon dioxide, dissolve in oil under pressure, reducing its viscosity and increasing its volume. An increase in volume of sufficient magnitude might free some oil from ganglia and the effect on oil viscosity might be useful.

4. *Reduction of oil viscosity by degradation of large molecules.* Singer *et al.* (1983) isolated an aerobic organism, designated H13, which reduced the viscosity of heavy crude by as much as 98%. Part of this reduction was due to formation of an oil–water emulsion under the influence of a glycolipid surfactant. The surfactant alone reduced viscosity by about half, the remaining effect being apparently due to changes in oil composition, as yet only partly characterized (Donaldson, 1982). Similar results have been reported by Zhang and Quin (1983). These reports serve to demonstrate that bacteria can reduce oil viscosity, apparently by selective degradation as well as by emulsion formation, but as yet there is no demonstration that this is possible under the anaerobic conditions

pertaining in the reservoir and the persistent emulsions generated would be disadvantageous.

5. *Reduction of oil–water interfacial tension by surfactants.* The introduction of chemical surfactants into the reservoir together with the waterflood forms the basis for a well-known EOR technique. The production of surfactants of different chemical types by a variety of bacteria is well documented. Many of the cultures used for MEOR operations are reported to reduce the surface tension of culture broths. However, where interfacial tension measurements have been reported for bacterial surfactants, the values with one exception fall far short of that required to effect oil mobilization under reservoir conditions (*see below*). The operation of this mechanism is thus hypothetical.

6. *Dissolution of oil by solvents e.g. alcohols and ketones.* It seems unlikely that appropriate substances would be produced in high enough concentrations to dissolve oil but they might act as co-surfactants, modifying the behaviour of the surfactant itself (*see below*).

7. *Sulphonation of oil by bacteria.* Sulphonated petroleum fractions are used as surfactants for EOR. No evidence exists of bacterial surfactant production in this way.

8. *Displacement of oil from rock surfaces by bacterial growth.* Most reservoir rocks are water-wet. The oil is trapped by pore throats rather than stuck to a surface. The mechanism might operate in oil-wet rocks.

9. *Increase in viscosity of the aqueous phase caused by bacterial polymers.* Polymers, mainly of bacterial origin, are already used to increase the viscosity of injected water. Polymers can be produced under anaerobic conditions (*see below*) so it is feasible to consider their production *in situ.*

10. *Selective plugging of regions of high permeability.* If permeability variations are a serious problem this is a highly plausible mechanism. Bacterial cells can block pores either directly or by producing extracellular slimes (*see below*).

CONDUCTING MEOR OPERATIONS

As a result of early laboratory experiments a picture emerged on the way in which an MEOR operation might be conducted in the field. The work of ZoBell (1946, 1947a, b, 1953) had placed emphasis on sulphate-reducing bacteria and on utilization of crude oil as a substrate for growth. Later workers had difficulty in demonstrating oil utilization under anaerobic conditions and sometimes found that sulphate-reducing bacteria were ineffective at removing oil (Beck, 1947; Updegraff and Wren, 1954) and were likely to cause a variety of problems (*see below*). Emphasis therefore shifted to the use of other bacteria, either strict or facultative anaerobes, and to the provision of a carbohydrate substrate, usually molasses, or an organic acid such as lactate, for growth together with a range of mineral nutrients (Updegraff and Wren, 1953, 1954, 1957). Hitzman (1962) pointed out that certain problems might arise if vegetative cells, injected together with nutrients, were used as the inoculum. Growth would begin in the vicinity of the well where, for geometrical reasons, large volumes of fluid would pass through a very small cross-sectional area of rock. Plugging of the formation at this point by cells or metabolic products (*see below*) would be particularly detrimental (Crawford, 1983), whereas the beneficial effects of bacterial growth and metabolism would be most effective further into the reservoir. Vegetative cells are susceptible to environmental hazards, such as high shearing forces, especially in the absence of nutrients, and Hitzman proposed to overcome these problems by first injecting the inoculum in the form of spores, introducing nutrients only after the spores had penetrated deep into the formation. He pointed out that spores penetrated sandpacks more readily than vegetative cells, an observation confirmed by later reports (e.g. Jang *et al.*, 1983). Later, as an alternative way of avoiding plugging, he proposed injecting the inoculum into the brine layer below the oil-bearing formation, the injection well being shut off from the brine layer. The inoculum would, it was claimed, grow at the oil–water interface, repressurizing the reservoir by producing methane for a 10–30-year period, but causing no dangers of plugging because bacterial growth would be outside the main oil-bearing layer (Hitzman, 1965).

Generally, the use of aerobic organisms has not been seriously considered because of the low oxygen status of reservoirs, because of the difficulties of introducing large quantities of oxygen without interfering with the waterflood and because engineers normally go to much trouble and expense to exclude oxygen to avoid corrosion of metal pipework. However, Jones (1967) proposed the use of aerobic bacteria, injected along with oxygen or air, to metabolize oil hydrocarbons *in situ*. Fracturing the rock formation to facilitate bacterial penetration was also proposed.

In many cases MEOR operations have proceeded as part of the waterflooding operations with nutrients and inoculum injected at one point and oil produced elsewhere (water-drive method). Alternatively the 'huff-and-puff' method has been used in which nutrients and inoculum are injected into a well which is then sealed for a period of months. Oil is produced by opening the well and reversing the flow.

POTENTIAL ADVANTAGES OF MEOR

MEOR offers, in principle, three advantages over other EOR systems. It is cheap, because the main raw material is an inexpensive carbohydrate source such as molasses, with or without inorganic nutrients. It is further expected that MEOR might be more effective than some types of EOR operation because the active materials would be generated throughout the reservoir, and their effectiveness would not be reduced by adsorption during migration to the site of action. It is also thought that bacteria, by diffusing out of the main lines of fluid flow, could mobilize oil in regions not accessible to other methods. The latter two advantages have yet to be proved.

LABORATORY MODELS OF MEOR

The techniques used to study oil release on a laboratory scale are of critical importance. Most reservoir rock is water-wettable and the surface is covered with a layer of water even when oil fills most of the pore-space. Oil is held in place by surface forces (*see later*). Thus the displacement of oil from oil-wet surfaces is not likely to be of importance in most reservoirs.

In order to simulate the porous reservoir for displacement studies, two main devices—sandpacks and cylindrical rock cores—usually derived from outcrop rocks, have been used. Sandpacks consist of glass or plastic tubes containing fine sand or other material such as glass beads. They are easy to make in a fairly uniform fashion but the grains are not normally fused as with a rock core, and it can be difficult to obtain permeabilities in the lower range. Yarbrough (personal communication) has used lucite (perspex) tubes packed with sand and then heated to fuse the sand to the tube wall in order to prevent edge channelling. Rock cores have been used in a variety of shapes and sizes. Berea sandstone has been a favourite material because of its fairly uniform structure and easy availability in the US in a range of permeabilities. Cylindrical shapes are usually used with fluid flow induced across the ends. To prevent the escape of fluids from along the length, the surfaces can be coated with an epoxy resin or the core can be sealed in a tightly fitting rubber sleeve (Hassler cell). In the latter case it is comparatively easy to apply uniform pressure to the core independent of the pressure differential across the ends. Occasional use has been made of other materials to simulate porous rock. Thus Bubela (1983a) described the use of concrete, which can be cast in a wide range of permeabilities, and cores of sintered alundum have also been used.

In some instances visual monitoring of the progress of an experiment may be useful. For studies of oil displacement a variety of systems have been devised in which a thin layer of the porous matrix is sandwiched between sheets of plastic or glass so that flows of oil and water can be observed directly (Dawe, personal communication). Similar arrangements have been used for bacterial experiments by the group at Queen Mary College (Andrew McKay, personal communication) and by Bubela (1983a).

The preparation of cores or sandpacks for an experiment is critical. Several workers steam-clean outcrop rocks before use to remove humic acids (Hank Yarbrough, personal communication, Jenneman *et al.* 1983). Sterilization may be carried out at this stage but is difficult to achieve. Jenneman *et al.* (1983) used a commercial chlorine dioxide preparation. This, they said, reduced the counts of indigenous bacteria for 24–48 hours but the population became re-established after this time or upon introducing nutrients. Even after steaming cores for two weeks, autoclaving for 12 hours and drying at 121°C, variable numbers of indigenous bacteria remained, presumably due to failure of steam fully to penetrate the pore structure. Jang *et al.* (1983) and Jang and Yen (1983) reported the use of 70% ethanol for sterilization. This, too, is unlikely to have been fully effective. The use of mercuric chloride or of low temperature to inhibit bacterial action in controls is not acceptable as both treatments reduce oil mobility (Davis and Updegraff, 1954).

Cores or sandpacks are first flooded with water (brine). Vacuum or pressure infiltration is necessary to get good penetration. Crude oil is then pumped in until no more water is displaced. In this way the rock surfaces become water-wet while the capillaries are full of oil as in a reservoir. If water is now injected at a steady rate, oil will be expelled. It is essential to continue with this process until no more oil emerges. This represents waterflooding to exhaustion and can be used both as a control and as a preliminary to bacterial injection. The lack of an explicit mention that this operation has been properly performed, makes it difficult or impossible to evaluate some of the reports in the literature. It is of no practical use to demonstrate that a bacterial culture (plus the waterflood) will release a certain percentage of the retained oil if the waterflood alone will produce as much or more.

Oil released can be measured volumetrically. Jang and Yen (1983) extracted it into toluene, evaporated the solvent and determined the oil by weighing. This seems advantageous if oil is released as an emulsion, as it may be if bacteria or synthetic surfactants are used, or if the quantities released are small. Small cores (say 7·5 cm × 2·5 cm) are frequently used commercially to test the effects of surfactants or polymers, but larger cores may be useful to facilitate accurate measurements of released oil and to avoid the possibility of end effects. Fluctuations in back pressure are almost inevitable: these are easily monitored by solid-state differential-pressure transducers connected across the column. Peristaltic pumps are widely available and convenient, but flow rates are very susceptible to fluctuations in back pressure. Continuous-action motor-driven syringe pumps are more expensive but can be virtually pulse-free and are much less sensitive to back pressure. Electronic flow-measuring devices are available but expensive.

Microbial activities and MEOR

PLUGGING AND PENETRATION OF RESERVOIR ROCK BY BACTERIA

Many authors have described the presence of a wide variety of bacteria in oil samples, formation rock and produced water. With some justification a range of undesirable activities has been attributed to bacteria and these include degradation of oil in reservoirs (Kuznetsov, Ivanov and Lyalikova, 1963; Anonymous, 1972; Westlake, 1983) the promotion of corrosion (Allred, 1976; Miller, 1971), souring by the production of H_2S, and formation plugging. The sulphate-reducing bacteria are considered to be especially undesirable (for general discussions *see* Beerstecher, 1954; Davis and Updegraff, 1954; Kuznetsov, Ivanov and Lyalikova, 1963; Davis, 1967; Moses and Springham, 1982).

Pressure maintenance and waterflooding operations involve the injection of enormous volumes of water into reservoirs, and micro-organisms present in the water may cause serious plugging problems. Crawford (1983) has calculated the effects of different degrees of plugging on waterflooding operations. Reservoir engineers have devised methods of treatment to alleviate this and other problems, and water treatment prior to injection may involve the use of flocculating agents to remove solids, de-aeration, addition of biocide to prevent bacterial growth on filters, passage through various filters and, finally, the addition of substances such as oxygen scavengers, corrosion inhibitors, scale inhibitors and biocides. Two dangers arise in respect of MEOR operations: the suspension of treatment to permit bacterial inoculation and growth may permit undesirable organisms to enter, and the inoculum itself may have a detrimental effect.

It is not easy to generalize on the dangers of plugging. Some of the early work demonstrated that serious plugging of cores could occur very rapidly (Plummer *et al.*, 1944) but the waters used for the study contained a variety of bacteria, fungi, algae, protozoa, as well as precipitates of calcium carbonate and metal sulphides. Beck (1947) studied injection waters which had given rise to plugging problems and noted the presence of a variety of bacteria together with ferrous sulphide and ferric hydroxide precipitates. He recommended pretreatment of the water by filtration and the addition of germicides. It is clearly important to distinguish between plugging caused by cells and that caused by metabolic products, and to ask to what extent different types of cells cause different degrees of plugging, so that bacteria intended for MEOR can be chosen to eliminate or minimize these problems. Particularly serious problems are likely to be caused by iron bacteria and sulphate-reducing bacteria producing ferric hydroxide and ferrous sulphide precipitates respectively, and by slime-forming organisms; these would normally be avoided in MEOR operations or used only for special purposes.

Many investigators have used killed or non-growing suspensions of bacteria, injecting them into a variety of rock cores and measuring the pressure drop as an index of plugging. In some cases pressure was measured at a serious of points along the core. There is general agreement that the injection of large volumes of dense bacterial suspensions produces a progressive reduction in permeability (Hart, Fekete and Flock, 1960; Kalish *et al.*, 1964; Raleigh and Flock, 1965; Jenneman *et al.*, 1983). The most severe plugging occurs at, or close to, the injection face. The greater the concentration of cells injected, the greater the degree of plugging. There is a concentration effect: a given number of cells causes more plugging if injected at low concentration than at high concentration (Hart *et al.*, 1960; Kalish *et al.*, 1964). Kalish *et al.* (1964) found that permeability values tended to stabilize after the injection of large volumes of suspension: the proportional reductions in permeability at this point were greatest in the formations of highest permeability. They attributed this to the ability of cells to penetrate further into the more permeable formations and thus cause plugging in depth. By contrast Hart, Fekete and Flock (1960) found that in some cases permeability was reduced almost to zero and were unable to relate permeability reductions to the initial permeability values of their rock samples. Using different bacterial species, Kalish *et al.* (1964) concluded that large cells cause more plugging than small cells and clumps cause more plugging than single cells. Jack, Thompson and DiBlasio (1983) isolated an anaerobic rod-shaped isolate which grew as discrete cells on sucrose medium but as chains on glucose/fructose medium. The chains were much more effective at plugging than the discrete cells. Bubela (1983b) found that rod-shaped organisms caused greater plugging than did cocci and that the plugs are more difficult to shift by the application of pressure.

Kalish *et al.* (1964) found that the adverse effects of bacteria on core permeability could be overcome readily by increasing the applied pressure, which increased permeability apparently by dislodging cell plugs, or by acid treatment followed by reverse flooding. Chlorine treatment is an alternative (Crawford, 1983). Sharpley (1961) discussed the various problems of plugging in waterflood operations and specified safe limits for the numbers of bacterial cells in injection water. The general validity of these limits may be open to question. Allred (1976), also discussing field operations, suggested that no such generalizations could be made.

The risks of serious plugging by the inoculum itself do not seem very high. The cells are an inoculum, not a reagent, and very high numbers should not be necessary. According to Kalish *et al.* (1964) the injection of 10^{11} cells of *Pseudomonas aeruginosa* into a core of 323 md permeability, reduced the permeability by only about 15% and, even with a 32 md core, 5×10^{10} cells reduced the permeability only by about half. The most vulnerable part of the formation, for geometrical reasons, is the part nearest to the face of the injection well, but that is the part easiest to clean.

Cells actually growing in the formation are likely to present a more serious problem (Jenneman *et al.*, 1983). Hitzman (1962) recommended introducing the bacteria first in the form of spores, nutrient being pumped in later when the spores had moved into the formation. This suggestion does not seem to have been followed up in field trials; in most instances the cells and medium are injected together. Despite this, plugging does not seem to have caused any serious problems. Yarbrough (personal communication) found that injectivity decreased as a result of inoculation but the effect could be overcome by pumping instead of letting fluids flow in by gravity.

One would expect that a number of factors might influence the ability to penetrate and the rates of movement. These include characteristics of the bacteria such as size, shape, deformability, the nature of the cell surface, whether or not cells aggregate into chains, production of extracellular slimes, and motility; reservoir characteristics such as permeability, pore structure, the chemical nature of the rock, nutrient and salt content of the connate water or waterflood, and temperature and pressure (which may affect motility) and also the fluid flow rate and the concentration of cells. Effects of cell size and shape on plugging have already been mentioned and these factors presumably have equivalent effects on penetration. Penetration can be studied by total or viable counts on cells emerging from rock cores or sandpacks, by staining cells and examining the appearance of rock to detect penetration (Kalish *et al.*, 1964) or by labelling cells with radioactive isotopes (Myers and McReady, 1965). Springham *et al.* (1983) have pointed out the problems which may arise in using ^{32}P-labelled cells due to release of soluble label and to isotope-induced lysis.

Failure of cells to penetrate could be caused by settling of cells out of suspension, by adsorption to rock surfaces or by sieving. Some idea of the likely importance of the latter mechanism can be obtained from data on so-called pore size distribution. This is routinely obtained by forcing mercury into rock samples and measuring the relationship between applied pressure and the volume of mercury which penetrates into the rock pores (Ritter and Drake,

1945). Strictly speaking, it measures the distribution of the constrictions controlling entry, rather than the pores proper. Burdine, Gournay and Reichertz (1950), demonstrated a relationship between pore throat size distribution and the permeability of reservoir rocks. Davis and Updegraff (1954) found that pore throats must be at least twice the diameter of cocci or short bacilli for cells to pass through readily and concluded that this would require permeabilities in excess of 100 md. Several workers have suggested figures in this region as representing minimum values for MEOR. A summary is given by Moses and Springham (1982).

Quantitative studies of penetration rates have been made by Yarbrough (personal communication) who found negative semilogarithmic relationships between spore numbers and distance moved, using a series of sample points along the length of sandpacks. Jang *et al.* (1983), using *Pseudomonas putida*, *Clostridium* sp. and *Bacillus subtilis*, found that, in general, cells were detectable in the fluid emerging from their rock cores after one pore volume of fluid had passed through. For several more pore volumes of fluid, cell concentrations remained more or less constant at a fraction of the input concentration. Eventually cell concentrations rose to equal the input level. They interpreted their data in terms of a deep bed filtration model, calculating the filtration coefficient (K_0) for cells and spores as:

$$K_0 = \frac{\ln\left(\frac{C_I}{C_L}\right)}{L}$$

where C_I = initial cell concentration, C_L = emerging cell concentration, L = length of core.

They studied the effects of ions and chelating agents and concluded that, at low cell concentrations, filtration was mainly due to adsorption on to rock surfaces. Spores had lower filtration coefficients than cells and the presence of residual oil lowered the filtration coefficient, suggesting that cells might penetrate waterflooded reservoirs more readily than most experiments with rock cores would suggest.

In contrast to the results of Jang *et al.* (1983), Jenneman *et al.* (1983), measuring the emergence of *Pseudomonas* cells from a rock core, found that the concentration of cells emerging fluctuated with time, with no discernible pattern. The cell concentration in the effluent never exceeded 1% of the input concentration. Unfortunately Jang *et al.* (1983) presented only a generalized curve, and not the results of individual experiments, so an exact comparison is not possible.

Linear rates of cell movement can be calculated from laboratory data and the values obtained are not wholly incompatible with MEOR operations, even where considerable distances are involved between wells (Moses and Springham, 1982). The reports of field trials generally indicate that bacteria can become distributed fairly rapidly (see the compilation of data presented by Hitzman, 1983) and, subject to the permeability limits mentioned above, there is no sign so far that penetration is a serious limitation to MEOR.

SELECTIVE PLUGGING BY BACTERIAL CELLS

Under some circumstances, plugging by bacterial cells may be positively advantageous. Von Heiningen, Jan de Haan and Jensen (1958) injected cultures of slime-forming bacteria, together with carbohydrate media, with the deliberate intention of restricting flow in the more permeable regions of the formation, thus redistributing the waterflood into the less permeable regions. A similar rationale lay behind the trial described by Yarbrough and Coty (1983) who intended to achieve a similar end by the generation of gas bubbles. Crawford (1961, 1962) pointed out that the proportional reduction in permeability due to bacterial cells was greatest in rocks of highest permeability and suggested that this might result in redistributions of flow in favour of the less permeable zones.

In some formations, permeability variations can be considerable. Fox (1983) mentions the problems caused by rapid water movement along the Brent Sands layers in many of the North Sea fields, Jack, Thompson and DiBlasio (1983) refer to the problems of the Lloydminster reservoir where bypassing is exacerbated by high oil viscosity, and renewed interest is being shown generally in selective plugging as a possible MEOR mechanism. Cells actively forming slime cause very severe plugging, whereas the same cells injected under nutrient conditions which prevent slime formation have little effect (Jack, Thompson and DiBlasio, 1983). Injections of cells can thus be made deep into the formation together with nutrients which do not support slime formation, and followed by modified nutrient solutions which promote slime formation and plugging. Jenneman *et al.* (1983) used two cores of different permeabilities connected in parallel, and by injecting cells of *Bacillus subtilis* followed by nutrients were able to achieve selective plugging of the more permeable core. Initially, the low-permeability core had accepted 24% of the liquid flow; after selective plugging it accepted 90%.

In my opinion, selective plugging is a mechanism of great potential importance. Its usefulness will depend on the degree to which plugging can be confined to the more permeable zones. It may be the case that it will not be limited to instances where there are obvious and major differences in rock permeability. It is likely that even rocks whose structure is apparently uniform have permeability variations on the micro scale. This is suggested by the data on pore size distribution in reservoir rocks (Burdine, Gournay and Reichertz, 1950) and by the results of Kalish *et al.* (1964) who examined cores which had been flushed with suspensions of stained bacteria and noted the very uneven extent to which cells had penetrated the matrix. It might also account for the rather surprising fact reported by Myers and McReady (1965) that bacteria appeared to penetrate cores of very low permeability ($<0{\cdot}1$ md).

Jang and Yen (1983) examined the release of oil from sandpacks which had been saturated with oil and then waterflooded. A culture of *Bacillus subtilis* injected into the rock caused release of about 35% of the residual oil, whereas sterile medium released only 12%. If sterile medium was pumped into a column after a bacterial inoculum, more oil was released and this coincided with an increase in the pressure drop across the column. From the brief details published, selective plugging seems the most likely reason for oil release in this instance.

PRODUCTION OF POLYMERS AND SURFACTANTS

We have seen that polymers and surfactants can play a part either in chemical or microbial EOR. Where such materials are produced by bacteria they could either be produced above ground and injected into the reservoir, or the organisms could be injected to produce them *in situ*.

The properties required of a polymer for EOR are fairly well understood. Water viscosity must be increased up to about the same value as that of the oil being displaced. The solution must be resistant to shear degradation as high shearing forces are normally experienced when water is injected into the formation. Shear thinning has the advantage that the viscosity of the aqueous phase drops (temporarily) on injection. Reservoir brines frequently contain high concentrations of salts so these must not seriously affect viscosity. The internal surface area of reservoir rocks is enormous: Hesselink and Teeuw (1981) mention $0{\cdot}1$–$1{\cdot}0\,m^2$ surface area per gram of sandstone. It is therefore important that adsorption is minimal. The two polymers most used (by injection from the surface) are partly hydrolysed polyacrylamides and a polysaccharide, xanthan. Polyacrylamides are good viscosifiers at low salt concentrations but lose most of their viscosity above 1 g/dm^3 NaCl and are very sensitive to shear degradation. Xanthan, a bacterial product, is a good viscosifier and is relatively insensitive to shear degradation and salt content.

Little work has been published on the generation of polysaccharide *in situ* as opposed to its injection from the surface. Xanthan itself is synthesized under aerobic conditions but some polysaccharides such as dextrans and levans can be produced anaerobically. A valuable body of experience concerning the selection and strain movement of suitable organisms is available (Sutherland, 1983). Improvements include the selection of high-yield mutants, modification of polysaccharide properties, and mutation to eliminate unwanted products. Successful *in situ* polysaccharide generation will also require an appropriate nutrient balance. In general, for good polysaccharide production, carbohydrate must not be limiting and high carbohydrate:nitrogen ratios are generally favourable. There is some evidence that different limiting nutrients may affect the precise chemical nature of the product.

Rather less is known in general terms about the factors controlling surfactant production but it appears that these, too, would probably require control of the nutrient status in the reservoir to achieve high levels of production. The properties required of a microbial surfactant can also be predicted from a consideration of oil displacement.

Under normal waterflooding conditions a large proportion of the oil is trapped by capillary (surface) forces, even in regions which are fully swept by the water-flood. Oil globules which are surrounded by water are unable to move through constrictions because this would require an increase in the interfacial surface area as the globule is distorted. If the flow rate of water is made sufficiently high, the viscous (flow) force tending to displace oil will overcome the capillary forces resisting displacement. Taber (1968) and Melrose and Brandner (1974) have stressed the utility of a single dimensionless parameter representing the balance between viscous and capillary forces. The one most frequently used is the capillary number:

$$N_C = \frac{\mu_W V_W}{\sigma}$$

where μ_W = water viscosity, V_W = water velocity and σ = oil:water interfacial tension.

After waterflooding, the local displacement efficiency of oil (in well-swept zones) is likely to be about 50% and the capillary number about 10^{-6}. To exceed significantly this displacement efficiency, capillary values above 10^{-3} are required, and to approach 100% displacement, the capillary number has to be increased to about 10^{-2} (Shah, 1981). The applied pressure gradient is limited by the operating equipment and by the need to avoid fracturing the reservoir rock and cannot be increased to the necessary values (Taber, 1968). Reduction of the interfacial tension from 30 mN/m to 10^{-2} or 10^{-3} mN/m can be achieved by the use of surfactants, thus increasing the capillary number to the required value.

The attainment of ultralow values of interfacial tension is by no means straightforward. Shah (1981) provides a useful survey of some of the complexities. Interfacial tension is affected by several factors, including the properties of both the surfactant and the oil, surfactant concentration, salt concentration, temperature, and the nature of the cosurfactant. Surfactant formulations for chemical EOR are thus designed for the specific oil type, salinity and temperature of a particular reservoir. The surfactant slug injected may contain one or more surfactants, salt, a cosurfactant such as a short-chain alcohol (which has a number of desirable effects), a sacrificial agent which will preferentially adsorb to rock surfaces and minimize loss of the surfactant, and perhaps a chelating agent to remove ions which may interfere with the surfactant.

The surfactant slug moves through the reservoir displacing trapped oil droplets which are pushed ahead and merge together to form the so-called 'oil bank'. Formation and maintenance of an oil bank is considered to be essential for efficient oil recovery. Displaced oil droplets must coalesce readily so the interfacial viscosity must be low. The surfactant slug must be of an appropriate viscosity to minimize fingering, the materials must not be unstable or insoluble under reservoir conditions, and loss by adsorption on to the enormous rock surface area must be minimized. Usually a polymer slug will follow the surfactant flood to optimize the flow pattern. It can be seen that, once ideal conditions have been achieved in the surfactant slug, there will be a tendency for the composition to change by adsorption, partitioning and degradation of components and for the slug itself to break up. These factors will tend to limit the efficiency of surfactant flooding operations over any but the shortest distances.

For micro-organisms to release oil by surfactant production requires more than just production of a surfactant. An effective product must have certain specific properties, especially the ability to lower oil:water interfacial tension to 10^{-2} or 10^{-3} mN/m, and it must be produced in sufficient concentration.

Cationic surfactants are unlikely to be useful because of adsorption on to negatively charged rock surfaces. Low values of interfacial viscosity are essential so that displaced oil ganglia can coalesce to form an oil bank. A wide variety of surface-active agents is known to be produced by micro-organisms (*see* Zajic and Panchal, 1976; Finnerty and Singer, 1983).

Where interfacial tension has been measured, the values have usually been too high to effect oil mobilization, although it should be remembered that even the best synthetic surfactants give ultralow interfacial tensions only in the presence of critical concentrations of salt. Singer *et al.* (1983) have isolated a bacterial strain (H13) which produces a glycolipid surfactant when grown on crude oil. The crude glycolipid solution gave an interfacial tension of 2×10^{-2} mN/m against hexadecane, and Finnerty and Singer (1983) mention a minimum interfacial tension of less than 10^{-4} mN/m with 0·5% pentanol as cosurfactant. The material has apparently not yet been tested for oil mobilization. As it is reported to stabilize oil-in-water emulsions it would presumably inhibit oil-bank formation and would therefore not mobilize oil effectively in a conventional recovery procedure.

However, there are recent indications from other sources that microbial surfactants can give low interfacial tensions and are very effective at mobilizing oil from sandpacks and rock cores without the addition of cosurfactants. The latter property would be particularly useful for *in situ* MEOR as it would simplify the attainment of critical conditions within the reservoir.

FIELD TRIALS

Over the past 30 years, field trials of MEOR processes have been conducted in Czechoslovakia, Hungary, Poland, the USSR, Romania, the Netherlands and the USA. Over 40 separate trials have been reported in varying degrees of detail. The reports are scattered and most are not in English. Moses and Springham (1982) have discussed the more important trials and Hitzman (1983) has published a tabulated summary. A single example discussed in detail, together with briefer consideration of others, will be used to illustrate the way in which such trials have been conducted and the difficulties of interpretation they pose.

In 1954, Coty, Updegraff and Yarbrough conducted one of the first-ever field trials in Union Country, Arkansas. A detailed account has recently been presented (Yarbrough and Coty, 1983). The rationale was as follows: it was intended that the injected bacteria should produce free gas *in situ*, thereby blocking the most permeable channels and diverting flow into unswept regions of the reservoir. It was hoped that generating gas *in situ* would require much less energy than forcing in gas from the surface and that the bacteria, because they could migrate outside the main streamlines, would generate gas and release oil from areas not accessible to processes operated direct from the surface.

A strain of *Clostridium acetobutylicum* was chosen as the test organism because of its ability to generate hydrogen, carbon dioxide and organic acids under anaerobic conditions from media containing sugars and mineral salts. Hydrogen and carbon dioxide were produced in approximately a 1:1 ratio and, in liquid-filled systems, pressures up to 8·8 MPa could be generated. The organism grew well at about 40°C and could tolerate high pressures of carbon dioxide as well as sodium chloride concentrations up to 30 g/dm^3. It was not inhibited by crude oil and produced no detectable changes in oil properties. Laboratory trials with sandpacks and rock cores demonstrated its ability to release oil; the most important variables appeared to be the nature of the rock and the pressure generated.

The wells chosen were in the Lisbon field, producing from the Nacatoch sand, a layer about 7 m thick at a depth of about 650 m. The formation was loosely consolidated sand containing about 8% carbonate and with high porosity and permeability (30% and up to 5·7 d respectively). Down-hole temperature was about 34°C and the formation water contained 42 g/dm^3 sodium chloride. The temperature was suitable for growth of the test organism, and the high permeability was expected to permit easy penetration, but the field was far from ideal in most other respects. Thus the salt content of the formation water was sufficient to inhibit bacterial growth and the amount of oil remaining in the formation was extremely low. The field had been discovered in 1925. After many years of primary production, waterflooding was started in 1949. Rock core samples, taken when extra wells were drilled for the waterflood operation, showed residual oil saturations of only 5–9% of the pore space. A single well, No. 30, was chosen as the injection well for the experiment and well 31, 123 metres to the south, which had been abandoned but was re-opened for the trial, was used as the production well.

Waterflooding operations in the field, resulting in a general flow of water from north to south, continued throughout the trial. It was considered that the predominant flow of water between the two wells was determined by the main waterflooding operation rather than by the injections into well 30.

Before the experiment started, water containing 20–25 g/dm^3 NaCl had been injected into well 30 to lower the salt concentration in the reservoir. In May 1954, fresh water was injected for two months and in July 1954 the experiment was started by the injection of dense bacterial suspensions, 18 separate batches totalling 18 m^3 being injected over a 4-month period. Simultaneously, injection of beet molasses was started, a concentration of 2% being maintained in the injection water for $5\frac{1}{2}$ months. At first the substrate solution was allowed to flow in by gravity: later, the resistance to flow increased slightly and a pump was installed to force the solution into the formation. Samples of gas, water and oil were taken at intervals from all the production wells in the area.

Fresh water injected at well 30 appeared 70 days later at well 31. From 80 to 90 days after inoculation until the termination of sampling in May 1955, unused sugars, seven different short-chain fatty acids, carbon dioxide and methane, together with traces of ethanol, butanol and acetone appeared at well 31. The gas produced from well 31 contained 38–82% carbon dioxide: gas from wells not affected contained 0·3–8·8%. By measuring the ^{14}C:^{12}C ratio of the carbon dioxide it was estimated that 20% came from acid dissolution of rock carbonate. Carbon dioxide produced by fermentation was equivalent to 19% of the sugar injected. The remaining gas produced was mainly methane: from ^{14}C:^{12}C ratios it was calculated that 20% of this originated from sugars.

Virtually no hydrogen was detected. Total acid production was equivalent to 59% of the sugar injected.

Neither the spectrum nor the ratios of fermentation products were those expected from a pure *Clostridium acetobutylicum* fermentation, which would produce only formic, acetic and butyric acids equivalent to 35–40%, carbon dioxide equivalent to about 47%, and hydrogen equivalent to 2% of the sugar used, but no methane. This suggests that the bacteria active in the formation were not exclusively derived from those injected. Bacteria, (not *Clostridium acetobutylicum*) were detected at the production well, but counts were low and erratic and no correlation with the production of fermentation products could be made.

The production of oil from well 31 dropped rapidly from 3·5 barrels per day when it was reopened to less than 1 barrel per day at inoculation (*Figure 2*). The dotted line shows the field engineers' estimate of continued oil production had no disturbance occurred. Oil production rose sharply, if not steadily, after July. Water production varied between about 400–500 barrels per day, the water:oil ratio fluctuating wildly. Yarbrough and Coty (1983) estimated that oil production from November to May, when measurements terminated, was 250% more than that projected. No change could be detected in oil properties. Since the test wells were part of a larger pattern of wells subjected to a major water-flooding operation, it was vital to decide whether increases in oil production had occurred at other producing wells. Changes in oil production during the period of the experiment at the nearest three production wells, nos. 33 and 34 to the south-west and no. 25 to the north-east were monitored. No significant increases were observed, showing that the increased production at well 31 was not a result of any general change in that part of the field.

The monitoring of fermentation products showed that, despite the absence of increased bacterial activity at the production well, an underground fermentation had been induced, and all those connected with the trial, both engineers and microbiologists, seem to have been convinced that a genuine stimulation of oil production had been obtained.

Two Dutch trials reported by Von Heiningen, Jan de Haan and Jensen (1958) are of interest because of the explicit intention to enhance oil production by selective plugging. Few details are provided but *Betacoccus dextranicus* was used in one case and a mixture of *Leuconostoc mesenteroides*, *L. dextranicum*, *Bacillus polymyxa* and *Clostridium gelatinosum* was used in the other. Molasses was the nutrient. An increase in oil production was reported in the first case and an increase in the oil:water ratio in the second.

Dostalek's group in Czechoslovakia (Dostalek and Spurny, 1957) reported the isolation of a soil *Clostridium* which would grow on petroleum, or on carbohydrate and yeast extract, with the production of large quantities of gas. Laboratory trials indicated that gas production was the critical factor in releasing oil. In a series of field trials starting in 1954, cultures of *Desulfovibrio* and *Pseudomonas* were injected together with molasses. Bacterial counts in the formation water were increased in every case. In three of the seven trials, increased production of oil was reported. In one case individual wells showed increases of 12–36% and the whole formation showed an increase of 7% over the 6-month trial. Trials in which either inoculum or molasses were omitted gave no increases in oil production. In the successful trials, permeabilities were in excess of 3 d.

A Soviet trial was described by Kusnetsov, Ivanov and Lyalikova (1963). A mixed inoculum, grown anaerobically on molasses, was injected together with 54 m^3 of 4% molasses solution. The well was closed for 6 months. On reopening, oil production increased by about 3 tonnes, from 37 to 40 t/day. Wellhead pressure increased and the proportion of co-produced water decreased. Oil viscosity increased by about 20%. Four months later, oil production had fallen to 36·5 t/day. The brief increase in production cannot have compensated for the loss during the 6-month shutdown. This trial is chiefly remarkable for the high productivity of the well chosen. In another Russian trial the inoculum consisted of a complex mixture of aerobic and anaerobic bacteria (including sulphate-reducing, denitrifying, putrefactive, butyric acid-producing, and cellulose-digesting activities), was grown up in oil and formation water and contained nutrient substances derived from peat and silt but no molasses. In this instance substantial increases in oil production were recorded, together with a reduction in oil viscosity, an increase in gas production, changes in gas composition, and

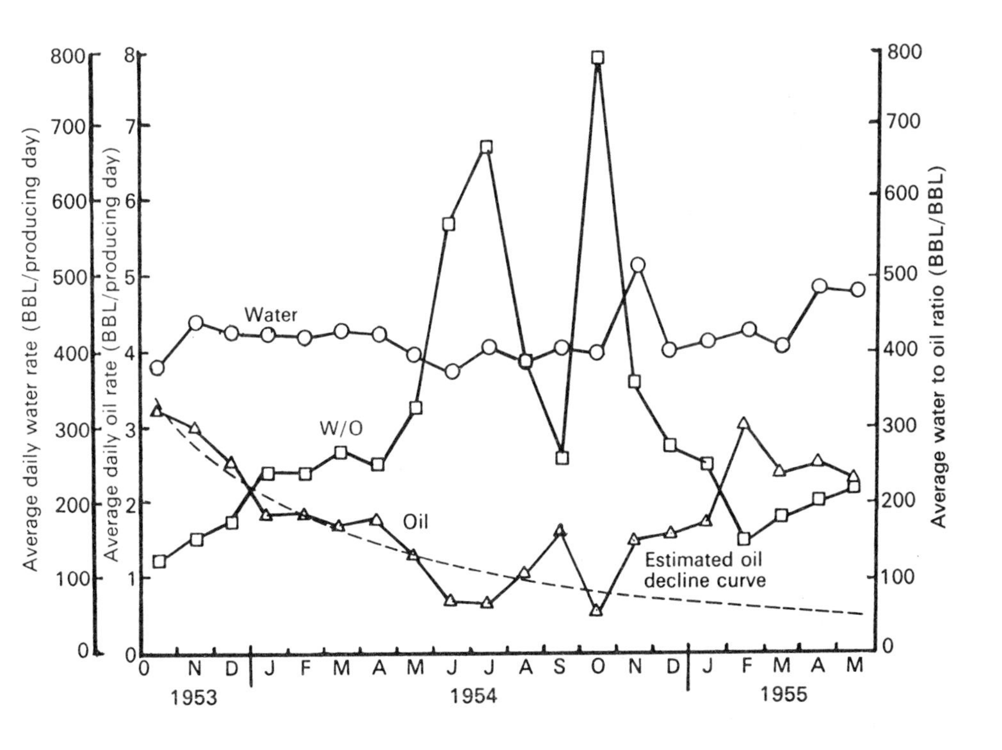

Figure 2. Production of oil and water from well 31 in the Arkansas field trial. The first bacterial inoculation was in July, 1954. Reproduced from Yarbrough and Coty (1983). (BBL = barrels).

an increase in bacterial counts in produced water. Russian workers have shown some interest in stimulating and controlling the natural population of the reservoir. This approach would appear to be useful only if the natural population has properties which can lead to oil mobilization. Particular attention has been paid to methanogens (Ivanov and Belyaev, 1983).

In a series of trials in Hungary (Jaranyi *et al.*, 1963; Dienes and Jaranyi, 1973) the inoculum consisted of mixed cultures of *Pseudomonas*, *Clostridium* and *Desulfovibrio* or of mixed sewage sludge cultures containing *Desulfovibrio desulfuricans*, and the complex injection protocols involved the addition of molasses, potassium nitrate, inorganic phosphate and sucrose. In some cases the injection well was closed for a period and operated on a huff-and-puff basis. In others, the injection well remained open, oil being produced from nearby wells. Sometimes supplementary charges of nutrient were made at a later date. Among the effects reported were reductions in oil viscosity, reductions in water pH, increases in gas production, increases in bacterial counts (in one case $60000\,m^3$ of formation was said to be affected). Increases of oil production varying from 0 to 60% were reported over periods up to 18 months. In two trials, the failure to increase oil production was attributed to low permeability (stated to be 10–70 md and 'low' respectively) but stimulation of oil production occurred in one instance where permeabilities as low as 67–104 md were recorded. One trial is of particular interest: it was performed on a reservoir nearly 2500 m deep with a temperature of 97°C and pressure of 22·8 MPa. Well spacings (injection to production) were 300–1700 metres. Increases in the numbers of *Desulfovibrio desulfuricans* and of the indigenous population were recorded throughout the formation, with a decrease in water pH from 9 to 6, a halving of oil viscosity and a 60% increase in oil production. This remarkable result indicates the potential of MEOR techniques even in deep hot reservoirs with widely spaced wells.

The series of trials conducted in Poland from 1961 (Karaskiewicz, 1974) include relatively detailed documentation. Some twenty wells in sandstone reservoirs between 500 and 1200 m deep were treated. Inocula consisted of complex mixtures derived from soil, crude oil, formation water, industrial wastes and sewage sludge and included strains of *Arthrobacter*, *Clostridium*, *Mycobacterium*, *Peptococcus* and *Pseudomonas*. Cultures were adapted and tested for the ability to mobilize oil by methods similar to those described by Lazar (1983a, b) and discussed below. Typically, $500\,dm^3$ of culture, containing about 2×10^{12} cells, 2 tonnes of molasses, and $50\,m^3$ of production water were injected followed by $150\,dm^3$ of crude oil and the well was sealed for a period of months. In each case the injection well was reopened and oil produced from it by lift pump (i.e. huff-and-puff operation). Data for oil production appear to have been kept carefully, production being logged every 12 h, and detailed records were available for at least 12 months prior to injection, sometimes for much longer.

In most trials, increases in oil production were reported: often a peak occurred after reopening, which quickly compensated for the loss during closure. Enhanced production then continued for periods as long as 9 years at declining rates. Increases of up to 300% occurred in some cases after allowing for lost production, the figures depending very much on the period over which the increase is calculated, and the projected oil-production curves used, which obviously became less reliable the longer the experiment continued. However, the projections appear to be conservative and the stimulations convincing. Some trials gave no increase in oil production. Sometimes oil viscosity decreased and fractional distillations revealed changes in oil composition. Increases in gas pressure were recorded in some cases. Water pH decreased in most, if not all, cases.

Bacterial counts increased in every case reported. In some trials, increased oil production was also observed in neighbouring wells. This was interpreted as a stimulation due to treatment and not to a general improvement due to other causes. Sometimes bacterial injection was combined with hydraulic fracturing to increase permeability and it is not always clear how much of a particular increase in oil production was due to bacterial action and how much to the fracturing. In one instance, a previous fracturing operation was performed without effect, suggesting a synergistic effect of the two treatments. An appreciation of the significance of these results demands a study of the oil-production graphs and the interested reader is strongly recommended to consult either the original monograph (Karasckiewicz, 1974) or the compilation presented by Hitzman (1983) which reproduces many of the original figures with English legends.

Lazar (1983a, b) has conducted trials on seven fields in Romania. Increases in bacterial counts were reported in all cases but only two fields showed overall increases in oil production, up to 200% increases being obtained for periods of 1–5 years. The successful trials appear to have given very convincing demonstrations of microbial stimulation. Good baselines are available for comparison as detailed records were kept for 2–3 years before inoculation. Some changes in oil viscosity, decreases in pH and alterations in composition of produced water were observed. Lazar regards the unsuccessful results as demonstrating the need for the correct choice of formation for treatment.

Detailed accounts have been given by Lazar (1983a) of the procedure for obtaining, adapting, testing and injecting bacterial cultures. His procedure is developed from those used by the Hungarian and Polish groups. Bacterial cultures were obtained from a wide variety of sources, the two which proved best in laboratory trials being formation water, which contained bacteria adapted to the reservoir conditions, and the solid sludge resulting from sugar refining, which contained vigorously fermenting populations. The organisms were cultured at reservoir temperature in formation water, supplemented with molasses and in contact with crude oil. The various 'adapted' populations which resulted were then tested in 'collectors': these are items of equipment of varying design in which the ability of growing bacteria to release oil from saturated sandpacks can be measured. Experience showed that strains isolated in pure culture released a maximum of 3·5% of the oil; mixtures of pure strains released up to 14·5%, whereas undefined but adapted mixed populations released from 18 to 25%. The best cultures contained predominantly strains of the following genera: *Pseudomonas*, *Escherichia*, *Arthrobacter*, *Mycobacterium*, *Micrococcus*, *Peptococcus*, *Bacillus* and *Clostridium*.

Recently Johnson (1979) carried out MEOR operations on over 150 wells in the United States. His approach was highly empirical, taking the view that field trials carried out on stripper wells, (those producing less than 10 barrels per day), are cheaper and more conclusive than laboratory simulations. He used various *Bacillus* and *Clostridum* species. The cultures were mixed with a molasses solution containing unspecified mineral nutrients and immediately injected into the well. About 10–14 days were needed for growth.

Operations were conducted on either a huff-and-puff or a water-drive basis. Johnson (1979) claimed that, given an appropriate choice of oil type and reservoir conditions, 'an average increase of oil production in excess of 350% may be expected' at a cost roughly in the range of 15–30 cents per additional barrel (1979 prices). For comparison, at the time of writing (in 1983) the spot price of Saudi Arabian light crude on the Amsterdam market is US $29·80 per barrel.

Johnson suggested that several mechanisms could be involved in oil release according to circumstances. The production of organic acids which react with reservoir carbonates (and incidentally thus produce more carbon dioxide) was reckoned to clean up formation damage, i.e. partial plugging in the region of the well bore. The copious production of gases, especially carbon dioxide and methane, could increase reservoir pressures and help to force out more oil. A variety of surfactants could help to reduce oil:water interfacial tension and thus aid oil release. No data were presented. It seems unlikely that Johnson was able to make very detailed collections of data either before or after well treatment. What makes his work particularly impressive is the number of trials carried out and the fact that he was apparently able to make an overall commercial success of his operations. This year (1983) Johnson has joined forces with other workers including Updegraff and the group at QMC Industrial Research Ltd in London, under the aegis of Petroleum Sciences Inc., to test a variety of newly isolated strains on twelve wells in Illinois (Anonymous 1983). In this instance detailed monitoring is being performed and the results are awaited with interest. It is reported in the same source that several other groups in the United States, including those at Oklahoma State University (E. Grula), the University of Texas, El Paso (J. Zajic), the University of Calgary, Alberta (T. Jack) and Phillips Petroleum (J. Norell), are also planning field trials.

In the earlier Romanian experiments, sources of inorganic nitrogen and phosphorus were injected together with molasses but Lazar (1983b) remarks that the mineral supplement was omitted in later trials since laboratory experiments showed that molasses and formation water provided adequate mineral nutrition.

About half of the trials reported throughout the world seem to have used mineral nutrients. Updegraff and Wren (1953) reported deficiencies in phosphate or nitrogen in many formation waters. Clark, Munnecke and Jenneman (1981), summarizing a survey in the top ten oil-producing states of the United States, reported general deficiencies in sulphur, nitrogen and phosphorus. Many successful trials have been conducted without mineral supplements, and Grula *et al.* (1983) reported that limestone or sandstone outcrop rock would support good growth of several clostridia supplied with molasses and ammonia. Analysis of formation water and molasses is clearly essential before a trial. The ability of nutrients to move, in solution, through rocks has been examined by Jang *et al.* (1983) and by Jenneman *et al.* (1983).

GENERAL CONCLUSIONS FROM FIELD TRIALS

The data which are readily available are imperfect: essential facts on methods or conditions are frequently omitted; results are often given in vague terms; sometimes, errors appear to have crept in during translation. The most important result, namely the extent to which oil production has increased (if at all), is not easy to decide, even if a full set of data is available, because an adequate control in the normal sense is difficult or impossible. Although controls are frequently mentioned in connection with field trials, the term usually refers simply to a nearby well. The fact that this is not a control in the normal sense is demonstrated when an increase in 'control production' is accepted as further evidence of the effectiveness of the bacterial inoculation. One of the best controls explicitly mentioned is in the work of Coty, Updegraff and Yarbrough (Yarbrough and Coty, 1983). Here control wells were chosen aligned fairly closely with the test wells in the overall waterflood but positioned in such a way that the bacterial inoculation would not be likely to affect them during the period of the experiment. The absence of a response in these wells was good evidence against a significant change in that part of the field generally. Dostalek's group omitted either inoculum or molasses in some of their trials and reported no oil stimulation. The significance of these controls clearly depends on how comparable they were with the wells receiving full treatment, since negative results are not uncommon, even with normal injections.

Generally, stimulations in oil production have to be detected by extrapolating the oil-production curve forward into the experimental period. In many cases the results of this exercise show a fairly convincing stimulation of oil production, but production curves are not straight lines (nor even very smooth curves in many cases) and the hazards of interpretation must be appreciated. Where stimulations appear to continue for several years but at a slowly declining rate, calculation of the enhanced recovery depends critically on the slope of the projection curve and exact figures become meaningless. One particularly convincing demonstration of increased production, not yet described in a published report, arose where a series of injections was made. Oil production rose for a period of months after each injection and then decreased until the next injection was made.

Changes in flow patterns attributable to events in other parts of the formation, or even disturbances such as the temporary closing of a well, can affect oil production and reports of field trials should make it clear that this has not happened. Where disturbances are inevitable, the possible effects should be considered carefully. For instance, the author is not aware of any instance reported in the literature where a well has been capped for a period with no other treatment, and the results monitored as a prelude to a similar treatment coupled with bacterial inoculation, although personal conversations make it clear that this test has, in fact, been made in at least one instance. Hydraulic fracturing of formation rock has sometimes been practised in conjunction with bacterial inoculation, to enhance overall permeability. In one of the Polish field trials it was reported that, previously, fracturing alone had been completely ineffective: more often, the results of the two treatments are indistinguishable.

Despite all these reservations, I am now convinced that genuine stimulations of oil production have occurred in a number of trials, if not in all cases where success has been claimed. Many trials were unsuccessful in that no stimulation of oil production resulted, but no disastrous effects of microbial inoculation were reported. In one of the Polish trials, slimy oil resulted, but no serious plugging or decline in oil production seems to have been experienced. The cynic might suspect that any such event would not quickly find its way into print.

In most cases it is difficult to draw reliable conclusions about the mechanism responsible for enhanced production, because cultures are usually chosen on the basis of empirical criteria. Most authors describe stimulation of bacterial counts in the reservoirs, encompassing both the strains inoculated and the indigenous flora. Two conclusions follow: (1) it is not always easy to be sure that a given effect results solely, or even partly, from the bacteria inoculated; (2) it may not be easy to produce a bacterial monoculture in the reservoir, and results therefore may be difficult to predict from laboratory-scale trials.

SPECIAL PROBLEMS OF MEOR OPERATIONS IN DEEP FIELDS

As the shallower oil fields on the earth's surface become insufficient to satisfy demand, increasing attention must be paid to deeper deposits. Temperature and pressure increase with depth and some of the fields now being exploited have ambient temperatures and pressures well beyond the limit tolerated by most forms of life. Thus, in the North Sea fields where formation depths exceed 2000 m, pressures range from 20 to 50 MPa, temperatures from 80° to 110°C and formation water salinities exceed 30 g/dm^3 dissolved solids (Fox, 1983). Additional difficulties in the North Sea are attributable to the offshore location, the depth of the sea (up to 200 m) and the bad weather which makes supplying the platforms, and even access, difficult for much of the year. The North Sea fields represent the bulk of the known oil deposits in the United Kingdom and the ambient conditions are by no means unique. As a result of the cost of drilling, well spacings are in excess of 600 m. It is thus of some interest to consider whether MEOR operations might be feasible in such fields.

The question of whether life was possible at all under these conditions was, until recently, open to doubt. Most bacteria will tolerate pressures up to 20 MPa but above this value an increasing proportion become inhibited by pressure, even when other conditions are optimal (for reviews *see* ZoBell and Kim, 1972; Marquis, 1976; Marquis and Matsumura, 1978; Marquis, 1983). Even if growth is possible, undesirable changes often occur, such as loss of motility (Meganathan and Marquis, 1973) and a tendency to filamentous growth. High temperature also prevents growth of most bacteria and, until recently, the highest temperatures reported for growth in culture were about 85°C, although ZoBell (1958) reported growth of an organism at 104°C and 100 MPa but was unable to subculture it. There were signs that growth at up to 100°C occurred in certain natural environments (Brock, 1967; Bott and Brock, 1969; Brock and Darland, 1970; Brock *et al.*, 1971, 1972; Jackson, Ramaley and Meinschein, 1973; Tansey and Brock, 1978). Even if bacteria could grow at reservoir temperatures, the effect of combinations of different adverse conditions, especially temperature and pressure (Marquis, 1983), and the effects of heavy metals (Bubela, 1983b) are unknown.

Recently, the known limits for growth have been extended substantially. Thus Stetter *et al.* (1981) isolated a new species of methanogen from an Icelandic hot spring with an optimum growth temperature of 83°C and a maximum of 97°C. Shortly afterwards Stetter (1982) reported the isolation, from submarine volcanic areas, of an organism which required hydrogen, carbon dioxide and elemental sulphur for growth, generating hydrogen sulphide. This organism had an optimum growth temperature of 105°C. This year Baross and Deming (1983) have reported the growth under laboratory conditions of still more remarkable organisms which were cultured from submarine volcanic vents at a temperature of 306°C. These organisms grew at temperatures from 100°C at atmospheric pressure up to 250°C at a pressure of 26·5 MPa. Some care is necessary when attempting to demonstrate bacterial growth at very high temperatures as it is easy to obtain misleading results if parts of the apparatus are below the design temperature. In the three cases mentioned there is no reason to suspect the validity of the authors' conclusions.

Thus temperature and pressure alone do not appear to rule out the growth of bacteria even in the hottest reservoirs, although their effects in combination with salinity, heavy metals and other oil components are, as yet, unknown. Little is known about the newly discovered extreme thermophiles and no information is available to assess their suitabilities for MEOR operations. The difficulty of access to wells such as those of the North Sea, together with their wide spacing, give rise to further difficulties. A substrate-fed MEOR operation would require the provision of substrate at very high rates (because of the high injection rates) for very long periods. Estimates of the time taken for fluid to pass from one well to another vary from 5 to 12 years, posing enormous problems of supply, investment cost and system stability.

There are strong arguments that the only way in which a large-scale MEOR operation could be conducted would be for crude oil to provide the substrate. Molecular oxygen is not available in the reservoir, so some other electron acceptor would be necessary. Moses *et al.* (1983) described bacterial cultures which, under strictly anaerobic conditions, grew on crude oil as sole carbon source in the presence of inorganic nutrients. Growth was measured by increases in absorbance and protein concentrations: both methanogenic and nitrate-reducing cultures were described, the latter producing nitrite, molecular nitrogen and carbon dioxide. Injection of oxygen into the gas phase of nitrate-reducing cultures completely inhibited gas production, whereas injection of nitrogen had no detectable effect. Although anaerobic growth at the expense of crude oil has been demonstrated, the rates of growth were very low, the oil components being utilized were unknown, and no thermophilic activity has so far been detected. Other workers have also reported signs of anaerobic growth on hydrocarbons (*see* Moses and Springham, 1982). When Bubela (1983a) incubated a bacterial culture with a crude oil sample for 3 months at 60°C under strictly anaerobic conditions, an increase was found in the asphaltene fraction. No change was detected in oil samples incubated at 30°C or at 60°C without bacteria, suggesting that the change was due to anaerobic metabolism of hydrocarbons.

Much remains to be done for a workable system to be developed and it may well be that gene manipulation will eventually have a role in transferring essential attributes into an extreme thermophile. Despite the recent spectacular advances in gene manipulation, the difficulties of doing this must not be underestimated since entire pathways rather than single enzymes may need to be transferred, and the problems of getting mesophilic enzymes to function at high temperatures (if that should be necessary) have only recently been addressed. However, one of the Hungarian field trials claimed success under reservoir conditions remarkably similar to those in the North Sea fields: perhaps the problem is not as difficult as it seems!

Summary and conclusions

The plethora of possible mechanisms for oil release has sometimes diverted attention from the vital necessity to determine which one is operating in a particular case. The temptation is strong to inject a mixture of organisms, offering between them the possibilities of several mechanisms, hoping that one or more will operate successfully. The author feels that this empirical 'eye of newt and toe of frog' approach must now give way to a determination to define the specific mechanism which is to function, and its critical parameters.

Gene manipulation may have a part to play both in increasing product yields and in producing desirable combinations of properties in a single strain. The problems will be much more difficult than those currently being tackled to produce pharmaceutical products, because the sort of property which may have to be transferred, e.g. polymer production, surfactant production, is likely to involve the products of a whole series of genes. Rational manipulation will require the use of single strains or, at most, of simple, well-defined mixtures, despite the present advantages which more complex cultures seem to enjoy (Lazar, 1983a).

For a number of reasons, an MEOR system which has been optimized in the laboratory may perform in a very different way in the reservoir. It is important to be aware of the problems, although in some cases effective solutions are not immediately obvious. Indigenous organisms may compete for substrate or, by supplying metabolic products or removing the products of the inoculum, may alter the expected metabolic balance. Thus the fermentation balance of *Ruminococcus albus* can be altered if the partial pressure of hydrogen is lowered by growing it together with *Vibrio succinogenes*, which removes the hydrogen by oxidation (Thauer, Jungermann and Decker, 1977; Tewes and Thauer, 1980). Interactions of this general sort could occur in the reservoir and their effects would not be easy to predict. Evidence of metabolic interactions between inoculum and indigenous organisms was presented by Yarbrough and Coty (1983).

The bacterial population which develops in the reservoir may not be the one injected. It is certain that any attempt to eliminate indigenous organisms from the reservoir would be futile since it has proved extremely difficult to sterilize rock cores in the laboratory, but numbers might be reduced locally by a biocide injected prior to inoculation. Reservoir populations are likely to have been selected for their abilities to survive for long periods rather than for rapid growth, and the effects of competition could be minimized by using a large inoculum of organisms with a high maximum growth rate.

Mutation is unlikely to be a problem except where distances between wells are great and very long periods are involved. The growth of ineffective variants would be a particular danger if strong selection pressures had been applied to obtain high-yielding mutants (e.g. surfactant producers) and it will be necessary to check strains carefully for genetic stability. One favourable aspect arises from the highly directional flows within the reservoir. Since there is little back-mixing and cross-mixing, a mutant, unless it arises very early, could affect only a comparatively small sector of the reservoir.

The provision of nutrients by injection from the surface raises another type of problem. Before substrate can reach bacteria growing in distant parts of the formation it must pass through established populations in the vicinity of the injection well. There is a danger that the latter will grow sufficiently fast to utilize all the substrate, thus localizing the region of bacterial activity. I have constructed a simple computer model to predict how substrate concentration, bacterial concentration and growth rate would vary with distance from the injection well, using a deep-bed filtration equation to describe bacterial movement and the Monod equation to describe growth rate. The model is still being checked and refined, but preliminary results indicate that the outcome is highly dependent on the choice of parameters such as the concentration of substrate injected, flow rate and bacterial growth rate. According to circumstances, growth can occur deep in the formation or it can be confined to a region close to the injection well. It is, of course, only the concentration of the limiting nutrient component that is critical. Data from field trials should provide useful information but are not easy to interpret. Thus Yarbrough and Coty (1983) measured appreciable concentrations of carbohydrate at the production well, but this does not prove that bacterial growth was possible in this region because the limiting nutrient was not known. Similarly, where increased numbers of bacteria are found in parts of the reservoir remote from the injection well it does not follow that substrate is also present to permit their growth.

The weight of published information suggests that real increases in oil production can be achieved by MEOR under appropriate circumstances. With the wide dissemination of information and the considerable research efforts which have been exerted in recent years, the chances of success in the future should be greatly improved.

The experience of Johnson (1979) suggests that, with the easiest targets, a high percentage of success is already attainable, but full exploitation of the potentialities requires considerable development work. A high degree of collaboration between disciplines is required for success: the properties of the reservoir are defined by geologists, geochemists and geophysicists; extraction of the oil requires the skills of engineers, chemists and mathematical modellers; microbiologists and geneticists must provide the organisms and the means of developing and controlling their activities. All parties involved must be fully aware of the requirements and limitations imposed by the other disciplines. Fortunately, the oil industry has a long experience of interdisciplinary collaboration.

The enormous potential benefits of MEOR were mentioned in the opening paragraphs of this review. MEOR is particularly attractive in comparison with other EOR techniques in that, despite the difficulties and uncertainties, the operating costs are low. Essentially all the cells and active substances are generated in the reservoir from relatively cheap raw materials and no expensive manufacturing facilities are required on the surface. The substrates are cheap sources of carbohydrate and, in some cases, mineral nutrients: if workable systems can be developed depending on anaerobic hydrocarbon degradation, no carbohydrate would be needed, although there would be a requirement for an electron-acceptor such as nitrate.

Almost all MEOR operations so far have been confined to stripper wells, so success in individual cases has resulted in only small quantities of extra oil. However, there are of the order of 400 000 stripper wells in the US alone. In 1981 these produced less than three barrels a day on average but accounted for about 13% of the US production, so the widespread and successful application of MEOR to these alone would be of some significance. The application of MEOR to more productive reservoirs has to face two sorts of hurdle: the first is that there may be technical difficulties not yet appreciated; the second is a matter of credibility and understanding. Many engineers still regard MEOR as unreliable and potentially harmful. Part of this mistrust will disappear if MEOR is widely used and regularly successful, but the technique will not take its place along other EOR techniques until the mechanisms are fully understood and the processes can, to some degree, be modelled successfully by computer.

The application to really difficult targets, as exemplified by the North Sea fields, will face the problems of extreme reservoir conditions, widely spaced wells and difficulty of access together with very high costs of maintaining equipment in working order. The high temperatures and pressures, although still serious problems, do not seem as formidable as they did a few years ago, but the economics and logistics of feeding organic substrates from the surface in such situations still look intractable. It remains to be seen if rapid anaerobic growth on crude oil components is possible. A considerable faith in the biological stability of MEOR systems working over long distances and extended times will be essential if depleted fields are to be kept operating for the long periods before results can be expected. It is indeed to be hoped that techniques, understanding and confidence develop to the necessary point before too many fields are finally abandoned.

Acknowledgements

Thanks are due to the many scientists and engineers who have provided facts, ideas and encouragement, especially Vivian Moses and John Robinson at Queen Mary College, my many colleagues at QMC Industrial Research Ltd. and to Donald Dunlop, Ion Lazar, Hank Yarbrough and Claude ZoBell. Grants from the Marine Technology Directorate of the Science and Engineering Research Council are acknowledged. The opinions expressed are the sole responsibility of the author. Figure 2 is reproduced from *Proc. Int. Conf. on Microbial Enhancement of Oil Recovery* with permission of the US Department of Energy, of Mobil Research and Development Corporation and of the authors.

References

ALLRED, R.C. (1976). Microbial problems in secondary oil recovery. In *The Role of Microorganisms in the Recovery of Oil*, pp. 133–135. National Science Foundation, Washington.

ANONYMOUS (1972). Bacteria have destroyed ten per cent of world's crude. *World Oil*, Feb. 1, pp. 28–29.

ANONYMOUS (1983). *McGraw-Hill's Biotechnology Newswatch* **3,** 1–2.

BAROSS, J.A. AND DEMING, J.W. (1983). Growth of 'black smoker' bacteria at temperatures of at least 250°C. *Nature* **303,** 423–426.

BECK, J.V. (1947). Penn Grade Process on use of bacteria for releasing oil from sands. *Producers Monthly* **11,** 13–19.

BEERSTECHER, E. (1954). *Petroleum Microbiology*. Elsevier, New York.

BOTT, T.L. AND BROCK, T.D. (1969). Bacterial growth rates above 90°C in Yellowstone hot springs. *Science* **164,** 1411–1412.

BROCK, T.D. (1967). Life at high temperatures. *Science* **158,** 1012–1019.

BROCK, T.D. AND DARLAND, G.K. (1970). Limits of microbial existence: temperature and pH. *Science* **169,** 1316–1318.

BROCK, T.D., BROCK, K.M., BELLY, R.T. AND WEISS, R.L. (1972). Sulfolobus: a new genus of sulfur-oxidising bacteria living at low pH. *Archiv für Mikrobiologie* **84,** 54–68.

BROCK, T.D., BROCK, M.L., BOTT, T.L. AND EDWARDS, M.R. (1971). Microbial life at 90°C: the sulfur bacteria of Boulder Spring. *Journal of Bacteriology* **107**, 303–314.

BROWN, J., ED. (1979). *European Symposium on Enhanced Oil Recovery*, 1978. Heriot-Watt University, Edinburgh.

BUBELA, B. (1983a). Physical simulation of microbiologically enhanced oil recovery. In *Microbial Enhanced Oil Recovery* (J.E. Zajic, D.C. Cooper, T.R. Jack and N. Kosaric, Eds), pp. 1–7. Penn Well Books, Tulsa.

BUBELA, B. (1983b). Combined effects of temperature and other environmental stresses on microbiologically enhanced recovery. In *Proceedings of the International Conference On Microbial Enhancement of Oil Recovery 1982*, (E.C. Donaldson and J.B. Clark, Eds), pp. 118–119. US Department of Energy, Bartlesville.

BURDINE, N.T., GOURNAY, L.S. AND REICHERTZ, P.P. (1950). Pore size distribution of petroleum reservoir rocks. *Petroleum Transactions, American Institute of Mining Engineers* **189,** 195–204.

CLARK, J.B., MUNNECKE, D.M. AND JENNEMAN, G.E. (1981). *In situ* microbial enhancement of oil recovery. *Developments in Industrial Microbiology* **22,** 695–701.

CRAWFORD, P.B. (1961). Possible bacterial correction of stratification problems. *Producers Monthly* (Dec. 1961), 10–11.

CRAWFORD, P.B. (1962). Continual changes observed in bacterial stratification rectification. *Producers Monthly* (Feb. 1962), 12.

CRAWFORD, P.B. (1983). Possible reservoir damage from microbial enhanced oil recovery. In *Proceedings of the International Conference on Microbial Enhanced Oil Recovery 1982*, (E.C. Donaldson and J.B. Clark, Eds), pp. 76–79. US Department of Energy, Bartlesville.

DAFTER, R. (1980). *Scraping the Barrel.* Financial Times Ltd, London.

DAVIS, J.B. (1967). *Petroleum Microbiology.* Elsevier, New York.

DAVIS, J.B. AND UPDEGRAFF, D.M. (1954). Microbiology in the petroleum industry. *Bacteriological Reviews* **18,** 215–238.

DIENES, M. AND JARANY, I. (1973). (Increase of oil recovery introducing anaerobic bacteria into the formation, Denjen Field, Hungary. In Hungarian). *Köolaj es Földgaz* **6,** 205–208.

DONALDSON, E.C. (1982). BETC research. *DOE/BETC Progress Review* **31,** 118–119.

DOSCHER, T.M. AND KUUSKRAA, V.A. (1979). Carbon dioxide for enhanced recovery of crude oil. In *European Symposium on Enhanced Oil Recovery*, 1978 (J. Brown, Ed.), pp. 225–252. Heriot-Watt University, Edinburgh.

DOSTALEK, M. AND SPURNY, M. (1957). Release of oil by microorganisms. I. Pilot experiment in an oil deposit. *Ceskoslovenska Mikrobiologie* **2,** 300–306.

FINNERTY, W.R. AND SINGER, M.E. (1983). Microbial enhancement of oil recovery. *Bio/Technology* **1,** 47–54.

FOX, W.N. (1983). Winfrith-enhanced oil recovery. *Atom* **319,** 95–100.

GEFFEN, T.M. (1976). Present technology for oil recovery. In *Conference on the Role of Microorganisms in the Recovery of Oil, 1975*, pp. 23–37. National Science Foundation, Washington.

GROSZEK, A.J., ED. (1977) *Symposium on Enhanced Oil Recovery by Displacement with Saline Solutions.* B.P. Educational Services, London.

GRULA, E.A., RUSSELL, H.H., BRYANT, D., KENAGA, M. AND HART M. (1983). Isolation and screening of clostridia for possible use in microbially enhanced oil recovery. In *Proceedings of International Conference on Microbial Enhancement of Oil Recovery, 1982* (E.C. Donaldson and J.B. Clark, Eds), pp. 43–47. US Department of Energy, Bartlesville.

HART, R.T., FEKETE, T. AND FLOCK, D.L. (1960). The plugging effect of bacteria in sandstone systems. *Canadian Mining and Metallurgical Bulletin* **53,** 495–501.

HESSELINK, F.T. AND TEEUW, D. (1981). Scope for polymers in enhanced oil recovery. *Shell Polymers* **5,** 82–86.

HITZMAN, D.O. (1962). *Microbiological Secondary Recovery.* US Patent 3,032,472.

HITZMAN, D.O. (1965). *Use of Bacteria in the Recovery of Petroleum from Underground Deposits.* US Patent 3,185,216.

HITZMAN, D.O. (1983). Petroleum microbiology and the history of its role in enhanced oil recovery. In *Proceedings of the International Conference on Microbial Enhancement of Oil Recovery, 1982* (E.C. Donaldson and J.B. Clark, Eds), pp. 162–218. US Department of Energy, Bartlesville.

IVANOFF, M.V. AND BELYAEV, S.S. (1983). Microbial activity in waterflooded oil fields and its possible regulation. In *Proceedings of the International Conference on Microbial Enhancement of Oil Recovery, 1982* (E.C. Donaldson and J.B. Clark, Eds), pp. 48–57. US Department of Energy, Bartlesville.

JACK, T.R., THOMPSON, B.G. AND DIBLASIO, E. (1983). The potential for use of microbes in the production of heavy oil. In *Proceedings of the International Conference on Microbial Enhancement of Oil Recovery, 1982* (E.C. Donaldson and J.B. Clark, Eds), pp. 88–93. US Department of Energy, Bartlesville.

JACKSON, T.J., RAMALEY, R.F. AND MEINSCHEIN, W.G. (1973). Thermomicrobium: a new genus of extremely thermophilic bacteria. *International Journal of Systematic Bacteriology* **23,** 28–36.

JANG, L.K. AND YEN, T.F. (1983). An experimental investigation on the role of bacterial growth and bacterial transport in MEOR processes. Presented at the *Symposium on Biological Processes Related to Petroleum Recovery, Seattle 1983*, pp. 789–799. American Chemical Society, Washington.

JANG, L.K., SHARMA, M.M., FINDLEY, J.E., CHANG, P.W. AND YEN, T.F. (1983). An investigation of the transport of bacteria through porous media. In *Proceedings of the International Conference in Microbial Enhancement of Oil Recovery, 1982* (E.C. Donaldson and J.B. Clark, Eds), pp. 60–70. US Department of Energy, Bartlesville.

JARANYI, I., KISS, L., SZALANCZY, G. AND SZOLNOKI, J. (1963). Veränderung einiger Charakteristiken von Erdölsonden durch Einwirkung von mikrobiologischer Behandlung. *Wissenschaftliche Tagung für Erdölbergbau, 1963*, pp. 633–650.

JENNEMAN, G.E., KNAPP, R.M., MENZIE, D.E., MCINERNEY, M.J., REVUS, D.E., CLARK, J.B. AND MUNNECKE, D.M. (1983). Transport phenomena and plugging in Berea sandstone using microorganisms. In *Proceedings of the International Conference on Microbial Enhancement of Oil Recovery, 1982* (E.C. Donaldson and J.B. Clark, Eds), pp. 71–75. US Department of Energy, Bartlesville.

JOHNSON, A.C. (1979). Microbial oil release technique for enhanced recovery. In *Conference on Microbial Processes Useful in Enhanced Oil Recovery, 1979*, pp. 30–34. National Technical Information Service, Springfield.

JONES, L.W. (1967). *Aerobic Bacteria in Oil Recovery.* US Patent, 3,332,487.

KALISH, P.J., STEWART, J.E., ROGERS, W.F. AND BENNET, E.O. (1964). The effect of bacteria on sandstone permeability. *Journal of Petroleum Technology* **16,** 805–814.

KARASKIEWICZ, J. (1974). Zastosowanie metod microbiologicznych w intensificacji karpakich ztoz ropy naftowego, pp. 1–67. Institutu Naftowego, Prace.

KUSNETSOV, S.I., IVANOV, M.V. AND LYALIKOVA, N.N. (1963). *Introduction to Geological Microbiology*, (English translation, ed. C.H. Oppenheimer). McGraw-Hill, New York and London.

LA RIVIÈRE, J.W.M. (1955). The production of surface active compounds by microorganisms and its possible significance in oil recovery. I. Some general observations on the change in surface tension in microbial cultures. II. On the release of oil from oil–sand mixtures with the acid of sulphate-reducing bacteria. *Antonie van Leeuwenhoek Journal of Microbiology and Serology* **21,** 1–8, 9–15.

LAZAR, I. (1983a). Some characteristics of the bacterial inoculum used for oil release from reservoirs. In *Microbial Enhanced Oil Recovery* (J.E. Zajic, D.C. Cooper, T.R. Jack and N. Kosaric, Eds), pp. 73–82. Penn Well Books, Tulsa.

LAZAR, I. (1983b). Microbial enhancement of oil recovery in Romania. In *Proceedings of the International Conference on Microbial Enhancement of Oil Recovery, 1982* (E.C. Donaldson and J.B. Clark, Eds), pp. 140–148. US Department of Energy, Bartlesville.

MACKENZIE, K. (1952). The metabolism of *Vibrio desulfuricans* in anaerobic petroliferous formations. *Biochemical Journal* **51,** 24–25.

MARQUIS, R.E. (1976). High pressure microbial physiology. *Advances in Microbial Physiology* **14,** 159–241.

MARQUIS, R.E. (1983). Barobiology of deep oil formations. In *Proceedings of the International Conference on Microbial Enhancement of Oil Recovery, 1982* (E.C. Donaldson and J.B. Clark, Eds), pp. 124–128. US Department of Energy, Bartlesville.

MARQUIS, R.E. AND MATSUMURA, P. (1978). Microbial life under pressure. In *Microbial Life in Extreme Environments* (D.J. Kushner, Ed.), pp. 105–158. Academic Press, New York.

MEGANATHAN, R. AND MARQUIS, R.E. (1973). Loss of bacterial motility under pressure. *Nature* **246,** 525–527.

MELROSE, J.C. AND BRANDNER, C.F. (1974). Role of capillary forces in determining microscopic displacement efficiency for oil recovery by waterflooding. *Journal of Canadian Petroleum Technology* **13,** 54–66.

MEYER, R.F., ED. (1977). *Future Supply of Nature-Made Petroleum and Gas.* Pergamon Press, Oxford.

MILLER, J.D.A. (1971). *Microbial Aspects of Metallurgy.* Medical and Technical Publishing Co., Aylesbury.

MOSES, V. AND SPRINGHAM, D.G. (1982). *Bacteria and the Enhancement of Oil Recovery.* Applied Science Publishers, London and New Jersey.

MOSES, V., ROBINSON, J.P., SPRINGHAM, D.G., BROWN, M.J., FOSTER, M., HUME, J., MAY, C.W., MCROBERTS, T.S. AND WESTON, A. (1983). Microbial enhancement of oil recovery in North Sea reservoirs: a requirement for anaerobic growth on crude oil. In *Proceedings of the International Conference on Microbial Enhancement of Oil Recovery, 1982* (E.C. Donaldson and J.B. Clark, Eds), pp 154–157. US Department of Energy, Bartlesville.

MYERS, G.E. AND MCREADY, R.G.L. (1965). Bacteria can penetrate rock. *Canadian Journal of Microbiology* **12,** 477–484.

MYERS, G.E. AND SAMIRODEN, W.D. (1967). Bacterial penetration in petroliferous rocks. *Producers Monthly* **31,** 22–25.

PLUMMER, F.B., MERKT, E.E., POWER, H.H., SAVIN, H.J. AND TAPP, P. (1944). Effect of certain microorganisms on the injection of water into sand. *Petroleum Technology Publications, AIME, No. 1678*, pp. 1–13.

RALEIGH, J.T. AND FLOCK, D.L. (1965). A study of formation plugging with bacteria. *Journal of Petroleum Technology* **17,** 201–206.

RITTER, H.L. AND DRAKE, L.C. (1945). Pore size distribution in porous material. *Industrial and Engineering Chemistry, Analytical Edition* **17,** 782–785.

SHAH, D.O., ED. (1981). *Surface Phenomena in Enhanced Oil Recovery.* Plenum Press, New York.

SHAH, R.P. AND WITTMEYER, E.E. (1978). A study of CO_2 recovery and tertiary oil production enhancement in the Los Angeles basin. *Fourth Annual DOE Symposium: Enhanced Oil and Gas Recovery and Improved Drilling Methods*, pp. C1/1–8. Petroleum Publishing Company, Tulsa.

SHARPLEY, J.M. (1961). Bacteria in floodwater: what they are—what they mean. *Petroleum Engineer* **33,** B55–B67.

SINGER, M.E., FINNERTY, W.R., BOLDEN, P. AND KING, A.D. (1983). Microbial processes in the recovery of heavy petroleum. *Proceedings of the International Conference on Microbial Enhancement of Oil Recovery, 1982* (E.C. Donaldson and J.B. Clark, Eds), pp. 94–101. US Department of Energy, Bartlesville.

SPRINGHAM, D.G., MCKAY, A., MOSES, V., ROBINSON, J.P., BROWN, M.J., FOSTER, M., HUME, J., MAY, C.W., MCROBERTS, T.S. AND WESTON, A. (1983). Some constraints on the use of bacteria in enhanced oil recovery. In *Proceedings of the International Conference on Microbial Enhancement of Oil Recovery, 1982* (E.C. Donaldson and J.B. Clark, Eds), pp. 158–161. US Department of Energy, Bartlesville.

STETTER, K.O. (1982). Ultrathin mycelia-forming organisms from submarine volcanic areas having an optimum growth temperature of 105°C. *Nature* **300,** 258–260.

STETTER, K.O., THOMM, M., WINTER, J., WILDGRUBER, G., HUBER, H., ZILLIG, W., JANÉCOVIC, D., KÖNIG, H., PALM, P. AND WUNDERL, S. (1981). *Methanothermus fervidus*, sp. nov., a novel, extremely thermophilic methanogen isolated from an Icelandic hot spring. *Zentralblatt für Băkteriologie, Mikrobiologie und Hygiene* **1,** 166–178.

STEWART, G. (1977). *Enhanced Oil Recovery.* Heriot-Watt University, Edinburgh.

STOCKIL, P.A., ED. (1977). *Our Industry Petroleum*, 5th edn. British Petroleum Company Ltd, London.

SUTHERLAND, I.W. (1983). Extracellular polysaccharides. In *Biotechnology* (H. Dellweg, Ed.) volume 3 (H.J. Rehm and G. Reed, Eds), pp. 531–574. Verlag Chemie, Weinheim.

TABER, J.J. (1968). Dynamic and static forces required to remove a discontinuous oil phase from porous media containing both oil and water. *Society of Petroleum Engineers Journal* (March, 1969), 3–12.

TANSEY, M.R. AND BROCK, T.D. (1978). Microbial life at high temperatures. In *Microbial Life in Extreme Environments* (D.J. Kushner, Ed.), pp. 159–216. Academic Press, New York.

TEWES, F.J. AND THAUER, R.K. (1980). Regulation of ATP synthesis in glucose-fermenting bacteria involved in interspecies hydrogen transfer. In *Anaerobes and Anaerobic Infections* (G. Gottschalk, N. Pfennig and H. Werner, Eds), pp. 907–1004. Gustav Fischer Verlag, New York and Stuttgart.

THAUER, R.K., JUNGERMANN, K. AND DECKER, K. (1977). Energy conservation in chemotrophic anaerobic bacteria. *Bacteriological Reviews* **41,** 100–180.

UPDEGRAFF, D.M. (1957). *Recovery of Petroleum Oil.* US Patent 2,807,570.

UPDEGRAFF, D.M. AND WREN, G.B. (1953). *Secondary Recovery of Petroleum Oil by* Desulfovibrio. US Patent 2,660,550.

UPDEGRAFF, D.M. AND WREN, G.B. (1954). The release of oil from petroleum-bearing materials by sulfate-reducing bacteria. *Applied Microbiology* **2,** 309–322.

VAN POOLLEN, H. (1980). *Fundamentals of Enhanced Oil Recovery.* Penn Well Books, Tulsa.

VON HEININGEN, J., JAN DE HAAN, H. AND JENSEN, J.D. (1958). *Process for the Recovery of Petroleum from Rocks.* Netherlands Patent, 89,580.

WESTLAKE, W.S. (1983). Microbial activities and changes in the chemical and physical properties of oil. In *Proceedings of the International Conference on Microbial Enhancement of Oil Recovery, 1982* (E.C. Donaldson and J.B. Clark, Eds), pp. 102–111. US Department of Energy, Bartlesville.

YARBROUGH, H.F. AND COTY, V.F. (1983). Microbially enhanced oil recovery from the Upper Cretaceous Nacatoch formation, Union County, Arkansas. In *Proceedings of the International Conference on Microbial Enhancement of Oil Recovery, 1982* (E.C. Donaldson and J.B. Clark, Eds), pp. 149–153. US Department of Energy, Bartlesville.

ZAJIC, J.E. AND PANCHAL, C.J. (1976). Bioemulsifiers. *CRC Critical Reviews in Microbiology* **5,** 39–66.

ZHANG, Z. AND QUIN, T. (1983). A survey of research on the application of microbial techniques to the petroleum production in China. In *Proceedings of the International Symposium on Microbial Enhancement of Oil Recovery, 1982* (E.C. Donaldson and J.B. Clark, Eds), pp. 135–139. US Department of Energy, Bartlesville.

ZOBELL, C.E. (1946). *Bacteriological Process for Treatment of Fluid-Bearing Earth Formations.* US Patent 2,413,278.

ZOBELL, C.E. (1947a). Bacterial release of oil from oil-bearing materials. I and II. *World Oil* **126,** 36–44 and **127,** 35–40.

ZOBELL, C.E. (1947b) Bacterial release of oil from sedimentary materials. *Oil and Gas Journal* **46,** 62–65.

ZOBELL, C.E. (1953). *Recovery of Hydrocarbons.* US Patent 2,641,566.

ZOBELL, C.E. (1958). Ecology of sulphate-reducing bacteria. *Producers Monthly* **22,** 12–19.

ZOBELL, C.E. AND KIM, J. (1972). Effects of deep-sea pressures on microbial enzyme systems. *Society for Experimental Biology Symposium XXVI, The Effects of Pressure on Organisms* (M.A. Sleigh and A.G. MacDonald, Eds), pp. 125–146. Cambridge University Press.

9

Biotechnological Approach to a new Foot-and-Mouth Disease Virus Vaccine

ANDREW K. CHEUNG* AND HANS KÜPPER

Biogen SA., 46, Route des Acacias, CAROUGE/Geneva, Switzerland

Introduction

Advances in biotechnology in the past decade have opened up new ways of manufacturing specific proteins with micro-organisms. This novel approach is the result of several major observations and discoveries.

Cohen, Chang and Hsu (1972) observed that bacterial cells can be transformed or transfected with plasmid or viral deoxyribose nucleic acid (DNA), allowing the isolation of individual DNA sequences. Smith and Wilcox (1970) and Kelly and Smith (1970) discovered Class II restriction enzymes that recognize specific nucleotide sequences and cleave phosphodiester bonds within the recognition sites, permitting the purification of defined segments of a genome. With the use of ligase, chimeric DNA molecules can be created *in vitro* from DNAs of diverse origins and replicated in a bacterial host after transformation. Maxam and Gilbert (1977) and Sanger, Nicklen and Coulson (1977) developed rapid methods for determining the nucleotide sequence of DNA molecules, thus enabling the deciphering of genetic information encoded by the cloned DNA molecules. In recent years, the degree of sophistication and the range of application of this technology has increased considerably; it is now a routine procedure for studying the structure and organization of individual genes. Once a desired gene is identified, it can be manipulated to be expressed in an appropriate host, where the gene product can be produced in large quantities. With the progress of recombinant DNA technology, a revolutionary concept of vaccine production has been envisaged.

Conventional vaccines comprise (1) specific compounds purified from the pathogenic agent, (2) the whole killed pathogenic agent, or (3) an avirulent strain of the pathogenic agent as a live vaccine. There are intrinsic problems associated with conventional vaccines. For killed vaccines, complete inactivation of the pathogenic organisms is of the greatest concern, because application of an inadequately inactivated vaccine will lead to infection with, instead of immunization against, the disease. With attenuated live organisms there is the danger

* To whom correspondence should be addressed.

Biotechnology and Genetic Engineering Reviews—Vol. 1, February 1984
0264–8725/84/01/223–37$10.00 + $0.00 © Intercept Ltd

that they may mutate and revert to the virulent form and so infect the recipient that they are intended to protect.

The use of genetic engineering technology to manufacture vaccines has several potential advantages over conventional production methods. For all practical purposes, genetically engineered vaccines are safe because they are free from the causative agent. There is no fear of inadequate inactivation or reversion to the virulent form, since only parts of the genome of the pathogenic organism are cloned and used for vaccine production; the antigens are produced in the absence of the whole infectious agent. There is no risk of generating infectious DNA, RNA or virion. It is possible to manipulate bacterial cells to express a recombinant polypeptide at levels up to 20–30% of the total cellular proteins. Bacterial cultures require less space, time, cost and effort to maintain than tissue cultures, embryonic eggs or experimental animals; therefore, it is more cost effective to produce specific antigens in bacteria. Hence, even at the expense of low yield in certain purification schemes to eliminate contaminants, one can well afford to increase the purity of the preparation. In cases where only limited quantities of certain polypeptides are synthesized in their natural environment, production by recombinant DNA technology is definitely a more practical approach.

Unfortunately, it appears that foreign polypeptides synthesized in bacterial hosts in some instances do not possess the same antigenic activity as that normally displayed by the 'natural' polypeptides and, therefore, are less effective antigens. This may be because cloned polypeptides formed in a different environment, i.e. the bacterial rather than the normal host cell, may not be folded in the same way to generate the same active sites or antigenic determinants as those that are present on the native polypeptides. Consequently, these cloned proteins will not perform the desired functions or will exhibit reduced activities. It is also possible that contamination with materials of bacterial origin, not separable from the cloned product, might cause undesirable side-effects.

Current control programmes against FMD

Foot-and-mouth disease (FMD) is one of the most feared and devastating diseases of animals because it is highly contagious. Transmission of the FMD virus (FMDV) has at various times been attributed to wild and domesticated animals, animal products, human beings, rodents, birds, insects and the wind (Westergaard, 1982). The virus infects cloven-hoofed animals which include economically important livestock such as cattle, swine, sheep and goats. The disease is characterized by vesicular lesions on the feet and the oral area of the infected animal; other parts of the body e.g. snout, teats, skin, rumen and myocardium, may also be involved. Although the disease, in general, only lowers the productivity of the adult animals because of poor body condition, mortality among young animals can be as high as 50%. In practice, infected animals are slaughtered and their carcasses disposed of safely. The damaging effects of an epizootic are considerable: not only is there a direct loss of meat, dairy products, wool, hides and other animal products; there is also an extremely costly indirect loss due to the shut-down of international markets.

A few countries (e.g. Australia, New Zealand, the United States, Canada and Japan) have been free of FMD for long periods. The last outbreaks in the United States and Australia were in 1929 and 1872, respectively. Denmark was free of FMD from 1970 until an outbreak occurred in March 1982 on the islands of Funen and Zeeland. These countries, and others such as the UK, do not immunize their livestock against FMDV. If an outbreak occurs, the disease is eradicated by slaughtering the infected and exposed animals and disposing of the carcasses. The major factor contributing to the fact that these countries remain FMD free is probably due to their effective policies in restricting importation of animals and animal products from countries where FMD is a continuing problem.

Most European countries started vaccination programmes in the early 1950s and have continued their efforts systematically. These countries have been able to keep FMD to low levels, and outbreaks occur only sporadically. At present, three categories of vaccination policies are practised in Europe: (1) no vaccination in the northern and north-eastern regions; (2) vaccination programmes in border areas of eastern countries; and (3) regular prophylactic immunization in western and central regions.

The principal reasons for not vaccinating the animals are: (1) susceptible herds are the best indicators for the presence of FMDV; (2) eliminating the risk of generating carriers for the disease through vaccination of the animals. However, when an outbreak occurs, the situation can be devastating in regions where vaccination is not practised. During the 1967–68 epizootic in England, approximately half a million head of infected or exposed cattle, swine, sheep and goats were slaughtered.

In South America, even though there has been a long practice of vaccination programmes, FMD has not been reduced to an acceptable level. In 1981, more than 500 million doses of vaccines were produced for South America, which makes this part of the world the largest market for FMDV vaccine. It is encouraging that the FMD incidence has declined significantly in recent years since the introduction of strict quality control for the vaccines. In Asian and African countries with strict vaccination programmes, the disease has been brought to low levels, whereas in countries with limited vaccination programmes, FMD continues to be endemic.

Current FMDV vaccines

At present, live attenuated and inactivated whole-virus vaccines are used for the control of FMD, with the inactivated vaccine being the most commonly used. These vaccines are multivalent and they include combinations of the seven serotypes prevalent in certain regions of the world. In Europe and South America, trivalent or quadrivalent vaccines which are composed of chemically inactivated viruses of serotypes A, O and C are used. Varying proportions of the virus types are incorporated into the vaccines. The composition of the vaccine depends on the immunogenicity of the viral particles used and the response in target animals, because some serotypes are better immunogens than others and because some animals respond better than others. It has been reported that larger amounts of serotype O viruses than serotype A or C viruses are required to protect swine, indicating that serotype O virus is intrinsically a poorer immunogen (McKercher and Farris, 1967; Bachrach and McKercher, 1972). The vaccine dosage for different animals also varies, and more viral antigen is needed to immunize swine than cattle (Morgan, McKercher and Bachrach, 1970). In some cases, subtype-specific vaccines are necessary because some subtype viruses within the same serotype differ significantly in their immunogenicity and give unsatisfactory levels of cross-protection. Duration of immunity induced by two inoculations of whole-virus vaccine lasts about one year in cattle but not as long in swine. In Europe, where FMD is sporadic, animals are vaccinated once or twice a year; in South America, where the disease is frequent, animals are vaccinated two or three times a year.

LIVE ATTENUATED VACCINES

The general practice for generation of avirulent viral strains has been serial passage of virulent strains in an 'unnatural' host, e.g. in tissue cultures or in animals other than the usual host. Such procedures have been very successful in generating attenuated strains for several animal diseases including canine distemper, rinderpest, equine influenza and vesicular stomatitis. There had been considerable effort to obtain avirulent strains of FMDV (Skinner, 1959, 1960; Mowat, Barr and Bennett, 1969; Dietzschold, Kaaden and Ahl, 1972). However, those strains that elicit a good immune response generally give low but significant levels of lesions and side-effects; those that do not provoke lesions elicit poor immunity in animals. There is also the danger that an avirulent strain may revert to its virulent form, and that the vaccine thus may cause a serious epizootic instead of preventing the disease. As a result, modified live-virus vaccines have not been widely used.

INACTIVATED WHOLE-VIRUS VACCINES

Chemically inactivated virus is the most commonly used vaccine against FMD at present. The viruses are produced in suspension or monolayer cultures of baby hamster kidney cells (BHK-21, clone 13) (Mowat and Chapman, 1962; Capstick *et al.*, 1965) or bovine tongue epithelial explants (Frenkel, 1951). Chemicals that have been used to inactivate the harvested virus include formalin, glutaraldehyde, β-propriolacton (BPL) or aziridine derivatives such as acetylethylenimine (AEI) or binary ethylenimine (BEI) (Brown *et al.*, 1963; Brown, Cartwright and Stewart, 1963; Sangar *et al.*, 1973; Bahnemann *et al.*, 1974; Bahnemann, 1975).

There has been a gradual decrease in the use of formalin and an increase in the use of AEI or BEI as the inactivating agent (Olascoaga *et al.*, 1982) in vaccine production, because formalin inactivation does not follow first-order kinetics and the aziridine derivatives are safe and easy to handle. Recent outbreaks (1979, 1981) in France (Brittany and Normandy) and in the United

Kingdom (Jersey and the Isle of Wight) have been associated with the incomplete inactivation of FMDV with formalin and subsequent virus spread from areas where vaccination was practised (King *et al.*, 1981). New methods of virus inactivation are still being explored and preliminary results for virus inactivation through the activation of virion-associated endonuclease at alkaline pH in the presence of monovalent ions are promising (Scodeller *et al.*, 1982).

There are several disadvantages associated with the production and use of killed vaccines. These include the handling of large volumes of infected cultures; a requirement for biological containment facilities; a requirement for refrigeration during storage and distribution, especially in tropical countries, because of the heat-labile nature of the viral particle; and the poor growth of some subtypes in culture to yield adequate amounts of antigen. Nevertheless, the present chemically killed virus vaccine is quite safe and effective, provided that complete inactivation is achieved.

General structure and molecular organization of FMDV

FMDV is an aphthovirus of the family Picornaviridae; other genera of the family include enteroviruses, rhinoviruses and cardioviruses. The genetic information of FMDV is encoded in a single-stranded RNA of molecular weight $2{\cdot}8 \times 10^6$ daltons, which corresponds to approximately 8000 nucleotides (*Figure 1a*). The viral RNA is infectious and can serve as messenger RNA (mRNA). The viral RNA is not capped; instead, a small viral-specific polypeptide (VPg) of molecular weight 4000 daltons is covalently linked to the uridylic acid at the 5′ terminus of the RNA via a tyrosine residue (Sangar *et al.*, 1977; Grubman, 1980; Forss and Schaller, 1982). A polycytidylic acid tract, poly (C), of 100–170 bases (Brown *et al.*, 1974) is located approximately 400–500 bases from the 5′ end of the RNA molecule (Harris and Brown, 1976; Rowlands, Harris and Brown, 1978). The major protein synthesis initiation site is present on the 3′ side of the poly (C) tract (Harris, 1979; Sangar *et al.*, 1980). A single polyprotein translation product is produced from the RNA and is subsequently cleaved. The intermediate cleavage products are then further processed to give the mature non-capsid and capsid viral proteins (VP) by host-specific and viral-specific proteases (Sangar *et al.*, 1977, 1980; for review *see* Sangar, 1979). In common with most eukaryotic mRNAs, the viral RNA carries a poly (A) tail at its 3′ terminus.

As in all other picornaviruses, the protein shell is composed of four major polypeptides (VP1, 2, 3 and 4). The viral particle is icosahedral, 22 nm in diameter, and consists of 60 copies of each structural protein and one or two copies of VP0, the uncleaved precursor of VP2 and VP4 (Vande Woude, Swaney and Bachrach, 1972). VP1, 2 and 3 have molecular weights of around 30 kilodaltons (kd) and VP4 is approximately 13·5 kd, as estimated by polyacrylamide gel electrophoresis (PAGE) (Burroughs *et al.*, 1971). All four are phosphorylated (La Torre *et al.*, 1980). The native virion sediments at 146S. By lowering the pH to 6·5 or heating at 56°C, the virion dissociates into 12S particles which consist of five copies each of VP1, VP2 and VP3, and an aggregate of VP4 molecules (Burroughs *et al.*, 1971; Vasquez *et al.*, 1979).

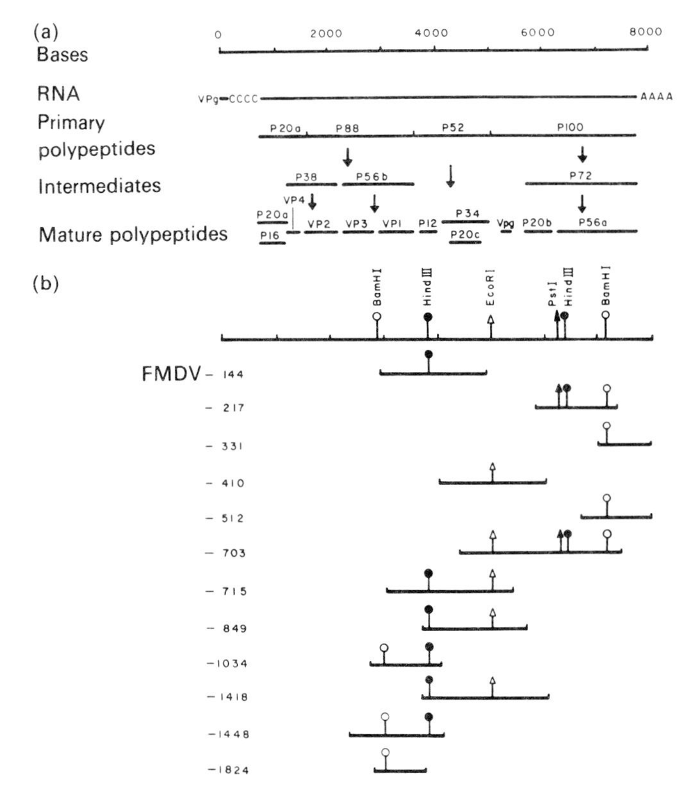

Figure 1(a) Biochemical map of foot-and-mouth disease virus. (Based on data of Sangar *et al.*, 1977 and Sangar, 1979).
(b) Restriction map and cDNA clones of FMDV strain O_1K. (*See* text for details. Courtesy of Küpper *et al.*, 1981).

Considerable confusion has been caused by the behaviour of the major capsid proteins of FMDV in various electrophoretic resolution systems. VP1, VP2 and VP3 in SDS-PAGE (Rowlands, Sangar and Brown, 1971) correspond to VP3, VP2 and VP1, respectively, as defined by urea-containing gel (Matheka and Bachrach, 1975). To resolve this confusion, each polypeptide has been named after its *N*-terminal amino acid residue, which is the same in all viral strains so far analysed, i.e. VP_{Thr}, VP_{Asp} and VP_{Gly} (Strohmaier, Wittmann-Liebold and Geissler, 1978) or VP_T, VP_D and VP_G (Bachrach, Morgan and Moore, 1979) which correspond to VP1, VP2 and VP3 in SDS-PAGE, respectively. The most recent designation is in favour of the SDS-PAGE terminology (American Society of Virology Meeting at Ithaca, New York, 1982), in conformity with the picornaviridae nomenclature. According to this system, VP1 of FMDV corresponds to VP1 of poliovirus and α of cardioviruses. In this review we have adopted the latest nomenclature.

At present, different strains of FMDV are categorized by serological assays such as complement fixation and neutralization tests. Seven serotypes, namely European (A, O and C), South African Territories (SAT1, SAT2 and SAT3) and Asian (Asia1) are recognized. Antibodies elicited by a particular serotype do not neutralize viruses of a different serotype; consequently, animals immunized with a particular FMDV serotype are not protected from infection by any of the other six heterotypic viruses. In addition, variation within each serotype has resulted in over 65 known subtypes. Reduction in cross-reactivity among subtypes of the same serotype can be quite significant. Animals immunized against one subtype may not be protected against infection by a different subtype.

Identification of the immunizing viral protein—VP1

REDUCTION OF ACTIVITIES OF PROTEASE-TREATED VIRUS

When serotype O viruses are incubated with trypsin (Wild and Brown, 1967; Rowlands, Sangar and Brown, 1971; Strohmaier and Adam, 1974; Strohmaier, Franze and Adam, 1982) or upon long incubation with chymotrypsin (Cavanagh *et al.*, 1977; Barteling *et al.*, 1979), the ability of viral particles to adsorb to host cells is greatly decreased and infectivity is reduced 100–1000-fold. The ability of the protease-treated virus to elicit neutralizing antibodies is also decreased 100–1000-fold. It was concluded that the cell attachment site and immunizing properties of FMDV serotype O viruses are closely associated with this enzyme-sensitive peptide.

The above conclusion does not extend to serotype A or C viruses. There is only a very slight reduction of infectivity and immunogenicity of type C-strain 997 viruses treated with trypsin (Rowlands, Sangar and Brown, 1971). Trypsinized type A_{12}-strain 119 virus is just as immunogenic as the native virus although infectivity is reduced 100–1000-fold (Bachrach *et al.*, 1975; Cavanagh *et al.*, 1977; Moore and Cowan, 1978). Recent experiments showed that absorption of A_{12}119 virus to cell membrane is abolished by trypsin (Baxt and Bachrach, 1982); hence 99·9% of the infectivity is lost (Bachrach, 1977).

The antibodies elicited by trypsinized virus, both serotypes C and O, differ from those elicited by the native virion (Rowlands, Sangar and Brown, 1971). When trypsin-treated virus was used to absorb serum induced by native virion, the neutralizing activity of the serum was only slightly diminished. In the same series of experiments, trypsinized virus induced only a low level of neutralizing antibodies, all of which could be absorbed by the trypsin-treated virus. These data suggest that the complete virion induces at least two classes of neutralizing antibodies: the major class neutralizes the intact virion and the minor type neutralizes only the trypsinized virus.

VP1—THE PROTEASE-SENSITIVE POLYPEPTIDE

From the trypsin-sensitive properties of O_1K viruses, where the infectivity and immunogenicity decrease dramatically, it was reasoned that the cell attachment and immunogenic sites are present on the surface of the virion. Electron-microscopy studies showed that neutralizing IgM antibodies bind specifically to the vertices of native virions, but not to trypsin-treated virus (Brown and Smale, 1970). Although there is no gross observable alteration (e.g. size, morphology, sedimentation constant and density in caesium chloride) in the virus treated with trypsin or chymotrypsin, the treated viruses do not react with neutralizing antibodies in immunodiffusion experiments, and migrate considerably faster in electrophoretic analysis (Wild, Burroughs and Brown, 1969; Brown and Smale, 1970; Meloen, 1976). When intact virions were iodinated to identify surface proteins (Laporte and Lenoir, 1973; Talbot *et al.*, 1973), more than 90% of the label was associated with VP1, even though in strain O_1K, VP2 and VP3 have twice as many tyrosines as VP1 (Bachrach, Swaney and Vande Woude, 1973). When the proteins of trypsin-treated (Wild and Brown, 1967; Talbot *et al.*, 1973) and chymotrypsin-treated (Cavanagh *et al.*, 1977) O_1K viruses were analysed on PAGE, VP1 was found to be cleaved into two lower molecular weight peptides, whereas VP2, VP3 and VP4 remained unaltered. Trypsinized $A_{12}119$ viruses also gave a similar PAGE pattern with VP1 being cleaved into two 16 kd polypeptides (VP1a and VP1b). It appears that trypsin cleaves VP1 only once in both cases and the effect of proteolytic enzymes on the infectivity and immunogenicity is strain specific.

VP1—THE IMMUNIZING POLYPEPTIDE

It was first reported by Laporte *et al.* (1973) that VP1 carries the major antigenic determinants of FMDV. Of the four purified viral capsid proteins, only VP1 was able to induce neutralizing antibody in guinea pigs.

This observation was confirmed by Bachrach *et al.* (1975) using the $A_{12}119$ virus. When guinea pigs were vaccinated twice with 100 micrograms (μg) of VP1, VP2 or VP3 or with 5 μg of AEI-inactivated serotype $A_{12}119$ virus, protective antibodies were present only in animals that received VP1 or AEI-inactivated virus. Protective antibodies were not observed in animals immunized with VP2 or VP3. The anti-VP1 sera were type specific since they did not precipitate type O1 or C3 viruses. The titres of the anti-VP1 sera were lower than antiviral sera (0·5–2·8 vs. 2·6–3·4) even though a greater amount of VP1 polypeptide was used. Qualitatively, the anti-VP1 sera recognized fewer antigenic determinants than did the antiviral sera in immunodiffusion precipitation analyses. No precipitin line was evident between anti-VP1 serum and the 12S particles, although a reaction between the antiviral serum and the 12S particles was seen. Apparently, the major antigenic determinants on VP1 are different whether they are present in the viral particle, 12S particle or the isolated protein.

Experiments in which swine were immunized either with VP1 or with VP1, VP2 and VP3 each at a separate site of the same animal, gave the same titre of neutralizing antibody and the same degree of protection after three inoculations (Bachrach *et al.*, 1977). When a mixture of VP1, VP2 and VP3 was used for vaccination, a lesser degree of protection was observed. The authors' explanation is that, upon mixing the three structural polypeptides, intermolecular association occurs and the antigenic determinants on VP1 are perturbed. Experiments with larger groups of swine (Bachrach *et al.*, 1977) showed that one vaccination with 100 μg VP1 is not sufficient to impart adequate protection to the animals (one out of 10), whereas two vaccinations significantly increased the degree of protection (eight out of 10). Although the AEI-treated virus is a more potent immunogen (five out of five swine were fully protected when vaccinated once with 5 μg of inactivated virus in the same experiment), it is possible to protect swine against FMDV with VP1 after multiple injections.

IMMUNIZING ACTIVITY OF VP1 FRAGMENTS

Kaaden, Adam and Strohmaier (1977) showed that cleavage fragments of VP1 are capable of inducing neutralizing antibodies which can protect animals against FMDV. Cyanogen bromide cleaves O_1K VP1 into five fragments, each of molecular weight less than 14 kd. After four injections of approximately 100 μg of either intact or cyanogen-bromide-cleaved VP1 of O_1K, guinea pigs showed similar levels of neutralizing antibody titre with both antigens; however, the level is much lower than that elicited by native virions. Upon challenge with virulent virus, only a fraction of the animals vaccinated with either the intact VP1 or the cyanogen-bromide-cleaved polypeptide proved to be resistant to the disease.

Similar experiments using the $A_{12}119$ virus revealed that the 13 kd peptide generated by cyanogen bromide cleavage of VP1 is the major peptide responsible for eliciting neutralizing antibodies in guinea pigs (Bachrach, Morgan and Moore, 1979). These studies demonstrated that the complete VP1 polypeptide is not necessary for the induction of neutralizing antibodies against FMDV. It is possible to use VP1 fragments for vaccination as long as they present the immunogenic determinants in a form recognized by the immune system.

In summary, all the experiments described above show that VP1 is the only polypeptide of FMDV capable of inducing neutralizing antibodies. This protein seems to carry all the antigenic determinants necessary to protect animals against infection, and therefore makes a good candidate for the application of recombinant DNA technology in vaccine production.

Application of biotechnology

CLONING OF VP1-SPECIFIC SEQUENCES IN *ESCHERICHIA COLI*

The introduction of genetic materials of exogenous origin into a prokaryotic host by recombinant DNA techniques has become part of the routine repertoire of a molecular biology laboratory. A specific gene can be isolated, and then propagated and studied in a 'controlled' environment.

With the realization that isolated VP1 could protect animals against FMD (LaPorte *et al.*, 1973; Bachrach *et al.*, 1975; Meloen, Rowlands and Brown, 1979), the advantage of a subunit vaccine over conventional vaccine became self-evident. A subunit vaccine produced by recombinant DNA is absolutely safe because neither the viral particle nor the viral RNA is involved in any step of vaccine production. Furthermore, in comparison with conventional methods, larger quantities of antigens can be manufactured more quickly and economically.

By 1981, three groups had succeeded in cloning and determining the nucleotide sequences of DNA copies of the genomes of three different FMDV strains; subsequently, two groups were able to express the VP1 coding sequences in *E. coli*. The same standard complementary DNA (cDNA) cloning procedures

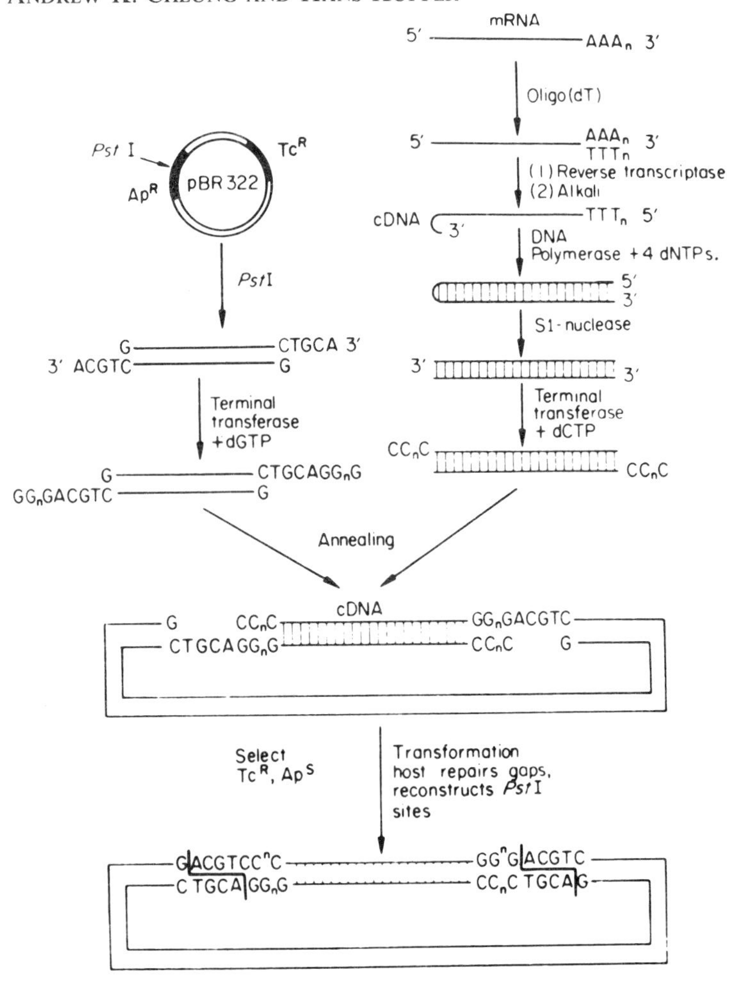

Figure 2. Schematic diagram of cDNA cloning. *See text* for details.

were used by all the investigators to produce bacteria that contain FMDV sequences. The general method for introducing foreign DNA sequences into a bacterial cell is outlined in *Figure 2*. Küpper *et al.* (1981) were the first to report the cloning and expression of FMDV VP1 cDNA sequences of O_1K. Their procedures and results are used as a prototype in the present review. The vector used was plasmid pBR322 which carries resistance markers for ampicillin and tetracycline. The *Pst*I endonuclease restriction site located within the β-lactamase gene was selected for the introduction of double-stranded (ds) cDNA sequences by the G:C tailing method (Villa-Komaroff *et al.*, 1978).

Since the genetic material of FMDV is a single-stranded RNA, a ds-cDNA copy has to be produced in order to introduce FMDV specific sequence into *E. coli*. The usual process is to prime the first strand cDNA with oligo(dT), which is complementary to the poly (A) tail at the 3′ end of the genomic RNA. DNA synthesis is carried out by reverse transcriptase. After removal of the viral RNA, the cDNA folds back on itself and thus primes synthesis of the second cDNA strand by DNA polymerase I. The hair-pin structure at the 5′ end of the ds-cDNA is cleaved with S1 nuclease and oligo(dC) tails are then added to both ends of the ds-cDNA molecule. These molecules are annealed to *Pst*I cut plasmid pBR322 DNA which had had oligo(dG) tails added to its ends. After annealing, the mixture is used to transform competent *E. coli* host cells. Because insertion of DNA at the *Pst*I site inactivates the β-lactamase gene responsible for ampicillin resistance, tetracycline-resistant and ampicillin-sensitive bacterial colonies were selected for further analysis. Thus a FMDV strain O_1K-specific cDNA library was established.

So far, none of the investigators have reported the cloning of the entire FMDV genome as a contiguous cDNA sequence in *E. coli*. Rather, they all obtained inserted fragments of variable length (up to 4000 base pairs) covering various parts of the viral genome (*Figure 1b*). In particular, the poly (C) track or sequences 5′ to the poly (C) track were not observed in any of the clones. It appears that the long homopolymer region is unclonable by the conventional methods described above. Consequently, the chance of generating infectious RNA or DNA molecules by introducing the entire FMDV genome into a bacterium appears highly unlikely.

IDENTIFICATION OF THE VP1-SPECIFIC SEQUENCE

Initially, Küpper *et al.* (1981) constructed a physical map (*Figure 1b*) of the FMDV genome based on the size and intensity of the DNA fragments generated by digestion of ^{32}P-labelled ds-cDNA with various restriction enzymes. Assuming that the cDNA sequences corresponding to the 3′ end of the viral RNA will occur more frequently because of the oligo(dT) priming method, the restriction fragments were ordered. This physical map from restriction enzyme analyses was aligned with the known biological map constructed from the order of translation and the molecular weights of the viral polypeptides (Doel *et al.*, 1978). By this comparison, the VP1 coding sequence was predicted to be present in a 850 base pair *Bam*H1/*Hin*dIII fragment which corresponds to a region of the viral RNA located 5000 nucleotides from the poly (A) tail. DNA fragments covering this area of O_1K (FMDV −115, −144 and −1034) were first analysed for their nucleotide sequence by the method of Maxam and Gilbert (1977).

By that time, the first 40 *N*-terminal amino acid residues of O_1K VP1 polypeptide had been elucidated (Adam and Strohmaier, 1974; Strohmaier, Wittmann-Liebold and Geissler, 1978). Computer analysis was then conducted to locate these 40 residues within the deduced amino acid sequences from the cDNA fragments; thus the coding sequence of capsid protein VP1 was located (Küpper *et al.*, 1981). Consequently, the complete VP1 coding sequence was established.

As the primary translation product from the viral RNA is a polyprotein from which the viral capsid proteins are derived, the correct reading frame should allow complete read-through of the nucleotide sequence. The reading frame was unequivocally established since only one frame allows translation without interruption, whereas the other two frames contain multiple termination codons (Kurz *et al.*, 1981; *Figure 3*).

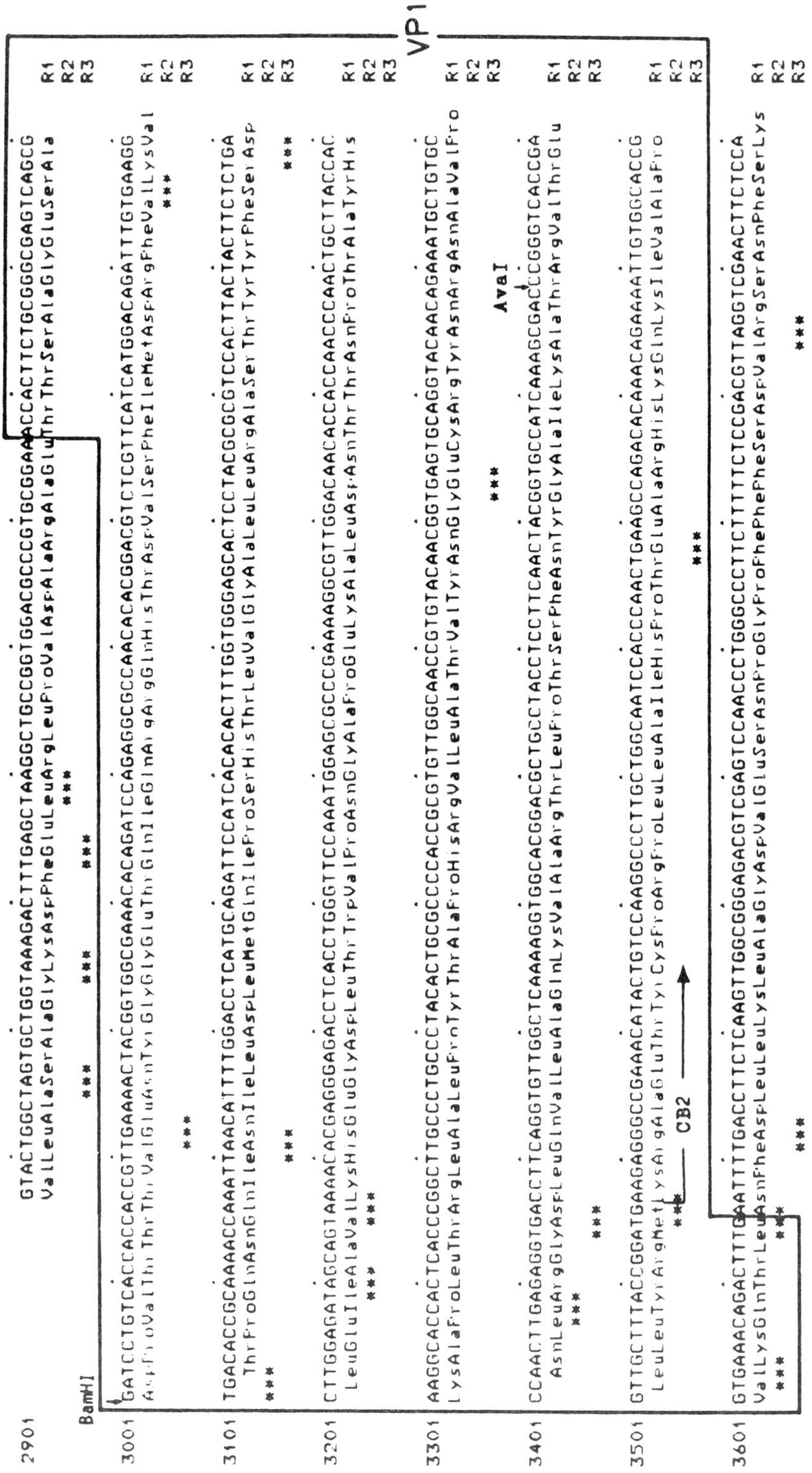

```
2901           GTACTGGCTAGTGCTGGTAAAGACTTTGAGCTAAGGCTGCCGGTGGACGCCCGTGCGGAAACCACTTCTGCGGGCGAGTCAGCG
               ValLeuAlaSerAlaGlyLysAspPheGluLeuArgLeuProValAspAlaArgAlaGluThrThrSerAlaGlyGluSerAla   R1
                                             ***                                                     R2
     BamHI              ***      ***      ***                                                        R3

3001 GATCCTGTCACCACCACCGTTGAAAACTACGGTGGCGAAACACAGATCCAGAGGCGCCAACACACGGACGTCTCGTTCATCATGGACAGATTTGTGAAGG
     AspProValThrThrThrValGluAsnTyrGlyGlyGluThrGlnIleGlnArgArgGlnHisThrAspValSerPheIleMetAspArgPheValLysVal   R1
                                                                                               ***           R2
                   ***                                                                                       R3

3101 TGACACCGCAAAACCAAATTAACATTTTGGACCTCATGCAGATTCCATCACACACTTTGGTGGGAGCACTCCTACGCGCGTCCACTTACTACTTCTCTGA
      ThrProGlnAsnGlnIleAsnIleLeuAspLeuMetGlnIleProSerHisThrLeuValGlyAlaLeuLeuArgAlaSerThrTyrTyrPheSerAsp   R1
     ***                                                                                                     R2
                  ***                                                                               ***      R3

3201 CTTGGAGATAGCAGTAAAACACGAGGGAGACCTCACCTGGGTTCCAAATGGAGCGCCCGAAAAGGCGTTGGACAACACCACCAACCCAACTGCTTACCAC
      LeuGluIleAlaValLysHisGluGlyAspLeuThrTrpValProAsnGlyAlaProGluLysAlaLeuAspAsnThrThrAsnProThrAlaTyrHis   R1
            ***   ***                                                                                        R2
                                                                                                             R3

3301 AAGGCACCACTCACCCGGCTTGCCCTGCCCTACACTGCGCCCCACCGCGTGTTGGCAACCGTGTACAACGGTGAGTGCAGGTACAACAGAAATGCTGTGC
     LysAlaProLeuThrArgLeuAlaLeuProTyrThrAlaProHisArgValLeuAlaThrValTyrAsnGlyGluCysArgTyrAsnArgAsnAlaValPro   R1
                                                                                                             R2
                                                                            ***                              R3
                                                                                          AvaI
3401 CCAACTTGAGAGGTGACCTTCAGGTGTTGGCTCAAAAGGTGGCACGGACGCTGCCTACCTCCTTCAACTACGGTGCCATCAAAGCGACCCGGGTCACCGA
       AsnLeuArgGlyAspLeuGlnValLeuAlaGlnLysValAlaArgThrLeuProThrSerPheAsnTyrGlyAlaIleLysAlaThrArgValThrGlu   R1
        ***                                                                                                  R2
              ***                                                                                            R3

3501 GTTGCTTTACCGGATGAAGAGGGCCGAAACATACTGTCCAAGGCCCTTGCTGGCAATCCACCCAACTGAAGCCAGACACAAACAGAAAATTGTGGCACCG
      LeuLeuTyrArgMetLysArgAlaGluThrTyrCysProArgProLeuLeuAlaIleHisProThrGluAlaArgHisLysGlnLysIleValAlaPro   R1
               ***  CB2 →                                                                                    R2
                                                                         ***                                 R3
                                                                                                          VP1
3601 GTGAAACAGACTTTGAATTTTGACCTTCTCAAGTTGGCGGGAGACGTCGAGTCCAACCCTGGGCCCTTCTTTTTCTCCGACGTTAGGTCGAACTTCTCCA
     ValLysGlnThrLeuAsnPheAspLeuLeuLysLeuAlaGlyAspValGluSerAsnProGlyProPhePhePheSerAspValArgSerAsnPheSerLys   R1
      ***        ***                                                                                         R2
                   ***                                                            ***                        R3
```

```
3701 AACTGGTGGAAACCATCAACCAGATGCAGGAGGACATGTCAACAAAACACGGGCCTGACTTTAACCGGTTAGTGTCCGCATTTGAGGAGTTGGCCATTGG
       LeuValGluThrIleAsnGlnMetGlnGluAspMetSerThrLysHisGlyProAspPheAsnArgLeuValSerAlaPheGluGluLeuAlaIleGly    R1
                                                                  ***                                          R2
                                              HindIII    ***    ***              ***                          R3

3801 AGTGAAAGCCATCAGAACCGGTCTCGACGAAGCCAAACCCTGGTACAAGCTTATCAAGCTCCTAAGCCGCCTGTCGTGCATGGCCGCTGTGGCAGCACGG
      ValLysAlaIleArgThrGlyLeuAspGluAlaLysProTrpTyrLysLeuIleLysLeuLeuSerArgLeuSerCysMetAlaAlaValAlaAlaArg     R1
      ***                                                    ***                                               R2
                                                                                                                R3

3901 TCCAAGGACCCAGTCCTTGTGGCCATCATGCTGGCCGACACCGGTCTCGAGATTCTGGACAGCACCTTCGTCGTGAAGAAGATCTCCGACTCGCTCTCCA
     SerLysAspProValLeuValAlaIleMetLeuAlaAspThrGlyLeuGluIleLeuAspSerThrPheValValLysLysIleSerAspSerLeuSerSer   R1
                                                                             ***                               R2
                                                                            XhoI                               R3

4001 GTCTCTTTCACGTGCCGGCCCCCGTCTTCAGTTTCGGAGCACCGGTCCTGTTGGCCGGGTTGGTCAAAGTTGCCTCGAG
       LeuPheHisValProAlaProValPheSerPheGlyAlaProValLeuLeuAlaGlyLeuValLysValAla                               R1
                                                                                                               R2
                                                                                                               R3
```

Figure 3. Nucleotide sequence of cloned FMDV–O_1K cDNA. The deduced amino acid sequence of the translation product is shown for the first phase, potential stop codons (***) in the other two phases. The sequence that codes for VP1 is boxed. (Courtesy of Kurz *et al.*, 1981).

With the current amount of FMDV sequence data available, the identification of VP1-specific sequences of any desired strain is relatively easy. Comparison of data from several FMDV strains showed that the 3′ flanking sequences, i.e. those coding for the P52 protein, are highly conserved. Cheung *et al.* (1983) have successfully used sequences from this region as probes to identify VP1-containing clones of other FMDV strains.

DETERMINATION OF THE GENE BOUNDARIES OF VP1

The amino acid sequence derived from the DNA sequence was compared with the partial amino acid sequence of VP1 obtained by protein sequencing (Strohmaier, Wittmann-Liebold and Geissler, 1978). Except for discrepancies at positions 7 and 33, the derived protein sequence is identical to the peptide-sequencing results. Thus the *N*-terminal boundary of VP1 could be assigned to the appropriate nucleotide sequence. The carboxyl-terminal boundary was determined by digestion both of the purified VP1 and of the *C*-terminal fragment of the cyanogen-bromide-cleaved VP1 with carboxypeptidase Y (EC 3.4.16.1). Leucine, threonine and glutamine were preferentially released in that order. In conjunction with the estimated molecular weight of VP1, the carboxyl end, Glu-Thr-Leu, was assigned to the corresponding codons (Kurz *et al.*, 1981; Strohmaier, Franz and Adam, 1982). Thus the VP1 polypeptide of O_1K contains 213 amino acid residues.

The VP1 gene boundaries for $A_{12}119$ were determined in an analogous way. The *N*-terminal residues for $A_{12}119$ were reported to be Thr-Thr-Ala-Thr (Matheka and Bachrach, 1975). Subsequent experiments extended this sequence to the first 26 amino acid residues (Bachrach, Morgan and Moore, 1979). Although there is a discrepancy between the deduced and the directly obtained amino acid sequences, there is enough agreement to assign the first codon of the $A_{12}119$ VP1 polypeptide. Since the $A_{12}119$ *C*-terminal residues, Gln-Thr-Leu, obtained by direct amino acid determination were available (Bachrach, Swaney and Vande Woude, 1973), the *C*-terminal codon could be designated. Thus $A_{12}119$ VP1 polypeptide contains 212 amino acid residues.

Both the *N*- and *C*-terminal gene boundaries for VP1 of $A_{10}61$ were first estimated based on the partial amino acid sequences published for other FMDV strains as described above. Subsequently, the *N*-terminal amino acid residues were confirmed by direct sequencing (Boothroyd *et al.*, 1981, 1982). The VP1 of $A_{10}61$ contains 212 amino acid residues.

Today, due to the availability of enough sequence information, the newly derived VP1 sequences of additional FMDV strains can be directly aligned with the already established boundaries by analogy.

EXPRESSION OF THE VP1 SEQUENCE IN *E. COLI*

In order to produce large quantities of a given protein, its synthesis in a fast-growing bacterium such as *E. coli* is desirable. The regulation of gene expression in this organism is well understood. Once the structural gene of the protein is isolated, it can be placed under the control of prokaryotic sequences required for efficient transcription and translation.

In general, it is desirable to regulate the expression of the inserted coding sequence because the expressed protein might be toxic to the cell and, therefore, interfere with cell growth even when the gene product is present at low intracellular concentration. Of the many regulated promoters known, two have been applied successfully for larger-scale fermentation of genetically engineered bacteria containing VP1 of FMDV: (1) the pL promoter of bacteriophage lambda which is inducible by high temperature if the cell contains a temperature-sensitive mutant repressor protein; (2) the bacterial *trp* promoter which can be induced by tryptophan starvation. Both of these are strong promoters and can be easily regulated. If the pL promoter is used, the corresponding repressor protein has to be present in the host cell. The gene for the repressor protein can be present either on the chromosome or on a separate plasmid.

No matter how strong the promoter, the final production of the proteins is further dependent on efficient translation of the mRNA and the stability of the product. In a number of cases, it appears that proteins which are naturally secreted are unstable inside the cell. One possible explanation for this is that the reducing conditions inside inhibit the proper folding of the protein which normally occurs in the relatively more oxidizing conditions outside the cell; the protein thus becomes an efficient substrate for the host's degradation mechanisms. Sometimes the problem of stability can be overcome by fusing the protein to part of a bacterial protein. This method of stabilization has been successfully applied to somatostatin (Itakura *et al.*, 1977) by fusing the gene for the small hormone to that of β-galactosidase; the relatively large fusion protein which resulted was resistant to degradation by bacterial proteases.

To express the VP1 gene of FMDV strain O_1K, Küpper *et al.* (1981) chose an expression vector which had already been used to express the human fibroblast interferon gene (Derynck *et al.*, 1980). This plasmid (pPL c24, *Figure 4*) carries the strong pL promoter together with the DNA sequence coding for the first 99 amino acids of bacteriophage MS2 replicase, followed by unique sites for the restriction endonucleases *Bam*H1 and *Hin*dIII. Using these two sites, the 847 bp *Bam*H1/*Hin*dIII fragment of clone pFMDV-1034 (*Figure 1b*) was inserted into the vector where only one orientation is possible. The insertion places the coding sequence of FMDV, starting with amino acid 9 of capsid protein VP1, in phase with the MS2 replicase; thus the ribosome binding site and the beginning of the replicase gene are used to initiate translation of the VP1 gene, which as a processing product does not have its own initiation signals. Translation should produce a fusion protein of 396 amino acids beginning with the 99 *N*-terminal amino acids of the MS2 replicase, followed by 284 residues specific for FMDV and then 13 residues encoded by the vector. The inserted FMDV fragment codes for 205 amino acids of VP1 plus the first 79 amino acids of protein P52, the gene for which is adjacent to the gene for VP1 in the FMDV genome (*Figure 1a*). This plasmid (pPL-VP1, *Figure 4*) was transformed into an *E. coli* C600 strain which already carried a gene specifying the temperature-sensitive lambda repressor on a second plasmid, pcI857, carrying the kanamycin resistance gene (Remaut, Stanssens and Fiers, 1981). This allowed the expression of the pL promoter to be controlled by temperature. Cultures grown at 28°C do not make the fusion protein. When the temperature is shifted to 42°C the

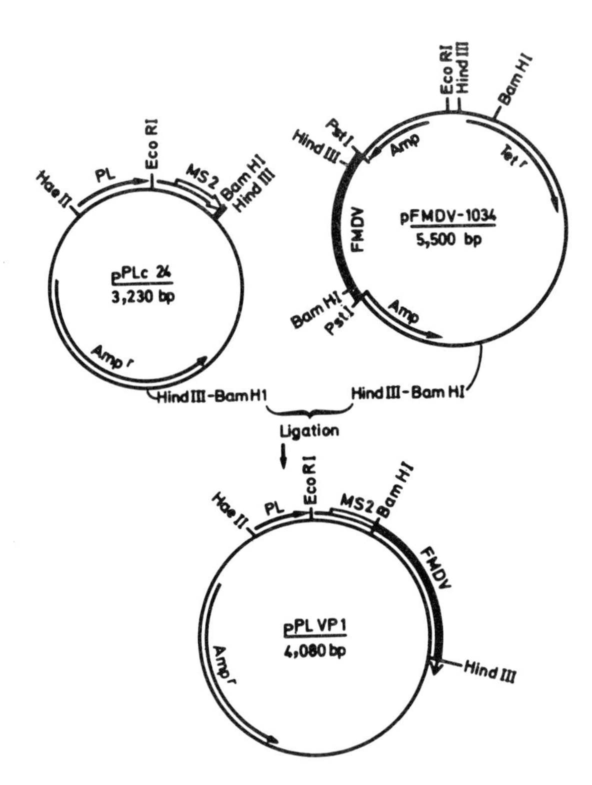

Figure 4. Construction of plasmid pPL VP1. The schematic outline shows the vector plasmid pPL c24 and the donor plasmid pFMDV–1034. Amp: β-lactamase region coding for resistance to ampicillin; Tet: coding region for resistance to tetracycline; PL: λ P_L control element; MS2: DNA fragment derived from bacteriophage MS2 where the boxed part indicates the coding sequence for the first 99 amino acids of the MS2 polymerase; FMDV: insert derived from FMDV cDNA, the shadowed area shows the region coding for VP1. The predicted fusion polypeptide from the constructed plasmid pPL VP1 is indicated. Only restriction sites relevant for this construction are shown. (Courtesy of Küpper *et al.*, 1981).

repressor is inactivated and expression of a 43 kd protein can clearly be demonstrated by SDS-PAGE (*Figure 5*). This protein accounts for up to 20% of the total cellular protein, which is equivalent to about 1×10^6 molecules per bacterial cell (Küpper *et al.*, 1982). The newly synthesized protein was identified as the MS2-VP1 fusion protein by its reaction with murine antibodies against capsid protein VP1. This test shows that, indeed, the correct protein is synthesized in larger amounts and that the protein is not degraded inside the bacterial cell. When smaller portions of the MS2 replicase (e.g. only the first 15 amino acids) were used, the amount of protein found in the cell was reduced 20-fold (Küpper, unpublished observation). It is possible that the hydrophobic part of the MS2 replicase stabilizes the fusion protein by forming a precipitate within the cell and thus making it insoluble and inaccessible to cellular proteases.

Phase contrast microscopy and thin-section electron microscopy revealed that intracellular inclusion bodies are present in bacteria (C600 (pcI857)/pPL-VP1)

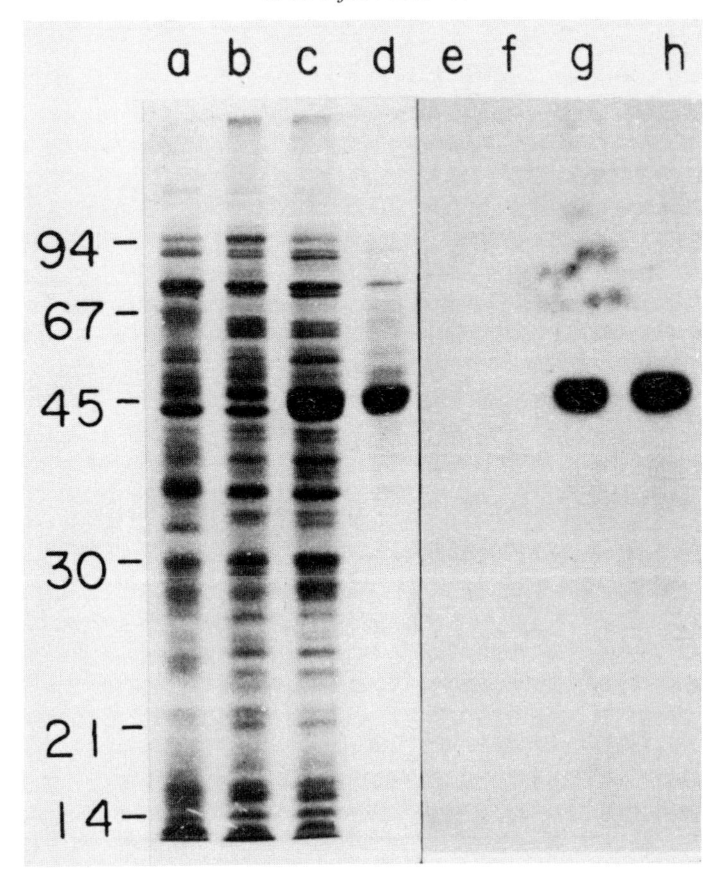

Figure 5. Expression of cloned FMDV–O_1K VP1 gene. SDS–PAGE of total cellular proteins. (a) C 600 (pcI 857) induced for 3 hours at 42°C; (b) C 600 (pcI 857)/pPL VP1 not induced, and (c) induced for 3 hours at 42°C. (d) Proteins extracted from the membrane pellet by 6 M urea. Lanes (e–h) are autoradiograms of same proteins as lane (a–d) but transferred to nitrocellulose and reacted with ^{125}I-labelled VP1-specific antibodies. (Courtesy of Küpper *et al.*, 1982).

that produce large quantities of the fusion protein but absent from cells not carrying the expression plasmid (*Figure 6*).

Such electron-dense inclusion bodies have been observed before in genetically manipulated bacterial cells producing a fusion protein consisting of β-galactosidase and human insulin (Williams *et al.*, 1982).

Hardy, Stahl and Küpper (1981) transferred the O_1K VP1 gene from the pPL-VP1 expression plasmid into the *Bacillus subtilis* pBD9 plasmid. In this way, VP1 is expressed in *Bacillus subtilis* as a chimeric protein containing 73 amino acid residues encoded by the erythromycin resistance gene, 205 residues by FMDV VP1, 79 residues by P52 and 13 residues by the bacterial vector. The fusion protein represented about 2% of total cell protein and was stable in the *Bacillus* host.

Kleid *et al.* (1981) inserted a ds-cDNA copy of the subtype A_{12}119 genome into the *Pst*I site of pBR322 and then transferred the VP1 cDNA sequences to a site in an expression vector downstream of a region containing the trypto-

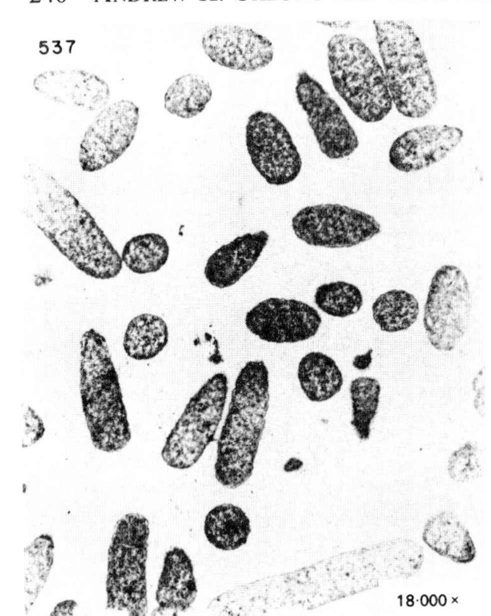

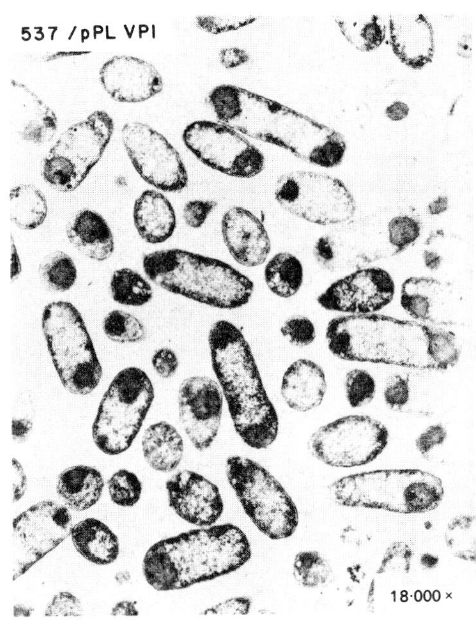

Figure 6. Electron micrograph of *E. coli* carrying pPL-VP1 that expressed VP1 fusion protein at high levels. 537 is the parent bacterium C600(pCI857) and 537/pPL-VP1 carries the expression plasmid for FMDV VP1. Both preparations were induced at 42°C.

phan promoter, the *trp* leader (L) gene and part of the *trp* E protein gene. This plasmid was transformed into *E. coli* and propagated in low concentration of tryptophan. When tryptophan in the medium is depleted, the tryptophan promoter is de-repressed and allows expression of a 394-residue VP1 fusion protein. This protein contains 190 amino acid residues from the (L + E) sequences and 204 amino acids of VP1 (the 6 *N*-terminal and the last *C*-terminal amino acids are missing). This chimeric protein is insoluble inside the host cell and is found in amounts corresponding to approximately 17% of total *E. coli* protein by SDS-PAGE. The protein reacts with anti-VP1 (A_{12}119) antibody.

PURIFICATION OF THE BACTERIAL VP1 FUSION PROTEIN

To prepare VP1 fusion protein for vaccination, it is necessary to extract the antigen from the bacterial host.

The bacterial strain C600 (pcI857)/pPL-VP1 is grown in L-broth containing ampicillin and kanamycin at 28°C overnight (approximately $A_{550\ nm} = 3$). Next day, the culture is diluted fourfold into a pre-warmed medium (42°C) and further incubated for 3 hours. During this time the cells divide once and accumulate approximately 20% of total cellular proteins as the VP1 fusion protein. The cells are collected by centrifugation, resuspended in phosphate-buffered saline (PBS) and broken by sonication. Because of the insoluble nature of the fusion protein, all soluble proteins can be washed away by simple washing procedures, leaving the desired protein and membrane particles in the precipitate. The VP1 protein is extracted from this pellet with high concentrations of urea to give a product of more than 50% purity (*Figure 5, lane d*). The protein is further purified by column chromatography, but must be kept in urea buffer to prevent aggregation and precipitation. These steps involved in the production of the VP1-fusion protein can be carried out on large volumes of cells. In fact, production in fermenters often gives better results than shake flask experiments because bacterial growth can be optimized by monitoring the pH and oxygen concentration of the culture.

EFFICACY OF BACTERIAL-PRODUCED VP1 IN ANIMALS

Although several laboratories are actively engaged in research in this field, relatively few animal experiments of the genetically engineered vaccine have been reported. This is because the vaccine has become available only recently and the animal experiments are time-consuming.

Kleid *et al.* (1981) injected LE-VP1 (A_{12}119), purified by SDS-PAGE, into cattle and swine using an oil adjuvant. Neutralizing antibodies were detected after two injections (250 μg each). This immunization procedure protected both cattle (five out of six fully protected, one animal with minor foot lesions) and swine (two out of two); non-vaccinated control animals (two cattle and two swine) came down with severe foot and mouth lesions (*Table 1*). Hofschneider *et al.* (1981) immunized goats with crude bacterial extracts of MS2-VP1 fusion protein (O_1K) and found that two out of three goats had anti-VP1 antibodies when tested by enzyme-linked immunosorbent assay (ELISA). When a purer preparation of MS2-VP1 was used for immunization, two out of three goats produced neutralizing antibodies. Upon challenge both goats had less fever and lower viraemia than non-vaccinated control animals. The most successful fusion-protein vaccine experiment until now has been the immunization of cattle using the LE-VP1 fusion protein of A_{12}119 by J. J. Callis *et al.* (personal communication). After two inoculations of 10, 50, 250 or 1250 μg each of fusion protein, five, seven, eight or nine out of nine cattle could be protected, respectively (*Table 2*).

It was also reported that high neutralizing antibody titres could be elicited in guinea pigs by a single injection of 100 μg of the LE-VP1 fusion protein

Table 1. Neutralizing antibody and immune responses of cattle and swine vaccinated with LE′–VP3 fusion protein or VP3 isolated from virus. VP3 of this table is identical to VP1 of this review. (Courtesy of Kleid *et al.*, 1981).

		Neutralizing antibody day after vaccination*								Challenge of immunity†		
Subject	Antigen	7	14	21	28	35	42	46	56	Mouth lesions	Foot lesions	VIA anti-body‡
Cattle												
1	LE′–VP_3	<0·3	0·5	0·3	1·2	2·3	2·7		2·9	0	0	–
2	LE′–VP_3	0·2	0·9	1·3	2·0	2·4	2·7		2·9	0	0	–
3	LE′–VP_3	0·2	0·4	1·1	1·7	1·9	2·1		2·9	0	0	–
4	LE′–VP_3	0·2	0·7	0·7	1·1	2·2	2·6		2·9	0	0	–
5	LE′–VP_3	<0·3	0·9	1·0	1·1	2·6	2·5		2·6	0	+	–
6	LE′–VP_3	<0·3	1·0	1·3	1·3	2·1	2·1		2·3	0	0	–
7	VP_3	0·1	1·0	2·1	1·9	2·2	2·6		2·1	0	0	–
8	VP_3	<0·3	0·4	0·4	0·7	1·7	1·6		4·1	+ + +	+ + +	+
9	None†								3·4	+ + +	+ + +	+
10	None								3·5	+ + +	+ + +	+
11	None								3·3	+ + +	+ + +	+
12	None								3·3	+ + +	+ + +	+
Swine												
13	LE′–VP_3	<0·3	<0·3	0·5	1·5			3·0		0	0	–
14	LE′–VP_3	<0·3	0·3	0·6	1·4			3·0		0	0	–
15	None†							3·5		+ + +	+ + +	+
16	None							3·4		+ + +	+ + +	+

* Revaccinations on day 24 for swine and day 28 for cattle; titre is $-\log_{10}$ of serum dilution that protects 50% of suckling mice. The 46- and 56-day sera were collected 14 days after challenge.
† Challenged on day 32 for swine and day 42 for cattle. Non-vaccinated animals 9 to 12 and 15 and 16 constituted the challenge groups; half were inoculated with virulent type A_{12}119ab virus and half were contact transmission controls; +, single small lesion, not generalized disease; + + +, numerous lesions and generalized infection.
‡ Presence, +, of virus-infection-associated antigen (VIA) antibody in sera collected 14 days after challenge indicated animals experienced FMD; –, VIA antibody absent.

Table 2. Neutralizing antibody and immunity in cattle vaccinated with biosynthetic A12 VP1 vaccine. (Courtesy of J. J. Callis *et al.*, personal communication).

	Weeks after vaccination											
Antigen (μg)	2	8	12	15	17	21	30	32	34	38	42	45
10	0·9	0·9	0·9	1·7†	2·0*	1·7	1·7‡	—	—	—	—	—
50	1·0	1·2	1·0	1·8†	2·1*	1·9	2·0‡	—	—	—	—	—
250	1·1	1·1	1·0	2·0	2·0	1·8	1·8	1·9†	2·3*	2·6	2·5	2·4‡
1250	1·2	1·3	1·3	2·0	2·6	2·0	1·9	2·3	1·9*	2·9	2·4	2·7

* Titre 2 weeks after revaccination.
† Revaccination with 10, 50, 250, or 1250 μg dose respectively.
‡ Challenge of immunity: 10 μg, 30 weeks, 5/9 immune; 50 μg, 30 weeks, 7/9 immune; 250 μg, 45 weeks, 8/9 immune; 1250 μg, 45 weeks, 9/9 immune.

(Moore, 1983; *Table 3*). This result is very important, because the efficacy of FMDV vaccines is currently judged by their ability to elicit a high titre of neutralizing antibody after a single injection. However, in a parallel experiment with the O_1Campos strain, injection of 500 μg of a LE-VP1 fusion protein did not give as high a neutralizing antibody titre as the fusion protein from A_{12}119. This is another indication that serotype O VP1 polypeptide is intrinsically less immunogenic (as described above), whether it is present as a fusion protein or in the virion when compared with serotype A. In addition, the same series of

Table 3. Neutralizing antibody responses of guinea pigs LE′–VP3 fusion proteins or FMDV VP3. VP3 of this table is identical to VP1 of this review (Courtesy of Moore, 1983).

Group	Antigen	Buffer†	Neutralizing antibody response at given dose (μg)			
			4·0	20	100	500
1	FMDV $A_{12}VP_3$	Urea–Tris	3·2	3·9	4·8	—
2	Long leader $A_{12}LE'–VP_3$	Urea–Tris	3·5	3·9	3·6	—
3	Short leader $A_{12}LE'–VP_3$	Urea–Tris	2·3	3·6	3·6	—
4	Short leader $A_{12}LE'–VP_3$	Gel slurry SDS–Tris	<0·3	1·7	4·2	—
5	Short leader $A_{12}LE'–VP_3$	Urea–Tris	—	—	3·2	3·6
6	Short leader $O_1LE'–VP_3$	Urea–Tris	—	—	1·5	2·2
7	Short leader $O_1LE'–VP_3$	Urea–Tris	—	—	1·8‡	3·2‡
8	Long leader $A_{12}LE'–VP_3$ tandem peptide Amino Acids #137–167	Urea–Tris	—	—	4·6§	—

* Neutralizing antibody titres determined by the suckling mouse serum neutralization test (Skinner, 1952) of guinea pig sera, taken 28–35 days after vaccination. Values are the $-\log_{10}$ serum dilution 50% end-point of a pool of 5 sera protecting mice from 100 LD_{50} doses of the homologous virus. For vaccination appropriate doses in a 2 ml aqueous-Freund incomplete oil adjuvant emulsion were given subcutaneously in the skin of the neck. — indicates not tested.

† Indicates the aqueous phase of the vaccines: Urea–Tris = 6 M urea, 0·05 M mercaptoethanol 0·014 M Tris–HCl, pH 8·6; gel slurry SDS–Tris = pulverized polyacrylamide preparative gel slice diluted in 0·1% SDS, 0·05 M Tris–glycine, pH 8·1.

‡ Neutralizing antibody titre 21 days after revaccination at 35 days with the same vaccine preparation.

§ Neutralizing antibody titre on day 28, revaccinated on day 14 with same preparation.

experiments showed that the length of the leader amino acid sequence does not affect the immunogenicity of the fusion protein.

In all the experiments reported, the ‘vaccine’ was freshly prepared in urea–tris buffer or with acrylamide particles present. However, for a commercial vaccine, it is necessary to formulate the vaccine with the appropriate adjuvants, e.g. oil, aluminum hydroxide or others, acceptable to the veterinarians in the field. Further experimentation is required to assess the duration of protection imparted to the animals, the shelf-life under various climatic conditions and the side-effects, if any. In addition, it should be stressed that all fusion proteins tested to date have employed protein of only one serotype at a time, whereas all commercial vaccines now in use are multivalent.

Large field trials have not yet been reported for the biochemically engineered vaccine. However, if extensive research, especially in the field of vaccine formulation, shows encouraging results the fusion protein will be tested in the field in the near future.

The above summary has shown that large quantities of VP1 fusion proteins can be produced in bacteria. The chimeric polypeptides are stable, with no detectable changes in molecular weight or antigenicity when treated for 10 minutes at 100°C in the presence of denaturing agents, e.g. β-mercaptoethanol and SDS (Kleid *et al.*, 1981; Küpper *et al.*, 1982). This genetically engineered protein is capable of protecting guinea pigs, swine, and cattle against FMD.

However, there are problems which remain unsolved. Protective antibodies elicited by viral or genetically engineered VP1 differ quantitatively and qualitatively from those elicited by the native virion (Cartwright, Chapman and Brown, 1980). On an equimolar basis, several orders of magnitude more VP1 protein

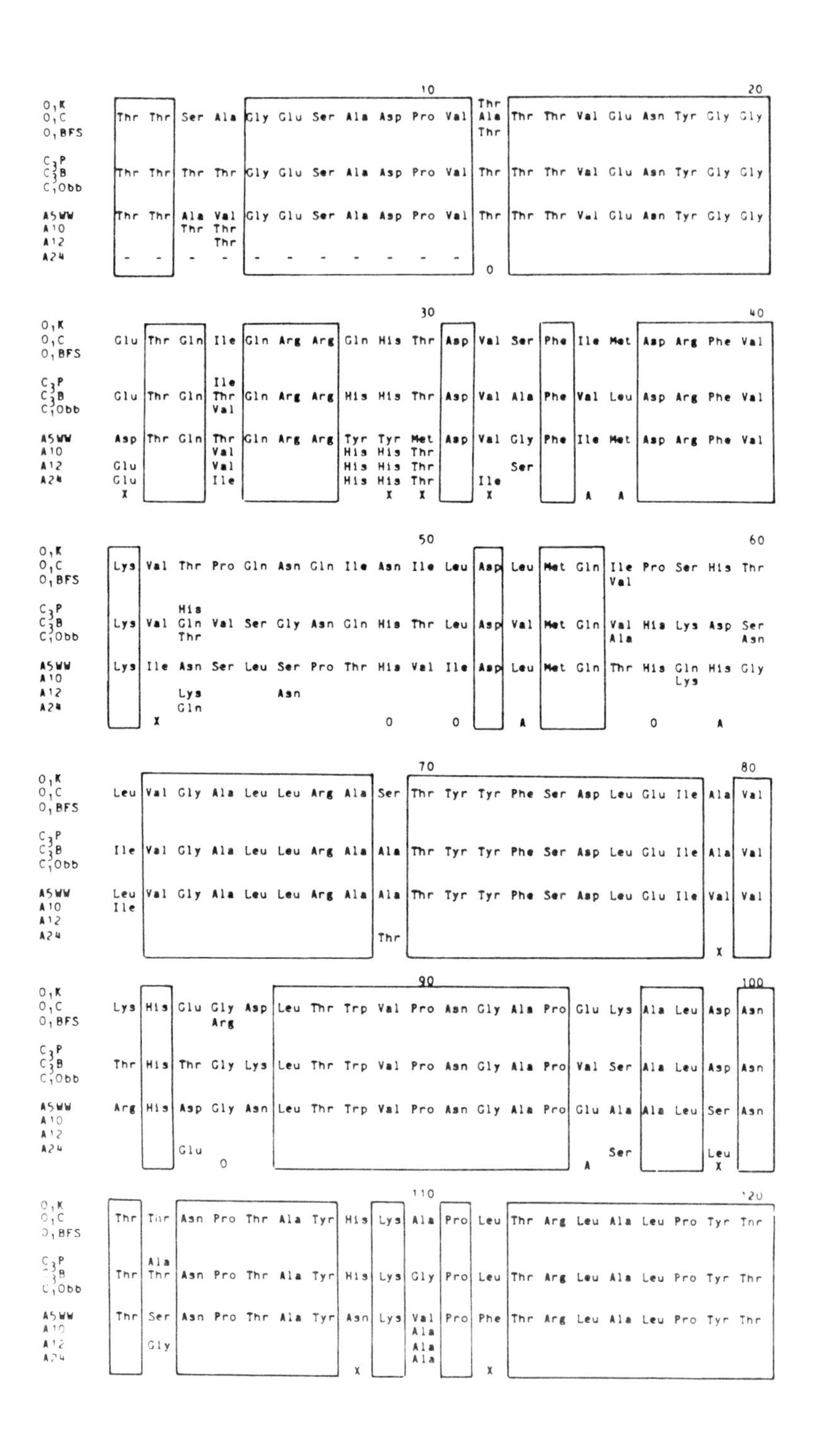
10
20
O1K
Thr
O1C
Thr Thr Ser Ala Gly Glu Ser Ala Asp Pro Val Ala Thr Thr Val Glu Asn Tyr Gly Gly
O1BFS
Thr
C3P
C3B
Thr Thr Thr Thr Gly Glu Ser Ala Asp Pro Val Thr Thr Thr Val Glu Asn Tyr Gly Gly
C1Obb
A5WW
Thr Thr Ala Val Gly Glu Ser Ala Asp Pro Val Thr Thr Thr Val Glu Asn Tyr Gly Gly
A10
Thr Thr
A12
Thr
A24
- - - - - - - - - - -
0
30
40
O1K
O1C
Glu Thr Gln Ile Gln Arg Arg Gln His Thr Asp Val Ser Phe Ile Met Asp Arg Phe Val
O1BFS
C3P
Ile
C3B
Glu Thr Gln Thr Gln Arg Arg His His Thr Asp Val Ala Phe Val Leu Asp Arg Phe Val
C1Obb
Val
A5WW
Asp Thr Gln Thr Gln Arg Arg Tyr Tyr Met Asp Val Gly Phe Ile Met Asp Arg Phe Val
A10
Val His His Thr
A12
Glu Val His His Thr Ser
A24
Glu Ile His His Thr Ile
X X X X A A
50
60
O1K
O1C
Lys Val Thr Pro Gln Asn Gln Ile Asn Ile Leu Asp Leu Met Gln Ile Pro Ser His Thr
O1BFS
Val
C3P
His
C3B
Lys Val Gln Val Ser Gly Asn Gln His Thr Leu Asp Val Met Gln Val His Lys Asp Ser
C1Obb
Thr
Ala Asn
A5WW
Lys Ile Asn Ser Leu Ser Pro Thr His Val Ile Asp Leu Met Gln Thr His Gln His Gly
A10
Lys
A12
Lys Asn
A24
Gln
X 0 0 A 0 A
70
80
O1K
O1C
Leu Val Gly Ala Leu Leu Arg Ala Ser Thr Tyr Tyr Phe Ser Asp Leu Glu Ile Ala Val
O1BFS
C3P
C3B
Ile Val Gly Ala Leu Leu Arg Ala Ala Thr Tyr Tyr Phe Ser Asp Leu Glu Ile Ala Val
C1Obb
A5WW
Leu Val Gly Ala Leu Leu Arg Ala Ala Thr Tyr Tyr Phe Ser Asp Leu Glu Ile Val Val
A10
Ile
A12
A24
Thr
X
90
100
O1K
O1C
Lys His Glu Gly Asp Leu Thr Trp Val Pro Asn Gly Ala Pro Glu Lys Ala Leu Asp Asn
O1BFS
Arg
C3P
C3B
Thr His Thr Gly Lys Leu Thr Trp Val Pro Asn Gly Ala Pro Val Ser Ala Leu Asp Asn
C1Obb
A5WW
Arg His Asp Gly Asn Leu Thr Trp Val Pro Asn Gly Ala Pro Glu Ala Ala Leu Ser Asn
A10
A12
A24
Glu Ser Leu
0 A X
110
120
O1K
O1C
Thr Thr Asn Pro Thr Ala Tyr His Lys Ala Pro Leu Thr Arg Leu Ala Leu Pro Tyr Thr
O1BFS
C3P
Ala
C3B
Thr Thr Asn Pro Thr Ala Tyr His Lys Gly Pro Leu Thr Arg Leu Ala Leu Pro Tyr Thr
C1Obb
A5WW
Thr Ser Asn Pro Thr Ala Tyr Asn Lys Val Pro Phe Thr Arg Leu Ala Leu Pro Tyr Thr
A10
Ala
A12
Gly Ala
A24
Ala
X X

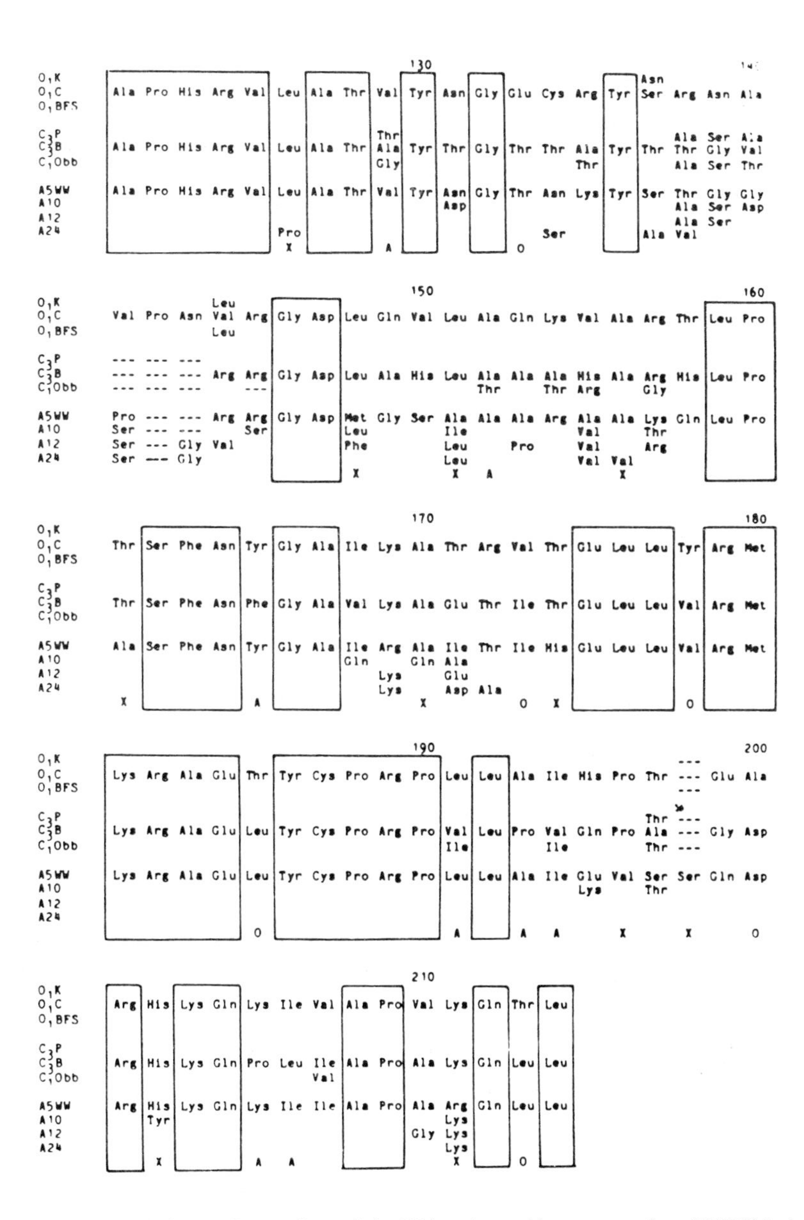

Figure 7. (Left and above). Comparison of the VP1 amino acid sequence of ten FMDV isolates. O_1Campos, C_3Indaial-B and A_5Westerwald sequences are used as prototype for comparison. Amino acid residues present in other isolates that differ from the prototype residues are replaced with the appropriate residues accordingly. Amino acids residues shared by all 10 isolates are enclosed in boxes. Additional residues common to serotypes (O and C), (C and A) and (O and A) are indicated by X, O and A respectively at the bottom of the appropriate positions, and residue deletion is indicated by (–––). (Based on data of Boothroyd *et al.*, 1981; Kleid *et al.*, 1981; Kurz *et al.*, 1981; Makoff *et al.*, 1982; Beck, Feil and Strohmaier, 1983 and Cheung *et al.*, 1983).

than complete virus is required to elicit neutralizing antibodies in animals. In addition, Meleon *et al.* (1982) have demonstrated that neutralizing monoclonal antibodies raised against the native virus do not recognize the isolated VP1 polypeptide.

Oligopeptide vaccines

The ready availability of protein sequences through DNA sequencing and the progress in oligopeptide synthesis has opened a totally new approach to making vaccines chemically. For efficient application of this approach, prior knowledge of those amino acid sequences corresponding to the antigenic determinants of the pathogen are useful.

LOCALIZATION OF IMMUNOGENIC EPITOPES OF VP1

Amino acid sequence approach

As discussed above, there are seven distinct serotypes and more than 65 known subtypes of FMDV based on serological assays such as complement fixation and virus neutralizing tests (Pereira, 1977). The fact that antisera raised with some subtypes do not always neutralize other subtypes of the same serotype suggests that there is considerable variation in the antigenic determinants of different strains of FMDV.

Of all the capsid proteins, only VP1 has been shown to carry the major antigenic determinant(s) responsible for inducing neutralizing antibodies in animals. By comparing the primary amino acid sequence of VP1 of various strains, it is possible to identify the regions where the amino acid sequences differ between serotypes and subtypes. At present, sequence determination of DNA is easier and faster than sequence determination of protein; very limited amino acid sequences of the VP1 polypeptide are obtained by direct residue determination. In contrast, complete VP1 nucleotide sequences of more than 10 isolates have been reported, allowing the amino acid sequences to be deduced according to the genetic code.

The currently available VP1 amino acid sequences have been aligned for maximum homology (Cheung *et al.*, 1983; *Figure 7*). The residues common to all strains are enclosed in boxes; those shared by all isolates of any two different serotypes are indicated by A, O or X at the bottom of the amino acid residues. This alignment indicates that approximately 60% of VP1 amino acid sequence is conserved when any two FMDV strains are compared. In general, three conserved regions are present: these lie between positions 1–41, 61–130 and 159 to the carboxyl terminus. Two variable regions are located within positions 42–61 and 134–158. Since the amino acid residues present at 42–61 are highly homologous among strains of the same serotype, it has been suggested that both variable regions contribute to serotype specificity while only the 134–158 variable region is likely to be associated with subtype specificity (Makoff *et al.*, 1982; Beck, Feil and Strohmaier, 1983; Cheung *et al.*, 1983).

Polypeptide fragmentation approach

The fact that the entire VP1 is not necessary for eliciting protective antibodies in animals encouraged Strohmaier, Franze and Adam (1982) to localize the major antigenic determinants along the VP1 polypeptide. O_1K virions were incubated with various proteolytic enzymes of different specificities and the virions were then re-purified. Fragments derived from VP1 were analysed on SDS–PAGE and the amino acid residues at the *N*-terminus and the *C*-terminus of each fragment were also determined. In conjunction with the amino acid sequence deduced from DNA sequence analysis, the cleavage sites and peptide fragment length can be determined unequivocally.

The results showed that trypsin (EC 3.4.21.4) cleaves O_1K at multiple sites, a finding that helps to clarify results from earlier investigations of the effect of proteases on infectivity (*above*). Although the virion remains visually intact, two small peptide regions (residue positions Asn^{139}–Lys^{154} and His^{202}–Leu^{214}) are cleaved off, and the infectivity and immunizing activity of the trypsin-treated O_1K virus is reduced by 100–1000-fold (as described above). Mouse submaxillary gland protease (MSGP; endoproteinase Arg-C) cleaves VP1 between Arg^{145} and Gly^{146} but there is no loss of any amino acid residues, and the infectivity of the virion is not affected. Initially, endoproteinase Lys-C nicks O_1K VP1 between Lys^{154}–Val^{155} and a fragment containing the Gln^{204}–Leu^{214} is cleaved from the virion. At this point, infectivity of the Lys-C treated virus is not affected. Upon further incubation, the large fragment (i.e. Thr^{1}–Lys^{154}) is shortened from 21 kd to 20 kd and the infectivity of the virion is reduced. A similar result was obtained with chymotrypsin (EC 3.4.21.1) (Cavanagh *et al.*, 1977). Here, the ability of virus to attach to host cells as well as its immunogenicity was greatly reduced after long periods of digestion. Collectively, these results showed that the loss of amino acid residues between Asn^{139} and Lys^{154} destroys the ability of the O_1K virus to attach to cells.

On the basis of the immunogenic properties of the peptide fragments generated by various procedures, Strohmaier, Franze and Adam (1982) localized two potential immunogenic sites on VP1 of O_1K. The peptide fragments Gln^{55}–Arg^{179} and Lys^{181}–Leu^{214} (generated by cyanogen bromide cleavage) and the Gly^{146}–Leu^{214} (generated by MSGP digestion) can elicit significant levels of neutralizing antibodies; a small peptide, Val^{154}–Arg^{201} (generated by trypsin digestion) does not possess any immunogenic activity. This suggests that Gly^{146}–Lys^{154} and His^{202}–Leu^{214} are the only non-overlapping amino acid sequences in VP1 capable of inducing neutralizing antibodies (*Figure 8*). However, these serological data were obtained from mice that had been injected four times, each time with 10 μg of peptides. On a molar basis, this is several hundred times more protein than is required to elicit similar neutralizing antibody levels with native virions.

IMMUNIZING ACTIVITY OF OLIGOPEPTIDES

On the basis of the deduced amino acid sequence of three FMDV strains ($A_{10}61$, $A_{12}119$ and O_1K) and data from the polypeptide fragmentation experiments

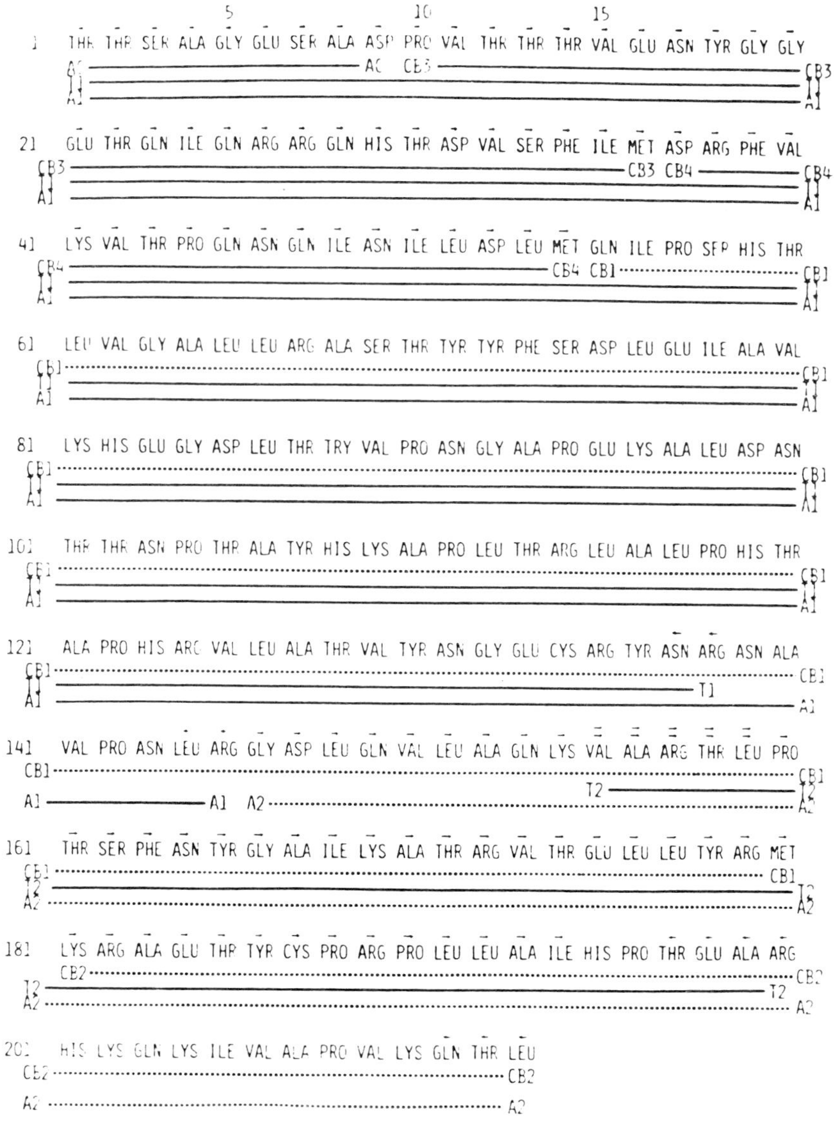

1 THR THR SER ALA GLY GLU SER ALA ASP PRO VAL THR THR THR VAL GLU ASN TYR GLY GLY

21 GLU THR GLN ILE GLN ARG ARG GLN HIS THR ASP VAL SER PHE ILE MET ASP ARG PHE VAL

41 LYS VAL THR PRO GLN ASN GLN ILE ASN ILE LEU ASP LEU MET GLN ILE PRO SER HIS THR

61 LEU VAL GLY ALA LEU LEU ARG ALA SER THR TYR TYR PHE SER ASP LEU GLU ILE ALA VAL

81 LYS HIS GLU GLY ASP LEU THR TRY VAL PRO ASN GLY ALA PRO GLU LYS ALA LEU ASP ASN

101 THR THR ASN PRO THR ALA TYR HIS LYS ALA PRO LEU THR ARG LEU ALA LEU PRO HIS THR

121 ALA PRO HIS ARG VAL LEU ALA THR VAL TYR ASN GLY GLU CYS ARG TYR ASN ARG ASN ALA

141 VAL PRO ASN LEU ARG GLY ASP LEU GLN VAL LEU ALA GLN LYS VAL ALA ARG THR LEU PRO

161 THR SER PHE ASN TYR GLY ALA ILE LYS ALA THR ARG VAL THR GLU LEU LEU TYR ARG MET

181 LYS ARG ALA GLU THR TYR CYS PRO ARG PRO LEU LEU ALA ILE HIS PRO THR GLU ALA ARG

201 HIS LYS GLN LYS ILE VAL ALA PRO VAL LYS GLN THR LEU

Figure 8. Immunizing activities of O_1K peptide fragments generated by cyanogen bromide or peptidases. CB = cyanogen bromide; T = trypsin; A = mouse submaxillary gland protease; ······, neutralizing antibody-producing peptides; ———, non-inducing peptides; → → ← ←, estimated by protein sequencing. (Courtesy of Strohmaier, Franze and Adam, 1982).

(Strohmaier, Franze and Adam, 1982), Bittle *et al.* (1982) chemically synthesized three oligopeptides (i.e. Thr^{1}–Lys^{41}, Val^{141}–Pro^{160} and Arg^{201}–Leu^{214}) of O_1K, the latter two of which correspond to the highly variable regions of VP1. These were coupled separately to keyhole limpet haemocyanin (KLH) and were tested for immunogenicity by injection (four injections of 200 μg peptide) into rabbits. Virus neutralizing activity was detected only in sera of animals immunized with the Val^{141}–Pro^{160} and Arg^{201}–Leu^{214} peptides coupled to the KLH carrier (*Table 4*). Subsequently, a single inoculation of 200 μg of peptide Val^{141}–Pro^{160} was shown to protect guinea pigs from homologous virus challenge. These data suggest that the oligopeptide–KLH molecules mimic, at least in part, the configuration of the VP1 epitopes on the virus. Low levels (1%) of cross-neutralization activity against serotype C (C_3Indaial) and serotype A (A_{10}61) viruses were detected.

Table 4. Antibody response of rabbits to different peptides of VP1 of FMDV, type 0, strain Kaufbeuren. (Courtesy of Bittle *et al.*, 1982).

Rabbit no.	Peptide	Anti-peptide antibody titre	Neutralization index (log_{10})
1	9–24	80–160	≦0·3
2	9–24	40–80	≦0·3
3	17–32	80–160	≦0·5
4	17–32	20–40	≦0·9
5	25–41	40–80	≦0·5
6	25–41	640–1280	≦0·9
7	1–41	320–640	≦0·9
8	1–41	320–640	≦0·7
9	141–160	320–640	≦3·9
10	141–160	320–640	≦3·7
11	151–160	80–160	2·9
12	151–160	160–320	1·1
13	200–213	>1280	3·5
14	200–213	160–320	3·1

An alternative approach for defining the immunogenic epitope of O_1K virus was attempted by Pfaff *et al.* (1982). They used various available methods of predicting the regions of α-helical structure in proteins to locate possible immunogenic determinants along the deduced O_1K VP1 sequence. Combining this information with the experimental data of Strohmaier, Franze and Adam (1982), they chose three regions (Leu^{144}–Leu^{159}, Ala^{167}–Lys^{181}, and Lys^{205}–Leu^{214}) for testing. Again, synthetic peptides covalently linked to the KLH carrier were used to immunize rabbits. After three inoculations, all three peptides elicited antibodies to the corresponding peptide but only Leu^{144}–Leu^{159} induced a high titre of antibody that neutralized virus when assayed in suckling mice (*Table 5*). Again, a low but measurable level of neutralizing activity against serotype C_1Obb virus (1%) was seen, although the serotype A_5WW virus was not affected. Furthermore the peptide Leu^{144}–Leu^{159} can absorb out 99% of the anti-VP1 antibody in serum raised with killed virus, suggesting that Leu^{144}–Leu^{159} is the dominant immunizing epitope of VP1.

Support for the idea that the region Asn^{134}–Arg^{154} of strain A_{12}119 is actually present on the surface of the virion was provided by Robertson *et al.* (1983).

Table 5. Reactivity of antipeptide antisera with the antigens, with coat protein VP1, and with FMDV serotypes O_1K, C_1O, and A_5. VP1–A = 144–159; VP1–B = 167–181; VP1–C = 206–214; VP1–A_1 = 144–151 and VP1–A_2 = 152–159. (Courtesy of Pfaff *et al.*, 1982).

		Titre in the ELISA*					
	Antigen:	Peptide		VP1	FMDV serotype		
Serum or IgG	Antiserum specificity	Free	Conjugate	O_1K FMDV	O_1K	C_1O	A_5
1	VP1–A/KLH†	3·2	3·5	3·3	3·8	1·8	NEG
2	VP1–B/KLH†	2·8	3·1	1·0	1·0	NEG	NEG
3	VP1–A/KLH†	3·6	4·4	4·4	5·1	2·3	NEG
	VP1–B/KLH†	3·3	4·0				
3A IgG§	VP1–A/KLH	3·5		3·9			
	[VP1–B/KLH]	[1·7]					
3B IgG#	[VP1–A/KLH]	[1·5]		1·2			
	VP1–B/KLH	3·1					
4	VP1–A/KLH‡	2·8		4·2	4·9	2·2	NEG
	VP1–B/KLH‡	2·3					
5	VP1–A_1/KLH†	2·0	2·2	1·7	2·2	NEG	NEG
	VP1–A_2/KLH†	2·3	2·7				
	VP1–C/KLH†	2·7	3·0				
6	VP1–A_1/KLH‡	1·0	2·5	NEG			
7	VP1–A_2/KLH‡	1·1	2·2	NEG			
8	VP1–C/KLH‡	1·3	2·7	1·9			

* $-\log_{10}$ values; NEG: values identical to those of pre-immune sera (extinction value $\varepsilon > 0{\cdot}05$ at a 1 : 10 dilution). The values shown in this table were obtained after the third injection.
† Coupling of the peptides to the carrier protein using glutaraldehyde.
‡ Coupling of the peptides to the carrier protein using *N*-ethyl-*N′*-(3-dimethylaminopropyl)carbodiimide.
§ IgG purified by an affinity chromatography on a Sepharose–VP1–B column.
# IgG purified by an affinity chromatography on a Sepharose–VP1–A column.

These investigators demonstrated that Tyr^{136} of VP1 is preferentially iodinated *in situ*, whether native, trypsinized or bacterial-protease-cleaved viral particles are used.

Experiments with chemically synthesized oligopeptides including positions 141–160 of $A_{10}61$ and $A_{12}119$ viruses have also been reported (Clarke *et al.*, 1983). Neutralizing antibodies against the native virion can be elicited in guinea pigs immunized with these oligopeptides. In fact, the neutralizing antibodies elicited by the oligopeptides exhibited subtype specificity. Anti-$A_{10}61$ peptide (position 141–160) serum does not neutralize the $A_{12}119$ virus or vice versa. Thus, subtype specificities of serotype A viruses are likely to be associated with the 141–160 variable region.

Although the major immunogenic domains of O_1K (144–159) (Bittle *et al.*, 1982; Pfaff *et al.*, 1982) and of $A_{10}61$ or $A_{12}119$ (141–160) (Clarke *et al.*, 1983) are located within the 134–159 variable region (Cheung *et al.*, 1983), there is a VP1 immunogenic domain present on $A_{12}119$ which is absent from O_1K. VP1 fragments of O_1K (Thr^1–Arg^{138} or Thr^1–Arg^{145}) do not (Strohmaier, Franze and Adam, 1982), whereas analogous fragments of $A_{12}119$ (Thr^1–Ala^{138} or Thr^1–Arg^{145}) do (Robertson *et al.*, 1983), elicit neutralizing antibodies.

In all the above experiments, sera raised in either rabbits or guinea pigs were studied. Even though guinea pigs are used as test animals for FMDV vaccines, results obtained with these animals do not always correlate with those obtained with cattle. Thus, the efficacy of any vaccine has to be evaluated in the field.

Some future research areas

HETEROTYPIC ANTIGENIC ACTIVITIES

Traditionally, FMDV vaccine development emphasizes the importance of antigen specificity, because animals immunized against a different serotype or subtype are not protected when challenged with heterologous viruses. Recently, several groups of investigators have reported that low but clearly detectable levels of heterologous neutralizing activity could be demonstrated in sera of animals immunized with the FMDV 12S particles or with chemically synthesized peptides (Meloen and Briare, 1980; Cartwright, Chapman and Brown, 1980; Bittle *et al.*, 1982; Cartwright, Chapman and Sharpe, 1982; Cartwright, Morrell and Brown, 1982; Pfaff *et al.*, 1982). This phenomenon may be helpful in the formulation of multivalent FMDV vaccines.

The 12S particle has two kinds of activities when used to elicit neutralizing antibodies: on the one hand, it is an ineffective primary immunogen for eliciting neutralizing antibodies, although neutralizing antibody titre does increase upon multiple inoculations; on the other hand, it is effective in stimulating homotypic or heterotypic neutralizing antibodies in animals primed with killed virus of the same or different serotype (Cartwright, Morrell and Brown, 1982).

ADJUVANTS

The most important factor in the effectiveness of a vaccine is the presentation of the desired antigenic determinants in the 'correct' conformation to the host immune system. Confusing and, at times, contradictory results have been obtained by various groups of investigators using the same immunogen. Some of this can be attributed to differences in the methods of preparing the antigens involved. Bachrach *et al.* (1982) demonstrated that A_{12}119 VP1 purified by three different conventional methods showed varying ability to elicit neutralizing antibodies in guinea pigs, cattle and swine. In particular, the use of lyophilized A_{12}119 VP1 eluted from SDS–PAGE gave a substantially lower neutralizing antibodies titre and imparted less protection to animals than did the same VP1 polypeptide without lyophilization or after DEAE-chromatography.

Vaccines are usually prepared by mixing the desired antigens with an adjuvant which helps to enhance the humoral and/or cell-mediated immune response. The stimulating effect of the adjuvant may or may not be specific in enhancing the immunogenicity of the antigen. However, regardless of whether the immune stimulation is specific or not, the antigens present in the vaccine preparation must be in the native conformation. Without exception, different adjuvants show varying degrees of enhancement for immunization with FMDV vaccines. At present, aluminum hydroxide is used as the adjuvant in FMDV vaccines. With the addition of saponin, the immune response in cattle is enhanced both after primary and after secondary vaccinations (Frenkel *et al.*, 1982). In another study, FMDV vaccines mixed with oil adjuvant (emulsion of water in mineral oil) gave longer-lasting immunity in cattle and enhanced the immunity in swine (Olascoaga *et al.*, 1982) than the aluminum hydroxide–saponin vaccine. Synthetic compounds are also being tested as adjuvants. Knudsen (1982) demonstrated that the synthetic lipid amine (CP20961) increases the antibody levels and enhances protection against FMDV in guinea pigs. Other immune stimulants, e.g. muramyl dipeptide and its derivatives or liposomes, should also be examined.

In general, the serum levels of FMDV neutralizing antibodies correlate well with the degree of protection against a particular serotype or subtype. The role of cellular immunity in protection from FMD has not been fully investigated. In two series of experiments using peptide fragments of A_{12}119 VP1, Bachrach *et al.* (1982) showed that both cattle and swine can be adequately protected against FMDV in the absence of high levels of neutralizing antibodies; presumably the protection is imparted by the host cellular immune response. Research work involving selective stimulation of the humoral and cellular immune responses would be extremely valuable in the development of other vaccines as well as that of FMDV vaccines.

CARRIERS

High-molecular-weight carriers are usually helpful in potentiating the immune response to low-molecular-weight antigens. Bittle *et al.* (1982), Pfaff *et al.* (1982) and Clarke *et al.* (1983) successfully immunized animals with chemically synthesized oligopeptides coupled to KLH. Of the carriers tested with the synthetic oligopeptides (bovine serum albumin, KLH, tetanus toxoid and thyroglobulin), KLH gave the best results. Unfortunately, KLH is quite expensive, and the chance of it being used in vaccine preparations is small. Synthetic carriers designed to stimulate the host immune systems or to enhance the immunogenicity of the antigens should be examined and applied to facilitate work in this challenging task of producing an absolutely safe and more stable FMDV vaccine.

HYBRID VIRUSES

Smith, Mackett and Moss (1983) showed that a hybrid virus, in which the gene for hepatitis B surface antigen had been inserted into the vaccinia virus genome, could raise high levels of hepatitis B neutralizing antibodies in test animals. The hepatitis antigen is expressed during the replicative cycle of the vaccinia virus and released into the bloodstream of the host animal. This method of applying an antigen, as a component of a modified live vaccine, would appear to have the potential ability to impart a longer-lasting immunity than could a killed vaccine and avoids purification of the antigen. The same approach could be applied to the development of a hybrid virus vaccine for FMD.

Summary

Major contributions towards the development of an absolutely safe FMDV vaccine are evident. With the identification of VP1 as the immunogenic protein, it is possible to manufacture a subunit vaccine via biotechnology. DNA sequences encoding the VP1 protein can be introduced into a bacterium with ease; under the appropriate conditions, large amounts of VP1 can be produced in a short time. The accumulation of amino acid sequences generated by recombinant DNA techniques allows identification of antigenic domains, which are the basis of variability among serotype and subtype viruses. As a result, vaccine production by chemical synthesis of short peptides corresponding to the antigenic determinants is greatly facilitated. At present, results from experimental vaccines employing genetically engineered or chemically synthesized VP1 antigens against homologous virus infection are encouraging.

The current approach of preparing vaccine is to utilize the antigenic specificity of the virus. Since FMDV undergoes antigenic drift, variants not neutralized by type-specific serum will arise. An alternative approach is to prepare vaccines based on antigenic sites shared among all serotype and subtype viruses.

Acknowledgements

We thank Drs Gerald Selzer and John DeLamarter for their suggestions. We also thank Ms Ruth Haehni, Deborah Peterson and Maija Cheung for their co-operation in preparing this manuscript. This work is supported by Biogen S.A., Geneva, Switzerland.

References

ADAM, K.H. AND STROHMAIER, K. (1974). Isolation of the coat proteins of foot-and-mouth disease virus and analysis of the composition and N-terminal end groups. *Biochemical and Biophysical Research Communications* **61**, 185–192.

BACHRACH, H.L. (1977). Foot-and-mouth disease virus: properties, molecular biology and immunology. In *Virology in Agriculture* (J.A. Romberger, Ed.), pp. 3–32. Allanheld, Osmun, Montclair.

BACHRACH, H.L. AND MCKERCHER, P.D. (1972) Immunology of foot-and-mouth disease in swine-experimental inactivated virus vaccines. *Journal of the American Veterinary Research Association* **160,** 521–526.

BACHRACH, H.L., MORGAN, D.O. AND MOORE, D.M. (1979). Foot-and-mouth disease virus immunogenic capsid protein VPT: N-terminal sequences and immunogenic peptides obtained by CNBr and tryptic cleavages. *Intervirology* **12,** 65–72.

BACHRACH, H.L., SWANEY, J.B. AND VANDE WOUDE, G.F. (1973). Isolation of the structural polypeptides of foot-and-mouth disease virus and analysis of their C-terminal sequences. *Virology* **52,** 520–528.

BACHRACH, H.L., MOORE, D.M., MCKERCHER, P.D. AND POLATNICK, J. (1975). Immune and antibody responses to an isolated capsid protein of foot-and-mouth disease virus. *Journal of Immunology* **115,** 1636–1641.

BACHRACH, H.L., MOORE, D.M., MCKERCHER, P.D. AND POLATNICK, J. (1977). An experimental subunit vaccine for foot-and-mouth disease. In *Proceedings, International Symposium on Foot-and-Mouth Disease (II), Lyon 1976.* (C. Mackowiak and R.H. Regamey, Eds). *Symposia Series in Developments in Biological Standardization,* volume 35, pp. 150–160. S. Karger, Basel.

BACHRACH, H.L., MORGAN, D.O., MCKERCHER, P.D., MOORE, D.M. AND ROBERTSTON, B.H. (1982). Foot-and-mouth disease virus: immunogenicity and structure of fragments derived from capsid protein VP3 and of virus containing cleaved VP3. *Veterinary Microbiology* **7,** 85–96.

BAHNEMANN, H.G. (1975). Binary ethylenimine as an inactivant for foot-and-mouth disease virus and its application for vaccine prodution. *Archives of Virology* **47,** 47–56.

BAHNEMANN, H.G. AUGE DE MELLO, P., ABARACON, D. AND GOMES, I. (1974). Immunogenicity in cattle of foot-and-mouth disease vaccines inactivated with binary ethylenimine. *Bulletin de l'Office International d'Epizooties, Paris* **81,** 1335–1343.

BARTELING, S.M., MELOEN, R.H., WAGENAAR, F. AND GIELKINS, A.L.J. (1979). Isolation and characterization of trypsin-resistant O_1 variants of foot-and-mouth disease virus. *Journal of General Virology* **43,** 383–393.

BAXT, B. AND BACHRACH, H.L. (1982). The adsorption and degradation of foot-and-mouth disease virus by isolated BHK-21 cell plasma membranes. *Virology* **116,** 391–405.

BECK, E., FEIL, G. AND STROHMAIER, K. (1983). The molecular basis of the antigenic variation of foot-and-mouth disease virus. *EMBO Journal* **2,** 555–559.

BITTLE, J.L., HOUGHTON, R.A., ALEXANDER, H., SHINNICK, T.M., SUTCLIFFE, J.G., LERNER, R.A., ROWLANDS, D.J. AND BROWN, F. (1982). Protection against FMDV by immunization with a chemically synthesized peptide predicted from the viral nucleotide sequence. *Nature* **298,** 30–34.

BOOTHROYD, J.C., HARRIS, T.J.R., ROWLANDS, D.J. AND LOWE, P.A., (1982). The nucleotide sequence of cDNA coding for the structural protein of FMDV. *Gene* **17,** 153–161.

BOOTHROYD, J.C., HIGHFIELD, P.E., CROSS, G.A.M., ROWLANDS, D.C., LOWE, P.A., BROWN, F. AND HARRIS, T.J.R. (1981). Molecular cloning of foot-and-mouth disease virus genome and the nucleotide sequences in the structural protein genes. *Nature* **290,** 800–802.

BROWN, F. AND SMALE, C.J. (1970). Demonstration of three specific sites on the surface of foot-and-mouth disease virus by antibody complexing. *Journal of General Virology* **7,** 115–127.

BROWN, F., CARTWRIGHT, B. AND STEWART, D.L. (1963). The effect of various inactivating agents on the viral and ribonucleic acid infectivities of foot-and-mouth disease virus and on its attachment to susceptible cells. *Journal of General Microbiology* **31,** 179–186.

BROWN, F., HYSLOP, N.St.G., CRICK, J. AND MORROW, A.W. (1963). The use of acetylethylenimine in the production of inactivated foot-and-mouth disease vaccines. *Journal of Hygiene* **61,** 337–344.

BROWN, F., NEWMAN, J., STOTT, J., PORTER, A., FRISBY, D., NEWTON, C., CAREY, N. AND FELLNER, P. (1974). Poly(C) in animal viral RNAs. *Nature* **251,** 342–344.

BURROUGHS, J.N., ROWLANDS, D.J., SANGAR, D.V., TALBOT, P. AND BROWN, F. (1971). Further evidence for multiple proteins in the foot-and-mouth disease virus particle. *Journal of General Virology* **13,** 73–84.

CAPSTICK, P.B., GARLAND, A.J., CHAPMAN, W.G. AND MASTERS, R.C. (1965). Production of foot-and-mouth disease virus antigen from BHK21 clone 13 cells grown and infected in deep suspension cultures. *Nature* **205,** 1135.

CARTWRIGHT, B., CHAPMAN, W.G. AND BROWN, F. (1980). Serological and immunological relationships between the 140S and 12S particles of FMDV. *Journal of General Virology* **50,** 369–375.

CARTWRIGHT, B., CHAPMAN, W.G. AND SHARPE, R.T. (1982). The stimulation by heterotypic antigens of foot-and-mouth disease virus antibodies in vaccinated cattle. *Research in Veterinary Science* **32,** 338–342.

CARTWRIGHT, B., MORRELL, D.J. AND BROWN, F. (1982). Nature of the antibody response to the foot-and-mouth disease virus particle, its 12S protein subunit and the isolated immunizing polypeptide VP1. *Journal of General Virology* **63,** 375–381.

CAVANAGH, D., SANGAR, D.V., ROWLANDS, D.F. AND BROWN, F. (1977). Immunogenic and cell attachment sites of FMDV: Further evidence for their location in a single capsid polypeptide. *Journal of General Virology* **35,** 149–158.

CHEUNG, A., DELAMARTER, J., WEISS, S. AND KÜPPER, H. (1983). Comparison of the major antigenic determinants of different serotypes of foot-and-mouth disease virus. *Journal of Virology*, in press.

CLARKE, B.E., CARROLL, A.P., ROWLANDS, D.J., NICHOLSON, X., HOUGHTON, R.A., LERNER, R.A. AND BROWN, F. (1983). Synthetic peptides mimic subtype specificity of foot-and-mouth disease virus. *FEBS Letters* **157,** 261–264.

COHEN, S., CHANG, A.C.Y. AND HSU, L. (1972). Nonchromosomal antibiotic resistance in bacteria: Genetic transformation of *E. coli* by R-factor DNA. *Proceedings of the National Academy of Sciences of the United States of America* **69,** 2110–2114.

DERYNCK, R., REMAUT, E., SAMAN, E., STANSSENS, P., DE CLERCG, E., CONTENT, J. AND FIERS, W. (1980). Expression of human fibroblast interferon gene in *Escherichia coli. Nature* **287,** 193–197.

DOEL, T.R., SANGAR, D.V., ROWLANDS, D.J. AND BROWN, F. (1978). A reappraisal of the biochemical map of foot-and-mouth disease virus RNA. *Journal of General Virology* **41,** 395–404.

DIETZSCHOLD, B., KAADEN, O.R. AND AHL, R. (1972). Hybridization studies with subtypes and mutants of foot-and-mouth disease virus type O. *Journal of General Virology* **15,** 171–174.

FORSS, S. AND SCHALLER, H. (1982). A tandem repeat gene in a picornavirus. *Nucleic Acids Research* **10,** 6441–6450.

FRENKEL, H.S. (1951). Research on foot-and-mouth disease III. The cultivation of the virus in explanation of tongue epithelium of bovine animals. *American Journal of Veterinary Research* **12,** 187.

FRENKEL, S., BARENDREGT, L.G., KLOOSTERMAN, E.G. AND TALMAN, F.P. (1982). Serological response of calves to aluminum hydroxide gel FMD vaccine with or without saponin. Influence of genetic differences on this response. In *Proceedings of the 16th Conference of the Foot-and-Mouth Disease Commission*, pp. 211–221. Office International des Epizooties, Paris, France.

GRUBMAN, M.J. (1980). The 5′ end of foot-and-mouth disease virion RNA contains a protein covalently linked to the nucleotide pUp. *Archives of Virology* **63,** 311–315.

HARDY, K., STAHL, S. AND KÜPPER, H. (1981). Production in *B. subtilis* of hepatitis B core antigen and of major antigen of foot-and-mouth disease virus. *Nature* **293,** 481–483.

HARRIS, T.J.R. (1979). The nucleotide sequence at the 5′ end of foot-and-mouth disease virus RNA. *Nucleic Acids Research* **7,** 1765–1785.

HARRIS, T.J.R. AND BROWN, F. (1976). The location of the poly(C) tract in the RNA of foot-and-mouth disease virus. *Journal of General Virology* **33,** 493–501.

Hofschneider, P.H., Buergelt, E., Kauzmann, M., Mussgay, M., Franz, R., Ahl, R., Boehm, H., Strohmaier, K., Küpper, H. and Otto, B. (1981). Studies on the antigenicity and immunogenicity of the foot-and-mouth disease viral protein VP1 expressed in *E. coli*. In *Prospective Biological Products for Viral Diseases. 6th Munich Symposium on Microbiology* (P.A. Bachman, Ed.), pp. 105–113. Taylor and Francis, London.

Itakura, K., Hirose, T., Crea, R., Riggs, A.D., Heyneker, H.L., Bolivar, F.P. and Boyer, H.W. (1977). Expression in *Escherichia coli* of a chemically synthesized gene for the hormone somatostatin. *Science* **198,** 1056–1063.

Kaaden, O.R., Adam, K.H. and Strohmaier, K. (1977). Induction of neutralizing antibodies and immunity in vaccinated guinea pigs by cyanogen bromide peptides of VP3 of foot-and-mouth disease virus. *Journal of General Virology* **34,** 397–400.

Kelly, T.J. and Smith, H.O. (1970). A restriction enzyme from *Haemophilus influenzae* II. Base sequence of the recognition site. *Journal of Molecular Biology* **51,** 379–392.

King, A.M.Q., Underwood, B.O., McCahon, D., Newman, J.W.I and Brown, F. Biochemical identification of viruses causing the 1981 outbreaks of foot-and-mouth disease in the UK. *Nature* **293,** 479–480.

Kleid, D.G., Yansura, D., Small, B., Dowbenko, D., Moore, D.M., Grubman, J.J., McKercher, P.D., Morgan, D.O., Robertson, B.H. and Bachrach, M.L. (1981). Cloned viral protein vaccine for foot-and-mouth disease: responses in cattle and swine. *Science* **214,** 1125–1129.

Knudsen, R.C. (1982). Adjuvant effects of two formulations of CP20961, a synthetic lipid amine, for FMD virus vaccine in guinea pigs. In *Proceedings of the 16th Conference of the Foot-and-Mouth Disease Commission*, pp. 184–197. Office International des Epizooties, Paris, France.

Küpper, H., Delamarter, J., Otto, B. and Schaller, H. (1982). Expression of major foot-and-mouth disease antigen in *E. coli*. In *Proceedings of the IVth International Symposium on Genetics of Industrial Microorganisms* (Y. Ikeda and T. Beppu, Eds), pp. 222–226. Kodansha Limited, Tokyo.

Küpper, H., Keller, W., Kurz, Ch., Forss, S., Schaller, H., Franze, R., Strohmaier, K., Marquardt, O., Zaslavsky, V.G. and Hofschneider, H. (1981). Cloning of cDNA of the major antigen of foot-and-mouth disease virus and expression in *E. coli*. *Nature* **289,** 555–559.

Kurz, C., Forss, S., Küpper, H., Strohmaier, K. and Schaller, H. (1981). Nucleotide sequence and corresponding amino acid sequence of the gene for the major antigenic sites of foot-and-mouth disease virus. *Nucleic Acids Research* **9,** 1919–1931.

Laporte, J. and Lenoir, G. (1973). Structural proteins of foot-and-mouth disease virus. *Journal of General Virology* **20,** 161–168.

Laporte, J., Grosclaude, J., Wantyghem, J., Bernard, S. and Rouze, P. (1973). Neutralisation en culture cellulaire du pouvoir infectieux du virus de la fièvre aphteuse par des serums provenant de porcs immunisés à l'aide d'une protéine virale purifiée. *Comptes rendus hebdomadaires des séances de l'Académie des sciences. Series D* **276,** 3399–3402.

La Torre, J.L., Grubman, M.J., Baxt, B. and Bachrach, H.L. (1980). Structural polypeptides of aphtovirus are phosphoproteins. *Proceedings of the National Academy of Sciences of the United States of America* **77,** 7444–7447.

McKercher, P.D. and Farris, H.E. (1967) Foot-and-mouth disease in swine: Response to inactivated vaccines. *Archiv für die gesamte Virusforschung* **22,** 451–461.

Makoff, A.J., Paynter, C.A., Rowlands, D.J. and Boothroyd, J.C. (1982). Comparison of the amino acid sequence of the major immunogen from three serotypes of foot-and-mouth disease virus. *Nucleic Acids Research* **10,** 8285–8295.

Maxam, A. and Gilbert, W. (1977). A new method for sequencing DNA. *Proceedings of the National Academy of Sciences of the United States of America* **74,** 560–564.

Matheka, H.D. and Bachrach, H.L. (1975). N-Terminal amino acid sequences in the major capsid proteins of foot-and-mouth disease virus types A, O and C. *Journal of Virology* **16,** 1248–1253.

MELOEN, R.H. (1976). Localisation on foot-and-mouth disease virus (FMDV) of an antigenic deficiency induced by passage in BHK cells. *Archives of Virology* **51,** 299–306.

MELOEN, R.H. AND BRIAIRE, J. (1980). A study of the cross-reacting antigens on the intact foot-and-mouth disease virus and its 12S subunits with antisera against the structural proteins. *Journal of General Virology* **51,** 107–116.

MELOEN, R.H., ROWLANDS, D.J. AND BROWN, F. (1979). Comparison of the antibodies elicited by the individual structural polypeptides of foot-and-mouth disease and polio viruses. *Journal of General Virology* **45,** 761–763.

MELOEN, R.H., BRIARE, J., WOORTMEYER, R.J. AND VAN ZAANE, D. (1983). The main antigenic determinant detected by neutralizing monoclonal antibodies on the intact foot-and-mouth disease virus particle is absent from isolated VP1. *Journal of General Virology* **64,** 1193–1198.

MOORE, D.M. (1983). Introduction of a vaccine for foot-and-mouth disease through gene cloning. *Beltsville Symposia in agricultural research, Genetic engineering in agriculture, Recombinant Technology, May 1982,* pp. 132–142. Allanheld, Osmun, Montclair.

MOORE, D.M. AND COWAN, K.M. (1978). Effect of trypsin and chymotrypsin on the polypeptides of large and small plague variants of foot-and-mouth disease virus: relationship to specific antigenicity and infectivity. *Journal of General Virology* **41,** 549–567.

MORGAN, D.O., MCKERCHER, P.D. AND BACHRACH, H.L. (1970). Quantitation of the antigenicity and immunogenicity of purified foot-and-mouth disease virus vaccine for swine and steers. *Applied Microbiology* **20,** 770–774.

MOWAT, G.N. AND CHAPMAN, W.G. (1962). Growth of foot-and-mouth disease virus in fibroblastic cell line derived from hamster kidneys. *Nature* **194,** 253–255.

MOWAT, G.N., BARR, D.A. AND BENNETT, J.H. (1969). The development of an attenuated foot-and-mouth disease virus vaccine by modification and cloning in tissue cultures of BHK21 cells. *Archiv für die gesamte Virusforschung* **26,** 341–354.

OLASCOAGA, P.C., SUTMOELLER, P., FERNANDEZ, A. AND ABARACON, D. (1982). FMD virus production and control of vaccines in South America. In *Proceedings of the 16th Conference of the Foot-and-Mouth Disease Commission,* pp. 139–152. Office International des Epizooties, Paris, France.

PEREIRA, H. (1977). Subtyping of foot-and-mouth disease virus. In *Proceedings, International Symposium on Foot-and-Mouth Disease (II) Lyon 1976* (C. Mackowiak and R.H. Regamey, Eds). *Symposia Series in Developments in Biological Standardization,* volume 35, pp. 167–174. S. Karger, Basel.

PFAFF, E., MUSSGAY, M., BOEHM, H.O., SCHULZ, G.E. AND SCHALLER, H. (1982). Antibodies against a preselected peptide recognize and neutralize foot-and-mouth disease. *EMBO Journal* **1,** 869–874.

REMAUT, E., STANSSENS, P. AND FIERS, W. (1981). Plasmid vectors for high-efficiency expression controlled by the pL promoter of coliphage lambda. *Gene* **15,** 81–88.

ROBERTSON, B.H., MOORE, D.M., GRUBMAN, M.J. AND KLEID, D.G. (1983). Identification of an exposed region of the immunogenic capsid polypeptide VP1 on foot-and-mouth disease virus. *Journal of Virology* **46,** 311–316.

ROWLANDS, D.J., HARRIS, T.J.R. AND BROWN, F. (1978). More precise location of the polycytidylic acid tract in foot-and-mouth disease virus RNA. *Journal of Virology* **26,** 335–343.

ROWLANDS, D.J., SANGAR, D.V. AND BROWN, F. (1971). Relationship of the antigenic structure of foot-and-mouth disease virus to the process of infection. *Journal of General Virology* **13,** 85–93.

SANGAR, D.V. (1979). The replication of picornaviruses. *Journal of General Virology* **45,** 1–13.

SANGER, F., NICKLEN, S. AND COULSON, A.R. (1977). DNA sequencing with chain-termination inhibitors. *Proceedings of the National Academy of Sciences of the United States of America* **74,** 5463–5467.

SANGAR, D.V., BLACK, D.N., ROWLANDS, D.J. AND BROWN, F. (1977). Biochemical mapping of the foot-and-mouth disease virus genome. *Journal of General Virology* **35,** 281–297.

SANGAR, D.V., ROWLANDS D.J., HARRIS, T.J.R. AND BROWN, F. (1977). Protein covalently linked to foot-and-mouth disease virus RNA. *Nature* **268,** 648–650.

SANGAR, D.V., ROWLANDS, D.J., SMALE, C.J. AND BROWN, F. (1973). Reaction of glutaraldehyde with foot-and-mouth disease virus. *Journal of General Virology* **21,** 399–406.

SANGAR, D.V., BLACK, D.M., ROWLANDS, D.J., HARRIS, T.J.R. AND BROWN, F. (1980). Location of initiation site for protein synthesis on foot-and-mouth disease virus RNA by in vitro translation of defined fragments of the RNA. *Journal of Virology* **33,** 59–68.

SCODELLER, E.A., LEBENDIKER, M.A., DUBRA, M.S., BASARAB, O., LA TORRE, J.L. AND VASQUEZ, C. (1982). Inactivation of FMD virus by activation of virion-associated endonuclease. In *Proceedings of the 16th Conference of the Foot-and-Mouth Disease Commission*, pp. 43–50. Office International des Epizooties, Paris, France.

SKINNER, H.H. (1952). One-week-old white mice as test animals in foot-and-mouth disease research. In *Proceedings of the 15th International Veterinary Congress, Stockholm 1952*, Part 1, pp. 195–199.

SKINNER, H.H. (1959). The immunogenicity of strains of the virus of foot-and-mouth disease modified by serial passage in white mice and chick embryos. In *Proceedings, 16th International Veterinary Congress, Madrid*, pp. 391–393.

SKINNER, H.H. (1960). Some techniques for producing and studying attenuated strains of the virus of foot-and-mouth disease. *Bulletin de l'Office International d'Epizooties, Paris* **53,** 634–650.

SMITH, G.L., MACKETT, M. AND MOSS, B. (1983). Infectious vaccinia virus recombinants that express hepatitis B virus surface antigen. *Nature* **302,** 490–495.

SMITH, H.O. AND WILCOX, K.W. (1970). A restriction enzyme from *Haemophilus influenzae* I. Purification and general properties. *Journal of Molecular Biology* **51,** 379–391.

STROHMAIER, K. AND ADAM, K.H. (1974). Comparative electrophoretic studies of foot-and-mouth disease virus proteins. *Journal of General Virology* **22,** 105–114.

STROHMAIER, K., FRANZE, R. AND ADAM, K.H. (1982). Location and characterization of the antigenic portion of the FMDV immunizing protein. *Journal of General Virology* **59,** 295–306.

STROHMAIER, K., WITTMANN-LIEBOLD, B. AND GEISSLER, A.-W. (1978). The N-terminal sequence of three coat proteins of foot-and-mouth disease virus. *Biochemical and Biophysical Research Communications* **85,** 1640–1645.

TALBOT, P., ROWLANDS, D.J., BURROUGHS, J.N., SANGAR, D.V. AND BROWN, F. (1973). Evidence for a group protein in foot-and-mouth disease virus particles. *Journal of General Virology* **19,** 369–380.

VANDE WOUDE, G.F., SWANEY, J.B. AND BACHRACH, H. (1972). Chemical and physical properties of foot-and-mouth disease virus: a comparison with Maus Elberfeld virus. *Biochemical and Biophysical Research Communications* **48,** 1222–1229.

VASQUEZ, C., DENOYA, C.D., LA TORRE, J.L. AND PALMA, E.L. (1979). Structure of foot-and-mouth disease virus capsid. *Virology* **97,** 195–200.

VILLA-KOMAROFF, L., EFSTRATIADIS, A., BROOME, S., LOMEDICO, P., TIZARD, R., NABER, S.P., CHICK, W.L. AND GILBERT, W. (1978). A bacterial clone synthesising proinsulin. *Proceedings of the National Academy of Sciences of the United States of America* **75,** 3727–3731.

WESTERGAARD, J.M. (1982). The epidemiology of FMD outbreaks on the Islands of Funen and Zeeland in Denmark. In *Proceedings of the 16th Conference of the Foot-and-Mouth Disease Commission*, pp. 527–550. Office International des Epizooties, Paris, France.

WILD, T.F. AND BROWN, F. (1967). Nature of the inactivating action of trypsin on foot-and-mouth disease virus. *Journal of General Virology* **1,** 247–250.

WILD, T.F., BURROUGHS, J.N. AND BROWN, F. (1969). Surface structure of foot-and-mouth disease virus. *Journal of General Virology* **4,** 313–320.

WILLIAMS, D.C., VAN FRANK, R.M., MUTH, W.L. AND BENNET, J.P. (1982). Cytoplasmic inclusion bodies in *Escherichia coli* producing biosynthetic human insulin proteins. *Science* **215**, 687–689.

10

Biotechnology and Effluent Treatment

A. D. WHEATLEY

Environmental Biotechnology Group, Department of Chemical Engineering, University of Manchester Institute of Science and Technology (UMIST), P.O. Box 88, Manchester M60 1QD, UK

Introduction

Effluent treatment is the largest and one of the most important controlled applications of micro-organisms in the manufacturing industries. *Table 1* compares the quantity of sewage treated on an annual basis with other biological and non-biological products in the UK. The Department of the Environment reports that Water Authority spending on effluent treatment and pollution control in the UK in 1978 was £1100 million (Water Data Unit, 1979). It is European Community policy (Ellington and Burke, 1981) that spending on pollution should continue to rise in accordance with increased expectations of environmental quality and inflation. A US Government technology report (Congressional Office of Technology Assessment, 1981) estimated that municipal, agricultural and industrial spending on pollution control will be $400 000 million between 1976 and 1986. The size and impact of this market for the combined use of microbiology, biochemistry and engineering will be second only to fine chemicals in the foreseeable future (Bull, Holt and Lilly, 1982).

Table 1. Quantities of sewage effluent treated compared with other common industrial products (UK figures) (modified from Dunnhill, 1981).

Product	Weight (tonnes/year)	Price (£/t)
Water as sewage	6000×10^6	0·10
Milk	16×10^6	25
Steel	12×10^6	300
Beer	$6{\cdot}6 \times 10^6$	280
Sugar	1×10^6	350
Cheese	$0{\cdot}2 \times 10^6$	1300
Bakers yeast	$0{\cdot}1 \times 10^6$	460
Citric acid	$0{\cdot}015 \times 10^6$	700
Penicillin	$0{\cdot}003 \times 10^6$	45 000

Abbreviations: ADF, alternating double filtration; BOD, biological oxygen demand; COD, chemical oxygen demand; DO, dissolved oxygen; RBC, Rotating Biological Contacter; TOC, total organic carbon; UASB, Upflow Anaerobic Sludge Blanket.

Biotechnology and Genetic Engineering Reviews—Vol. 1, February 1984
0264–8725/84/01/261–49$10.00 + $0.00 © Intercept Ltd

The traditional aims of effluent treatment have been to reduce the concentration of organic matter in waste water which otherwise exerts an oxygen demand expressed as BOD (biological oxygen demand), and to decrease the number of potential pathogens in the waste. This allows the effluent to be discharged into the environment without adverse effect.

The recent discoveries of how to alter the characteristics of micro-organisms by genetic manipulation, combined with new advances in reactor design, offer new opportunities for waste treatment technology, and broader aims for environmental biotechnology can now be identified such as (1) the recycling of materials for economic self-defence, (2) conservation of resources and (3) management of environmental problems. Not all of these strategic goals can be stimulated by normal market forces but the economic deficit can be rectified by legislation on pollution, government subsidy for conservation of vital resources and as a consequence of disruption in existing supplies. In Italy, for example, effluent treatment plant to recover biogas from wastes attracts a 70% grant because the country is short of indigenous energy (Alfani, 1983). This chapter reviews existing effluent treatment, emphasizing processes available for by-product recovery, and discusses the future prospects for environmental biotechnology for further recovery of useful materials and the breakdown of recalcitrants (used here in the sense of a persistent organic compound).

The characteristics of waste

Waste water is treated to prevent any adverse effects on the receiving water and to allow reuse. The polluting load is measured by a range of analyses to assess its physical, chemical and biological effects on the environment. The biggest problem with domestic, agricultural and a wide range of industrial wastes (i.e. food, drinks, and fermentation wastes) is caused by their organic content. The metabolism of this organic matter by the natural organisms contained in the receiving water causes rapid oxygen depletion and elimination of the normal aquatic flora and fauna. The organic content of waste water is determined by three tests: of chemical oxygen demand (COD), total organic carbon (TOC), and biological oxygen demand (BOD). COD and TOC are measures of total organic carbon determined by chemical oxidation and pyrolysis (with infra-red absorption of the carbon dioxide formed) respectively. BOD is the biologically oxidizable fraction measured by incubation over 5 days. These are standardized tests and the methods and procedures are described in several texts (Department of the Environment, 1972; American Public Health Association, 1975, Department of the Environment 1979). The amount of suspended solids, determined by filtration and turbidity tests, is a parameter which is used to assess the effects of pollution from wastes on light penetration in streams (which controls algal growth) and the amount of precipitation on to the river bed (which affects bottom fauna and fish eggs).

The ammonia (as nitrogen) content of wastes is also normally measured. In domestic wastes nitrogen is derived from urea and the deamination of protein. Residual ammonia is directly toxic to aquatic life and its oxidized products can cause nutrient enrichment and excessive algal growth or eutrophication.

An additional complication arises if the waste water is recycled for water supply: nitrate combines with haemoglobin and can be toxic to young growing children. Nitrogen is, however, an essential additional nutrient for any biological treatment process. *Table 2* compares the character and strength of a typical domestic agricultural and industrial waste water.

Table 2. A comparison of some different types of waste water.

Characteristics	Type of waste water		
	Domestic	Agricultural	Industrial
pH	7·8	8·5	4·5
BOD (mg/ℓ)	370	15 000	8000
COD (mg/ℓ)	670	25 000	17 370
TOC (mg/ℓ)	219	N.K.	N.K.
Total solids (mg/ℓ)	1309	22 000	3200
Suspended solids (mg/ℓ)	146	14 500	570
Ammonia (as N) (mg/ℓ)	46	2000	0·5

N.K., not known.

A detailed characterization of sewage has been carried out by Painter and co-workers (Painter 1958, 1971; Painter and Viney, 1959; Painter, Viney and Bywaters, 1961), who have identified 75% of the organic nutrient present. They found that 30–40% of the total organic material present was in solution; 10–15% was present as colloidal solids (0·1–1 μm); 20–25% as supra-colloidal solids (1–100 μm); and 30–35% as settleable solids (> 100 μm). The single largest organic constituents in both solution and suspension were the fatty acids, representing about 30% of the total organic carbon. The next largest constituents were the sugars (about 15% of the total organic carbon). It is probable that the sugars are progressively converted into the fatty acids and that the amount of fatty acid varies according to the age of the sewage. The total fat and grease content of sewage is 40–100 mg/ℓ and all the fatty acids up to C20 have been identified in sewage, including those with odd numbers of carbon. The most common are the simple acids acetic, butyric and propionic derived

Table 3. Concentration of inorganic materials (mg/ℓ) in domestic sewage.

Element	Soft water	Hard water
Cl	20·1	68
Si	3·9	N.K.
Fe	0·8	0·8
Al	0·13	N.K.
Ca	9·8	109
Mg	10·3	6·5
K	5·9	20·0
Na	23	100
Mn	0·47	0·05
Cu	1·56	0·2
Zn	0·36	0·65
Pb	0·48	0·08
S	10·3	22·0
PO_4 as P	6·6	22·0

N.K., not known.

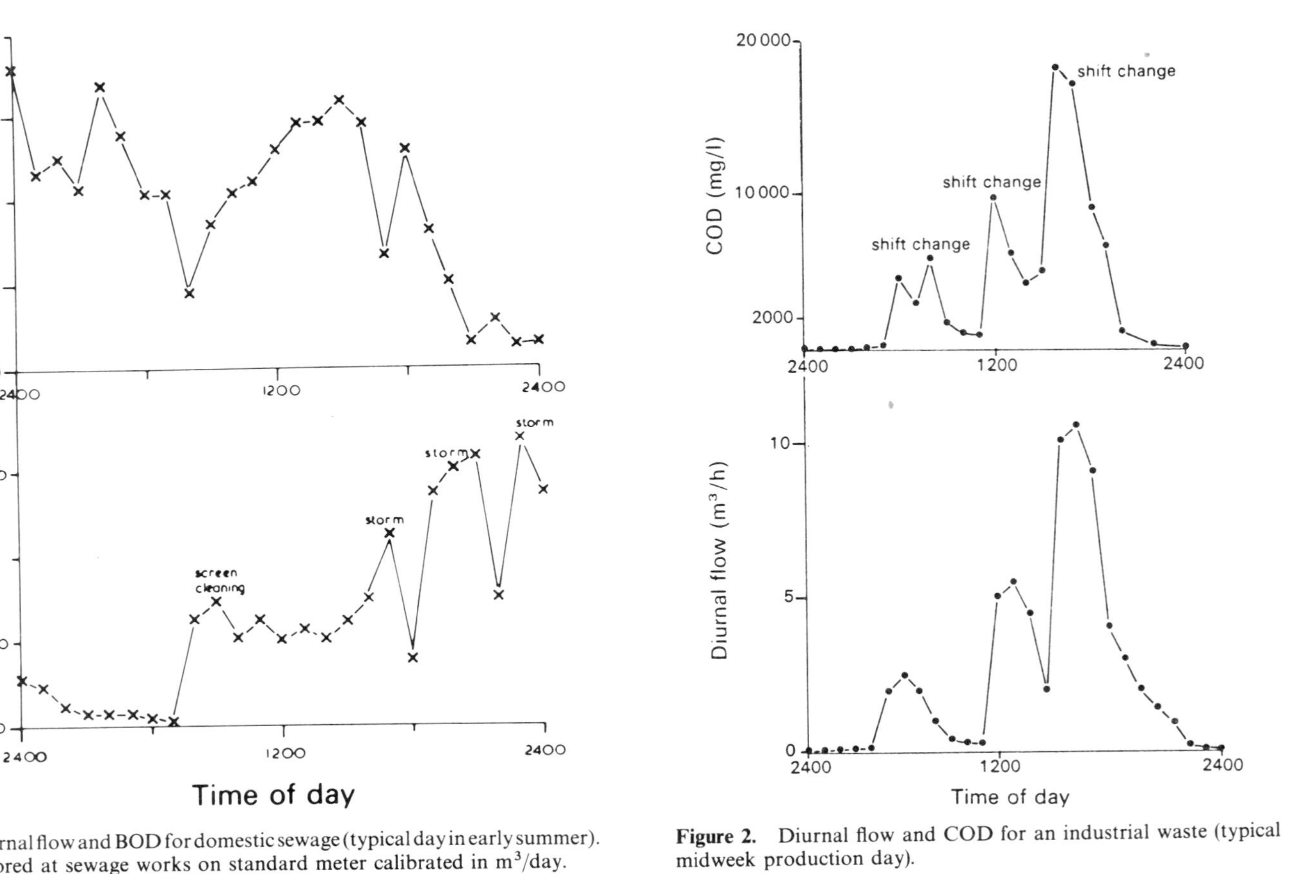

Figure 1. Diurnal flow and BOD for domestic sewage (typical day in early summer). Flow monitored at sewage works on standard meter calibrated in m^3/day.

Figure 2. Diurnal flow and COD for an industrial waste (typical midweek production day).

from the simple sugars. Glucose, sucrose and lactose are the predominant sugars, followed by galactose, fructose, xylose and arabinose; most of the sugar is glucose. Carbohydrate in suspension is mainly cellulose and starch.

The inorganic substances present in waste waters are influenced by the characteristics of the water supply, as well as by the nature of any industrial processing. The differences between soft- and hard-water domestic sewage are shown in *Table 3.*

There are also major changes in the flow and strength of waste waters according to the diurnal patterns of behaviour and production. *Figures 1 and 2* show the diurnal changes in flow and strength of a domestic and an industrial waste (Wheatley and Williams, 1976; Wheatley and Cassell, 1983). Changes of up to 300% of average are common with domestic works when affected by storm run-off. The flows never stop completely, because of the infiltration of ground water.

Present treatment technology

Waste waters are normally complex in composition, and treatment plants need to acquire a diverse range of micro-organisms with the metabolic capacity to degrade various types of wastes and to allow the effluent to be discharged to the environment without adverse effect. Effluent treatment depends on mixed-culture fermentation but also includes higher grazing organisms so that a complete ecosystem is formed with various trophic levels. There are two common types of fermenter—plug-flow fixed-film systems (referred to as biological or percolating filters) and completely mixed flocculant processes or activated sludge (*Figure 3*).

Most effluents are heterogeneous and contain both dissolved and suspended matter. The primary step in biological treatment is, therefore, adsorption of the substrate on to the biological surface. This is then followed by a sequence of of steps with the breakdown of the adsorped solids by extracellular enzymes, the absorption of dissolved materials into cells, growth, endogenous respiration, release of excretory products and the ingestion of the primary population by secondary grazers (*Figure 4 and 5*). The artifical ecosystem can be in complete balance in such a way that all the available substrate and subsequent primary growth are consumed or leave the system. Thus the only products of waste treatment are simple inorganic salts and gases.

The earliest type of treatment was the percolating filter, which developed directly from experiments carried out on the percolation of sewage through soil. The first biological filters were built at Salford in 1890 and despite nearly 100 years of use, the current basic design is virtually unchanged. They are circular or rectangular tanks made from brick or concrete containing a graded bio-support medium—usually clinker, slag, stone or gravel, depending on local availability. The components of the medium are normally 30–100 mm in size and the packed depth 1·8 m. Biological filters are classified by the organic and hydraulic load which they receive (*Table 4*). They are easy to maintain, incur low running costs, produce only small quantities of surplus biomass and have a plant life of 30–50 years. The process has proved to be very reliable

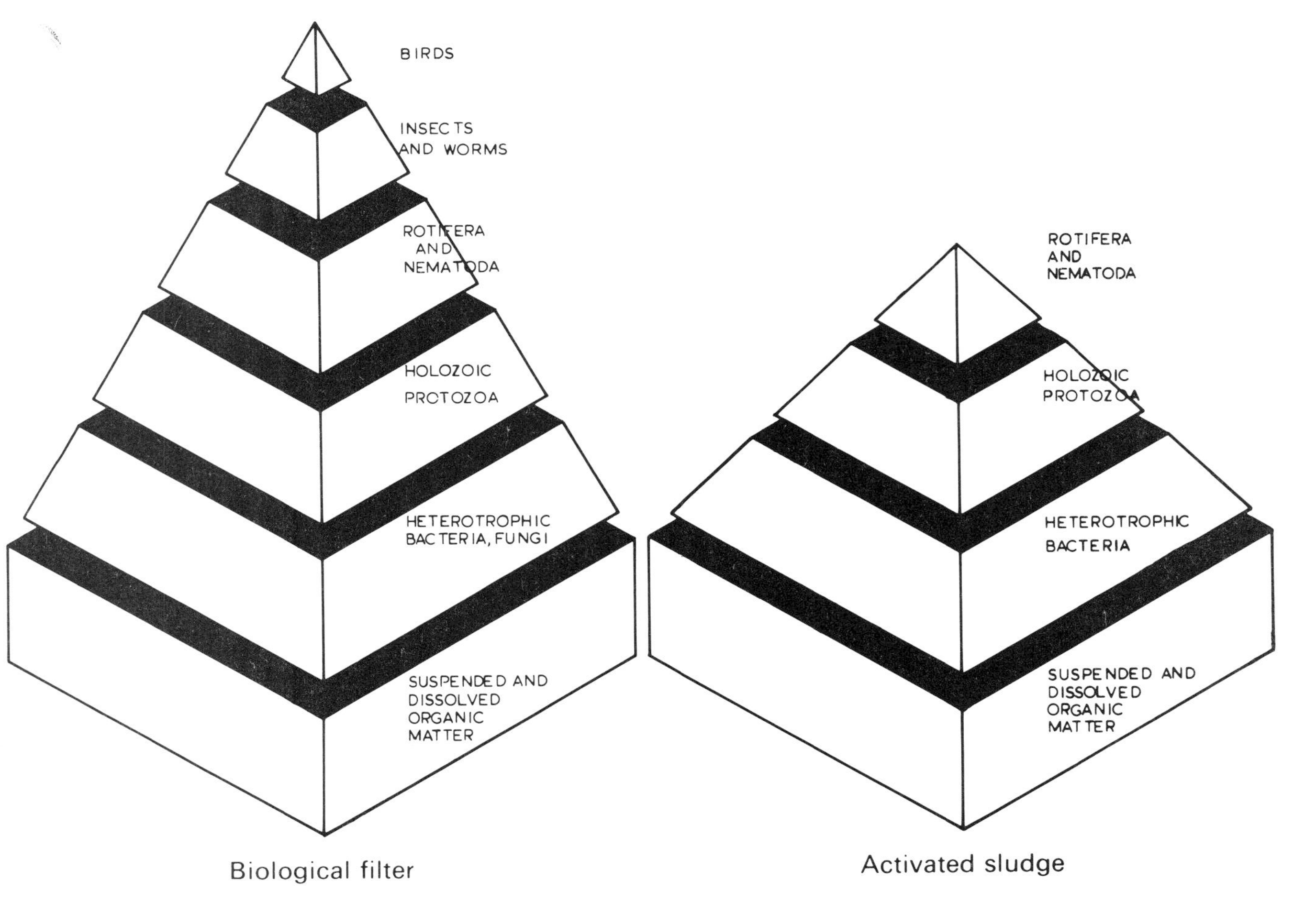

Figure 3. Diagram of the main food links in biological waste treatment.

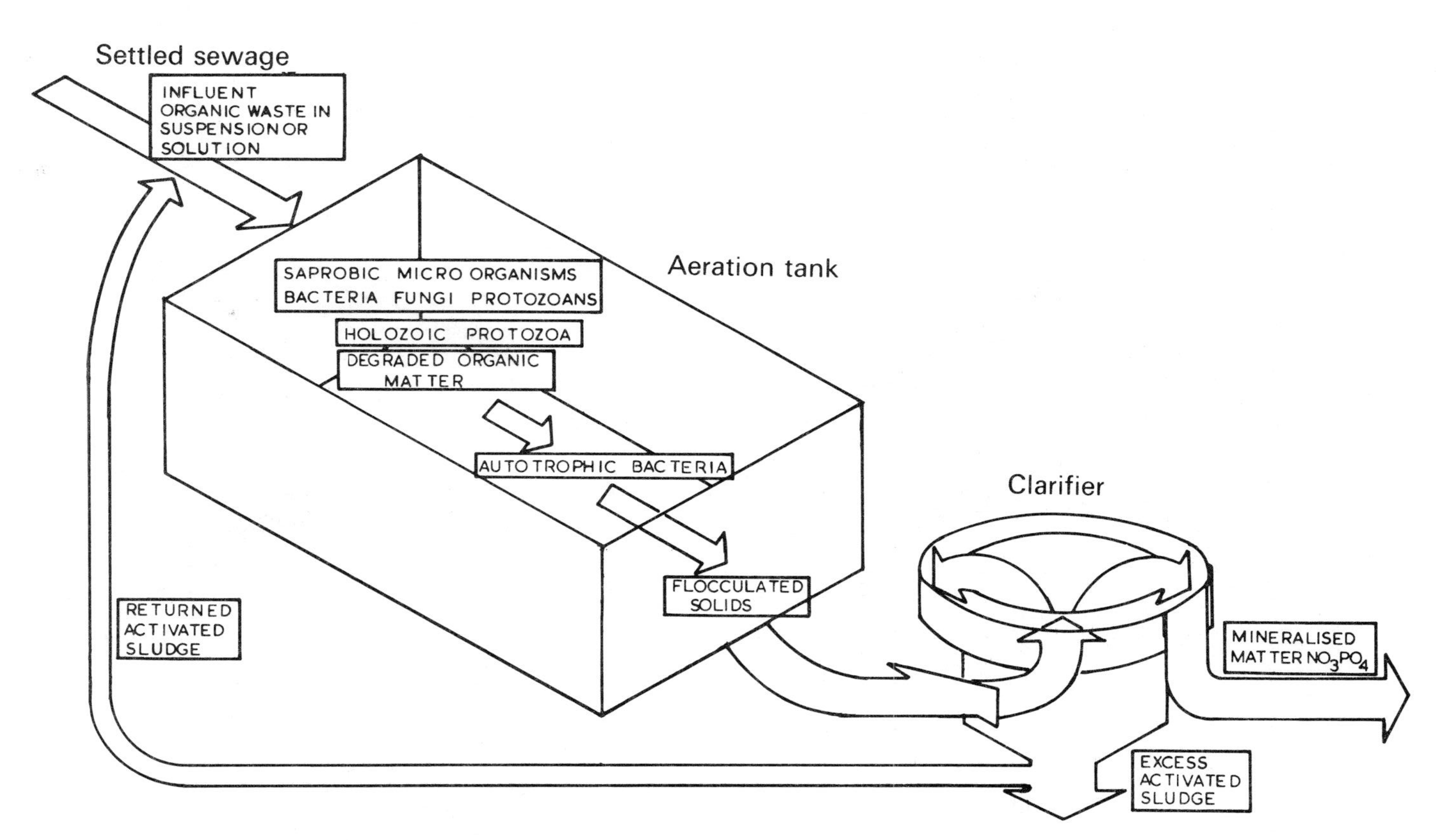

Figure 4. Mass transfer processes in activated sludge.

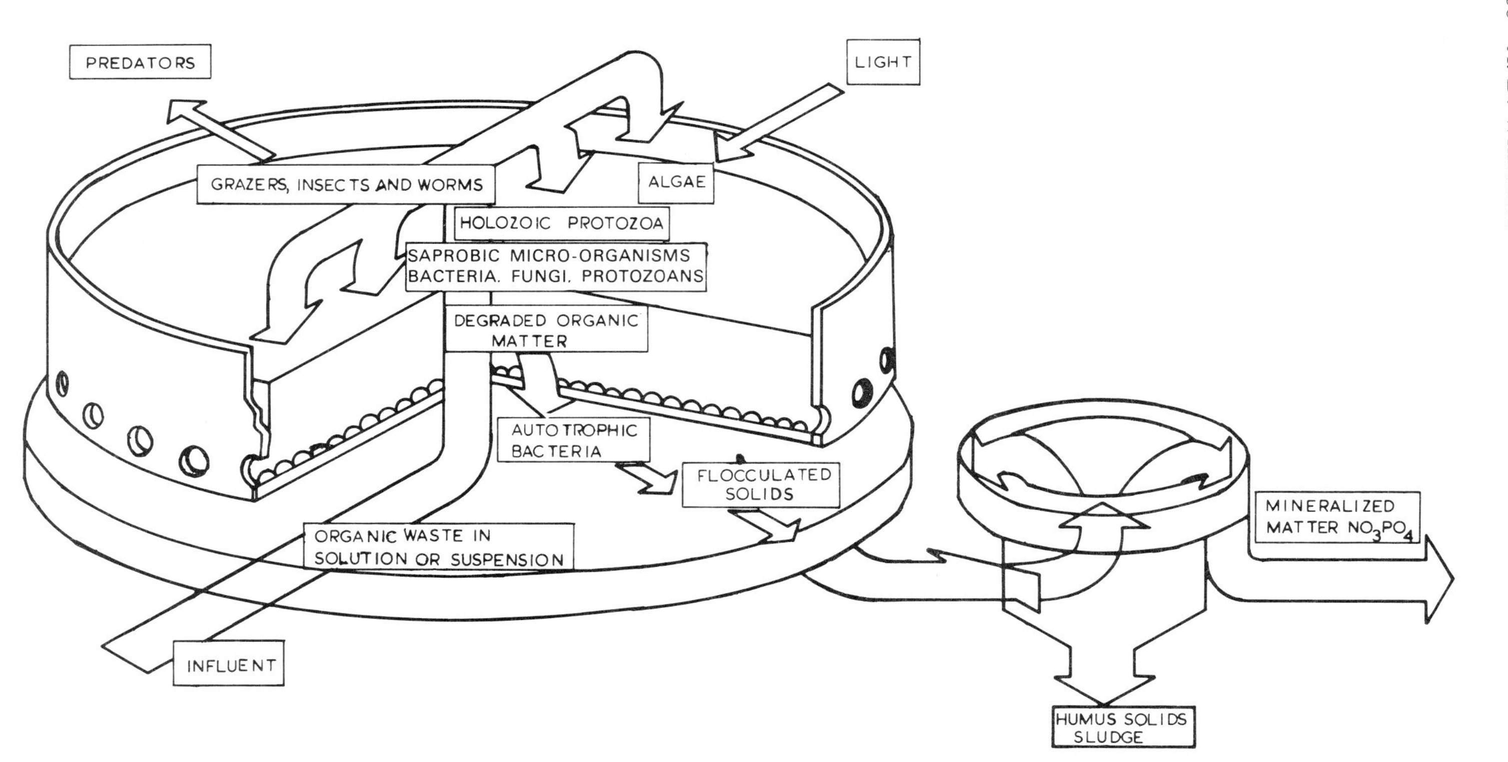

Figure 5. Mass transfer processes in biofiltration.

Table 4. Biological filter loadings.

Type of filter	BOD load (kg/m^3 of media)	Efficiency of BOD removal (%)	Sludge conversion (%)
Septic tank systems	0·05	95	5
Low-rate percolating	0·05–0·1	95	15–20
Mid-range double filtration, ADF, recirculation	0·1–0·5	85–90	20–50
High rate filtration	2·0–6·0	50–70	50–90

and it is estimated that 70% of European and American effluent treatment plants use this type of system.

The most common alternative is the activated sludge process which is a completely mixed process, aerated and stirred by bottom diffusion or surface agitation; it was first used at Manchester in 1914. The incoming screened degritted settled waste is aerated with active sludge or bios recycled from a settling tank. The residence time in the aeration tank is 4–10 hours, after which the final effluent is separated from the active sludge by settlement (1 hour) and most of the sludge is returned to mix with the incoming waste (*Figure 5*). The process is more intense than that of biological filters and is able to treat approximately 10 times the effluent per volume of reactor and is therefore much cheaper to build. It is, however, more difficult to operate and maintain, incurs high running costs through mixing and aeration, and produces large quantities of surplus biomass or sludge.

The extra operational difficulties result both from increased mechanical plant and from a reduced ecological trophic diversity. Growths of fungi or filamentous bacteria, both of which are natural organisms in polluted waters and normal components of biological filter film, seriously reduce the density of the activated sludge bios and interfere with the settlement and return of the active sludge to the aeration tank. A second effect of the reduced ecological diversity seems to be a reduction in the resilience of activated sludge (compared with that of biological filters) to shock or variable organic loads. Despite these operational difficulties, and because of the large areas of land required for equivalent biological filtration, activated sludge has been the preferred type of treatment for loads from population equivalents of more than 50 000. Normal BOD loads are 0·2–0·5 kg/kg dry biomass for a 90% removal of BOD (*Table 5*). The majority of effluent in the UK is treated by activated sludge; *Table 6* shows the size of the largest treatment works.

Table 5. Activated sludge loadings.

Type of plant	BOD load per kg MLSS* (kg)
Extended aeration	0·02–0·06
Conventional load	0·2–0·5
High rate	1·5–2·0

* Mixed liquor suspended solids.

Table 6. Sewage works in England, Wales, Northern Ireland and Scotland in 1978 with a theoretical dry-weather flow greater than 100 000 m^3/day.

Sewage treatment works	Water Authority or River Purification Board	Dry-weather flow (000's m^3/day)	Population (000's)
Beckton	Thames	912	2250
Crossness	Thames	505	1550
Minworth	Severn Trent	384	1050
Davyhulme	North West	314	714
Mogden, Isleworth	Thames	439	1330
Deephams	Thames	171	685
Stoke Bardolph	Severn Trent	146	465
Blackburn Meadows	Yorkshire	128	466
Derby (Raynesway)	Severn Trent	108	210
Knotstrop (High Level)	Yorkshire	150	481
Finham	Severn Trent	112	339
Maple Lodge	Thames	113	434
Avonmouth	Wessex	160	500
Rye Meads	Thames	78	322
Dalmuir	Clyde	219	440
Dalmarnock	Clyde	136	300
Belfast (Duncrue Street)	Northern Ireland	110	530

Source: Water Data Unit, 1979.

TREATMENT PLANT DESIGN

The basis of aerobic treatment design is to provide an environment in which the waste, the organisms responsible for purification, and the air, are brought into contact. Treatment efficiency is, in principle, therefore proportional to the amount of biomass and contact time between waste and biomass. Normally the limiting factor is aeration efficiency: excessive biomass rapidly occludes the voidage in biological filters and exerts too great an oxygen demand in activated sludge. Early work (Dunbar and Calvert, 1908; Royal Commission on Sewage Disposal, 1908) rapidly established empirically the amounts of biomass, the concentrations of waste and the aeration or ventilation required to ensure a treated effluent suitable for river discharge. Improvements followed as the understanding of the process grew. O'Shaughnessy (1931) conducted some early experiments on growth rate and noted that BOD removal was first order: dilution of the waste reduced the amount of biomass generated. Several practical methods of controlling biological treatment evolved from this early appreciation of the kinetics. A common problem at that time was excessive biological growth in filters, thus restricting both ventilation and effluent flow. This led to a deterioration in performance and ultimately, if the filter blocked, complete failure. It affected biological filters mainly in the winter when film autolysis and grazing activity was low. Wishart and Wilkinson (1941) described some experiments in which they successfully controlled biomass by washing the filter with biologically treated effluent. They then suggested a scheme of alternating double filtration (ADF) where the waste normally applied to two filters in parallel was instead applied to two filters in series. Biomass accumulates in the 1st filter as it receives and consumes virtually all of the applied substrate. The order of the filters is then reversed before the bios in the first filter prevents the free

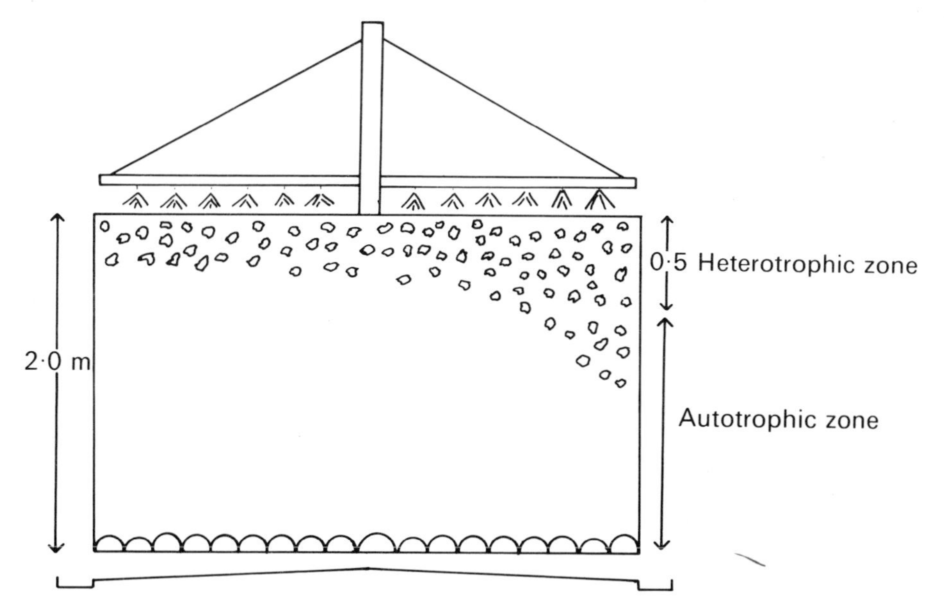

Figure 6. Ecological zones in biological filtration.

flow of waste and air through it. The original primary filter then receives treated effluent which has passed through the new primary filter. This results in the rapid autolysis and consumption of the starved biomass. ADF enables filters to be operated at more than twice the normal load with few operational problems. Additional experiments by Mills (1945a) established that the best frequency of alternation was weekly, and Tomlinson (1946) reported on the effects on ecology. ADF is now extensively used, particularly with industrial wastes. Mills (1945b) noted another change regarding the distribution of biomass through the depth of the filter. Normally the biomass was in two distinct zones because the system was plug flow and the surface of the filter received and removed the greatest amount of applied substrate. Thus most of the active organic-consuming organisms are in this upper 0·5 m (*Figure 6*), known as the heterotrophic layer. Below this is the autotrophic zone, which is about 1·5 m deep. The bios material in the base of the filter occurs in flocculated deposits rather than as prostrate slime adhering closely to the support media, as in the case of surface growth (Wheatley, 1981) (*Figure 7*). The material is the debris from the primary population or the surface micro-organisms. There is very little BOD removal in these lower layers but considerable autotrophic nitrification. Higher rates of instantaneous irrigation such as those from ADF, recirculation and intermittent dosing are thus used to increase the efficiency of biological filtration by dissipating the load deeper into the filter. Tomlinson (1946) showed that 62% of the biomass occurred in the top 0·5 m of a conventional filter, whereas there was only 44% in the top of an ADF filter. This improves total BOD removed by making more efficient use of the filter media but normally reduces the nitrifying activity. Dilution of the incoming waste by recirculation is also used with activated sludge and biological filters: the strength of the waste

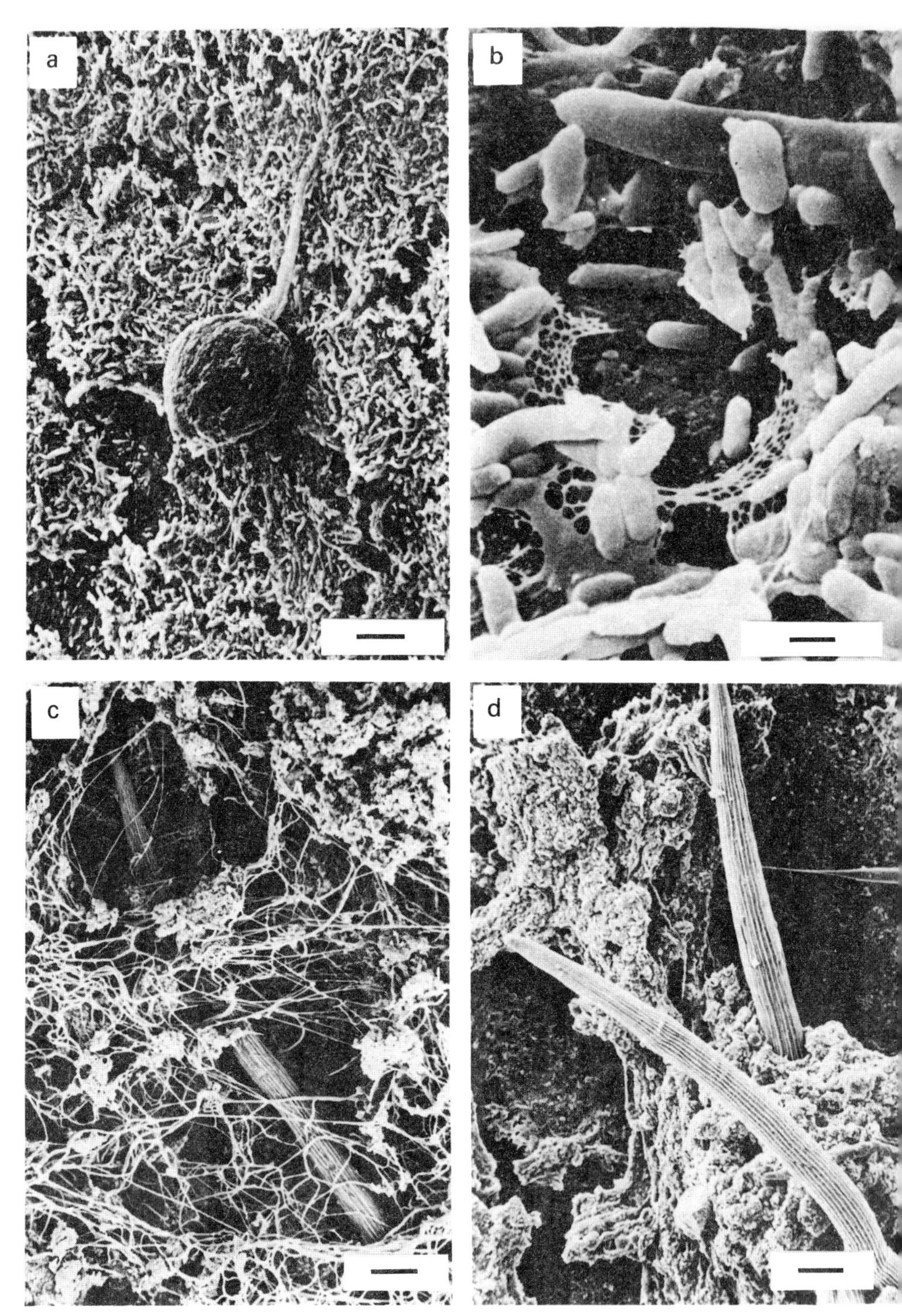

Figure 7. Electron micrographs of biological growth adhering to support media. (a) Bacterial growth and stalked protozoan *Opercularia* from a filter treating domestic sewage (surface growth) (bar = 1 μm); (b) Higher magnification of bacterial growth showing bacterial cells and extracellular polymer (bar = 0·73 μm); (c) Fungal growth from a biological filter treating dairy waste (also shown is a nematode worm) (bar = 22·5 μm; (d) Humus solids and debris from the base of a biological filter (two nematode worms are also visible) (bar = 22·5 μm).

is diluted, important trace nutrients are returned and the effluent is reaerated (Mills, 1945b; Lumb and Eastwood, 1958; Lumb, 1960). Other work (Lumb and Barnes, 1948; Tomlinson and Hall, 1955; Hawkes and Shephard, 1972) established that slowing down the distributor on biological filtration also improved performance by encouraging an even spread of biomass through the filter depth. One of the more recent techniques was direct double filtration with the first filter of a series of two containing a larger medium (150–300 mm) to accommodate more biological growth but at a lower BOD removal efficiency (about 70%). The second filter contains conventional media and completes the removal of organic waste to 95%. Overall loading is higher. Different combinations of processes are now also used, such as high-rate activated sludge followed by filters, or high-rate filters followed by activated sludge (Tebbutt, 1971). These are often simple extensions to existing plant. Similar modifications have been made to the activated sludge process to even out energy and aeration demand through the tank. A common method of operation is tapered aeration, which is designed to match oxygen demand to aeration capacity. Based on a plug flow through the aeration basin, the oxygen demand at the outlet is much less than that at the inlet, so that aeration can be progressively reduced along the length of the tank. This is normally accomplished by a reduction in the number of diffuser domes in the base of the tank. A similar modification, more applicable to surface aeration, is step aeration which introduces the waste at intervals throughout the length of the tank. Aeration of the return sludge without addition of further waste is also used to encourage the organisms to utilize any stored nutrient. The activated sludge is then able to assimilate a greater amount of substrate or waste when re-introduced into the main treatment tank. This process is known as contact stabilization.

ECOLOGY

The ecology of waste treatment by open mixed culture is complex and there are problems in differentiating between the organisms that are actively growing and those organisms that can be isolated from the fermenters. There is a fundamental difference between the two processes: activated sludge is a truly aquatic environment, whereas the biological filter contains only a thin film of water over the surface of the bios. The bacteria are the basis of both processes and the Zoogloea (Gram-negative, non-sporing, non-motile capsulated rods) are normally referred to as being the major group responsible for treatment (Wattie, 1943; Hawkes, 1963) (*Figures 8* and *9*). There are still some doubts as to whether Zoogloea are a defined group or a strain of *Pseudomonas* which develop a slime layer because of the particular conditions under which they are grown (McKinney, 1956). It is assumed (Pike and Carrington, 1972; Pike, 1975) that only a small proportion of the sludge floc and slime is actively growing but most of the bios could still carry out biochemical reactions by extracellular enzymes and adsorption of organic matter on to the floc. A number of other bacteria have also been shown to be active in waste-water treatment processes (*Chromobacter*, *Achromobacter*, *Flavobacter*, *Arthrobacter*) (Pike, 1975). Work has also been carried out on the filamentous bacteria because of their association

Figure 8. Common bacteria and algae in effluent treatment processes. (a) *Sphaerotilus* (actual filament diameter 2–3 μm); (b) *Leptothrix* (filament diameter 1–2 μm); (c) *Phormidium* (filament diameter 5 μm); (d) *Ulothrix* (filament diameter 15–17 μm); (e) *Zoogloea* (cells 0·5–1·0 μm in diameter and 1–2 μm long; finger-like projections normally approx. 50 μm in diameter).

with poorly operating activated sludge plants. The four common filamentous bacteria that have been identified and characterized are *Beggiatoa*, *Sphaerotilus*, *Leptothrix* and *Nocardia* (Eikelboom, 1975; van Veen, Mulder and Deinema, 1978; Dhaliwal, 1979; Beccari, Mappelli and Tandoi, 1980). A wide variety of enteric organisms, including some potential pathogens, can be detected in fresh sewage, but they have no role in the treatment process and their numbers fall rapidly (Pike and Carrington, 1979) (*Table 7*). The organisms responsible for treatment are derived from aerial inoculation and via infiltration water.

Six filamentous fungi have also been shown to grow actively in waste-water treatment systems: these are *Fusarium*, *Geotrichum*, *Subbaromyces*, *Saprolegnia*,

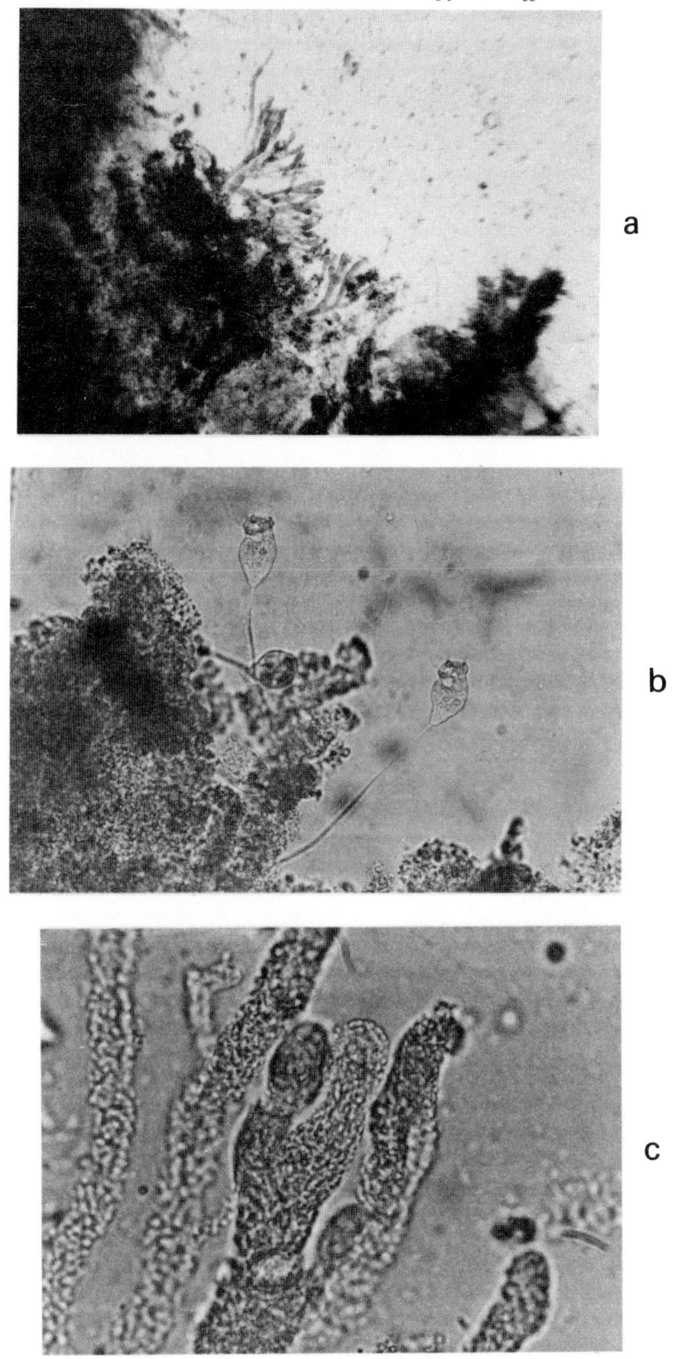

Figure 9. *Zoogloea* and *Sphaerotilus* bacteria growing in activated sludge. (a) Showing distinct finger-like projections (50 μm in diameter) of zoogloea; (b) Cells of zoogloea and peritrichous protozoa (*Vorticella*, cell approx. 50–90 μm in length); (c) Higher magnification of finger-like projections of zoogloea.

Table 7. Summary of pathogen removal by various sewage treatment processes†.

Organisms	Parameters*	Primary sedimentation	Trickling filter with primary and secondary sedimentation, sludge digestion and sludge drying	Activated sludge with primary and secondary sedimentation, digestion and sludge drying	Oxidation ditch with sedimentation and sludge drying	Septic tanks	Land application or slow sand filtration as tertiary treatment	Anaerobic digestion (30–40°C)
Enteric viruses	A	10^3–10^5	10^3–10^5	10^3–10^5	10^3–10^5	0–10^9	10–10^4	May survive for over 3 months
	B	10^3–10^5	10^2–10^4	10–10^4	10–10^4	10–10^3	0–10^2	
	C	0–30	90–95	90–99	90–99	50	99–100	
	D	Contaminated	Contaminated	Contaminated	Contaminated	Contaminated		
Salmonellae	A	10^3–10^4	10^3–10^4	10^3–10^4	10^3–10^4	0–10^9	10–10^3	May survive for several weeks
	B	10^2–10^3	10^2–10^3	10–10^3	10–10^3	0–10^8	0/1	
	C	50–90	90–95	90–99	90–99	50–90	100	
	D	Contaminated	Contaminated	Contaminated	Contaminated	Contaminated		
Shigellae	A	10^3–10^4	10^3–10^4	10^3–10^4	10^3–10^4	0–10^9	10–10^3	Unlikely to survive for more than a few days
	B	10^2–10^3	10^2–10^3	10–10^3	10–10^3	0–10^8	0/1	
	C	50–90	90–95	90–99	90–99	50–90	100	
	D	Contaminated	Contaminated	Contaminated	Contaminated	Contaminated		
E. coli	A	10^6–10^8	10^6–10^8	10^6–10^8	10^6–10^8	10^7–10^9	10^4–10^7	May survive for several weeks
	B	10^5–10^7	10^5–10^7	10^4–10^7	10^4–10^7	10^6–10^8	0–10^3	
	C	50–90	90–95	90–99	90–99	50–90	99·99–100	
	D	Contaminated	Contaminated	Contaminated	Contamined	Contaminated		
Cholera vibrio	A	10–10^3	10–10^3	10–10^3	10–10^3	0–10^9	0·1–10^2	May survive for 1 or 2 weeks
	B	1–10^2	1–10^2	0·1–10^2	0·1–10^2	0·10^8	0/	
	C	50–90	90–95	90–99	90–99	50–90	100	
	D	Contaminated	Contaminated	Contaminated	Contaminated	Contaminated		

Leptospires	A	Very few	Very few	Very few	Very few	Very few	Very few	Survive for not more than 2 days
	B	Very few	Very few	Very few	Very few	0/1?	0	
	C	0	0	0?	0	100	100	
	D	Safe	Safe	Safe	Safe	Safe	—	
Entamoeba histolytica cysts	A	$10–10^4$	$10–10^4$	$10–10^4$	$10–10^4$	$0–10^5$	$10–10^3$	May survive for 3 weeks
	B	$5–10^4$	$5–10^3$	$5–10^3$	$5–10^3$	$0–10^5$	0	
	C	10–50	50?	50?	50?	0?	100	
	D	Contaminated	Safe	Safe	Safe	Contaminated	—	
Hookworm ova	A	$10–10^3$	$10–10^3$	$10–10^3$	$10–10^3$	$0–10^4$	$10–10^2$	Ova will survive
	B	$10–10^2$	$10–10^2$	$10–10^2$	$10–10^2$	$0–10^3$	0	
	C	50	50–90	50–90	50–90	50–90	100	
	D	Contaminated	Contaminated	Contaminated	Contaminated	Contaminated	—	
Ascaris ova	A	$10–10^3$	$10–10^3$	$10–10^3$	$10–10^3$	$0–10^4$	$0–10^2$	Ova will survive for many months
	B	1–10	$0–10^2$	$0–10^2$	$0–10^2$	$0–10^3$	0	
	C	30–80	70–100	70–100	70–100	50–90	100	
	D	Contaminated	Contaminated	Contaminated	Contaminated	Contaminated	—	
Schistosome ova	A	1–100	1–100	1–100	1–100	1–100	1–10	Ova may survive up to 1 month
	B	1–10	1–10	1–10	1–10	1–10	0	
	C	80	50–99	50–99	50–99	50–90	100	
	D	Contaminated	Safe	Safe	Safe	Contaminated	—	
Taenia ova	A	1–100	1–100	1–100	1–100	$0–10^3$	0·1–50	Ova will survive for a few months
	B	0·1–50	0·1–50	0·1–50	0·5–50	0–50	0	
	C	50–90	50–95	50–95	50?	50–90	100	
	D	Contaminated	Contaminated	Contaminated	Contaminated	Contaminated	—	

* A: in typical inflow (no./ℓ); B: in typical outflow (no./ℓ); C: removal (%); D: final sludge.
† After Seachem *et al.* (1980).

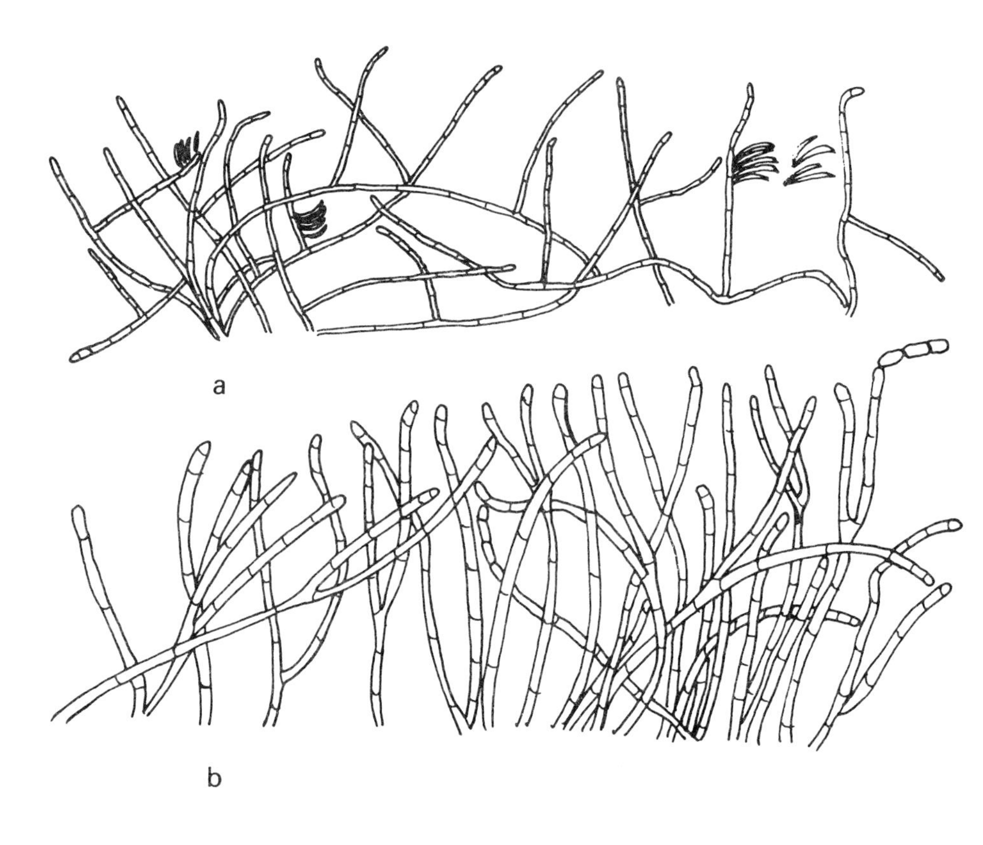

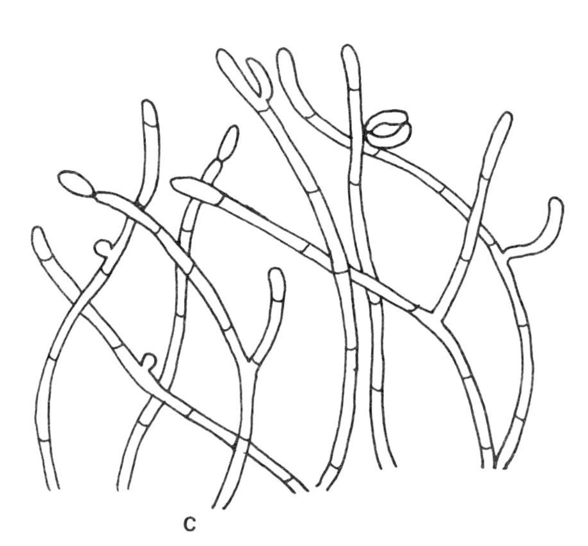

Figure 10. The most common fungi in effluent treatment processes. (a) *Fusarium aquaeductuum* (hyphal diameter 5 μm); (b) *Geotrichum candidum* (hyphal diameter 10 μm); (c) *Subbaromyces splendens* (hyphal diameter 10–12 μm).

and *Ascoidea* (Tomlinson, 1942; Cook, 1954; Painter, 1954; Wheatley, Mitra and Hawkes, 1982). The combination and dominance of the organisms which become established depends on local conditions, but in general the filamentous bacteria and fungi are more tolerant than the flocculated bacteria (Tomlinson and Williams, 1975). Thus, changes from a neutral pH, or less than optimum C:N:P ratios (i.e. 100:5:1), or poor concentrations of DO ($<2{\cdot}0\,mg/\ell$) have the effect of encouraging the filamentous groups. This is of little consequence in biological filtration but seriously interferes with settling in the activated sludge process. If the activated sludge floc contains a significant proportion of filamentous material, then the density of the sludge is reduced and there are major problems with regard to settling and returning the bios (Pipes, 1977; Chambers and Tomlinson, 1982) (*Figures 10 and 11*); this condition is known as 'bulking'.

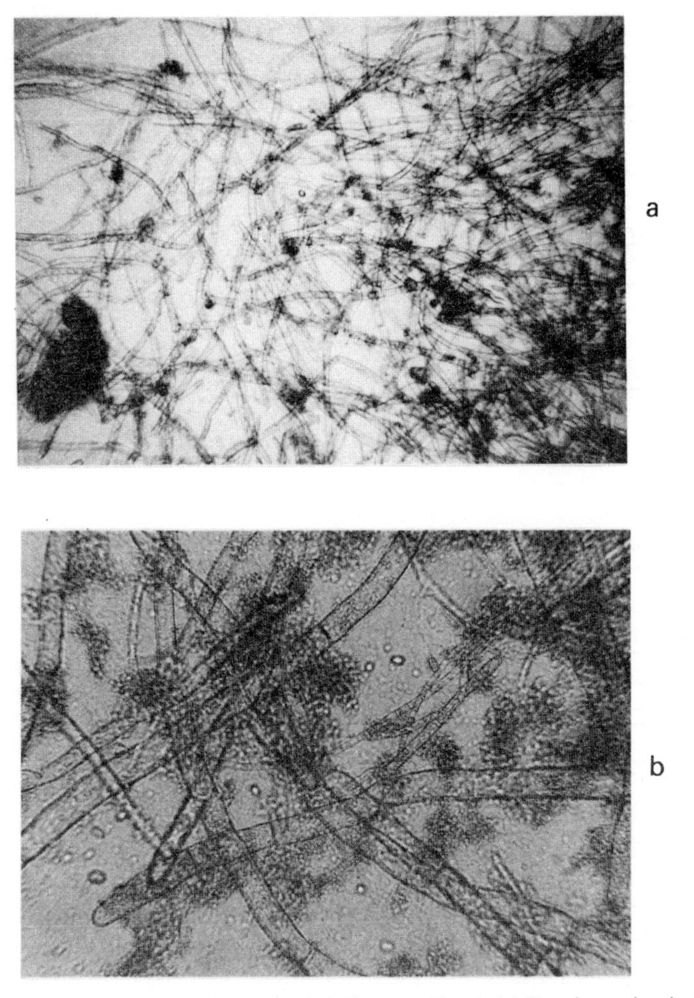

Figure 11. *Geotrichum* spp. in effluent (hyphal diameter $10\,\mu m$). (a) Fungi growing in high-rate biofilter film; (b) *Geotrichum* sp. at 5 × magnification of (a).

For the successive stages in the complete oxidation and mineralization of the organic waste, a number of other organisms also become established. The heterotrophic bacteria (*Figures 4 and 5*) are responsible for primary removal of organic matter. They are followed by a number of secondary communities feeding off the primary population and its breakdown products. One of the important groups are the autotrophic nitrifying bacteria *Nitrosomonas* and *Nitrobacter* which have been studied in some detail because of their importance in removing ammonia. Nitrifying bacteria are slow growing and sensitive to low concentrations of dissolved oxygen ($<2{\cdot}0\,mg/\ell$) (Painter, 1970; Bruce, Merkens and Haynes, 1975). In the activated sludge process, unlike biological filtration, the autotrophic bacteria occupy the same physical position in the floc as the heterotrophs and although not competing for substrate they do compete for oxygen. This means that if an activated sludge plant is to produce a well-nitrified effluent, residence times and oxygen concentrations have to be higher than those required simply to remove BOD. The algae are the other major autotrophic group present in waste-water treatment systems. They are normally restricted by the light available, and are not of primary importance in the UK. Further stages in treatment, such as oxidation ponds or lagooning, would generate large algal and diatom populations. The algae are always present in biological filters and the most commonly identified are the blue-green algae (Cyanophyceae), *Phormidium* and *Oscillatoria*, together with *Stigeoclonium*, *Ulothrix* and *Chlorella* from the Chlorophyceae (*Figures 8 and 12*).

Secondary grazers rapidly become established in waste-treatment reactors with the holozoic protozoa, rotifera and nematode worms occupying the second trophic level. The species and diversity of the ciliates also depends on the degree of treatment, with a succession to the stalked peritrichous protozoa from the holotrichia and flagellates (Baines *et al*., 1953; Curds and Cockburn, 1970) (*Figure 13*). It seems likely that most of the protozoa are scavengers feeding both saprobically (on organic detritus) and holozoically (on living organisms). Experiments conducted by Curds, Cockburn and Vandyke (1968) have shown that the protozoa play an important part in the treatment process, with a reduction in protozoa coinciding with a more turbid effluent with a higher BOD.

In biological filters there is also a wide range of Metazoa, absent from the truly aquatic activated sludge. *Table 8* shows the range present in a typical

Table 8. Common Metazoa in biological filters

Oligochaeta:	Diptera:
Lumbricillus	Psychoda
Enchytraeus	Chironomids
Eisenia	Anisopus
Dendobenea	Scatella
Collembola:	Acarina
Hypogastura	Platyseius
Crustacea:	
Daphnia	
Canthocamptus	

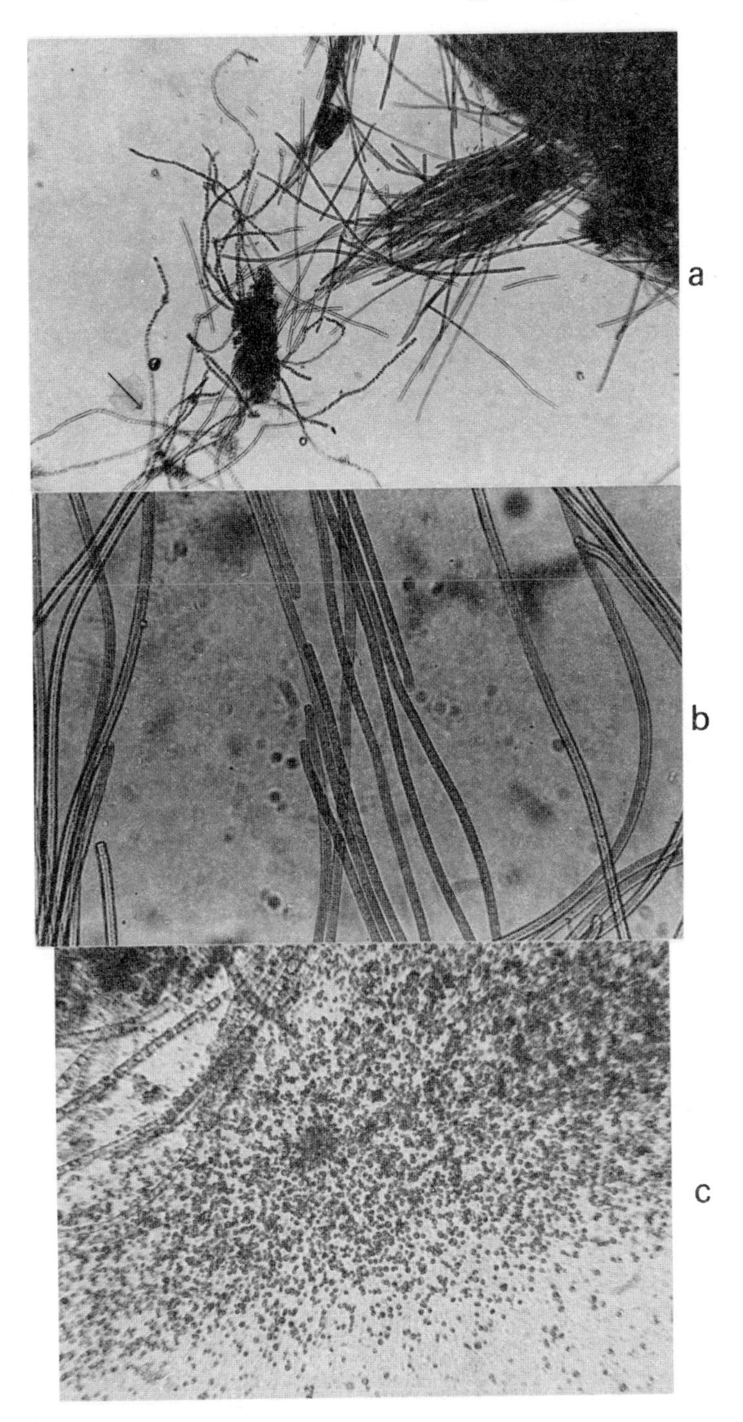

Figure 12. Algae in effluent treatment processes. (a) Algae growing in biological filter film, showing (arrowed) *Ulothrix* (Chlorophyceae) (with filament diameter 5 μm) and straight rods of *Phormidium* (Cyanophyceae) (5μm in diameter); (b) *Phormidium* at 4 × the magnification in (a): (c) *Chlorella* (3–5 μm in diameter)

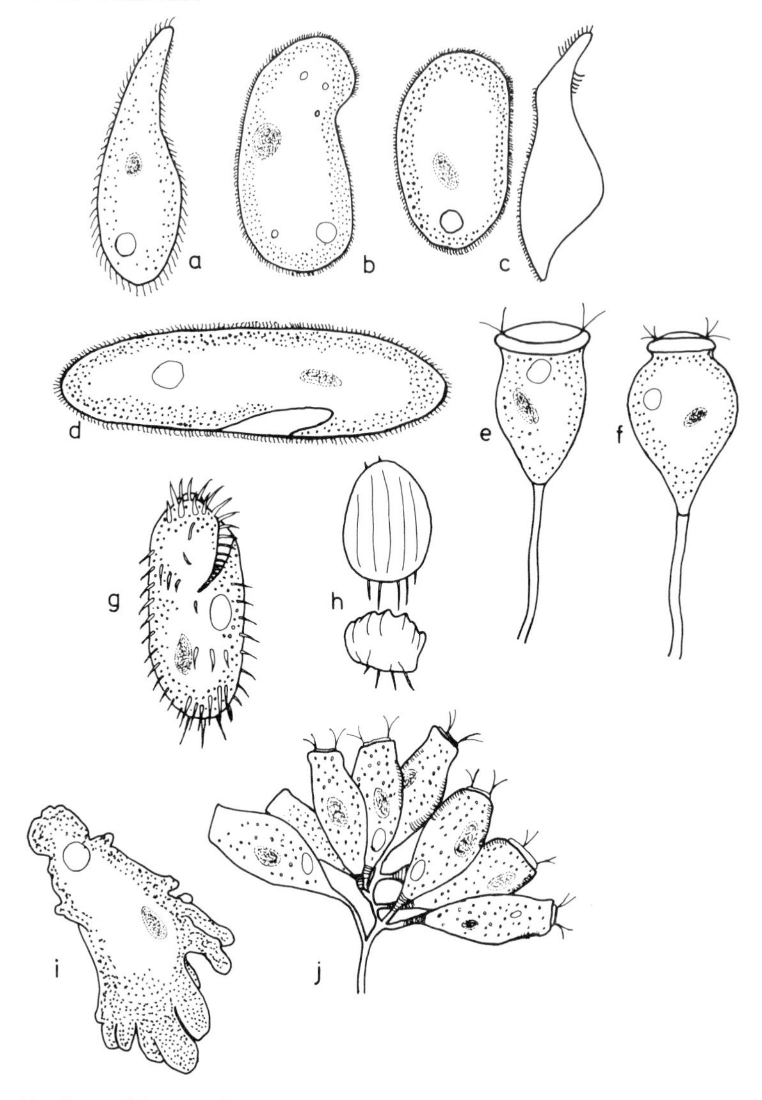

Figure 13. Some of the most important protozoa in waste treatment processes. (a) *Hemiophrys* sp. (length 150–200 μm); (b) *Colpidium colpoda* (100–120 μm); (c) *Chilodonella* sp. (100–120 μm); (d) *Paramoecium caudatum* (150–180 μm); (e) *Vorticella alba* (cell (not stalk) length 60–80 μm); (f) *Vorticella microstoma* (cell length 50–90 μm); (g) *Stylonychia* sp. (150–200 μm); (h) *Aspidisca costata* (25–40 μm); (i) *Amoeba proteus* (approx. 10–20 μm); (j) *Opercularia coarctata* (head, not stalk, each 45–50 μm).

filter; these organisms originate from mud flats. Strong wastes or high irrigation rates tend to limit the species diversity. The Psychoda flies are the most common and are present even in very high-rate filters. The flies can be a nuisance but fortunately, with one rare exception, are non-biting.

Because of their importance in controlling film accumulation, the flies and worms have been subject to a considerable amount of investigation. Lloyd (1945) has published a series of papers on the origins and the role of the flies, and Hawkes (1957), Terry (1956) and Hawkes and Shephard (1972) have given details about the preferences of each species. Reynoldson (1941, 1943, 1947, 1948) has published a similar account of the worms. Williams, Solbe and Edwards (1969) and Solbe (1971) have conducted work on the aspects of distribution, life history and amount of film metabolized by the worms. The grazing organisms are responsible for a major shedding of bios in the spring of each year, coincident with a consistent rise in temperature in late April and May. Sludge production can double during this period but fortunately the solids produced are easily settleable. Biological filters accumulate solids during the colder months from a combination of waste solids and partly autolysed film debris.

RECENT DEVELOPMENTS IN TECHNOLOGY

An important advance in biological filtration took place in 1970 with the introduction of plastic media. This has extended the range of applications to include high-concentration wastes typical of industrial effluents. Conventional mineral media become clogged with too much biomass when treating strong industrial wastes. Plastics are light compared with stone and it is now possible to build space-saving tall filters without the usual substantial retaining walls (*Figure 14*). The high rates of treatment possible with the plastic media (*Table 4*) have enabled major savings to be made in the cost of effluent disposal and its use to treat industrial wastes has become widespread. There are now about 1000 such plants in Europe.

Two characteristics of media which exert an important influence on the efficiency of biological treatment are the specific area available for biological growth and the void space between the pieces allowing ventilation, discharge of solids and drainage (Chipperfield, 1967). Providing a large specific surface while retaining adequate voidage has, in the past, been the problem in selecting suitable mineral media. Plastic materials can be fabricated into shapes with optimum surface area and voidage but they also have to be capable of promoting uniform utilization of the greater surface area. The modular synthetic media rely on a high irrigation rate through recirculation to ensure adequate wetting of the available surface, but at lower rates of load a random pack medium has to be used to give an even flow pattern through the filter depth.

Another innovation introduced into the UK in 1973 was the Rotating Biological Contacter (RBC). A honeycomb of plastic sheets is slowly rotated on a shaft through a tank containing the waste water. About 40% of the surface is submerged and a typical unit 7·5 m in length and 3·5 m in diameter will have a surface area of 9500 m^2 for biological growth. RBCs have characteristics similar to those of biological filters, with a large surface area for fixed biological culture which is alternately in contact with the waste water and air. Rotation of the media ensures that uniform concentrations of dissolved oxygen and substrate are available to the total biomass. The effluent tank may be baffled to produce a plug flow of waste and in this way high BOD removals of over 95% and nitrification can be achieved in the final compartments. There are about 4000 such units operating throughout the world, mostly as treatment systems for small communities or factories (Iggleden, 1981). Pike *et al.* (1982) have carried

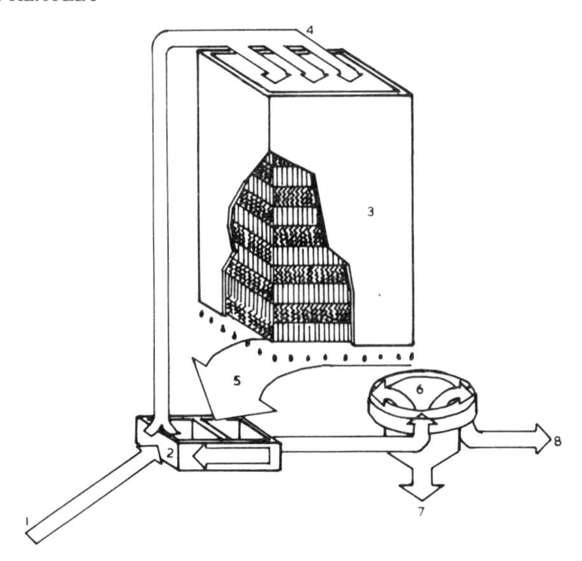

Figure 14. High rate biotower. 1. Incoming waste; 2. Recycle sump; 3. Biotower; 4. Distribution system; 5. Effluent; 6. Settlement tank; 7. Sludge; 8. Clarified effluent.

out a performance survey of some RBC treating domestic sewage.

Recent developments have also been made in the activated sludge process: the first was the use of pure oxygen in completely enclosed tanks in 1972. The system was based on the availability of cheap oxygen from pressure-swing adsorption equipment. The effluent treatment plant is very similar to conventional activated sludge but is able to operate at higher biomass concentrations and lower residence times without a loss in efficiency (*Figure 15*). The higher dissolved oxygen concentrations are reported to overcome the problem of bulking sludges suffered by many strong industrial wastes (Fuggle, 1981). About 150 plants have now been built world-wide, with five in the UK.

An innovative application to effluent treatment was the use of a deep air-lift fermenter in 1974 by Imperial Chemical Industries (Hemming *et al.*, 1977) (*Figure 16*). The concentric tube arrangement, the 'deep shaft', can be 50–100 m deep and 0·5–5 m in diameter. The shaft is partitioned into a downflow section, into which the incoming waste, returned active sludge and the process air are injected, and an upflow section. In the upflow section, bubbles of dissolved gases are released as the pressure decreases, thus reducing the density of the circulating water. The difference in density between the downflow and the upflow liquids above the point of air injection produces the net driving force for the shaft. The velocity generated is much higher than the rise rate of air bubbles in the downflow portion and the air is carried down. Dissolved oxygen and turbulence in the shaft are high, which permits a high concentration of very active biomass to be maintained. To avoid flotation, the effluent is vacuum degassed before settlement. The system is expected to be more economic than conventional activated sludge by virtue of its low residence time and low running cost. There are about five full-scale plants, one of which is in the UK.

A combination of fixed film and activated sludge, the fluidized bed, was

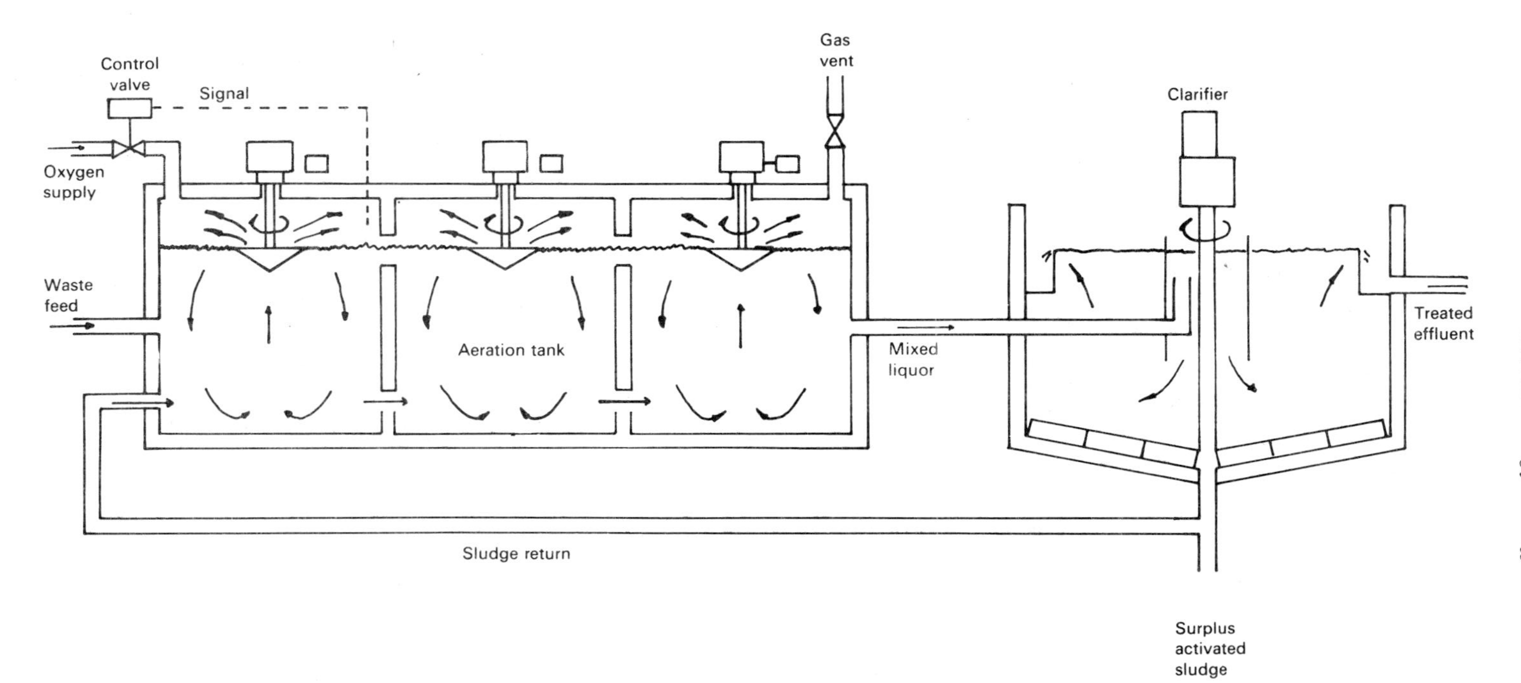

Figure 15. Schematic diagram of an oxygen-activated sludge plant.

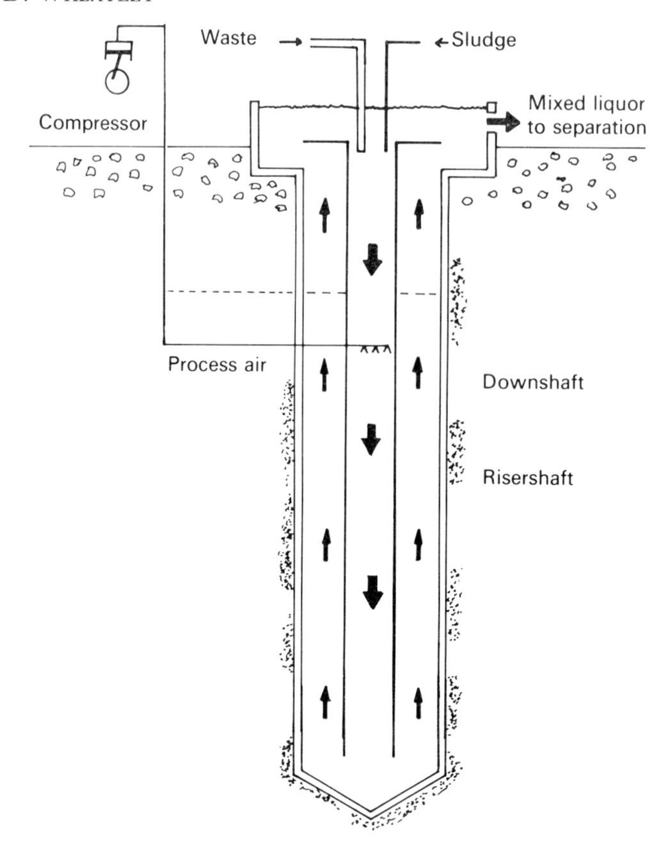

Figure 16. Diagram of the deep shaft.

introduced in 1980. The biological fluidized bed, like the deep shaft and pure-oxygen-activated sludge, is able to retain a high concentration of micro-organisms—about five times the concentration of conventional plant with consequent savings in capital cost. There is a further saving in capital because the plant does not require final settlement. The waste sludge from the reactor is highly concentrated and the effluent is reported to be suitable for direct discharge. There are two basic types: the Simon Hartley Captor with plastic support particles and compressed air for fluidization (Walker and Austin, 1981) and the Dorr-Oliver Oxitron, with sand support particles and oxygen injection for fluidization (Sutton *et al.*, 1981).

In the Oxitron system, the sand overflows the reactor and is recycled after cleaning in an agitator or vibrating screen (*Figure 17*). In the Captor system, the particles are retained in the reactor by a mesh but a portion is periodically cleaned by squeezing. They are then returned to the bottom of the reactor by an air lift (*Figure 17*).

An unwanted consequence of intensifying the aerobic treatment processes has been an increase in the amount of surplus biomass (sludge) to be disposed of.

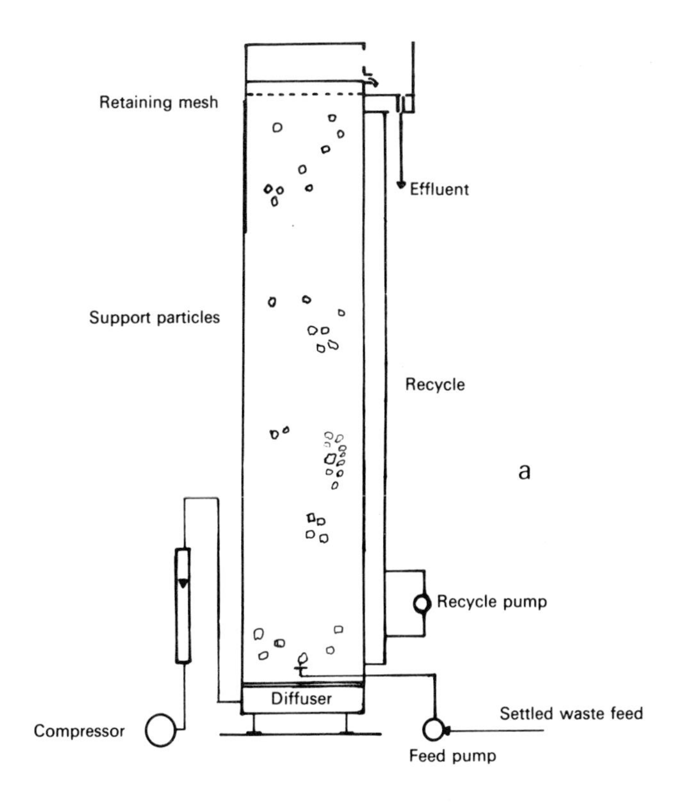

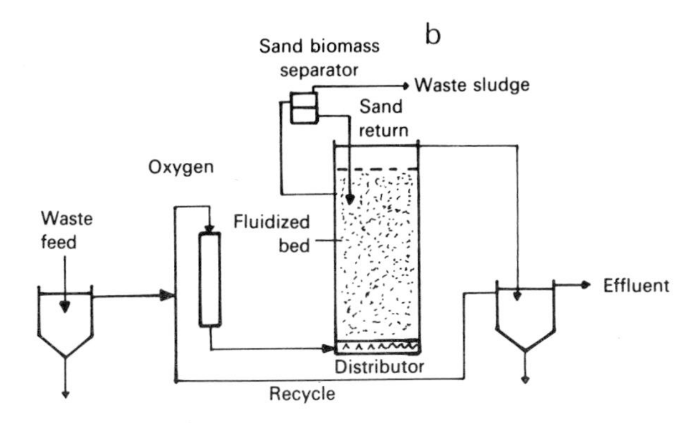

Figure 17. Fluidized-bed process. (a) Captor system; (b) Oxitron system.

Sludge disposal can represent 50% of effluent treatment costs (Porch, Bayley and Bruce, 1977). Thus one of the most important possibilities to be derived from a greater understanding of the metabolism of micro-organisms would be to uncouple microbial anabolism and catabolism. This would ensure an inefficient conversion of substrate to biomass. Empirical work already conducted (Hawkes and Shephard, 1972) indicates that certain trace nutrient deficiencies and intermittent feeding may have this effect.

By-product recovery

One of the largest potential sources of organic pollution in the UK is the biologically based industries. The scale of fermenters, the water consumption and the biological nature of the processes used tends to generate large volumes of strong wastes. *Table 9* shows some strengths of a few typical wastes from

Table 9. Strengths of common wastes from the food, drink and fermentation industries (approximate figures; actual values vary from site to site depending on the process).

Industry	Strength of waste (COD in mg/ℓ)
Pharmaceutical:	
Citric acid	20 000
Antibiotics	10 000
Distillery waste	25 000
Confectionery	15 000
Cheese, butter, cream	5000
Brewery	5000

the food, drink and fermentation industries. Most of these wastes are spent broths or wash waters which are very variable in composition. Traditionally, the only suitable treatment has been biological oxidation admixed with domestic sewage. Typically, the waste from antibiotics or citric acid fermentation, for example, can have a COD of 10 000–20 000 mg/ℓ, and although easily biodegradable, this type of waste is 20 times the strength of domestic sewage. Thus an antibiotics plant producing tetracylines with an effluent flow of 250 m^3/day and a COD of 10 000 mg/ℓ has a population equivalent of 36 000 and would cost about £200 000–300 000 a year in water charges to dispose of to sewers (*Table 10*). If the biologically based industry is considered as a whole, then a relatively minor improvement in the biotechnology of waste treatment would have a substantial financial impact. There is therefore considerable interest in reducing the volume or utilizing these wastes. There are two approaches to the recovery of materials from wastes: (1) direct recovery or concentration of valuable materials, and (2) transformation of wastes into useful materials. There are three widely used large-scale product-recovery biotechnologies concerned with, (1) water; (2) biomass and (3) biomethanation.

Table 10. Mogden formula for trade effluent charges (North West Area, 1979).

$$C = R + V + \frac{O_t}{O_s} B + \frac{S_t}{S_s} S$$

where:

C = daily cost of trade effluent (pence*/m^3)
R = reception and conveyance cost of sewage (1·11 p/m^3)
V = volumetric and primary treatment cost of sewage (1·37 p/m^3)
O_t = COD of trade effluent after 1 hours' settlement
O_s = the average COD of settled sewage
B = unit biological oxidation cost for settled sewage (2·537 p/m^3)
S_t = the total suspended solids in trade effluent (mg/ℓ)
S_s = total suspended solids in crude sewage (mg/ℓ)
S = treatment and disposal cost of primary sludges (1·463 p/m^3)

* One new penny = £0·001.

WATER RECLAMATION

A major problem with reclaiming organically polluted water to a high standard, suitable for reuse, is the capital cost of the equipment compared with the cost of mains water. Mains water costs 75–100 p/tonne and there are, in most cases, sufficient supplies of the required quality and price for most industrial needs. Recycling industrial water is economic only if a water quality inferior to that available from the supply is acceptable. This is often the case in the heavy industries such as power generation, steel making and coal preparation; water treatment can then be kept to a minimum—usually settlement and/or cooling.

In the water supply industry about 30% of raw water is obtained from recycled effluent (Water Data Unit, 1979) principally in areas without access to supplies of upland water, for example South-East England and the southern counties. The biologically treated effluent is discharged to lowland rivers which then serve as aqueducts and sources of water supply. The biological activity in the river further improves the quality of the recycled water (*Figure 18*). The rivers Thames and Lee can be 75% recycled sewage effluent in summer (Packham, 1983). Supplies of upland water are now scarce and it is envisaged that any expansion in supply will have to include a substantial quantity of reclaimed water (Water Resources Board, 1973). There are potential problems associated with this type of supply from the recycling of recalcitrant pollutants. All raw waters are normally treated by a combination of biological and physico-chemical processes. Solids are removed by chemically assisted precipitation; the residual liquid is filtered through sand and is then chemically sterilized before being pumped into the mains. Additional treatment with carbon, membranes or deionizing plant may be necessary if the raw water quality is low. The sand filter is partly biological—a layer of micro-organisms which grows on the surface of the sand particles, degrades organic matter and reduces some of the nitrate, phosphate and carbonate present in the water. In slow sand filtration this biologically active layer is known as the schmutzdecke and is responsible for the principal filtering action.

It is likely that there will be improvements in recycled water supply technology

a

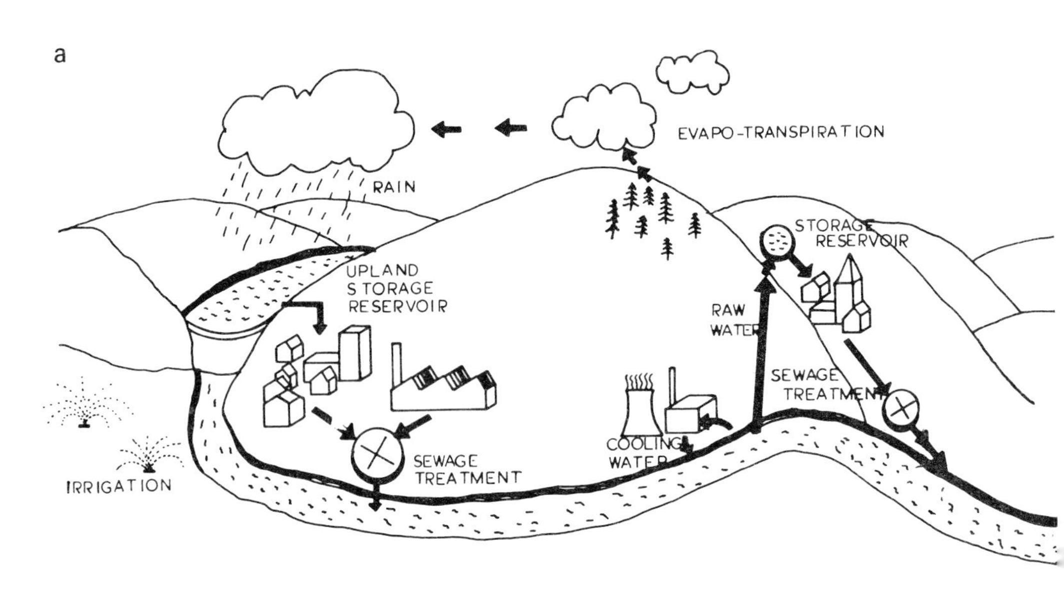

b

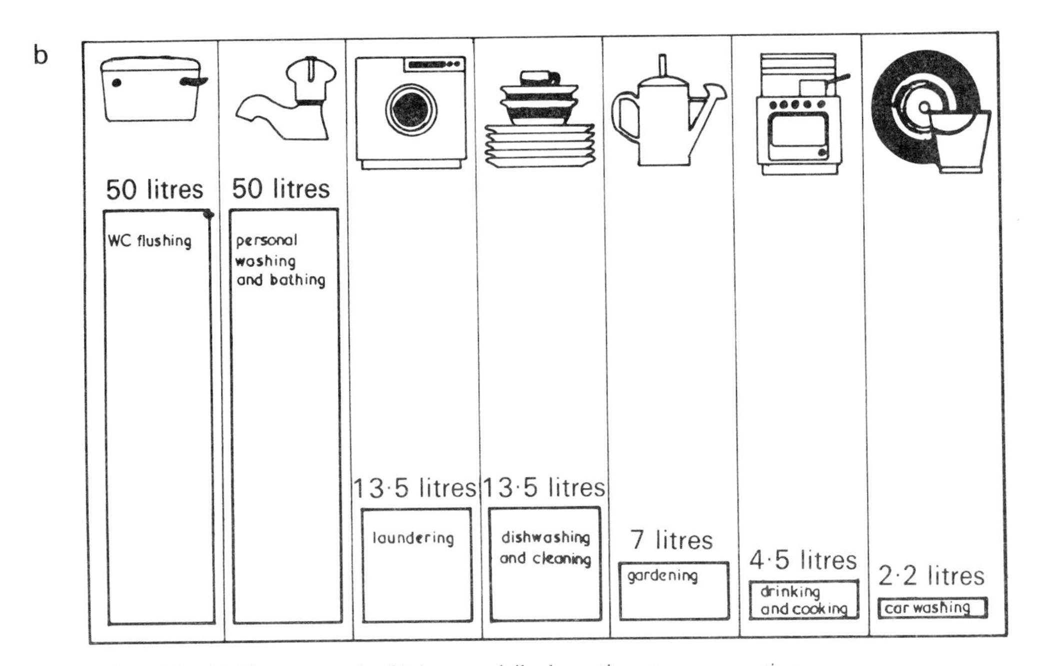

Figure 18. (a) The water cycle; (b) Average daily domestic water consumption.

as a result of the application of micro-organisms for the breakdown and scavenging of recalcitrants.

Recalcitrant compounds

According to Storck (1979) the total world production of synthetic organic chemicals (xenobiotics) is now estimated at 150×10^6 tonnes/year. One hundred and fifty chemicals are produced in quantities in excess of 50 000 tonnes/year and the use and losses of these compounds are fairly well known (Schmidt-Bleek, 1980). There is, however, little information on the fate of a much wider range of chemicals produced in smaller quantities. A large number of these synthetic compounds are recalcitrant or persistent, and considerable efforts are being made to understand how micro-organisms might acquire the metabolic potential to degrade these recalcitrant xenobiotics. The high affinity of biological catalysts for their substrates makes biodegradation particularly useful at low concentrations of the xenobiotic. This makes biodegradation complementary to the traditional methods for disposing of bulk waste products (oxidation, incineration and precipitation). Another advantage is that the xenobiotic is converted into natural products.

Most biodegradation research has been classic, investigating the metabolism of pure cultures; this reduces the complexity of the system and ensures reproducible results. The two basic methods used are enrichment culture and gene manipulation. In enrichment culture the compound to be degraded is added in progressively larger doses during an acclimatization period until the xenobiotic then becomes a source of essential nutrient. The starting inoculum is important and should be taken from an area where the organisms are most likely to have evolved, a habitat which has been exposed to the compound of interest. Batch culture is usually used to screen for suitable organisms, and continuous culture is used to exert continuous selection pressure for the desired organisms. Important additional information on the biochemistry of the microbial degradation, such as the rate-limiting steps, catabolic sequences and the substrate specificity, can also be obtained from these cultures. Harder (1981) in a recent review of enrichment techniques emphasized the role of mixed cultures. Previous work had shown that an interdependent microbial community was often necessary to metabolize the complex sequence of intermediates which resulted from the breakdown of xenobiotics. This biochemical approach can be complemented by work on the genetics of degradative pathways. Work has been reported on the analysis of plasmids, gene cloning, and transposon mutagenesis. Williams (1981), for example, has extended original work by Chakrabarty (1980) on the structural analysis of plasmids in *Pseudomonas*. The plasmids contain information for the breakdown of alkanes, toluene and xylene. Another group (Franklin, Bagdasarian and Timmis, 1981) have studied the use of gene cloning and transposon mutagenesis for the cleavage of a wide range of aromatic compounds.

The list of recalcitrants currently under investigation (Leisinger, 1983) include the azo dyes, stilbenes (optical brighteners), chloroaromatics, *S*-triazines, chlorinated hydrocarbons, DDT and lignin.

Although the results of microbiological, biochemical and genetic research on biodegradation are accumulating there are, as yet, no practical processes on an industrial scale. Nevertheless, three important areas of application are envisaged:

1. Controlled degradation of specific wastes with specialized cultures at the source of the waste. Cook *et al.* (1983) have described a specific example for the treatment of parathion insecticide wastes.
2. Improvement of waste treatment systems by inoculation with adapted laboratory strains. Cook (1983) has reported commercially available specialized organisms for improving BOD removal, for eliminating filamentous growths in 'bulking', and for improving methane generation in anaerobic treatment. Five companies (Biolyte, London; Bactozyme, Worksop, Nottingham; Agrico, Stafford; Ubichem, Middlesex; and Interbio, London) are now actively selling this type of preparation for supplementing existing effluent treatment systems, but no controlled experiments have been reported so far.
3. The clean-up of spills and decontamination of soils using specialized cultures.

Specific reactors for the breakdown of recalcitrants would be the easiest to operate and control because the waste stream could be better defined and applied to existing reactor designs. To supplement existing waste treatment systems, or to be effective on spills, the organisms and their enzymes will have to survive and function in the suboptimal conditions of a complex ecosystem.

Removal by adsorption

It has been observed for some time that biological effluent treatment effectively removes metals from sewage (Brown and Lester, 1979). The metals are precipitated with the sludge or surplus biomass and can reach a concentration sufficient to interfere with the subsequent processing of the sludge. Research at UMIST (Kiff and Brown, 1981) was directed at understanding the mechanism of this removal. Investigations on an acetate rayon waste which contained zinc in high concentrations (50 mg/litre) showed that the biomass could contain up to 12% by weight of zinc without interfering with BOD removal. Diffraction and microscopical examination showed that the zinc was bound in the extracellular slime of the zoogloeal growth. The zinc and other metals present in sewages could be elutriated by oxidative acid hydrolysis, a process which has been established by Simon Engineering and Richland Resources Ltd. Hydrochloric acid and hydrogen peroxide at pH 1·5 are used in conjunction with a polyelectrolyte to release the water. Recovery of the metal is 70–95% but the process is not economically viable unless subsidized by a high negative cost from the disposal of a metals-contaminated sludge. Examination of the biofilm (*Figure 7b*) shows that a large proportion of the bios is extracellular bacterial polymer rather than actual bacterial cells. This polymer has important metabolic functions with regard to the absorption and transfer of nutrient from solution to bacterial cells. Field data from operating plants have indicated that the

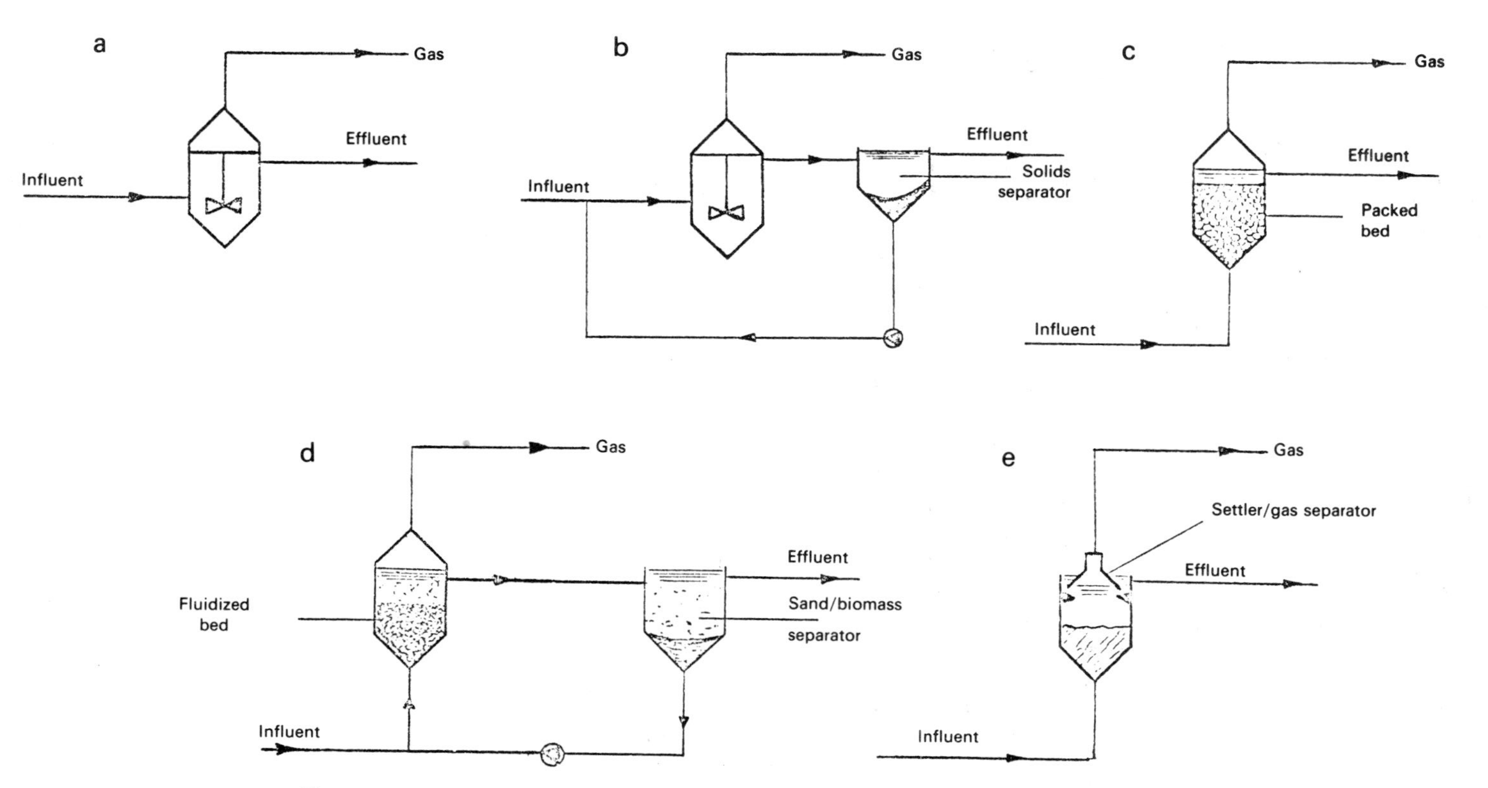

Figure 19. Schematic diagram of various anaerobic treatment processes. (a) Conventional digester; (b) Anaerobic contact process; (c) Anaerobic filter; (d) Anaerobic fluidized or expanded bed; (e) Upflow anaerobic sludge blanket. (*See* pages 300–302)

structural characteristics and quantity of polymer vary according to the substrate.

Carbohydrate wastes, particularly the oligosaccharides, generate the most polymer at similar concentrations and metabolic rates. Preformed polymers such as dextrans, starch and polyacrylamides also produce large quantities of extracellular polymer.

Little information has been derived from these early experiments on the chemical structure and nature of the polymer. Further work is required to determine its structure and how it effects adsorption. It is known to be a complex co-polymer held together by fibrils and to contain large amounts of bound water. It is also rich in a number of bound extracellular enzymes for the specific uptake of required nutrients. Long-term research is directed at altering the chemical structure of the polymer to recover a wide range of useful materials. Mercury tolerance and accumulation has also been isolated in a strain of the fungus *Chrysosporium* which could be adapted to detoxify mercury-bearing wastes. The mercury is adsorbed and fixed to the hyphal walls of the fungus (Williams and Pugh, 1975).

Bioscrubbing

Traditional odour and waste gas control processes, such as incineration, dispersion, catalytic oxidation, scrubbing and adsorption, are all very expensive, are best suited to large volumes of well-defined waste gases and are not able to adjust to varying complex mixtures of materials. Malodour problems particularly are normally caused by complex varying mixtures at very low concentrations: many of the mercaptans, for example, have odour thresholds below 1 ppb. Biological control of this type of problem, like that of the recalcitrants, may offer a simpler alternative treatment. Other advantages of the biological nature of the process include combined liquid and gas treatment, no chemical costs and low energy requirements, which are of general interest in all gas scrubbing. Patents for biological control systems were filed in 1930 but there are still few working systems. It seems likely that the idea arose from the observation that biological growths occurred on many water-scrubbing systems. The types of apparatus currently used resemble physico-chemical scrubbing units combined with a biological filter. The design requirements of bioscrubbing systems are similar to those of physico-chemical scrubbing units, the principal requirement being the best surface possible for gas–liquid–biomass interchange. Le Roux and Mehta (1980) have reported on the design of biological systems for the oxidation of malodour, using both a bioscrubber and percolation through a soil bed. Later, in 1982, Le Roux discussed a full-scale application for the treatment of malodours from a high-rate biological filter. The initial removal mechanism, as in a biological filter, is likely to be biophysical adsorption and solubilization of the material. In this form it will then be available for microbial breakdown. The packed bed serves both to provide a large gas–liquid surface area for adsorption and to immobilize the biomass.

Excessive amounts of biomass could occlude some of the surface area and the concentration of organic matter in the effluent should be controlled to maintain optimal efficiency. It is probable that the prototype bioscrubbing systems will be part of effluent treatment plants because this is the most likely source of substrate for biological growth. Specific inoculation would not, therefore, be required but later plants for the removal of xenobiotic materials from gases would require specific cultures and a controlled substrate. Completely novel chemicals may require a prolonged acclimatization period, but seeding of subsequent systems handling similar materials would then be simplified. Biological control of gases is as yet an under-researched area, although applications both for oxidizing biofilm, (for mercaptans, hydrogen sulphide and cyanide) and for reducing anaerobic biofilms (for sulphur dioxide and chlorinated hydrocarbons) can be envisaged.

RECOVERY OF BIOMASS

Biomass and single-cell protein recovery (on a small scale) from wastes is now very well documented with reviews in 1959 by Prescott and Dunn and in 1968 by Peppler. James and Addyman summarized the state of the technology in 1974 and described the most common recovery processes from molasses, whey and starch using pure culture to ensure product consistency. These well-defined materials are now, however, rarely wastes but are valuable commodities as substrates for other higher-value microbial products (Tomlinson, 1976a). The production of protein from the more common dilute wastes has met with little success despite a number of investigations in both Europe and the United States. The two most common groups of micro-organism investigated have been the yeasts and the Fungi Imperfecti. Important work on the yeasts has included that by Holderby and Moggio (1960) on paper mill wastes, by Wassermann, Hampson and Alvare (1961) on milk wastes, by Thanh and Simard (1973) on domestic sewage, by Righelato, Imrie and Vlitos (1976) and Tomlinson (1976a) on food processing wastes, by Smith and Bull (1976) on coconut waste water and by Braun *et al.* (1976) on citric acid wastes. The filamentous fungi are easier to separate and dry than the yeasts, and they also have a better food texture than other micro-organisms. Work by Church, Erikson and Widmer (1973) was conducted on full-scale plant trials using *Trichoderma* and *Geotrichum* for the treatment of vegetable processing wastes, Munden and King (1973) reported on the use of *Fusarium* for treatment of carbohydrate wastes; Quinn and Marchant (1980) described the use of *Geotrichum* for distillery wastes and Wheatley, Mitra and Hawkes (1982) used *Geotrichum* and *Fusarium* for the treatment of milk wastes. The utilization of bacteria (Grau, 1980), algae (McGarry, 1971), and water hyacinths (Cornwall *et al.*, 1977), have also been tested.

In general, it has been concluded that waste waters are not suitable for large-scale single-cell protein (SCP) production because of the difficulties of producing a biomass of consistent nutritive value and a material free from toxicological uncertainties or microbial contamination. Production of SCP becomes even less attractive when the cost of the fermenters and drying equipment are considered. Tomlinson (1976b) and Wheatley, Mitra and Hawkes (1982) report that protein recovery can be justified only if contamination-free wastes from the food, drinks and fermentation industry are used, if effluent treatment efficiency is unimpaired and if high sludge-treatment costs are reduced. The recovered SCP can then

Table 11. Comparison of costs of protein production* by fermentation† and biofiltration‡ in 1975.

Fermentation		Biofiltration	
Capital costs			
Civil Engineering	£45 000	Civil Engineering Tower	£11 000
Mechanical plant	£15 000	Base slab	£5000
Separation and drying costs	£50 000	Pumps and pipework	£10 000
		Sedimentation tank	£5000
		Plastic medium	£17 000
		Separation and drying costs	£20 000
Total	£110 000	Total	£68 000
Extras (10%)	£11 000	Extras (10%)	£7000
	£121 000		£75 000
Running costs			
Power required		4 pumps at 4 hp	
At 2·25 p/kWh	£3500	At 2·25 p/kWh	£1000
Manpower maintenance	£10 000	Manpower maintenance	£7500
Chemicals/nutrient	£3500	Chemicals/nutrient	£3000
Total	£17 000	Total	£11 500
Capital over 10 years at 15% interest	£24 200	Capital over 10 years at 15% interest	£15 000
Total annual costs	£41 000	Total annual costs	£26 500
Protein yield 106 t/year		Protein yield 50 t/year	

* Waste details: maltings liquor, hydraulic load 454 m³/day; BOD load 454 kg/day; BOD concentration 1000 mg/ℓ; final required BOD 75 mg/ℓ; BOD removal required 92% (925 mg/ℓ).
† Batch process 30 h (including let down); BOD removal 630 kg/batch; oxygen required 0.7 kg/kg BOD removed ∴ 441 kg O_2 per batch aeration; capacity 681 m^3; biomass production 295 kg/day.
‡ Three-stage treatment load 1·5 kg/m^3; 310 m^3 plastic medium; biomass production 170 kg/day.

be used as a protein additive to a mixed animal feed: *Table 11* shows some anticipated costings. Forage (1978) has described the full-scale recovery of *Candida* yeasts from confectionery wastes.

There has also been interest in the possibility of energy recovery from surplus biomass. So far, interest has been restricted to energy recovery from short-rotation forestry and from gasohol (ethanol) (Nyns and Naveau, 1983) but there has also been work on utilizing biomass, principally water hyacinths and grass, grown on effluent (Reddy, Hueston and McKim, 1983; Schwegler and Cynoweth, 1983). It has also recently been concluded that the European agricultural surpluses could be converted economically into energy (Verstraete, 1983). Processes have also been suggested for the production of more valuable materials such as ethanol and ketones (Amberg, Aspitante and Cormack, 1969) and the volatile acids and alcohol (Coombs, 1981 and Chapter 11 of this volume).

BIOGAS

The most common anaerobic treatment is the digestion of sewage sludge. This process has been successfully applied since 1901, to reduce sludge volume, eliminate pathogens, prevent smell nuisance and generate methane as a by-product. This is well established and widely publicized technology, but there are difficulties which have prevented its more widespread use and development. Most of these problems have been associated with the slow growth rate of the obligate anaerobic methanogenic bacteria, which makes the process susceptible to a wide range of interferences and less resilient to sudden changes in load. Until recently, the conversion of 50% of the organic matter into inorganic salts typically took 30 days (Ministry of Housing and Local Government, 1954; Stanbridge, 1976) and the process is, consequently, expensive to build.

The Water Pollution Research Laboratory has published the results of a national survey on digestion (Swanwick, Shurben, and Jackson, 1969), indicating that inhibition of the metabolic processes, for example by detergents, metals and chlorinated hydrocarbons, is not as common as was believed; most difficulties were attributed to poor design and operation. Recent work by Brade, Noone and colleagues (Brade and Noone, 1981; Brade *et al.*, 1982) has shown that the cost of building anaerobic digestion plants accounted for 20% of the capital of new works and represented £20–40 per capita served. They concluded that similar problems were encountered in all types of sludge treatment and that the cost of digestion could be reduced by some changes in the traditional civil engineering.

The rising costs of aerobic waste treatment and energy have revived interest in anaerobic treatment as an alternative and this has led to important developments in the microbiology and engineering of anaerobic treatment. The University of Louvain has recently completed a survey of anaerobic processes which includes data on over 400 operating plants (Demuynck, Naveau and Nyns, 1983).

Microbiology

Like aerobic waste treatment, the anaerobic digestion of wastes is based on a complex interdependent microbial community. Three different but interdependent groups of bacteria are involved (Zehnder, 1978; Mosey, 1978, 1982). The first group are the hydrolytic fermentative bacteria which hydrolyse complex polymeric substrates to organic acids, alcohols, esters and sugars, generating carbon dioxide and hydrogen. The second group are the hydrogen-producing and acetogenic bacteria; they convert the fermentation products of the first group into hydrogen, acetate and carbon dioxide. The third group are the methanogens which then convert the acetate and hydrogen into methane and carbon dioxide.

An important discovery, leading to a major advance in anaerobic treatment, was made by Bryant *et al.* (1967) who observed that the concentration of hydrogen exerted a crucial control on the anaerobic process. For the successful breakdown of carbohydrate under anaerobic conditions it was suggested that the partial pressure of hydrogen had to be less than 10^{-3} atm ($\approx$ 1 kPa). It is only at such low hydrogen concentrations that NADH formed by the breakdown of organic matter can be reoxidized by the release of hydrogen. These low hydrogen concentrations result from the activity of the hydrogenotrophic bacteria which constantly scavenge the hydrogen. Two groups—the methanogens (utilizing hydrogen and bicarbonate to form methane and water) and the sulphate reducers (forming hydrogen sulphide and water)—are the most important (*Table 12*). The hydrogenotrophic methanogens are thought to have doubling times of about 8–10 h (Schauer and Ferry, 1980) and are much slower growing than the acid-forming hydrolytic bacteria (doubling time about 30 minutes). Thus the hydro-

Table 12. Anaerobic digestion of waste and microbial groups involved.

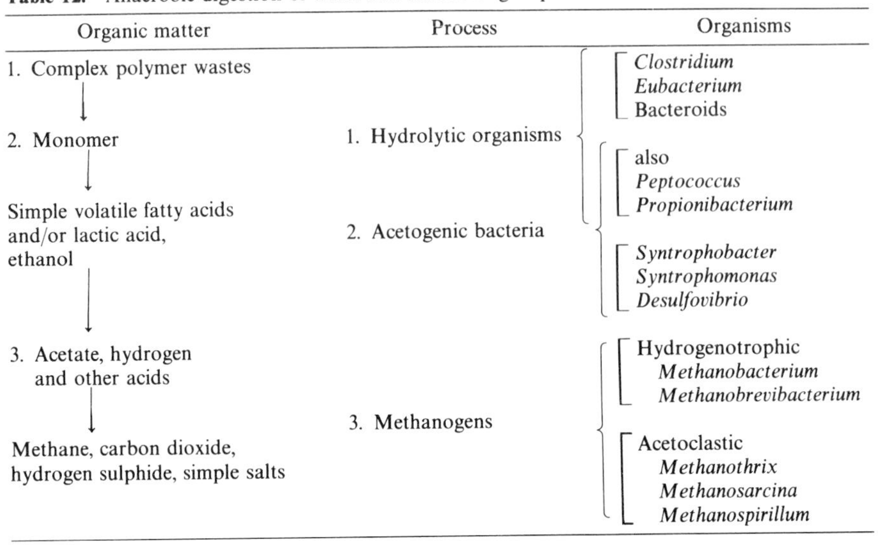

Organic matter	Process	Organisms
1. Complex polymer wastes ↓		*Clostridium* *Eubacterium* Bacteroids
2. Monomer ↓	1. Hydrolytic organisms	also *Peptococcus* *Propionibacterium*
Simple volatile fatty acids and/or lactic acid, ethanol ↓	2. Acetogenic bacteria	*Syntrophobacter* *Syntrophomonas* *Desulfovibrio*
3. Acetate, hydrogen and other acids ↓		Hydrogenotrophic *Methanobacterium* *Methanobrevibacterium*
Methane, carbon dioxide, hydrogen sulphide, simple salts	3. Methanogens	Acetoclastic *Methanothrix* *Methanosarcina* *Methanospirillum*

genotrophic methanogens are susceptible to stress as a result of sudden surges in load and an accumulation of hydrogen from the increased activity of the acid-forming hydrolytic bacteria. The acid-forming bacteria respond to the increased hydrogen concentration by altering their metabolism to form propionic, butyric, valeric, caproic and lactic acids instead of acetic. This decreases the formation of hydrogen:

$$C_6H_{12}O_6 + 2H_2O \rightarrow 2CH_3COOH + 2CO_2 + 4H_2$$
$$C_6H_{12}O_6 \rightarrow CH_3CH_2CH_2COOH + 2CO_2 + 2H_2$$
$$C_6H_{12}O_6 + 2H_2 \rightarrow 2CH_3CH_2COOH + 2H_2O$$

So far no methane bacteria capable of metabolizing propionic or butyric acids have been discovered and it is assumed that these acids have to be converted to acetic to form methane (McInerney, Bryant and Stafford, 1980). Although bacteria capable of this conversion have not yet been isolated, enrichment culture studies by Lawrence and McCarty (1969) have shown that the process does exist, although it is very slow. The reactions as explained by Heyes and Hall (1981) are energetically difficult (i.e. of low energy benefit) and theoretical doubling times of 1·5–4 days were suggested.

$$CH_3CH_2COOH + 2H_2O \rightarrow CH_3COOH + CO_2 + 3H_2$$
$$CH_3CH_2CH_2COOH + 2H_2O \rightarrow 2CH_3COOH + 2H_2$$

For benefit to be derived from these reactions the partial pressure of hydrogen must not exceed about 10^{-6} atm. ($\approx$ 1 Pa) (Sahm, 1983). The hydrogen-utilizing methane bacteria are therefore thought to regulate the formation of acetic acid, which is the vital substrate for the most economically important group—the acetoclastic methanogens. The acetoclastic methanogens are responsible for about 70% of the methane produced by the anaerobic digestion

Table 13. Characteristics of methanogens isolated in pure culture (Sahm, 1983).

Species	Morphology	Substrate	Cell wall
Methanobacterium spp.	Long rods or filaments	H_2, formate	Pseudomurein
M. formicium		H_2	
M. thermoautotrophicum		H_2	
(*Methanothix*) *soehngenii*		acetate	
Methanobrevibacter spp.	Lancet-shaped cocci or short rods	H_2	Pseudomurein
M. ruminantium		H_2, formate	
M. smithii		H_2, formate	
M. arboriphilus		H_2	
Methanococcus vannielii	Motile irregular small cocci	H_2, formate	Polypeptide subunits
M. voltae		H_2, formate	
M. thermolithotrophicus		H_2, formate	
M. mazei	Pseudosarcina	H_2, methanol, acetate	
Methanomicrobium mobile	Motile short rods	H_2, formate	Polypeptide subunits
Methanogenium carciaci	Motile irregular small cocci	H_2, formate	Polypeptide subunits
M. marisnigri		H_2, formate	
Methanospirillum hungatei	Motile regular curved rods	H_2, formate	Polypeptide
Methanosarcina barkeri	Irregular cocci as single cells or packets	H_2, acetate, methanol, methylamines	Heteropolysaccharide

process. They have a doubling time of 2–3 days. They are strict anaerobes and require a lower redox potential (≈ -330 mV) than most anaerobes, relying on the hydrolytic organisms to maintain this low redox potential. In comparison with the hydrogen-utilizing methanogens, relatively few species are known to degrade acetate (*Table 13*). Biochemical studies of the methanogenic bacteria have shown that they differ from other bacterial groups in several important respects. Their cell walls contain no muramic acid and they exhibit resistance to antibiotics such as the penicillins and cephalosporins which affect the cell wall. Their lipids also differ chemically from those of classic bacteria, and several new coenzymes and factors have been discovered in association with their metabolism; coenzyme F420 has been recommended by Nyns and co-workers (Binot, Naveau and Nyns, 1982) as a method of recognizing methane bacteria in mixed culture. The methanogens have also been shown to require several trace elements such as iron, manganese, molybdenum, zinc, copper, cobalt, selenium, tungsten and nickel: Sahm (1983) has demonstrated that *Methanosarcina bakeri* requires 1·0 nmol cobalt, 150 nmol nickel and 0·5 nmol molybdenum; the cobalt is required for corrinoid Factor 3 and the nickel for a nickel tetrapyrolle cofactor.

These discoveries have enabled changes to be made in the design and control of anaerobic systems. Any sudden large changes in load to the reactor must be avoided because these would encourage the hydrolytic and acid-forming bacteria (doubling times 30 min and 1–4 h respectively) to metabolize much more rapidly than the acetoclastic methanogèns (doubling time 2–3 days). Any overload will be expressed by an increase in the hydrogen concentration in the gas, together with more propionic and butyric acids in solution. The concentration of hydrogen and types of volatile acid can therefore be monitored to control the feed rate.

Design of Anaerobic Systems

The second major advance in anaerobic treatment has been in reactor design. Traditionally, because of the recalcitrance of particulate organic matter and the slow growth rate of anaerobic bacteria, reactors have been designed as stirred tanks with very long residence times. Unlike sludges, the organic materials in waste waters are often in solution and thus are more amenable to treatment. In a completely mixed system such as the traditional sludge digester, hydraulic retention time (HRT) and solids retention time (SRT) are the same and the minimum HRT is defined by microbial growth rate. To overcome the long generation time (several days) special reactors have been developed to retain most of the organisms inside the reactor or to recycle the bacteria after separation. Solids retention time is thus uncoupled from HRT and high bacterial concentrations can be obtained. There are four basic designs (*see Figure 19* on page 293): (1) the anaerobic contact process; (2) the Upflow Anaerobic Sludge Blanket (UASB) in which the biomass is held in the reactor by flocculation; (3) fluidized or expanded beds; (4) anaerobic filtration (by upflow or downflow). In types (3) and (4) the biomass is attached to an internal support medium. Three companies (Biotechnics, Sweden; ESMIL, Holland; Agip, Italy) have installed more than 10 large-scale plants for industrial waste treatment. The most common application is for the treatment of vegetable processing waters, particularly from sugar production; the volume of such reactors currently ranges from 2000 m^3 to 20 000 m^3.

Dorr-Oliver (USA) have built three full-scale fluidized-bed plants; American Celanese (USA) have built three upflow anaerobic filter plants, and Bacardi Rum have constructed one downflow filter plant of 1300 m^3. There are four full-scale anaerobic plants in the UK: two are completely mixed, one has external biomass separation, and one has an upflow anaerobic filter. In all cases, anaerobic treatment has been designed as a high-rate pre-treatment stage to remove large amounts of polluting load and to recover energy. The most successful applications have been for the treatment of warm strong soluble wastes from the food and drinks industry. Under the best conditions, loads with a COD of 25–30 kg/m^3/day can be applied (Verstraete, 1983, Wheatley and Cassell, 1983). The quantity of methane produced can be related to the quantity of COD degraded, based on the stoichiometry of conversion of organic matter into carbon dioxide and methane (Andrews and Graef, 1971). The equation demonstrates a disproportionate breakdown of organic materials: in theory, therefore, the methane content of the biogas is correlated with the chemical composition of the substrate;

when alcohols are converted, then the methane content of the biogas is about 75%, but when carbohydrates are metabolized the methane content is only about 50%. Thus,

$$C_cH_hO_oN_nS_s + \tfrac{1}{4}(4c - h - 2o + 3n + 2s)H_2O \rightarrow$$
$$\tfrac{1}{8}(4c - h + 2o + 3n + 2s)CO_2 + \tfrac{1}{8}(4c + h - 2o - 3n - 2s)CH_4$$
$$+ nNH_3 + sH_2S$$

where *c*, *h*, *o*, *n*, *s*, represent the number of atoms of carbon, hydrogen, oxygen, nitrogen and sulphur, respectively. In practice, each kilogram of COD (mixed substrate) will yield 350 ℓ of methane and this gives a useful estimate of the efficiency of the process. *Table 14* shows some results obtained from three different types of waste treated by anaerobic upflow filtration.

Table 14. Comparative summary of results from anaerobic filtration.

Type of waste	Date	COD load applied (kg/m³/day)	COD removed (%)	Gas ratio (m³/kg COD removed)	Methane (%)
Maltings	May–July	4	65	0·23	68
	Aug.–Oct.	6	72	0·16	70
	Nov.–Jan.	8	70	0·10	65
Confectionery	May	5·0	40	0·23	74
	June	7·5	72	0·13	70
	July	5·4	78	0·10	59
	August	4·6	78	0·15	61
	September	4·2	75	0·18	58
	October	10·5	32	0·31	62
Distillery	Jan.–March	13·5	45	0·3	57
	April–June	22·1	69	0·6	57
	July–Sept.	23·1	63	0·5	59
	Oct.–Dec.	25·6	56	0·5	59

Most investigations into anaerobic treatment have encountered problems associated with pH control. The difficulty arises from differences in the growth rates and optimal pH of the synergistic bacterial populations in anaerobic fermentation. Once a stable population of the groups has been established, then the mixed culture is able to operate without external pH control but the differential rates of growth of the bacterial populations makes the system susceptible to overload, especially when treating industrial wastes which have insufficient alkalinity or additional nutrients. Three methods of controlling this problem have been tried (Henze and Harremoes, 1982): these are (1) high recycle rates; (2) chemical buffering, and (3) buffering using domestic sewage. High rates of recycle are an integral part of some anaerobic reactor designs such as fluidized beds and downflow filters (Szendrey, 1983).

On the basis of the energy recovered and the savings in discharge costs, there is a good economic case for the anaerobic treatment of strong wastes, and a rapid expansion of the technology is probable. Some basic research and development is still required, to improve reliability and start-up procedures and to reduce capital costs.

Future work

There is now a better understanding of the characteristics of the micro-organisms involved in the treatment of effluent and of the interrelationships between them. High biomass concentrations can now be retained in anaerobic reactors and loading rates of COD up to 30 kg/m^3/day can be achieved, with the maximum expected amounts of gas produced. Better conversions of direct substrates such as acetates have been reported (Verstraete, 1983) and it may be possible to increase the methanogenic activity by conventional strain selection or genetic manipulation.

Further work may also assist the control of toxicity; problems with ammonia, sulphate and calcium are very common. Wastes containing large quantities of organic nitrogen and of a high pH may produce toxic concentrations of ammonia (ammonia concentration should be below 100 mg/ℓ). High concentrations of sulphate ($>$ 500 mg/ℓ) reduce methane production and generate large quantities of H_2S in the biogas (up to 4%). Most investigators (Sahm, 1983) have found that sulphide starts to inhibit methane formation at about 3 mmol/ℓ. Very hard waters (Ca $>$ 6 mmol/ℓ) can give rise to calcium carbonate and calcium phosphate precipitates which may then cause blockage and reduced activity.

Microbial methane production occurs over a wide temperature range (0–97°C) but two distinct optima exist, one at 35°C and one at 60°C. Most work has been carried out on the mesophilic group and more work is required on the thermophilic and psychrophilic methane bacteria, particularly to test whether a temperature increase improves performance. Psychrophilic methanogens have not been isolated nor have mixed cultures shown a clear psychrophilic optimum; much more work is required to confirm that the process could be used for cold wastes. The start-up of anaerobic treatment systems is still a problem: at present a 30–50% inoculum of actively digesting sludge is used and techniques need to be developed to reduce this volume. Most authors agree that the process is slow (possibly taking up to 6 months) but the differences in acclimatization period necessary for different combinations of the various substrates and the trace nutrients required, are not well understood.

Conclusions

Aerobic effluent treatment is the largest controlled use of micro-organisms in the biotechnological industries. There is no obvious alternative to this type of treatment and it is likely that the size of this market will continue to stimulate the development of more efficient processes.

Of the new technologies discussed, only biomethanation has a promising economic future. The market forces are already sufficient to ensure that this technology is applied and continues to develop. Confidence in the reliability of the process is growing and it is predicted that the majority of new effluent plant installed in the biological industry will be of this type. The research and development priorities have been identified and there is a high level of development activity in the United States and Europe, to improve the process and to expand the range of substrates to which it is applied.

The use of bioprocesses for the breakdown of aqueous and atmospheric recalcitrants has yet to be realized and most research is still at a fundamental stage. There is strong competition from the non-biological methods which do not need such large investments in research and development. This type of process could, however, make a significant contribution to removing recalcitrants from the environment and there is a case for centralized or concerted action by national and international institutions to fund development work.

Experience has shown that the recovery and marketing costs of large-scale biomass recovery cannot, at present, be justified by the value of the product. More valuable products such as important metals, special oils, peptides and vitamins need to be recoverable from specific wastes if this technology is to expand.

References

ALFANI, F. (1983). Full-scale plants for biological treatment of agricultural and industrial wastes in Italy. In *Environmental Biotechnology: Future Prospects* (R.S. Holdom, J.M. Sidwick and A.D. Wheatley, Eds), pp. 100–105. The European Federation of Biotechnology CEC, Brussels.

AMBERG, H.R., ASPITANT, T.R. AND CORMACK, J.F. (1969). Fermentation of spent sulphite liquor for the production of volatile acids. *Journal of the Water Pollution Control Federation*, **41,** R419–430.

AMERICAN PUBLIC HEALTH ASSOCIATION, AMERICAN WATER WORKS ASSOCIATION, WATER POLLUTION CONTROL FEDERATION (1975). *Standard Methods for the Examination of Water and Waste Water*, 14th edn. Office of the American Public Health Association, Washington.

ANDREWS, J.F. AND GRAEF, S.P. (1971). Dynamic modelling and simulation of the anaerobic digestion process. In *Anaerobic Biological Treatment Processes. Advances in Chemistry Series 105*, pp. 126–162. American Chemical Society, Washington.

BAINES, S., HAWKES, H.A., HEWITT, C.H. AND JENKINS, S.H. (1953). Protozoa as indicators in activated sludge treatment. *Sewage and Industrial Wastes* **25,** 1023–1033.

BECCARI, M., MAPPELLI, P. AND TANDOI, V. (1980). Relationship between bulking sludge and physicochemical properties of activated sludge. *Biotechnology and Bioengineering* **22,** 969–979.

BINOT, R.A., NAVEAU, H.P. AND NYNS, E.J. (1982). Biomethanation of soluble organic residues by immobilised fluidised cells. In *Proceedings, 2nd International Symposium on Anaerobic Digestion, Travemünde, Germany, 6–11 September 1981*, pp. 363–376. Elsevier Biomedical Press, Amsterdam, New York and Oxford.

BRADE, C.E. AND NOONE, G.P. (1981). Anaerobic sludge digestion, Need it be expensive? Making more of existing resources. *Water Pollution Control* **80,** 70–94.

BRADE, C.E., NOONE, G.P., POWELL, E., RUNDLE, H. AND WHYLEY, J. (1982). The application of developments in anaerobic digestion within Severn-Trent Water Authority. *Water Pollution Control* **81,** 200–219.

BRAUN, R., MEYRATH, J., STAPAREK, W. AND ZERLAUTH, G. (1979). Feed yeast production from citric acid waste. *Process Biochemistry* **14,** 16–18; 20.

BROWN, M.J. AND LESTER, J.N. (1979). Metal removal in activated sludge: the role of bacterial extracellular polymers. *Water Research* **13,** 817–837.

BRUCE, A.M., MERKENS, J.C. AND HAYNES, B.A.O. (1975). Pilot scale studies on the treatment of domestic sewage by two stage biological filtration with special reference to nitrification. *Journal of the Water Pollution Control Federation* **74,** 80–100.

BRYANT, M.P., WOLIN, E.A., WOLIN, M.J. AND WOLF, R.S. (1967). *Methanobacillus omelianskii*, a symbiotic association of two species of bacteria. *Archiv für Mikrobiologie* **59,** 20–31.

BULL, A.T., HOLT, G. AND LILLY, M.D. (1982). *Biotechnology: International Trends and Perspectives.* OECD, Paris, 84 pp.

CHAKRABARTY, A.M. (1980). Plasmids and dissimilation of synthetic environmental pollutants. In *Plasmids and Transposons: Environmental Effects and Maintenance Mechanisms* (C. Stuttard and K.R. Rozee, Eds), pp. 21–30. Academic Press, London.

CHAMBERS, B. AND TOMLINSON, E.J. (1982). *Bulking of Activated Sludge—Preventative and Remedial Methods*, 1st edn. Ellis Horwood, Chichester.

CHIPPERFIELD, P.N.J. (1967). Performance of plastic filter media in industrial and domestic waste treatment. *Journal of the Water Pollution Control Federation* **39,** 1860–1874.

CHURCH, B.D., ERIKSON, E.E. AND WIDMER, C.M. (1973). Fungal digestion of food processing wastes *Food Technology* **2,** 36–42.

CONGRESSIONAL OFFICE OF TECHNOLOGY ASSESSMENT (1981). *The Impacts of Genetics: Applications to Micro-organisms, Animals and Plants.* United States Government, Washington, DC.

COOK, A.M., GROSSENBACHER, H., HOGREFE, W. AND HUTTER, R. (1983). Biodegradation of xenobiotic industrial wastes: applied microbiology complementing chemical treatments. In *Biotech 83: Proceedings of the International Conference on the Commercial Applications and Implications of Biotechnology*, pp. 717–724. Online Publications, Northwood, London.

COOK J. (1983). Biotechnology in waste water treatment. *Water and Waste Treatment Journal* **26,** 24–29.

COOK, W.B. (1954). Fungi in polluted water: a review. *Sewage and Industrial Wastes* **26,** 539–549; 661–674.

COOMBS, J. (1981). Biogas and power alcohol. *Chemistry and Industry* **4,** 223–229.

CORNWALL, D.A., ZOLTEK, J., JR., PATRINELY, C.D., DES FURMAN, T. AND KIM, J.I. (1977). Nutrient removal by water-hyacinth. *Journal of the Water Pollution Control Federation* **49,** 57–65.

CURDS, C.R. AND COCKBURN, A. (1970). Protozoa in biological sewage treatment processes. I. A survey of the protozoan fauna of British percolating filters and activated sludge plants. *Water Research* **4,** 225–236.

CURDS, C.R., COCKBURN, A. AND VANDYKE, J.M. (1968). An experimental study of the role of ciliated protozoa in activated sludge processes. *Water Pollution Control* **67,** 312–329.

DEMUYNCK, M.G., NAVEAU, H.P. AND NYNS, E.J. (1983). Anaerobic fermentation technology in Europe. In *Environmental Biotechnology: Future Prospects* (R.S. Holdom, J.M. Sidwick, and A.D Wheatley, Eds), pp. 64–70. European Federation of Biotechnology, CEC, Brussels.

DEPARTMENT OF THE ENVIRONMENT (1972). *Analysis of Raw Potable and Waste Waters*, 1st edn. HMSO, London.

DEPARTMENT OF THE ENVIRONMENT, NATIONAL WATER COUNCIL (1979). *Methods of Examination of Waters and Associated Materials.* HMSO, London.

DHALIWAL, B.S. (1979). *Nocardia amarae* and activated sludge foaming. *Journal of the Water Pollution Control Federation* **51,** 344–350.

DUNBAR, D.R. AND CALVERT, H.T. (1908). *Principles of Sewage Treatment.* 1st edn. Griffen, London.

DUNNHILL, P. (1981). Biotechnology and industry. *Chemistry and Industry* (April), 204–217.

EIKELBOOM, D.H. (1975). Filamentous organisms observed in activated sludge. *Water Research* **9,** 365–388.

ELLINGTON, A., AND BURKE, T. (1981). *Europe: Environment. The European Communities Environmental Policy.* Commission of the European Communities, Brussels; Ecobooks, London, 23pp.

FORAGE, A.J. (1978). Recovery of yeast from confectionery effluent. *Process Biochemistry* **13,** 8–11.

FRANKLIN, F.C.H., BAGDASARIAN, M. AND TIMMIS, K.N. (1981). Manipulation of degradative genes of soil bacteria. In *Microbial Degradation of Xenobiotics and Recalcitrant*

Compounds. (T. Leisinger, R. Hutter, A.M. Cooke and J. Nuesch, Eds), pp. 109–130. Academic Press, London.

FUGGLE, R.W. (1981). Treatment of waste water from food processing and brewing. *Chemistry and Industry* **13,** 453–458.

GRAU, P. (1980). Utilisation of activated sludge as fodder supplement. *Environmental Technology Letters* **1,** 557–570.

HARDER, W. (1981). Enrichment and characterisation of degrading microbes. In *Microbial Degradation of Xenobiotics and Recalcitrant Compounds* (T. Leisinger, R. Hutter, A.M. Cook and J. Nuesch, Eds), pp. 77–96. Academic Press, London.

HAWKES, H.A. (1957). Film accumulation and grazing activity in the sewage filters at Birmingham. *Journal of the Proceedings of the Institute of Sewage Purification* (2), 88–110.

HAWKES, H.A. (1963). *The Ecology of Waste Water Treatment*, 1st edn. Pergamon Press, Oxford.

HAWKES, H.A. AND SHEPHARD, M.R.N. (1972). The effect of dosing frequency on the seasonal fluctuations and vertical distribution of solids and grazing fauna in sewage percolating filters. *Water Research* **6,** 721–730.

HEMMING, M.L., OUSBY, J.C., PLOWRIGHT, D.R. AND WALKER, J. (1977). 'Deep Shaft'—latest position. *Water Pollution Control* **76,** 441–451.

HENZE, M. AND HARREMOES, P. (1982). Review paper: anaerobic treatment of waste water in fixed-film reactors. *Presented at a Special Seminar on the Anaerobic Treatment of Waste Water in Fixed-Film Reactors, Copenhagen, 16–18 June 1982*, pp. 1–90. International Association for Water Pollution Research, London.

HEYES, R.H., AND HALL, R.J. (1981). Anaerobic digestion modelling—the role of H_2. *Biotechnology Letters* **3,** 431–436.

HOLDERBY, J.M. AND MOGGIO, W.A. (1960). Utilisation of spent sulphite liquor. *Journal of the Water Pollution Control Federation* **32,** 171–181.

IGGLEDEN, G.J. (1981). Rotating biological contactors. *Chemistry and Industry* **13,** 458–465.

JAMES, A. AND ADDYMAN, C.L. (1974). By-product recovery from food wastes by microbial protein production. In *A Symposium on Treatment of Wastes from the Food and Drink Industry, University of Newcastle, 8–10 January 1974*, pp. 149–155. Institute of Water Pollution Control, Maidstone.

KIFF, R.J. AND BROWN, S. (1981). The development of the oxidative acid hydrolysis process for sewage sludge detoxification. In *Proceedings, International Conference on Heavy Metals in the Environment, Amsterdam, September 1981*, pp. 159–163. Commission of the European Communities/WHO.

LAWRENCE, A.L. AND MCCARTY, P.L. (1969). Kinetics of methane fermentation in anaerobic treatment. *Journal of the Water Pollution Control Federation* **41,** R1–R7.

LEISINGER, T. (1983). Microbial degradation of recalcitrant compounds. In *Environmental Biotechnology: Future prospects* (R.S. Holdom, J.M. Sidwick and A.D. Wheatley, Eds), pp. 119–125. European Federation of Biotechnology, CEC, Brussels.

LE ROUX, N.W. (1982). Odour control at Turriff, STW. *Water Research News* **12,** 6–7.

LE ROUX, N.W. AND MEHTA, K.B. (1980). Biological oxidation of odours. In *Odour Control—A Concise Guide* (F.H.H. Valentine and A.A. North, Eds), pp. 161–162. Department of Industry, Warren Spring Laboratory, Stevenage.

LLOYD, L. (1945). Animal life in sewage purification processes. *Journal of the Proceedings of the Institute of Sewage Purification* (2), 119–139.

LUMB, C. (1960). The application of recirculation to the purification of sewage and trade wastes. In *Waste Treatment. Proceedings of the 2nd Symposium on the Treatment of Waste Waters, Department of Civil Engineering, Kings College, Newcastle upon Tyne, University of Durham, 1959* (P. Isaac, Ed.), pp. 188–205. Pergamon Press, Oxford.

LUMB, C. AND BARNES, J.P. (1948). The periodicity of dosing percolating filters. *Journal of the Proceedings of the Institute of Sewage Purification* (1), 83–98.

LUMB, C. AND EASTWOOD, P.K. (1958). The recirculation principle in filtration of settled sewage—some comments on its application. *Journal of the Proceedings of the Institute of Sewage Purification* (4), 380–398.

McGARRY, M.G. (1971). Water and protein reclamation from sewage. *Process Biochemistry* **6,** 50–53.

McINERNEY, M.J., BRYANT, M.P. AND STAFFORD, D.A. (1980). Metabolic stages and energetics of microbial anaerobic digestion. In *Anaerobic Digestion: Proceedings of the 1st International Symposium on Anaerobic Digestion, University College, Cardiff, 1979* (D.A. Stafford, B.I. Wheatley and D.E. Hughes, Eds), pp. 91–98. Applied Science, London.

McKINNEY, R.E. (1956). Biological flocculation. In *Biological Treatment of Sewage and Industrial Waste. Volume 1. Aerobic Oxidation* (B.J. McCabe and W.W. Eckenfelder, Eds), pp. 88–100. Reinhold Publishing Corporation, New York.

MILLS, E.V. (1945a). The treatment of settled sewage in percolating filters in series with periodic changes in the order of filters. *Journal of the Proceedings of the Institute of Sewage Purification* (2), 35–55.

MILLS, E.V. (1945b). The treatment of settled sewage by continuous filtration with recirculation. *Journal of the Proceedings of the Institute of Sewage Purification* (2), 94–110.

MINISTRY OF HOUSING AND LOCAL GOVERNMENT (1954). *Report of an Informal Working Party on the Treatment and Disposal of Sewage Sludge.* HMSO, London.

MOSEY, F.E. (1978). Anaerobic filtration: A biological treatment process for warm industrial effluents. *Water Pollution Control* **77,** 370–378.

MOSEY, F.E. (1982). New developments in the anaerobic treatment of industrial wastes. *Water Pollution Control* **81,** 540–552.

MUNDEN, J.E. AND KING, R.W. (1973). High rate biofiltration system for food pilot plants. *Effluent and Water Treatment Journal* **13,** 159–165.

NYNS, E.J. AND NAVEAU, H.P. (1983). Biogas from agricultural industrial and urban wastes. In *Energy from Biomass, 2nd EC conference* (A. Stub, P. Chartier and G. Schleser, Eds), pp. 364–371. Applied Science, London.

O'SHAUGNESSY, F.R. (1931). Some considerations in the oxidation of sewage. *Proceedings of the Association of Managers of Sewage Disposal Works* 74–92.

PACKHAM, R.F. (1983). Water quality and health. In *Pollution: Causes, Effects and Control* (R.M. Harrison, Ed.), pp. 19–32. The Royal Society of Chemistry, London.

PAINTER, H.A. (1954). Factors affecting the growth of some fungi associated with sewage purification. *Journal of General Microbiology* **10,** 177–190.

PAINTER, H.A. (1958). Some of the characteristics of domestic sewage. *Water and Waste Treatment Journal* **6,** 496–498.

PAINTER, H.A. (1970). A review of literature on inorganic nitrogen metabolism in microorganisms. *Water Research* **4,** 393–450.

PAINTER, H.A. (1971). Chemical, physical and biological characteristics of wastes and waste effluents. In *Water and Water Pollution Handbook* (L.L. Ciaccio, Ed.), volume 1, pp. 329–364. Marcel Dekker, New York.

PAINTER, H.A. AND VINEY, M. (1959). Composition of domestic sewage. *Journal of Biochemical and Microbiological Technology and Engineering* **1,** 15–21.

PAINTER, H.A., VINEY, M. AND BYWATERS, A. (1961). Composition of sewage and sewage effluents. *Journal of the Proceedings of the Institute of Sewage Purification* (4), 302–314.

PEPPLER, H.J. (1968). Industrial production of single-cell protein from carbohydrates. In *Single-Cell Protein* (R.I. Mateles and S.R. Tannenbaum, Eds), pp. 229–242. MIT Press, London.

PIKE, E.B. (1975). The aerobic bacteria. In *Ecological Aspects of Used-Water Treatment* (C.R. Curds, and H.A. Hawkes, Eds), volume 1, pp. 1–63. Academic Press, London.

PIKE, E.B. AND CARRINGTON, E.G. (1972). Recent developments in the study of bacteria in the activated sludge process. *Water Pollution Control* **71,** 583–605.

PIKE, E.B. AND CARRINGTON, E.G. (1979). The fate of enteric bacteria and pathogens during sewage treatment. In *Biological Indicators of Water Quality, a Symposium at the University of Newcastle upon Tyne, 12–15 September 1978* (A. James and L.M. Evison, Eds), pp. 20-1–20-32. John Wiley, Chichester.

PIKE, E.B., CARLTON-SMITH, C.H., EVANS, R.H. AND HARRINGTON, D.W. (1982). Performance of rotating biological contactors under field conditions. *Water Pollution Control* **81,** 10–27.

PIPES, W.O. (1977). Microbiology of dairy waste activated sludge separation problems. *Industrial Wastes* **236,** 6, 26, 29, 31.

PORCH, D., BAYLEY, R.W. AND BRUCE, A.M. (1977). *Economic Aspects of Sewage Sludge Disposal. Water Research Technical Report TR42.* Water Research Centre, Stevenage.

PRESCOTT, S.C. AND DUNN, G.G. (1959). *The Yeasts and the Propagation of Yeasts,* 3rd edn. McGraw Hill, London.

QUINN, J.P. AND MARCHANT, R. (1980). The treatment of malt whiskey distillery waste using the fungus *Geotrichum candidum. Water Research* **14,** 545–551.

REDDY, K.R., HUESTON, F.M. AND MCKIM, T. (1983). Water-hyacinth production in sewage effluent. In *Proceedings, Symposium on Energy from Biomass and Wastes VII, Lake Buena Vista, Florida, 24–28 January 1983* (D.L. Klass and H.H. Elliot, Eds), pp. 135–167. Institute of Gas Technology, Chicago.

REYNOLDSON, T.B. (1941). The biology of the macrofauna of high-rate double filtration at Huddersfield. *Journal of the Proceedings of the Institute of Sewage Purification* (1), 109–129.

REYNOLDSON, T.B. (1943). A comparative account of the life cycles of *Lumbricillus lineatus* and *Enchytraeus albidus* in relation to temperature. *Annals of Applied Biology* **30,** 60–66.

REYNOLDSON, T.B. (1947). An ecological study of the Enchytraeid worm population of sewage bacteria beds. *Annals of Applied Biology* **34,** 331–345.

REYNOLDSON, T.B. (1948). An ecological study of the Enchytraeid worm population of sewage bacteria beds—synthesis of field and laboratory data. *Journal of Animal Ecology* **17,** 27–38.

RIGHELATO, R.C., IMRIE, F.K.E. AND VLITOS, A.J. (1976). Production of single-cell protein from agricultural and food processing wastes. *Resource Recovery and Conservation* **1,** 257–269.

ROYAL COMMISSION ON SEWAGE DISPOSAL (1908). *5th Report.* HMSO, London.

SAHM, H. (1983). Anaerobic waste water treatment. In *Environmental Biotechnology: Future prospects* (R.S. Holdom, J.M. Sidwick and A.D. Wheatley, Eds), pp. 39–56. European Federation of Biotechnology CEC, Brussels.

SCHAUER, M.L. AND FERRY, J.G. (1980). Metabolism of formate in *Methanobacterium formicium. Journal of Bacteriology* **142,** 800–807.

SCHMIDT-BLEEK, F. (1980). Testing and Evaluation methods for environmental chemicals. *Toxicological and Environmental Chemistry Reviews* **3,** 265–290.

SCHWEGLER, B. AND CYNOWETH, D. (1983). Update on Walt Disney World water hyacinth treatment system. *Biomass Digest* **5,** (8), 4.

SEACHEM, R.G., BRADLEY, D.J., GARELICK, H. AND MARA, D.D. (1980). *Appropriate Technology for Water Supply and Sanitation. Health Aspects of Excreta and Sullage Management—a State-of-the-Art Review.* World Bank, Washington, DC.

SMITH, M.E. AND BULL, A.T. (1976). Studies on the utilisation of coconut water waste for the production of the food yeast *Saccharomyces fragilis. Journal of Applied Bacteriology* **41,** 81–95.

SOLBE, J.F. DE L.G. (1971). Aspects of the biology of the lumbricids *Eiseniella tetraedra* (Savigny) and *Dendrobaena rubida* (Savigny) *F. subrubicunda* (Eisen) in percolating filter. *Journal of Applied Ecology* **8,** 845–867.

STANBRIDGE, H.H. (1976). *The History of Sewage Treatment in Great Britain. 9. Anaerobic Digestion of Sewage Sludge.* Institute of Water Pollution Control, Maidstone.

STORCK, W.J. (1979). In *Chemical Engineering News* 7 May, pp. 21–29, cited by Hutzinger, O. and Veerkamp, W. (1981). Xenobiotic chemicals with pollution potential. In *Microbial Degradation of Xenobiotics and Recalcitrant Compounds* (T. Leisinger, R. Hutter, A.M. Cook and J. Nuesch, Eds), pp. 3–45. Academic Press, London.

SUTTON, P.M., SHIEH, W.K. KOS, P. AND DUNNING, P.R. (1981). Dorr-Olivers oxitron system in fluidised bed water and waste water treatment process. In *Biological*

Fluidised-Bed Treatment of Water and Waste Water (P.F. Cooper and B. Atkinson, Eds), pp. 285–304. Ellis Horwood, Chichester.

SWANWICK, J.D., SHURBEN, D.G. AND JACKSON, S. (1969). A survey of the performance of sewage sludge digestion in Great Britain. *Water Pollution Control* **68,** 639–661.

SZENDREY, L.M. (1983). The Bacardi Corporation digestion process for stabilizing rum distillery wastes and producing methane. In *Proceedings, Symposium on Energy from Biomass and Wastes VII, Lake Buena Vista, Florida, 24–28 January 1983* (D.L. Klass and H.H. Elliot, Eds), pp. 767–794. Institute of Gas Technology, Chicago.

TEBBUTT, T.H.Y. (1977). *Principles of Water Quality Control,* 1st edn. Pergamon, Oxford.

TERRY, R.J. (1956). The relations between bed medium and sewage filters and the flies breeding in them. *Journal of Animal Ecology* **25,** 6–14.

THANH, N.C. AND SIMARD, R.E. (1973). Biological treatment of waste water by yeasts. *Journal of the Water Pollution Control Federation* **45,** 674–680.

TOMLINSON, E.J. (1976a). The production of single-cell protein from strong organic waste waters from the food and drink processing industries. 1. Laboratory Cultures. *Water Research* **10,** 367–371.

TOMLINSON, E.J. (1976B). The production of single-cell protein from strong organic waste waters from the food and drink processing industries. 2. The practical and economic feasibility of a non-aseptic batch culture. *Water Research* **10,** 372–376.

TOMLINSON, T.G. (1942). Some aspects of microbiology in the treatment of sewage. *Journal of the Society of Chemical Industry* **62,** 53–58.

TOMLINSON, T.G. (1946). The growth and distribution of film in percolating filters treating sewage by single and alternate double filtration. *Journal of the Proceedings of the Institute of Sewage Purification* (1), 168–183.

TOMLINSON, T.G. AND HALL, H. (1955). The effect of periodicity of dosing on the efficiency of percolating filters. *Journal of the Proceedings of the Institute of Sewage Purification* (1), 40–47.

TOMLINSON, T.G. AND WILLIAMS, I.L. (1975). The Fungi. In *Ecological Aspects of Used-Water Treatment. Volume* 1. *The Organisms and Their Ecology* (C.R. Curds and H.A. Hawkes, Eds), pp. 93–152. Academic Press, London.

VAN VEEN, W.L., MULDER, E.G. AND DEINEMA, M.H. (1978). The Sphaerotilus–Leptothrix group of bacteria. *Microbiological Reviews* **42,** 329–356.

VERSTRAETE, W. (1983). Biomethanation of wastes: Perspectives and potentials. In *Biotech 83: Proceedings of the International Conference on the Commercial Applications and Implications of Biotechnology*, pp. 726–742. Online Publications, Northwood.

WALKER, I. AND AUSTIN, E.P. (1981). The use of plastic porous biomass supports in a pseudo fluidised bed for effluent treatment. In *Biological Fluidised-Bed Treatment of Water and Waste Water* (P.F. Cooper and B. Atkinson, Eds), pp. 272–284. Ellis Horwood, Chichester.

WASSERMANN, A.E., HAMPSON, J.W. AND ALVARE, N.F. (1961). Large-scale production of yeast in whey. *Journal of the Water Pollution Control Federation* **33,** 1090–1094.

WATER DATA UNIT (1979). *Department of the Environment Water Data, 1978.* Water Data Unit, Reading.

WATER RESOURCES BOARD (1973). *The Trent Research Programme,* volume 1. HMSO, London.

WATTIE, E. (1943). Cultural characteristics of zoogloea-forming bacteria isolated from activated sludge plants and trickling filters. *Sewage Works Journal* **15,** 476–490.

WHEATLEY, A.D. (1981). Investigations into the ecology of biofilms in waste treatment using scanning electron microscopy. *Environmental Technology Letters* **2,** 419–424.

WHEATLEY, A.D. AND CASSELL, L. (1983). Energy recovery and effluent treatment of strong industrial wastes by anaerobic biofiltration. In *Proceedings, Symposium on Energy from Biomass and Wastes VII, Lake Buena Vista, Florida, 24–28 January 1983* (D.L. Klass and H.H. Elliot, Eds), pp. 671–704. Institute of Gas Technology, Chicago.

WHEATLEY, A.D. AND WILLIAMS, I.L. (1976). Pilot scale investigations into the use of random pack plastics filter media in the complete treatment of sewage. *Water Pollution Control* **75,** 468–486.

WHEATLEY, A.D., MITRA, R.I. AND HAWKES, H.A. (1982). Protein recovery from dairy industry wastes with aerobic biofiltration. *Journal of Chemical Technology and Biotechnology* **32,** 203–212.

WILLIAMS, P.A. (1981). The genetics of biodegradation. In *Microbial Degradation of Xenobiotics and Recalcitrant Compounds* (T. Leisinger, R. Hutter, A.M. Cooke and J. Nuesch, Eds), pp. 97–107. Academic Press, London.

WILLIAMS, J.I. AND PUGH, G.J.F. (1975). Resistance of *Chrysosporium pannorum* to an organomercury fungicide. *Transactions of the British Mycological Society* **64,** 255–263.

WILLIAMS, N.V., SOLBE, J.F. DE L.G. AND EDWARDS, R.W. (1969). Aspects of the distribution, life history and metabolism of the Enchytraeid worms *Lumbricillus rivalis* (Levinsen) and *Enchytraeus coronatus* (N. and V.) in a percolating filter. *Journal of Applied Ecology* **6,** 171–183.

WISHART, J.M. AND WILKINSON, R. (1941). Purification of settled sewage in percolating filters in series with periodic change in the order of the filter. *Journal of the Proceedings of the Institute of Sewage Purification* (1), 15–38.

ZEHNDER, A.J.B. (1978). The ecology of methane formation. In *Water Pollution Microbiology, Vol. 2* (R. Mitchell, Ed.), pp. 349–376. Wiley, New York.

11
Sugar-cane as an Energy Crop

J. COOMBS

Bio-Services, King's College, University of London, 68 Half Moon Lane, London SE24 9JF, UK

Introduction

Sugar-cane is well known as a source of a wide variety of food and beverage products including white and speciality sugars, molasses and rum, as well as other fermentation products (Paturau, 1982). The term energy, as applied here to crop, is used (as so often) in a loose manner. Hence, the title of this review would perhaps be more correctly expressed as 'The potential of sugar-cane as a source of raw material for the production of high-grade fuels for transport, heat or power applications.' This in turn raises the question of why sugar-cane might be singled out in this context and what is the relevance to the rapidly expanding fields of biotechnology and genetic engineering. If these questions are considered in the reverse order, then it might be anticipated that developments in biotechnology will follow two eventually divergent paths: the group taking the first path is related to the production of relatively small quantities of high-value speciality chemicals and health-care products; the group taking the second path is concerned with the large-scale production of fuels and chemical feedstocks from renewable resources. As far as the first group is concerned, the cost and local availability of suitable substrates will not be of prime importance. On the other hand, as far as bulk chemicals and fuels are concerned, it is already clear that the major consideration which will decide the eventual size and geographical location of any new biotechnical energy industries will relate first to the nature, availability and cost of raw materials and second to the competitive advantages of biological conversion technologies over thermochemical alternatives. For instance, processes based on biological raw material (biomass) must compete now with oil and natural gas, and in the longer term with coal as raw material. In Europe it has been estimated (Anonymous, 1983) that the total energy which might be derived from biomass sources in the year 2000 will be of the order of 85×10^6 tonnes oil equivalent, i.e. of about the same order of magnitude as the present stockpile of coal in the UK. In the same way, if all the molasses available in Europe were converted to alcohol it would not be sufficient to provide even a 10% blend with petroleum as car fuel. As far as technology is concerned, the production of methanol from wood (not to mention natural gas or coal) has both a greater weight yield and a better energy efficiency than the

Biotechnology and Genetic Engineering Reviews—Vol. 1, February 1984
0264-8725/84/01/311-35$10.00 + $0.00 © Intercept Ltd

biological hydrolysis of cellulose and fermentation to ethanol of the sugars so released.

The challenge to biotechnology and genetic engineering in this area is to cause significant increases in production of energy crops, by increasing the yield per unit area of land harvested each year, and also to improve biological conversion processes so that fuels produced through biological routes can become competitive in real terms. The true assessment of such competition is important. Systems can be shown to be economically viable, at present, where raw materials represent biological wastes from agriculture or industry. Fermentation alcohol is being produced in a number of countries at a price which makes it competitive with alcohol derived from natural gas. However, this is due to a distortion of prices by artificial economic factors, resulting from surpluses of agricultural products in the USA and Europe, coupled with European (EC) prices which are maintained at about twice world prices for sugar and starch. Alternatively, in the less-developed regions of the world, apparently viable programmes aimed at the production of fuel alcohol may be based on very low incomes in the agricultural sector.

The term 'biomass energy system' has been adopted to describe systems where the raw materials are derived from plants, through fixation of carbon dioxide using solar energy in the process of photosynthesis. The use of plant material as a source of fuel is, of course, not new. Before the industrial revolution it was the major energy source throughout the world, and it remains so in rural regions of most tropical developing countries, representing over 15% of the world's total energy supply (Hall, Barnard and Moss, 1981). In many developing countries the use of such biological materials as fuel may represent over 80% of the total use: even in the USA, with the highest fuel consumption expressed both on a national and on a per capita basis, biological materials represent over 3% of the current energy use. However, until the last decade, little attention was paid either to improvements in such use of biological energy, or to considerations of true costs, energy balances, effects on the environment, rural populations or national economies. This situation changed during the 1970s in response to real or anticipated shortages of oil-derived fuels, as well as to rapid price increases. As a result, many research programmes were initiated and feasibility studies carried out, reaching a peak of activity at the end of the last decade. Peak activity to find alternative fuels coincided in 1979 with the peak of oil consumption (about 64 million barrels per day) (Anonymous, 1981a). Since then, demand for oil has dropped, resulting in both a decrease in the price of oil in real terms and a decrease in the amount of activity being devoted to schemes for the production of alternative fuels. This lull in activity permits an evaluation of existing biomass-based fuel programmes, as well as an assessment of what systems might be developed in order to meet future crises which may arise.

At the national level, the only significant use of biomass as a fuel is by direct combustion of wood, of agricultural residues, and of domestic solid waste (which may be regarded as a biomass energy resource because of its large wood and paper content). There is a world-wide increase in the use of such materials as a substitute for oil for the production of heat, and to a lesser extent for the production of power—this may be considered as a biomass activity rather than as biotechnology. In addition, in a few countries there is a significant production of ethanol by fermentation of sugar and starch (derived in the main from sugar-cane, molasses and maize) to provide material for use as a liquid transport fuel. The major fuel alcohol programmes are in Brazil (based on sugar cane: Rothman, Greenshields and Calle, 1983), in USA (based on maize: OTA, 1980) and in Zimbabwe and elsewhere (based on molasses and cane juice: Kovarik, 1982b). Anaerobic digestion, resulting in the production of methane, receives considerable attention and may be of importance in China in terms of the number of people who obtain some light and heat from the 7 million digestors which have been built there (Chen, 1980). However, in the rest of the world, both the number of digesters which have been built (about 500 in the EC, for example, discounting sewage digesters) and the amount of energy generated, are negligible. Hence, at present the only biotechnical energy system of any significance is ethanol production by fermentation.

Fuel alcohol programmes have already been subjected to extensive criticism on the basis of a wide range of factors, some related to economics, some to energy balances and some to environmental and social aspects. Numerous studies have shown that alcohol is expensive to produce in comparison with the use of oil: in many systems more oil is consumed in the production of alcohol than is gained, and with current technology there is competition between the use of agricultural raw materials for fuel or food production (Brown, 1980; Flaim and Hertzmark, 1981; Levinson, 1982). Where such food/fuel competition exists it has been suggested that crops for fuel could be produced by clearing more land, or by using marginal land—causing further concern about deforestation, desertification and other environmental issues, including worries about the effects of discharge of large volumes of stillage from the alcohol factories where suitable treatment is not available, or not economic. Above all, it is still not clear whether fuel alcohol can be produced on a sustainable basis, resulting in a significant net gain of energy, at a realistic cost under circumstances which also give a reasonable financial return to the grower. Hence, it would be of interest to consider an optimal system from a biological viewpoint and to see what are the limitations, to what extent these are biological, and where biotechnology might have an input which could help to realize the optimum potential.

Under ideal conditions, sugar-cane (for many reasons which will be detailed below) represents at present both an optimal agricultural system for trapping solar energy into plant biomass (Hudson, 1975; Thompson, 1978; Coombs, 1980; Lipinsky and Kresovich, 1982) and an ideal raw material for the biological production of higher-value fuels, because it can produce both fermentation substrate in the form of a solution of fermentable sugars in the juice, and a combustible fuel in the form of the fibrous parts of the plant (bagasse). In addition, breeding programmes, agricultural expertise, extensive research experience, available technology for both production and conversion, and established plantations capable of producing and processing very large quantities (approaching 10×10^3–15×10^3 tonnes/day in extreme cases) already exist. However, to quote from a recent publication from within the sugar industry (Bennett, 1983): 'It is simply a matter of fact that in the US, faced with an opportunity to produce and market fuel alcohol, the corn industry has been successful while the sugar industry has failed.' This is in spite of the fact that considerable opposition to the use of corn for this purpose, on the basis of energy considerations in particular, was voiced in the US.

The failure of the sugar industry (other than in Brazil) to exploit the fuel alcohol market in the late 1970s can be linked with a general depression in the sugar industry world-wide. This depression has been greater and more prolonged than that of the general world recession, and led to ill-advised projects (such as that in Kisumu, Kenya; Kovarik, 1982a) and inaccurate or exaggerated claims by some sectors of the sugar construction industry in an attempt to obtain contracts during the period of depression. At the same time, significant changes have occurred in the structure of the world sugar markets as a result of nationalization of previously colonial sugar estates following granting of independence of many nations in Africa and the Caribbean in particular, followed by periods of national instability and political problems. At the same time, entry of the UK into the European Economic Community led to the reshaping of the previous Commonwealth Sugar Agreement and its replacement by the Lohme Agreement, resulting in a reduction of about 50% in cane sugar imports to Europe and a disruption of sugar markets for Australia, followed by an increase in European beet sugar production and a slump in world sugar prices. Hence, it would be quite unrealistic to try to deal with the potential of sugar-cane as an energy crop in the abstract, considering only aspects of photosynthesis, fermentation and genetic engineering, without a realistic assessment of both the industry and the capability of producing countries to make the contributions necessary to enable such production to occur.

The sugar industry

Sugar-cane is grown in more than a hundred different countries located between latitudes of about 40 degrees North and South of the equator. The distribution reflects the availability of water and the need cane has for a high annual solar input averaging about 200 Wm^2. The major producers of cane sugar include Brazil, India, Cuba, China, Mexico, the Philippines, South Africa, Australia and the USA, as well as many smaller nations in the Caribbean, Latin America, Africa, the Far East, the Pacific and even in Europe at the southern tip of Spain and Portugal. Climatic differences, soil types and variations in farming practice result in crops with widely differing yields grown for anything from 9 months to almost 2 years between harvests. Yields, expressed in terms of tonnes (t) of green (natural moisture content) cane per hectare vary from as low as 20 t to over 200 t. This is material 'as harvested', which may represent about 60% of the above-ground biomass. An annual yield of 100 green tonnes (equivalent to 30 dry tonnes) would be regarded as a good average for almost any region, with current national averages of all developed countries at about 80 green tonnes per hectare and developing countries averaging about 54 tonnes per hectare (FAO, 1981); such yields have been more or less static over the last decade. This contrasts dramatically with the situation in respect of almost all other important agricultural crops, including sugar beet and the major grains, where steady annual rates of increase of about 2·5–3·5% have been recorded. As a result, increased production of cane reflects increased land area use rather than increased yield, again in contrast to most other crops, where yield increases reflect the introduction of new varieties, mechanization, and increased inputs in terms of fertilizer, irrigation and crop protection.

In spite of the apparent constant global production of cane sugar, very marked changes in regional production can be seen. Cane production has increased in Brazil, linked with the alcohol programme discussed in more detail below; most other major producers have remained static. However, dramatic decreases in production are seen in many of the smaller island states which were formerly colonies of European countries. Although these decreases may not be significant in terms of total world production, in many cases cane was a major agricultural crop in terms of land area harvested, and a major source of income to the islands. These falling production figures reflect major changes in the world pattern of sugar trade during the last decade.

The sugar market as a whole can be divided into three parts: (1) that intended for internal consumption by the producer; (2) that which is preferentially traded between countries at fixed price under specific quotas; (3) that part subject to free trade on the world market. This last part has been controlled by a series of International Sugar Agreements, aimed at reducing fluctuations in the world sugar price which arise as a result of periods of over-production or under-production. However, in spite of such agreements sugar prices rose in an uncontrolled manner in late 1974, to approach £500/t. This had serious consequences because it stimulated production of beet sugar within the European Community, resulting in over-production, flooding the free market and depressing prices to below costs of production in many tropical countries. The depression in the sugar industry was increased by the advent of immobilized enzyme technology leading to the development of high-fructose corn syrup in the USA, which has now displaced around 11 million tonnes of sucrose (*see* Chapter 5).

These more recent events followed on from important changes in sugar supply patterns which started to evolve in the the 1950s. Prior to this, most cane sugar was produced in colonies and exported thence as raw sugar for refining in Europe and the USA, often being re-exported back to the tropics as white sugar. However, since then there has been a gradual shift towards greater self-sufficiency, with the ratio of exports to production for such countries falling from 54% to 39%, resulting in a reduction in foreign exchange for purchase of goods and of oil-based fuels in particular. At the same time, problems of establishing stable management systems (both for cane plantations and for islands themselves, following independence) has contributed further to the decline. For instance, in 1971 cane sugar production in Barbados and in Trinidad were about 146×10^3 t and 229×10^3 t respectively. In June 1983 the estimated crop in Barbados is less than 86×10^3 t and the crop in Trinidad is expected to be around 70×10^3 t, with the cost of production substantially in excess of the current price. There is thus a paradoxical situation, that those areas which have lost foreign exchange income needed for fuel-oil purchase also have a depleted industry unable to take advantage of the possibilities of using cane without outside support. As a result, numerous feasibility studies have been carried out, funded by aid programmes and development banks, which tend to support the use of sugar cane as a source of fuel alcohol. However, the number of projects which have been initiated remain low.

Sugar cane

The world-wide distribution of cane is a result of man's intervention, as is the occurrence of present varieties with thick stems and high sugar content at maturity. Sugar cane is a large grass of the genus *Saccharum*, belonging to the family Gramineae in the tribe Andropogoneae. It is of relatively recent evolution with an origin in South-East Asia. As with most members of the tribe, the sugar cane is highly specialized and well adapted to dry subtropical savannahs. The subtribe Saccharineae includes *Saccharum* and the related genera *Erianthus*, *Sclerostachya* and *Narenga* (the so-called Saccharum complex), the common features of which are large size (with vegetative stems of over 5 metres tall), prominent nodes with root initials which will readily propagate, and the ability to form intergeneric crosses. The similarity of form, plus the ability to form hybrids and the fact that the initial distribution by man took place 200–300 years ago, can lead to problems in the identification of different types of 'sugar-canes'. Further difficulties have arisen from the splitting of what were different species into different genera and the division of what was the genus *S.officinarum* L. into various species or genera. Sugar-cane plants that have been selected for high sugar are in general natural or purpose-bred hybrids (Mukherjee, 1957).

It is probable that the major ancestor of the present cultivated cane is *S. spontaneum*, probably arising from *Erianthus* with germplasm contributions from *Miscanthus* originating in Indo-China (Simmonds, 1976). No true *Saccharum* species, or even a distant relative, has ever been found in the New World; it is believed that it was introduced to America by Columbus. On the other hand, the diversity of cane germplasm is greatest in Melanesia, with *S. robustum* as the dominant species. This species was cultivated by primitive man, in the region of New Guinea, on the basis of characteristics such as sweetness and thickness, for the purpose of chewing. As the commercial cane industry developed, the need for cane with characteristics such as increased resistance, higher sugar content, and erect habit, led to further hybridization with *S. spontaneum*, because this species was resistant to attack by mosaic virus. Two other forms, *S. sinense* and *S. bakeri*, which are probably natural hybrids, have also been used for breeding purposes. Early European-controlled cultivation in the West Indies was based on the Creole cane, a *sinense* derivative. Although questions of origin and nomenclature may appear to be academic, cultivated cane (like so many of the world's major crops) is based, in some areas, on very narrow germplasm resources. An understanding of these, coupled with an opportunity to introduce new genetic material from diverse natural populations by conventional crossing, remains the major method of genetic recombination, certainly in the near term at least, and is thus of paramount importance. Within the various species of *Saccharum*, wide variations of both sugar and fibre content occur. For instance Bull and Glasziou (1963) obtained a range of sucrose content, based on fresh cane weight, from 4% in *S. spontaneum*, through 8% in *robustum*, 14% in *sinense* and 17% in *officinarum*. Fibre content showed an opposite trend, dropping from around 35% in *spontaneum* to less than 10% in *officinarum*. This wide variation offers the opportunity to breed cane for either high sugar content or high fibre, as discussed below (page 332).

Cane production

Cane is a perennial crop taking 8–20 months to mature, depending on the region, after being planted as stem cuttings or sett pieces (Barnes, 1974). The first 'plant crop' is taken after about a year; thereafter, regrowth is followed by 'ratooning' until a reduction in yield indicates the need for replanting. Exactly how many crops are taken, the length of time between crops, and harvesting techniques, vary widely throughout the cane-growing regions. The highest sugar yields are obtained with a long warm growing season followed by a cooler and drier ripening period, free from frosts. Ripening (i.e. accumulation of sugar in the lower portion of the stem) may also be encouraged by deprivation of water, by low nitrogen or, under some circumstances, by application of plant growth regulating chemicals ('cane ripeners'). Ripening can thus be regarded as a 'stress' response because it is favoured by conditions which restrict vegetative growth. For this reason, if cane were bred and grown for total biomass, rather than for sucrose content and juice purity as at present, higher yields of total dry matter might be expected. However, to achieve high yields, considerable application of fertilizer is needed. A crop of 70 t/ha may require 100 kg N, 60 kg P_2O_5 and 300 kg K_2O. Part of this may be re-released into the field by burning dead leaves and trash on the crop before harvest, as well as returning boiler ash and/or filter mud to the field from factory operations. Cane will grow on a wide variety of soil types as long as adequate water is available (about 150 cm/year); the amount of water needed per year is about 1 t/kg sugar produced.

Production systems vary from vertically integrated plantations with agriculture, transport and processing controlled by a single management system—as often found in the developed countries and their previous colonies—to small groups of farmers selling to a central processing station. Traditionally, cane was harvested by hand and cane production was labour intensive. Over the last 10–15 years there has been a shift towards mechanical harvesting, but the cost of machinery is high. In 1980 it was estimated that about 15% of the 8×10^6 ha grown world-wide were being harvested mechanically and that to reach a figure of 25% of cane mechanically harvested by 1985 would take an investment of over US $\$1{\cdot}5 \times 10^9$.

Over the last 10 years, considerable attention has been paid to questions of the cost of cane production, yields and agricultural energy ratios. The reason for this is twofold: first, rising costs and other problems have, as detailed above, caused a decline in many of the traditional cane-producing regions and new methods of production will be necessary if the industries are to recover; second, such considerations are of particular relevance if sugar cane is to be used as a source of substitute fuel.

Because of the wide range of living conditions in the various countries which produce sugar cane, production costs may vary from less than US $10 t to over $60 t (Nathan, 1978; Anonymous, 1980a, 1980c; Bohall *et al.*, 1981).

the difference reflecting both labour costs (or farmer incomes) and the extent of inputs into the agriculture, in terms of fertilizer, crop protection and irrigation. The total energy input per hectare varies from the subsistence level to over 1500 MJ/t of cane produced. In the more developed regions the average input is about 40 GJ/ha for irrigated cane with the major inputs being fertilizer and fuel (Austin *et al.*, 1978). Of particular interest is the fact that studies both of highly mechanized systems in South Africa (Donovan, 1978) and of low-input systems in North Brazil (Khan and Fox, 1982) have indicated that high inputs do not necessarily give high yields in return. For instance, in the South African study a low-energy input group (at about 230 MJ/t yielded an average of 69 t/ha/year. In contrast, the high-energy group (at 1470 MJ/t) gave only 60 t/ha. Conditions on farms vary and in some cases added inputs represent an attempt to increase acreage to land which is not very suitable for cane growing. However, at the same time it would appear that, before the increase in fuel and fertilizer costs in 1973, insufficient attention may have been paid to such considerations, or to the alternative possibilities of breeding cane more suited to a specific use, soil type or climatic region.

Cane processing

The production of white refined cane sugar evolved as a two-stage process with raw sugar manufacture in the country of origin, followed by refining by European, North American, Japanese or other importers. Both stages of processing are similar, in that impurities are removed from the juice or melted raw sugar by precipitation following addition of various chemicals such as lime and carbon dioxide (carbonatation), phosphoric acid (phosphatation) or sulphite (sulphitation). (Baikow, 1982). Sucrose is then recovered by crystallization from a thick syrup, derived by concentration of the clarified and decolorized liquor under vacuum. Once the crystal sugar has been removed from the mother liquor by centrifugation, the residual syrup may be recycled until it is not economic to remove further sugar. This liquid residue—which contains the concentrated impurities, considerable ash derived from the original crop plus process chemicals, and varying amounts of residual sucrose as well as invert sugar (glucose and fructose)—becomes molasses when concentrated to about 80% solids. As far as the cane-sugar industry is concerned, molasses may be of three types: (1) raw factory (or blackstrap) molasses; (2) refinery molasses; (3) high test molasses. The third category represents concentrated, partly inverted cane juice from which sucrose has not been removed. This is of interest where cane is considered as an energy crop, because juice may be stored in this concentrated form for use outside the normal cane-harvesting season, or transported to other processing localities.

The raw sugar process is outlined in *Figure 1*, which also indicates an approximate mass balance for conventional sugar cane treatment. However, much of the process used for sugar production is of little relevance to the use of cane as an energy crop, because pure sugar does not represent a suitable fuel. Although it has been suggested that sugar surpluses might be converted to fuel alcohol, such direct use does not make sense from the energy viewpoint because the

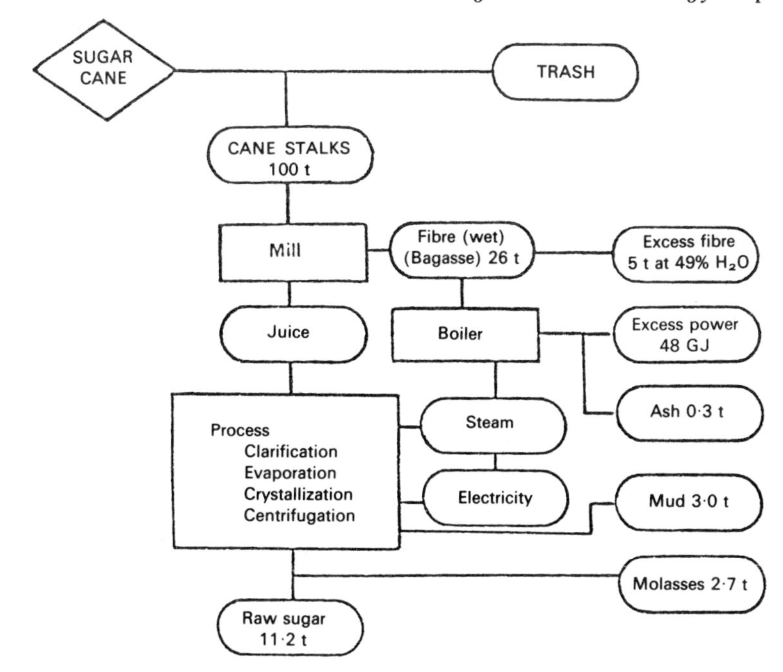

Figure 1. Processing of sugar cane.

energy input of the major process is associated with the removal of water (either in alcohol distillation or in sugar crystallization). There is thus no point in removing water to produce solid sugar and then adding it back to produce the liquid fermenter feed stream, other than as a means of reducing transport or storage costs. The use of high test molasses for this purpose has the advantage that the syrup will also contain invert sugar derived from the crop, and hence the total fermentable carbohydrates are recovered, rather than just sucrose.

Where possible, if the objective is to produce fuel alcohol from sugar-cane, it is better to ferment the juice directly. Thus, from the energy viewpoint, interest lies in the 'front end' of the process—juice extraction. Most cane is extracted by cutting and crushing the cane by passing it through large rollers (mills), which squeeze out the juice, residual sugar being removed by addition of water at the final stage. In addition, some cane is processed using counter-current diffusion techniques. Publicity has also been given to the Tilby separator, which separates the pith from the rind and wax; this was developed in order to obtain pith for cattle food (Pigden, 1974; James, 1975). It has been claimed that this technique has advantages in the processing of cane for energy purposes (Lipinsky, 1981). However, problems exist in extraction of the pulp, maintenance of the high-speed cutters, and the need to feed the machine with uniform straight billets of cane. These disadvantages probably outweigh any advantages which might accrue from an initial separation of rind and pith, as far as the use of cane for energy purposes is concerned; nevertheless, where the objective is to produce board or paper from the rind, the separator may be of value.

The mills and other equipment may be driven by steam or by electricity derived from bagasse-fed boilers and in-house electricity generation. The choice of power used to drive the mills may depend on the overall steam and energy balance of the factory. Where sugar is produced, process steam is condensed in the vacuum pans used in crystallization. However, where alcohol is the only product, surplus low-pressure steam may be available. In the same way, if all the bagasse available is burnt at high efficiency, then surplus electricity may be available for sale to outside consumers.

The objective in cane processing for sugar production is to obtain a sugar solution of high purity, without diluting the juice by addition of too much water (causing hydrolysis of sucrose), and without extracting unripe cane, which would contribute more impurities and colour precursors. In processing cane for energy, the objectives may be quite different: juice quality and purity can be sacrificed in order to obtain a higher yield of fermentable material. Nevertheless, the bagasse should have a low moisture content because the effective calorific value decreases as water content increases (Atchison, 1978; Paturau, 1982). However, because mechanical methods are the most energy-efficient for initial dewatering of cane, even if it is grown as a fibre crop it is probable that the initial stages of processing will remain the same (Alexander, 1980), and the juice produced will be used for fermentation or production of high test molasses. However, the efficiency with which juice can be extracted from high-fibre cane decreases with fibre content (*Figure 2*). Hence, an optimum balance exists between the total energy which can be recovered and the process sequence used.

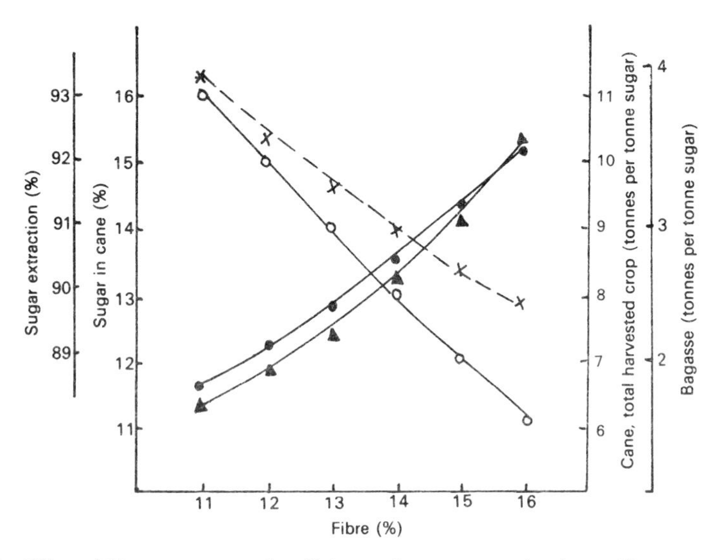

Figure 2. Effect of fibre content on the efficiency of sugar extraction in a mill. ×---×, sugar extraction (%); ○——○, sugar in cane (%); ●——●, cane, total harvested crop (tonnes per tonne sugar); ▲——▲, bagasse (tonnes per tonne sugar).

High-energy products

The only major raw materials available from cane are fermentable carbohydrate (as juice or molasses) and bagasse. The amount of fibre which can be harvested may be increased by inclusion of cane tops and leaves. The energy content available reflects the yield of millable cane, plus that of cane tops, so that a crop of 100 green tonnes could give between 650 and 750 GJ/ha of which about 40% would be in leaves and trash, 25% in each of fermentable solids and bagasse fractions, and the rest in molasses and waste streams. However, recoverable yields of power, electricity, alcohol or other up-graded fuels will be much lower, because of the relative efficiencies of conversion processes.

At the conventional assumed moisture content of 49%, bagasse as it comes from the mills has a gross calorific value of about 10 GJ/t (Atchison, 1978), and at the boiler efficiency of 60–70% of typical factory installations, will produce 2–2·5 kg of steam at 300°C per kg of bagasse. In a sugar factory, about 450 kg steam may be used per tonne of cane processed so that the ratio of fibre used to cane processed is about 0·1 to 1. Thus, with cane containing 13% fibre there would be a surplus of about 20% of bagasse: in an alcohol factory where the demand for steam is less, up to 30% bagasse may be surplus. This surplus may be used to generate electricity for the grid at a rate of about 0·25 kW/kg bagasse: alternatively, it may be used as fuel in order to process alternative sources of fermentable carbohydrate, such as cassava or maize, during the period when cane is not available, although storage may cause problems (Cusi, 1980) which may be overcome by pelletization (Bouvet and Suzor 1980).

Traditionally, the cane-sugar industry in Hawaii has supplied electricity to the local grid. In 1979 Murata reported that the plantations, with a total capacity of approximately 180 MW of electricity-generating capacity, generated a total of 669×10^6 kW, of which about 187×10^6 kW were sold to local utilities under various contractual arrangements. This electricity was produced by burning most of the bagasse and some additional leafy trash from a total of 14×10^6 t of cane grown on about 45 000 ha. This produced 9×10^6 t of prepared cane, yielding $2{\cdot}8 \times 10^6$ t of wet bagasse, 1×10^6 t of sugar and 310 000 t of molasses. Since then the efficiency of electricity generation has been improved by installation of bagasse driers powered by recycling flue gas as a means of increasing the calorific value of the bagasse as burnt, resulting in the provision of about 7% of the Hawaiian States total electricity use.

Although world-wide many factories sell some electricity, opportunities to do this vary, depending both on local demand and on the amount, seasonality and quality of bagasse produced on the other. Hawaii is in a unique position, with very high yields of cane because of the longer growing season and high inputs, as well as having an extended harvest with milling for 40 weeks of the year. Elsewhere, climatic restrictions on crop production and length of the milling season can result in a less favourable energy balance. This balance also depends on the range of products being manufactured in any particular factory, i.e. the proportions of raw sugar, factory white sugar, alcohol and/or molasses.

As far as the production of alcohol is concerned, the following alternatives exist using cane as the original raw material: (1) fermentation of molasses with an external fuel supply (e.g. European production from imported molasses); (2) fermentation of molasses in a distillery added on to an existing cane sugar factory, using bagasse as fuel; (3) an autonomous distillery fermenting cane juice using bagasse as fuel; (4) an integrated alcohol system in which cane is used to provide fermentable juice for part of the year and surplus bagasse is used as fuel with starch feedstocks obtained from cassava or maize for the non-crop season; (5) possible future systems in which the amount of ethanol produced would be increased by also using part of the bagasse, following hydrolysis, as fermentation substrate; (6) factory complexes in which the juice is fermented to ethanol, but the bagasse is converted to methanol via synthesis gas following thermal gasification.

At present, commercial fuel alcohol production is restricted to the first three categories, with the major activity in Brazil.

The Brazilian alcohol programme

The Brazilian national alcohol programme (Programa Naccional do Alcool) is at present by far the largest attempt to use biomass-derived liquid fuel as a substitute for petroleum products. The objective is to reduce foreign exchange spending; with current inflation running at about 160% and the largest national debt of any Third World country, the incentive to do this is great. Before initiation of the present programme in 1975, alcohol was produced in significant amounts as a by-product of the cane sugar industry. However, the programme had the specific objective of using alcohol to displace Brazil's petroleum demand, both by use of alcohol in blends and as a fuel in its own right in cars developed to run on hydrated (95%) alcohol. The target for 1985 is the production of about 11×10^9 ℓ, mostly derived from sugar cane (Anonymous, 1979c; Pimental, 1980; Anonymous, 1981d; Rothman, Greenshields and Calle, 1983).

The growth of alcohol production in Brazil during the last decade is shown in *Figure 3*, which shows both annual production (inset *a*) and monthly production for both the central and north-east region (inset *b*). The ability to produce the very large amounts of cane reflects the large areas of land available, but at the cost of destruction of wide areas of forest (inset *c*). The concentration of the industry in the central region can be seen from this Figure, as well as the variation in the length and season of the harvest campaign depending on the latitude, with a maximum of around 200 days of harvest of cane for alcohol production which contrasts with the shorter harvest of cane for sugar manufacture of around 160 days per annum.

Brazilian sugar cane production is currently approximately 153×10^6 t, with a low average yield of about 55 green tonnes per hectare grown over an area of some $2{\cdot}8 \times 10^6$ ha. This is equal to about one-fifth of all the land used to grow cane throughout the world. To achieve self-sufficiency in auto-fuel production will require almost five times as much land to be used; to reach even the 1985 objective will require 10% of all the Brazilian crop land to be devoted to cane.

Initial increase in alcohol production was based on the use of molasses in

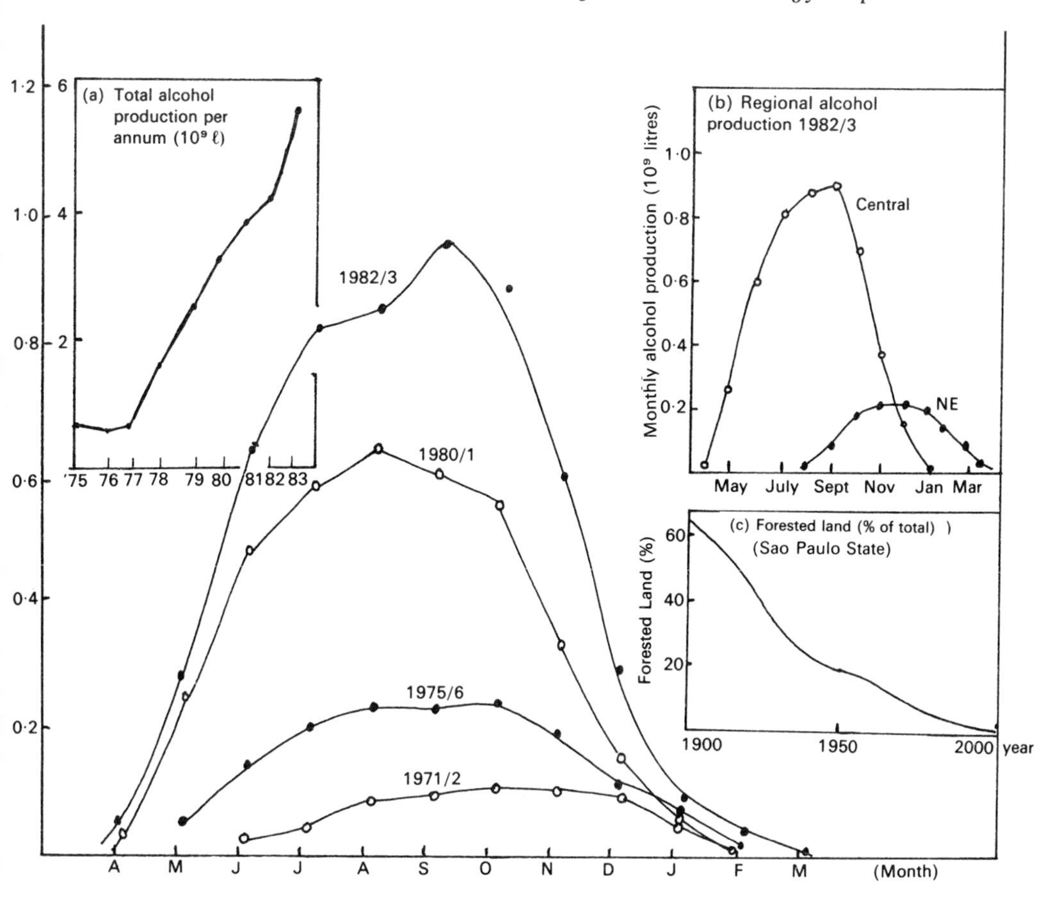

Figure 3. Growth of the production of fuel alcohol in Brazil: Source of data F. O. Licht, GmbH, International Molasses report, volume 12 (1983).

annexed distilleries. However, the rapid growth phase has been achieved by construction of several hundred autonomous distilleries with production rates of 30–250 m^3 per day. In general, the technology used has been conventional, based on traditional cane mills, followed by batch fermentation and distillation. Most of the equipment, for both agricultural production and processing, has been manufactured in Brazil, resulting in an expanding manufacturing industry which includes the production of specifically designed cars to run on alcohol fuel.

The programme has been widely criticized in terms of economics, environmental effects, diversion of food to fuel products, energy balances and lack of benefit to the poor rural work force. These arguments will not be repeated in detail here. Considerations of energy balance—a central theme of this chapter—

are dealt with on pages 327–329. Environmental issues relate in part to the effects of destruction of forest (*see Figure 3*, inset *c*) and in part to the effects of discharge of large amounts of stillage (*see* page 338). The economics of the programme are complex. If the objective is to produce foreign exchange savings, but at the same time an overseas market exists for sugar, then the programme is running at a net loss of revenue to Brazil. According to Kovarik (1982b) it would have been 30% cheaper to export sugar and buy oil on the proceeds in 1979, when sugar stood at a low of US $0·22 per kilo and oil was US $25 per barrel (160 ℓ). By 1981, with sugar prices up to US $0·33 per kilo and oil at US $40 a barrel, the sugar from one tonne of cane would be worth about $30 whereas the alcohol produced would directly replace only $17 worth of oil. Two other factors complicate the issue. First, substitution of ethanol for petroleum results in an imbalance within the oil refineries and a surplus of light fractions which have to be re-exported. The situation is made worse by the fact that alcohol is not suitable for use in diesel engines—diesel being the main fuel for agricultural machinery, road haulage and public transport. Second, the growth of distribution facilities and public demand has not matched the growth of production, resulting in stockpiles and the need also to export alcohol.

At the same time, if all the Brazilian cane were used to produce sugar, this would be equivalent to about 15% of the world total production of sugar (cane plus beet sugar), and equal to the amount traded on the free non-preferential markets. Hence, the international market would collapse if even a fraction of the Brazilian cane at present being used to produce alcohol were to be released on the world market. At present, therefore, an increasing proportion of the alcohol produced is being diverted to the chemical industry.

The Brazilian programme is characterized by having a single product (ethanol) derived from cane grown at relatively low yield and processed using traditional technology. The important questions thus relate to the extent to which innovations based on biotechnological advances might have an effect on the overall process and improve economics, energy balances and environmental effects. At present the greatest cost relates to cane production; at the same time, since the factory process can be energy self-sufficient, the major energy inputs which affect the net energy production also lie in the agriculture sector. Thus, what is needed is higher agricultural yields at lower energy inputs and costs.

Energy yields

Energy yields within the cane system may be considered from a number of viewpoints, which include the following: (1) the yield of total biomass, expressed in terms of the available solar radiation falling on the crop; (2) the yield of harvestable cane; (3) the yield of alcohol obtained by fermentation; (4) the yield of all fuels produced, i.e. for an alcohol distillery, not only alcohol but also surplus electricity and any excess bagasse. The energy content of the end-products (fuels) may also be expressed in terms of the energy content of the raw materials used, to give the efficiency (in terms of the ratio between fossil fuels used and the high-grade biological fuels produced, to give the net energy balance) or—in terms of all outputs divided by all inputs—to give the overall energy efficiency in thermodynamic terms. The problems with all such estimates lie both in definitions and in methods used to estimate the size of a given component within the system. A further problem arises from the fact that quite often the actual balance in question is not defined and it is often difficult, therefore, to compare data published in various sources. For this reason the various possible approaches to defining yields are discussed briefly, and some typical values for the complete process—from trapping of solar energy by photosynthesis to final alcohol production—are then considered.

THE EFFICIENCY OF TRAPPING OF SOLAR ENERGY

The total energy trapped into standing biomass before harvest may be defined in terms of the total energy incident on a unit area of land (hectare) during one growth season (which may be greater or less than one year) multiplied by the efficiency of conversion of solar energy to biomass. Because both land area and average annual solar irradiance are constants, the amount of energy trapped reflects the product of photosynthetic efficiency and effects of such factors as nutrient deficiency, water stress, pests and diseases which may prevent this theoretical efficiency being attained.

BIOMASS YIELD

Not all the biomass produced in the field will be harvested. The fraction used, or harvest index, may be expressed in terms of biomass harvested divided by total biomass. However, only some of this biomass will be recovered in the form of fuel. Hence, the gross fuel production can be derived from the weight of biomass harvested, multiplied by its average calorific value and by a factor which reflects the efficiency with which the harvested material is converted to the fuel under discussion.

NET FUEL GAIN

The net gain in fuel will be equal to the energy content of the final fuel product, less the energy put into the system for agriculture, processing and transport. These parameters may be further divided to distinguish, for instance, agricultural fuels, fertilizer, irrigation, transport of raw material to the factory, process fuels and so on. In general the inputs are considered in terms of fossil fuels only, because it is assumed that biomass feedstocks are free whether used as fuel or as raw material for the actual production of fuel exported from the factory. Such energy balances can be changed dramatically if an energy value is given to by-products or co-products formed at the same time. Such materials may include surplus bagasse, methane generated by anaerobic digestion of stillage, animal feed produced from yeast generated during fermentation or surplus electricity. Again, such by-products may be ascribed an energy value which reflects their actual calorific value, or they may be defined in terms of an energy equivalence. For example, the value given to animal feed need not be expressed in terms of the calorific value as cattle feed, but may be expressed in terms of the energy which would be needed to produce an equal protein source by growth of soybeans. Problems in interpretation of such data are compounded where the results are expressed in terms of ratios, often referred to as net energy ratios (NER).

NET ENERGY RATIOS

The simplest way to define the net energy ratio is in terms of the calorific value of the fuel produced divided by the energy content of all external (fossil) fuel inputs to the system. This may be defined as a simple ratio, or in terms of the number of joules put into the system in order to produce one joule of product. The problem here is that in the first case a number greater than one reflects an efficient system whereas in the second type of expression net fuel production will result in a value of less than one: i.e. an NER of 3 means that 0·33 joules of fossil energy are expended per joule of product energy achieved. Again, the end-product may be expressed in 'equivalence' terms rather than true calorific value. For example, the efficiency with which alcohol will serve as a fuel in a spark ignition engine is greater for a blend (where it confers an octane-boosting benefit) than for use in a 'pure ethanol' engine. However, even when used neat, the efficiency of the performance on a volumetric mileage basis is still higher than would be expected from a direct comparison of calorific value (remembering that the pure alcohol car is in fact running on 95% ethanol: water mix). Again, where the objective is petrochemical substitution and coal, hydroelectric or off-peak nuclear energy are available, these may not be regarded as significant inputs.

Considerable attention has been paid (particularly in the USA) to NER for alcohol production from grain, which may indicate an overall energy loss. However, NER should be combined with consideration of the overall process efficiency, as well as of the efficiency of use of the biomass raw materials.

EFFICIENCY OF RAW MATERIAL USE

The efficiency with which raw material is used may be defined in terms of the energy content in the final product as a fraction of the energy content of the biomass as brought to the factory, whereas the efficiency of the overall process can be derived by adding the energy content of all inputs to the denominator. The importance of looking at systems in this way can be demonstrated by considering the case of alcohol production (methanol or ethanol) either by fermentation of sugar, starch or cellulose, or by catalytic synthesis following thermal degradation to synthesis gas, a mixture of carbon monoxide and hydrogen. If wood is used both as fuel and as raw material for ethanol production, a positive NER is obtained; however only 18% of the biomass is recovered in fuel, and the overall efficiency on all energy inputs is 0·16. In contrast if wheat is used, combined with coal, 60% of the biomass energy ends up as fuel, with an overall efficiency of 0·34. The NER of the thermal chemical conversion system is lower, but the efficiency of the overall system is comparable to that of fermentation using sugar or starch as raw material. Hence, the 'best' route will reflect the relative availability and economic cost (rather than market price) both of the biomass raw material and of alternative energy inputs available at any particular site.

Overall cane energy balance

Estimates of the solar energy available to plants (photosynthetically active radiation, PAR, of wavelength about 400–700 nm), coupled with knowledge of the basic photochemical and biochemical processes and their thermodynamic efficiency, enable the theoretical maximum productivity at any given level of incident radiation to be calculated on a unit leaf area basis. Assuming that a leaf is fully formed and metabolically active, the maximum efficiency with which it can form carbohydrate can be estimated by giving numerical values to a series of constraints (Loomis and Williams, 1963; Bassham, 1977; Bolton and Hall, 1979; Coombs, 1983). The series of constraints usually considered are as follows: (1) the proportion of light which comprises PAR (given the value 0·43–0·5); (2) the proportion of PAR which is involved in useful work (0·7–0·9); (3) loss due to degradation of the absorbed quanta to excitation at 700 nm (given the value 0·86); (4) loss due to conversion of excitation energy to chemical energy of D-glucose. The value of this last factor varies depending on whether a value of 8 quanta per molecule of carbon dioxide reduced is assumed, or whether the more probable value of 10 is taken, giving factors of 0·26–0·33. Multiplication of these factors together gives an efficiency of production of carbohydrate of about 10%, which is further reduced by the need to use metabolic energy to convert carbohydrate to more reduced components such as protein, lipids or nucleic acids, so that the optimum efficiency of conversion of solar energy into plant biomass for a crop such as sugar cane is approximately 6%. However, this efficiency must be superimposed on the natural growth and development cycle of the crop. In this respect, the most important factor is the rate at which the developing crop forms a closed canopy capable of intercepting the available radiation at high efficiency and—once the crop has developed to its mature form—the availability of storage tissue. In general, the form of growth will follow the classic sigmoid growth curve, with the initial rate of the exponential phase depending on the size of the original seed material, and the final size reached reflecting the genotypic characteristics of the particular plant. An optimum 'ideotype' for biomass production has been suggested by Coombs, Hall and Chartier (1983) as one in which the plant has prolonged vegetative growth, indeterminate growth habit, and arises from a large initial root stock. The canopy should be erect in order to obtain a high leaf area index resulting in efficient solar energy trapping. In addition, an effective translocation system capable of removing assimilated carbon from the leaves at a high rate is necessary in order to reduce feedback inhibition of photosynthesis resulting from accumulation of sugars in the chloroplast. Sugar cane can be shown on this basis to be capable of very high rates of dry matter accumulation compared with other land plants, if grown under optimum conditions of nutrients, water and pest control (Alexander, 1973).

The overall flow of energy in a sugar-cane-based system producing ethanol is shown in *Table 1*. In this, average figures from various sources (da Silva *et al.*,

1978; Hopkinson and Day, 1980; Essien and Pyle, 1983; Johnson, 1983) have been rounded off. This Table can be considered in terms of the overall balance for trapping of solar energy into biomass, i.e. photosynthetic efficiency; agricultural energy inputs; process energy inputs.

Table 1. Energy balance for one hectare of sugar-cane receiving 200 W/m^2 per year.

(a) *Solar energy harvested by the crop*	(GJ/ha)	%
Total irradiation	30000	100·0
Available as photosynthetically active radiation	15000	50·0
Intercepted by canopy	13000	43·0
Converted through photosynthesis to carbohydrate	3000	10·0
Trapped as plant material, allowing for respiratory losses	1500	5·0
After losses due to water stress, nutrient deficiency and pests.	750	2·5
Material harvested	500	1·7
Fermentable carbohydrate	200	0·7
Ethanol	170	0·6

(b) *Agricultural energy inputs* (GJ/ha):

	Yield (t/ha/year)		
Input	Subsistence 60	Low 80	High 100
Machinery	1·4	2·2	10·0
Nitrogen	—	5·4	11·0
Fuel and labour	6·2	8·0	10·0
Other chemicals	0·5	1·5	2·0
Seed cane	1·1	2·5	2·5
Irrigation	—	—	15·0
Total	9·2	19·6	50·5

(c) *Energy equivalence of factory machinery:* 5·6 to 10 GJ/ha per year

Table 1a indicates the conversion of solar radiation at 200 Wm^2 per hectare of a hypothetical cane crop, which would receive approximately 30 000 GJ per annum of solar radiation. At the suggested optimum efficiency of 6%, this would give a total biomass yield of 1800 GJ or 100 dry tonnes (300 green tonnes). Theoretical maximum productivity could, in fact, be higher than this because for many regions where cane grows, light intensities may be higher than 200 Wm^2, with PAR of full sunlight approximately 400 Wm^2 (1800 μE/m^2/s). However, as is well known, actual yields are much lower than these optimum values because of the effects of water stress, lack of nutrients (nitrogen in particular), temperature, pests and disease as well as ineffective management of the production systems, in many cases.

The cane crop produces about 80 t of millable cane per hectare at the factory gate. This would represent about 60% of the standing above-ground biomass, which would thus be equivalent to approximately 500 GJ, or a 1·7% efficiency of solar energy capture (*Table 1a*); this can be compared with record high values for the efficiency of cane photosynthesis of about 4%. The energy content of alcohol produced from this cane (at 90% extraction efficiency and 90% fermentation efficiency) would be about 170 GJ. Assuming an agricultural energy input of 20 GJ/ha (*Table 1b*) and a factory machinery energy equivalence (based on a 20-year life time) of 6 GJ, the NER (ethanol; non-biomass energy inputs) would be 6·5, with a surplus of bagasse of 80 GJ/ha, having consumed 100 GJ of bagasse as process energy. It is important to note that most process energy is consumed in milling the cane. However, the high-pressure steam used to generate electricity may subsequently be used at lower pressure for distillation. If this energy comes from bagasse, then fuel imports are restricted to start-up and the energy equivalence of the capital plant manufacture, discounted over its life time. As

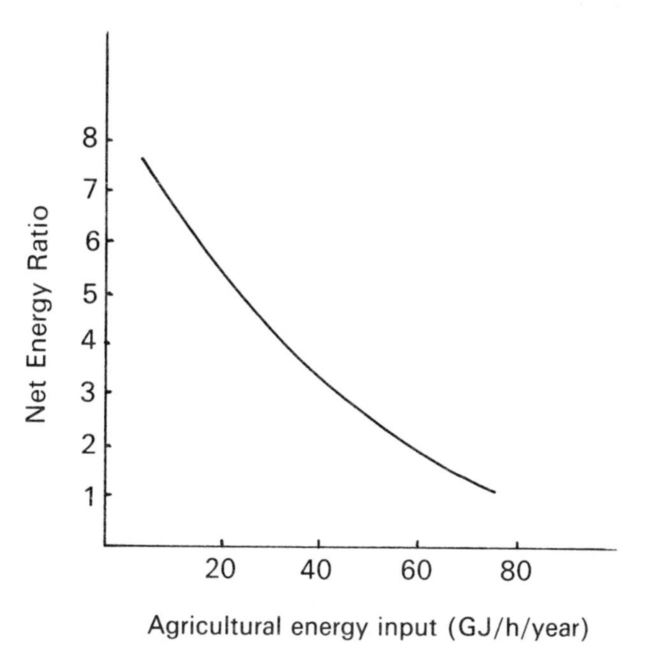

Figure 4. Decrease in Net Energy Ratio for alcohol production in a bagasse-fired distillery as a function of the agricultural energy input.

discussed on page 318, it has been estimated that energy inputs into cane production, for example for irrigation, fertilizer, fuel, may be as high as 150 GJ/ha. The extent of such inputs is a critical factor as far as alcohol production is concerned. Under these high inputs there will not be a significant net energy gain even if the process is fuelled using bagasse (*Figure 4*). Hence, the key factor in such systems is a high biomass yield at low agricultural energy input. This can be achieved if the best areas (in terms of such factors as soil, and water availability) are used for energy crops. However, it is often suggested that such crops are produced on marginal lands. Under such conditions marginal yields are to be expected, unless inputs into the system are increased, or unless the cane is bred specifically to survive adverse conditions.

Photosynthesis and physiology of sugar cane

Some of the figures quoted above will have indicated the high productivity of sugar cane compared with that of many temperate crops. When one considers parameters such as maximum recorded yields of dry matter per year, net rate of carbon assimilation, and rates of short-term carbon dioxide fixation for all the major cultivated crops, sugar-cane will be among the leaders. Other species showing high productivities include the related grasses sorghum and maize. It is now known that these high productivities are associated with a number of specific anatomical, ultrastructural, physiological and biochemical characteristics which together make up what has been referred to as the C4 syndrome (Laetsch, 1974). The discovery of the mechanism of C4 photosynthesis was a direct result of work carried out within the cane sugar industry, in Hawaii (Kortschak, Hartt and Burr, 1965), Australia (Hatch and Slack, 1970) and the UK (Coombs, 1976). In Hawaii and Australia this interest in photosynthesis followed detailed studies on the mechanism of sucrose synthesis and translocation (*see* Alexander, 1973), which established the highly efficient mechanism of sucrose accumulation during ripening. In particular, C4 plants show low compensation points (the compensation point being the equilibrium concentration of carbon dioxide reached in a closed atmosphere in the light, that is when carbon dioxide output from respiration equals carbon dioxide assimilation by photosynthesis), and do not lose carbon dioxide in the light, through photorespiration.

It is now clear that most temperate plants (now known as C3 species) do lose a significant proportion of the carbon fixed in the light, because of the process of photorespiration. This may result in an overall loss of yield, in crops such as wheat, of 10–30%, depending on factors such as light intensity and temperature. The reason for this is that the enzyme (ribulosebisphosphate carboxylase, EC 4.1.1.39) which catalyses the initial fixation of carbon dioxide in the reductive photosynthetic cycle is inhibited by oxygen in a competitive manner (Lorimer, 1981). As a result of this oxygen-dependent reaction, a two-carbon compound, glycollate, is produced. This contrasts with the normal carboxylation reaction in which two molecules of phosphoglyceric acid (a three-carbon compound, hence the term C3) are formed (Gibbs and Latzko, 1979): these are then reduced to three-carbon sugars, two molecules of which form fructose 1, 6-bisphosphate. In order to recover some of the carbon lost in photorespiration, glycollate is recycled through the C2 pathway (Tolbert, 1971) during which the three-carbon phosphoglycerate is re-formed with the loss of a molecule of carbon dioxide. As a result, in normal atmospheric concentrations of oxygen, photosynthesis in C3 plants is inhibited, in part because of loss of carbon from the carbon reduction cycle, and in part because of loss of carbon dioxide from the plant.

Sugar-cane and other C4 plants do not show photorespiration. Hence, it may be concluded that the high productivity of sugar-cane is associated with a distinct mechanism of photosynthesis which has evolved in order to overcome deficiencies associated with photorespiration. This is of particular importance in the tropics because the ratio of photorespiration to photosynthesis increases with increase in temperature and light intensity; the effects are therefore more deleterious in tropical latitudes and any species which develop a mechanism to overcome these effects have a competitive advantage.

The mechanism of C4 photosynthesis is of particular interest as far as improving productivity of C3 crops is concerned. The differences between C3 and C4 plants are sufficient to confirm the benefits of reduced photorespiration. For this reason extensive searches have been made for C3 plants with reduced photorespiration, lower compensation point, or a ribulosebisphosphate carboxylase which has a greater resistance to inhibition by oxygen (Somerville and Ogren, 1982). These studies include attempts to select mutants defective in enzymes of the C2 pathway, but so far without success. In the same way, attempts to obtain hybrids with higher photosynthetic capacity by crossing C3 and C4 species of *Atriplex* have also been unsuccessful (Bjorkmann, Gauhl and Nobbs, 1969). It seems, therefore, that it will not be easy to incorporate the characteristics of the C4 plants into other species: the reason for this can be seen if the actual mechanism of C4 photosynthesis is considered in more detail.

In the C4 plants, photosynthetic tissue is arranged in two layers around the vascular bundles, comprising an inner bundle sheath and an outer mesophyll layer. Chloroplasts in the outer layer appear to be normal: however, in the most advanced C4 plants such as sugar cane the bundle sheath chloroplasts lack some of the characteristic structures found in other higher plants, and also lack the ability to produce oxygen. Ribulosebisphosphate carboxylase is confined to this inner cell layer. The outer layer of photosynthetic cells lack this crucial enzyme of the carbon reduction cycle, but contain high levels of a second carboxylase (phosphoenolpyruvate carboxylase, EC 4.1.1.31) which catalyses the formation of the four-carbon acid, oxaloacetic acid (hence the term C4 plant). Oxaloacetic acid is reduced to malate in the mesophyll cells, and transported to the bundle sheath where decarboxylation results in release of carbon dioxide. This is refixed into phosphoglyceric acid which is then reduced to a sugar (dihydroxyacetone phosphate), using reductant derived from the activity of the mesophyll cells which possess complete non-cyclic photophosphorylation, are capable of splitting water, and do evolve oxygen. By this means the oxygen-sensitive reactions are separated in space from the light-dependent water-splitting reactions which generate oxygen. At the same time the C4 cycle acts as a pump to maintain a higher internal carbon dioxide concentration and thus to decrease the competitive inhibition by oxygen.

Although comparatively little work has been carried out on the molecular genetics of sugar cane, considerable information is now available concerning the structure of genes coding for the synthesis of ribulosebisphosphate carboxylase of maize (Link *et al*., 1978). This enzyme is of particular interest as it is composed both of large and of small subunits (Ellis, 1976), the former coded within the chloroplast and the latter being coded for by nuclear genes. Genes for both subunits have now been cloned and expressed in yeast. It is therefore possible (although not very probable) that some engineering of the genes controlling the reaction centre may bring about changes in the relative affinity for oxygen and carbon dioxide. However, the complex cytogenetics of sugar-cane offer not only advantages for breeding a wide variety of canes, but also the disadvantage of providing a very complex base on which to attempt to use *in vitro* genetic recombinant techniques.

Sugar-cane breeding

In the past, sugar-cane breeding has been directed towards varieties which produce good yields of juice of high purity, are resistant to major diseases and are of reasonable stem diameter and erect habit in order to facilitate harvesting. Experience has shown that the most profitable cane to grow is not necessarily that with the highest total carbohydrate content, but rather one in which a sweet juice with a very high percentage of sucrose and a low content of all other organic impurities occurs; this will reduce processing costs and the amount of residual molasses. However, if cane is to be bred for the production of fuels, a different approach may be taken: if the primary objective is the manufacture of ethanol, then breeding can be directed towards high production of total fermentable solids; if the objective is the production of combustible material, then cane can be bred for high fibre (Alexander, 1980; Giamalva, Clarke and Bischoff, 1981; Giamalva and Clarke, 1982). In an extensive programme aimed at production of 'energy cane' in Puerto Rico, Alexander has selected for high fibre cane which also gives high total dry-matter yields of approximately 80 ha. This has been achieved in part by breeding and selection, in part by increased inputs and in part by different management techniques (Alexander and Allison, 1981). In spite of the high fibre content, the actual yield per hectare of fermentable carbohydrate produced in the form of high test molasses from these canes is greater than that produced by local cane grown for the production of sugar. The variation in fibre and sugar content of the different *Saccharum* species has been discussed on pages 316–317. This wide variation gives ample scope for the production of cane suited to specific uses. However, cane breeding is complicated by factors related to high chromosome numbers (reflecting polyploidy), daylength responses of the various varieties, and low seed viability resulting from the complex cytogenetic base.

Chromosome numbers in cultivated canes vary between 60 and 200 (Stevenson, 1965). The basic chromosome number (x) of the Andropogoneae is 10. In fact no diploids ($2n = 2x = 20$) of *Saccharum* varieties are known. In addition, there is a large degree of variation in chromosome number, not only between varieties of the same species or hybrid parentage, but also between cells within the same plant, a considerable degree of aneuploidy being common. Of the various named species, *S. officinarum* is more stable at 2n = 80; ranges for the other recognized species are as follows: *S. robustum*, 2n = 60 or 80 in general but range from 63 to over 200; *S. spontaneum*, 2n = 40–128; *S. sinense*, 2n = 82–124. In addition, wild hybrids of *spontaneum* × *robustum* of 2n = 80–101 and intergeneric hybrids of *Saccharum* × *Miscanthus* of 2n = 114–205 are also known. A wide range of interspecific hybrids have also been produced, at the Indian cane breeding station in particular: these include crosses between *Saccharum* and *Erianthus*, *Miscanthus*, *Narenga*, *Imperata*, *Sorghum* and *Zea* (maize).

Sugar canes are wind pollinated and show inbreeding depression. Hence, crossing techniques are those characteristic of outbreeding clonally propagated crops (Simmonds, 1979). The progeny of a desired cross are planted out in large numbers and offspring are selected by visual inspection, survival, pest resistance, chemical analysis, etc., in order to obtain a single plant suitable for cloning through subsequent propagation of sett pieces derived from each mature node. The canes are daylength sensitive and flower in response to light periods greater than 12 hours, assuming that other conditions of nutrients, water supply and temperature are correct. The extent of flowering also depends on the variety, species and growth cycle in relation to season and latitude. Flowering is greater in the middle tropics and decreases towards the equator. In some regions, breeding and, in particular, interspecific crosses, may be facilitated if the parent material is subjected to the required conditions to promote flowering by use of chambers with artificial illumination, temperature control etc.

Crossing of noble canes with wild (*spontaneum*) types results in an unusual restitution of chromosome number; if a female of a noble variety is crossed with a wild-type male cane the viable progeny have the somatic complement of the female parent (80 chromosomes) plus the gamete number of the male. On the other hand, self-fertilization or crosses between noble canes yield progeny with 80 chromosomes. Restitution occurs for only two generations, but not at the third, so there is a limit to the increase in chromosome number, resulting in the higher values of about 200 mentioned previously.

During the breeding of a new variety, the rate of build-up of new clonal propagation material is limited by the number of sett pieces which can be obtained from each mature stem. Hence, the rate of multiplication of a new variety determines the time taken before it can be used commercially on a large scale. Because of the odd cytogenetic behaviour and low seed viability, similar material cannot be produced through a further cross of the same parents, nor will seed produced in the same crossing result in similar progeny. For this reason propagation by tissue culture has obvious attractions. A considerable amount of work has been carried out on *in vitro* culture of both callus and suspension tissues of cane, following the initial work of Nickell (1964). It has been suggested that such cultures are of value in studies of cane physiology and nutrition (Nickell and Maretzki, 1969), but that they are not suitable as a means of cloning, because of the wide variation which occurs in chromosome number of individual cultured cells, resulting in plants of diverse character. On the other hand, this diversity lends the technique both to production and to selection of new varieties (Heinz and Mee, 1969). Nevertheless, as pointed out by Heinz *et al.* (1977), 'the problems arising from the identification of disease and pest resistant clones that are also high-yielding are similar whether selecting from subclones derived from callus or from clones obtained by conventional breeding techniques ... the greatest gains in the use of tissue culture will be realized when directed genetic changes can be coupled with effective cellular selection methods'. Using *in vitro* techniques these authors were successful in screening for resistance to three diseases, namely eyespot disease (caused by *Helminthosporium sacchari*); Fiji disease (caused by a virus); downy mildew (*Sclerospora sacchari*), but not for smut (*Ustilago scitaminea*). Subclones derived from cell cultures produced about 80 t of cane per hectare, in one clone giving 12 t of sugar, which was similar to the parent cane. However, other clones were either less resistant, or gave a lower sugar yield than the parents. Although physiological responses to materials such as disease toxins can be screened at the cell level, other characteristics such as sucrose content, fibre, ash, juice purity, and acceptability for mechanical harvesting or milling, can be determined only after propagation of suitable quantities by conventional use of sett pieces and prolonged growth to maturity in the field.

Nitrogen fixation

The total energy input used in growing sugar-cane varies very considerably between the various countries where it is grown, and between subsistence level farms and modern fully mechanized plantations. Khan and Fox (1982) describe the energy cost of nitrogen fertilizer as 5·4 GJ/ha out of a total agricultural input of 16 GJ. However, this use is low: in some regions over 500 kg/ha (38·5 GJ) of ammonium nitrate may be applied, representing well over 50% of the total agricultural energy input. Such inputs of nitrogen represent an additional financial burden for those regions where fertilizer has to be imported. At the same time, such high inputs results in a significant reduction of the overall net energy balance. The observation of Dobereiner (1959), that populations of the nitrogen-fixing organism *Beijerinckia* were higher in the region of sugar cane roots (rhizosphere) than in the soil between the rows, is of particular significance; furthermore, these organisms were active in nitrogen fixation, as demonstrated by the acetylene reduction technique (Dobereiner, Day and Dart, 1972). Subsequently, various species of *Azotobacter*, *Beijerinckia*, *Derxia*, *Caulobacter Vibrio* and *Clostridia* have been isolated from free-living associative interactions with the roots of various grasses. Both acetylene reduction tests and the assimilation of heavy isotopic nitrogen (^{15}N) have been used in order to determine the significance of such fixation. Ruschel, Henis and Salati (1975) obtained rates of fixation of N_2 in sugar-cane roots of 523 μg N per gram of root per day, equivalent to about 20 g of nitrogen per hectare per day (Ruschel *et al.*, 1978). Sugar-cane, as well as a number of other C4 grasses, can also possess associative nitrogen-fixing organisms (e.g. *Spirillum lipoferum*) located within the root cortex (Dobereiner and Day, 1976) although *S. lipoferum* may be more abundant on the surface of the roots than within the tissue (Burris, Okon and Albrecht, 1978). The latter authors showed that sterile maize seeds could be inoculated with *Spirillum* and cultured in such a way that nitrogen reduction occurred and viable cells could be recovered; however, they reported that no benefit from such infection could be detected in field trials. At present there seems little doubt that an association can be formed between nitrogen-fixing organisms and roots of non-leguminous crops (Evans and Barber, 1977), in as much as organic material produced by the plant and excreted through the root represents a substrate for microbial growth. However, the extent to which any nitrogen assimilated by these bacteria is transferred to the plants appears to be small. It is possible that plants better suited to take advantage of the association could be selected or, alternatively, more effective bacterial strains could be selected from natural populations, or developed using genetic manipulation.

Fermentation

Because fermentation of sugars to ethanol remains the only important commercial process for the conversion of biomass to liquid fuel, it has been given particular attention. Actual or potential improvements may result from changes either in the organisms used or in the design of fermentation systems. The objectives are to improve efficiency (in terms both of substrate and of energy use), volumetric productivity and final net yields. However, fermentation cannot be looked at in isolation, as is sometimes done, because it represents only a small part of the overall capital and energy investment in its own right. Of greater importance are interactions between the techniques used for the fermentation step and their effect on the design, cost and energy balance of the system as a whole; this extends to the nature of primary raw materials, pretreatment to produce the fermenter feed stream, and downstream processing (including product separation, effluent treatment and recovery of by-products with cash or energy value).

Where the main substrate is soluble sugars, the organism used is a yeast (*Saccharomyces*), generally *S. cerevisiae*, which has the ability to use sucrose, glucose, fructose, maltose and maltotriose. This organism has the advantage that selection over hundreds of years in the beverage industry has resulted in known strains with stable behaviour attributable to their polyploid nature and low frequency of sporulation (Stewart and Russell, 1981). This is an advantage from the commercial viewpoint, but a disadvantage as far as genetic modification is concerned. The attraction of yeast as a means of converting biomass to a fuel of higher value lies in the fact that although the theoretical weight yield of alcohol is about 50%, over 90% of the energy in the substrate is retained in the ethanol, resulting in an increase from 16 kJ/g for glucose to 30 kJ/g for ethanol (Coombs, 1981a). In practice the yield is less, because part of the substrate is used for cell growth and part is diverted to by-products. Furthermore, as the final product is in dilute aqueous solution, considerable energy has to be expended to separate and dehydrate it. In batch fermentation systems, productivity is low on a volumetric basis and the actual yield may be only 80–85% of the theoretical yield. On a volumetric basis, productivity depends both on cell concentration and on substrate concentration and thus varies from approximately 1 g ethanol per litre of reactor volume per hour to over 40 g/ℓ/h for systems with cell recycle or as high as 80 g/ℓ/h for some experimental systems (Wilke *et al.*, 1983).

Most commercial processes, in developing countries, use batch fermentation or a cell recycle system in which each fermenter is inoculated with a high cell concentration obtained from washed yeast from the previous run. The objectives of the large number of experimental systems which have been investigated are as follows:

1. To increase efficiency of substrate utilization by removing the need to produce cells for each run, and decreasing the amount of by-products formed;
2. To increase productivity on a volumetric basis, in order to decrease capital cost;

3. To produce continuous systems more amenable to automatic control, in order to decrease labour costs;
4. To increase concentrations of alcohol in the product stream and to work at higher temperature, to improve energy efficiency by reducing the need for fermenter cooling and decreasing the energy needed for product recovery;
5. To extend the range of substrates which can be used, to cellulose in particular;
6. To decrease sensitivity of the organisms used to inhibition by alcohol and other metabolites.

As far as yeasts are concerned, factors which affect alcohol production include oxygen concentration, substrate concentration, cell density, ethanol concentration, temperature and the ratio of metabolism associated with cell growth to that associated with cell maintenance, alcohol production and the amount and nature of by-products. Where the cells are harvested by sedimentation or flotation, the flocculation characteristics are also important. Where practicable, substrate levels at 16–25% solids, giving alcohol concentrations of 8–12%, are desirable. Present yeasts work best at 28–34°C: thus, for alcohol production from cane, because of the problem of obtaining suitable fermenter cooling water in the tropics, a particular advantage would be availability of yeasts capable of the same alcohol tolerance and production rate at higher temperatures.

Thermophilic organisms are also of interest because they possess the following attractive characteristics: (1) higher cell growth and metabolic rates; (2) low cell biomass yields; (3) higher stability of enzymes and cells, in the absence of oxygen; (4) ability to use a wider variety of substrates. In particular, many thermophilic anaerobes have the ability to degrade lignocellulose or its hydrolysis products, to varying extents. These advantages in turn result in increased ethanol production rates, increased product recovery and improved energy balances. The problems associated with the use of organisms such as *Clostridium*, *Thermoanaerobacter*, or *Bacillus stearothermophilus*, relate to lower alcohol tolerance and continued production of by-products such as acetate. However, considerable improvements both in alcohol tolerance and in reduction in by-product formation have been made using mutagens and selection techniques (Wang *et al.*, 1981; Hartley *et al.*, 1983). At the present level of performance it is thought by Hartley and colleagues that the thermophilic process has advantages over the best-yielding yeast-based systems using cell recycle and vacuum fermentation (*see* Wilke *et al.*, 1983).

It is possible that further advances in development of these bacteria could be made using genetic engineering techniques. However, at present this approach has additional associated problems because little is known about the metabolism, metabolic pathways, life cycles or genetics of many of the cellulolytic fungi or bacteria which might be used. For yeasts, however, the life cycle is well established, as are details of the seven chromosomes and location of about 200 genes (Stewart and Russell, 1981) although the polyploid nature of commercial yeasts makes conventional genetic analysis and crossing difficult. On the other hand *in vitro* techniques of protoplast fusion and recombinant DNA technology (Hollenberg *et al.*, 1976; Beggs, 1978; Hinnen, Hicks and Fink, 1978) are being used with yeasts world-wide. In the short term these techniques can be used to determine further details of metabolic control and of the nature of alcohol tolerance, and to study the genetics of yeast in more depth. In the longer term it may be possible to incorporate genes coding for a greater range of hydrolytic enzymes and to obtain strains with ability to hydrolyse starch and cellulose. The use of plasmid vectors in yeasts is established (Beggs *et al.*, 1980). However, it remains to be seen whether introduced hydrolytic enzymes can be expressed in an efficient manner, and whether addition of such abilities would, in fact, confer a genuine commercial advantage. With regard to the use of cane, an advantage could be obtained if such genetic changes led to an efficient system for the hydrolysis of lignocellulose and fermentation of the sugars thus produced, including the pentoses derived from hemicellulose.

USE OF CELLULOSE AS FERMENTATION SUBSTRATE

Because of the extensive availability and low cost of waste wood, paper and agricultural residues, considerable research has been devoted to obtaining liquid fuels, by fermentation, from lignocellulose (Stone and Marshall, 1980). However, in spite of much excellent research, no viable commercial process exists which could be applied to excess bagasse from a cane distillery in order to increase the output of ethanol. What is needed is a process with the following characteristics: (1) high yield of alcohol per tonne of bagasse: (2) low capital cost; (3) low energy input; (4) either no necessity for, or ease of efficient recovery of, expensive chemicals, solvents or enzymes; (5) no production of large amounts of solid wastes, such as are formed during acid neutralization; (6) high NER and high efficiency overall.

Various partial solutions to such problems have been achieved, (Levy *et al.*, 1981; Wilke *et al.*, 1981; Tsao *et al.*, 1982; Ladisch *et al.*, 1983), but at present extensive pre-treatment to reduce particle size of raw materials must be followed by acid or enzyme hydrolysis, resulting in dilute impure feed streams with considerable loss or destruction of sugars if acid systems are used, or high enzyme costs if this route is followed. In spite of substantial increase in the specific activity of enzymes, from the fungus *Trichoderma ressei* in particular, specific activities of cellulases are still several orders of magnitude lower than those of commercial amylases used in the production of alcohol from corn.

Process plant

In the production of fuel alcohol from cane, batch fermentation systems associated with distillation and azeotropic dehydration are universal. In spite of the development of a large number of improved fermentation systems (Wilke *et al.*, 1983) these have not been accepted by the cane-based alcohol industry to any great extent, although pilot and demonstration plant have been run by a number of major engineering and sugar companies such as John Brown Engineering (Alcon Biotechnology), Alpha Laval, Tate and Lyle, Uhde, W. S. Atkins, etc. In the same way, the belief that distillation contributes a major part of the energy demand in a fuel alcohol system has led to extensive research into alternative methods of separation, including the use of selective membranes, adsorption, supercritical liquids and reverse osmosis. However, at the same time developments in the efficiency of steam distillation by techniques such as multiple effect evaporators and vapour recompression have improved steam consumption so that demand may be less than 1–2 kg steam per litre of alcohol (Keim, 1983). The variation reflects the way in which stillage is handled. With present plant, the use of energy in crushing sugar-cane in order to produce the juice is about twice that of distillation, although the development of alternatives for the former procedure attracts much less attention than that for the latter. One approach which has been suggested is the direct fermentation of fresh pith or chopped cane (Rolz, de Cabrea and Garcia 1979), in packed-bed reactors. However, this remains at the development stage.

The problem with distillation is the large amount of waste liquor produced at stillage. At an 8% alcohol concentration in the stream leaving the fermenter, 12 volumes of stillage will be produced per volume of alcohol; in other words, by 1985 Brazil will produce 110×10^9 litres of stillage of high biological oxygen demand (BOD). This can be spread on the land, but discharge into rivers or leaching can cause problems. Alternatives include drying for use as animal feed or fuel, aerobic treatment, or anaerobic digestion, the relative costs and merit of which have recently been reviewed by Maiorella, Blanch and Wilke (1983). The lower protein content in stillage from a cane distillery, compared with one based on maize as raw material, makes the production of animal feed of less interest. However, if an effective anaerobic digestion process can be developed, this would have the advantage of also providing some process energy. In spite of extensive research on anaerobic digestion and the development of a wide variety of systems (Callander and Barford, 1983) aimed at the treatment of industrial effluents of various strengths and composition cost (*see* Chapter 10), effective systems have yet to be proved for use with cane stillage.

As an alternative approach to the use of sugar-cane as an energy crop, attempts are being made in Brazil to produce methane by anaerobic digestion of entire cane. This has an advantage that the high-value fuel product (methane) does not have to be separated from the liquor. However, if the gas is to be used for generation of electricity, or in larger installations fed to pipe lines or compressed for use as a transport fuel, gas cleaning (removal of hydrogen sulphide and carbon dioxide) and compression costs may negate this advantage. A further problem relates to the long generation times needed in order to digest the lignocellulose (bagasse) fraction of the entire cane. Digestibility of agricultural residues such as straw can be increased by mechanical or chemical treatment, but the costs are again, high (Mardon, 1981).

ECONOMICS

Detailed site-specific studies have been carried out for many countries or regions in attempts to determine the actual costs of producing ethanol as a liquid transport fuel (e.g. Japan: Anonymous, 1981b; Italy: Anonymous, 1979b; India: Anonymous, 1979a; New Zealand: Harris *et al.*, 1979; Australia: Stewart *et al.*, 1979; USA: Nathan, 1978; Anonymous, 1980a; OTA, 1980; South Africa: Ravno

1979; Sweden: Anonymous, 1980b; Brazil: Anonymous, 1979c; Rothman, Greenshields and Calle, 1983; developing countries: Anonymous, 1980c; Thailand: Anonymous, 1981c). In general, the costs of producing ethanol by fermentation of feed derived from sugar or starch crops exceed the alternative cost of buying oil. In cost terms, value of the crop as a source of food or animal feed can be compared with the value of the volume of petrol (gasoline) which can be saved if ethanol is used as substitute. However, this is justified only for small countries which do not have their own oil refineries, because substitution of the light end of the barrel causes disruption of established petroleum technology. For this reason, ethanol looks most attractive as an alternative octane booster for use in unleaded petrol.

At present it is claimed that the cost of fermentation ethanol produced from maize in the United States is less than that of ethanol produced from natural gas (Ng *et al.*, 1983). Again, this comparison relates the selling price of the alcohol from the two sources, but does not take into account the underlying economic reasons. In the USA, corn prices are low because of over-production; it is also possible to produce a valuable by-product in terms of animal feed protein which lowers the actual raw material costs to less than $1 per bushel (about $40/t). A considerable proportion of maize-based protein feed is currently being sold into the European Communities (EC), where it commands a higher price due to the effects of the Common Agricultural Policy (Pearce, 1981), which results in grain and sugar prices which are twice those of the world market. These high prices have, in turn, encouraged the over-production of beet sugar, resulting in the export of EC sugar on to the world market and depressing the price of sugar, molasses and, eventually, the value of sugar-cane-based ethanol, if considered in terms of international trade. However, the production of alcohol can be justified, for any specific country, on the basis of self-sufficiency, foreign exchange savings and improvement of rural economies. Such alternatives may be viewed as shown in *Figure 5*. However rigorous application of such a flow chart analysis (Coombs, 1981b) in most countries at present gives a negative answer as far as the true value of producing fermentation alcohol as fuel is concerned.

In general, the greater proportion of the cost of alcohol production relates to the production of raw materials. Where inputs are low, sugar cane may be produced at fairly low cost, but also at low yield. If local supplies of fertilizer are not available, import costs may outweigh the benefits of fuel production. The prime objectives of any biotechnological programme therefore should be to improve productivity and to decrease the need for inputs of synthetic nitrogen.

Conclusions

Sugar cane represents one of the most efficient land-based systems for converting solar energy into biomass. This biomass can provide both a solid fuel for combustion to produce process heat, steam or electricity (as well as power or heat which may be sold), and an easily fermentable juice which may be upgraded to ethanol which is of value as a liquid transport fuel and, in particular, as an octane booster for use in lead-free petrol. Fibre and juice may be separated easily by established mechanical means; both agricultural and process expertise

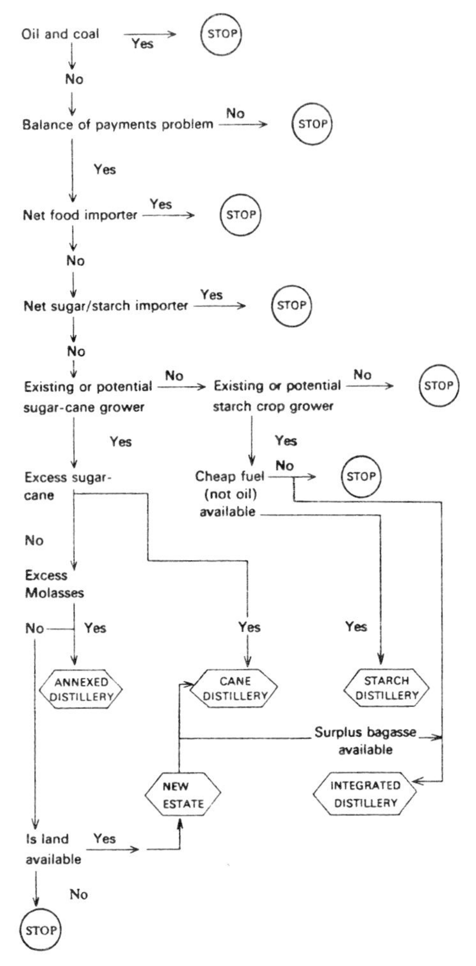

Figure 5. Options for the production of fuel alcohol from sugar and starch crops, using bagasse as fuel. An annexed distillery is one associated with a sugar factory: an integrated distillery is one which uses more than one feedstock.

exist to enable 10000–15000 t/day of raw material to be handled and, as long as fertilizer inputs are not excessive, the system will show a positive net energy gain. However, at present, developments of such systems are restricted by the economics of cane production in those countries with established plantation systems. Where cane is produced on a smaller scale by peasant farmers, apparent economic systems may reflect low farmer incomes. If biotechnology is to have a significant effect on the use of cane as an energy crop, it will have to relate to the overall production process, including cane breeding and waste treatment, rather than to fermentation alone. In the short term, it is possible that the most rapid advance will be in the growing of high-fibre canes for use as combustion fuel, rather than in the continued production of alcohol.

References

ALEXANDER, A. (1973). *Sugar Cane Physiology*. Elsevier, Amsterdam.

ALEXANDER, A.G. (1980). The potentials of sugarcane as a renewable energy resource for developing tropical nations. In *Bioresources for Development* (A. King, H. Cleveland and G. Streatfield, Eds), pp. 223–236. Pergamon Press, New York.

ALEXANDER, A.G. AND ALLISON, W. (1981). Cane management for energy in Puerto Rico. In *Proceedings of Condensed Papers, 4th Miami International Conference on Alternative Energy Sources* (T.N. Veziroglu, Ed.), pp. 332–334. Clean Energy Research Institute, Miami.

ANONYMOUS (1979a). *Biological Production of Ethanol, an Alternative Chemical Feedstock and Liquid Fuel.* Biochemical Engineering Research Centre, Indian Institute of Technology, New Delhi.

ANONYMOUS (1979b). *Etanolo per Via Fermentativa. Possibilita e Convenienza della Produzione in Italia per l'Uso Nel Settore dell'Autotrazione.* Progetto Finalizzata Energetica, Consiglio Nazionale Richerche, Milano.

ANONYMOUS (1979c). The Brazilian Alcohol Program and its objectives. In *International Molasses Report, Special Edition*, pp. 3–20. F.O. Licht, Ratzeburg.

ANONYMOUS (1980a). *A Manual on Ethanol & Gasohol Production in Louisiana.* Louisiana Department of Natural Resources, Baton Rouge.

ANONYMOUS (1980b). *Etanol ur Jordbruksproduckter. Publication DsJo 1980:7.* Ministry of Agriculture, Stockholm.

ANONYMOUS (1980c). *Alcohol Production from Biomass in Developing Countries.* The World Bank, Washington, DC.

ANONYMOUS (1981a). *BP Statistical Review of World Energy, 1981*, The British Petroleum Company p.l.c., London.

ANONYMOUS (1981b). *The Alcohol Fuels From Energy Farming.* The Institute of Energy Economics, Tokyo.

ANONYMOUS (1981c). *Thailand–Biomass Alcohol Subsector Review.* The World Bank, Washington, DC.

ANONYMOUS (1981d). *Assessment of Brazil's National Alcohol Programme.* Ministry of Industry and Commerce, Secretariat of Industrial Technology, Brasilia.

ANONYMOUS (1983). Energy from biomass in Europe. In *Energy from Biomass; 2nd E.C. Conference* (A. Strub, P. Chartier and G. Schleser, Eds), pp. 51–62. Applied Science Publishers, London.

ATCHISON, J.E. (1978). The industrial uses of bagasse and its fuel value as compared to fossil fuels. *Sugar y Azucar* **73,** 39–44.

AUSTIN, R.B., KINGSTON, G., LONGDEN, P.C. AND DONOVAN, P.A. (1978). Gross energy yields and the support energy requirements of sugar from beet and cane: a study of four production areas. *Journal of Agricultural Science* **91,** 667–675.

BAIKOW, U.E. (1982). *Manufacture and Refining of Raw Cane Sugar*, 2nd edn (2 volumes). Elsevier, Amsterdam.

BARNES, A.C. (1974). *The Sugar Cane*, 2nd edn. Leonard Hill Books, Aylesbury.

BASSHAM, J.A. (1977). Increasing crop production through more controlled photosynthesis. *Science* **197,** 630–632.

BEGGS, J.D. (1978). Transformation of yeast by a replicating hybrid plasmid. *Nature* **275,** 104–109.

BEGGS, J.D., VAN DEN BERG, J. VAN OOOYEN, A. AND WEISSMAN, C. (1980). Abnormal expression of chromosomal rabbit B-globin gene in *Saccharomyces cerevisiae. Nature* **283,** 835–840.

BENNETT, M. (1983). What price technology? *Tate and Lyle Times* **15,** 11–12.

BJORKMANN, O., GAUHL, E. AND NOBBS, M.A. (1969). Comparative studies of *Atriplex* species with and without beta carboxylation photosynthesis and their first generation hybrids. *Year Book, Carnegie Institute, Washington, California* **68,** 620–624.

BOHALL, R., SHAPOURI, H. AND D'ANGELO, L. (1981). *Cost of Producing and Processing Sugar Cane and Sugar Beets in the United States including Projections for the 1981/82 Crop.*

National Economics Division, United States Department of Agriculture, Washington.

BOLTON, J.R. AND HALL, D.O. (1979), Photochemical conversion and storage of solar energy. *Annual Review of Energy* **4,** 353–401.

BOUVET, P.E. AND SUZOR, N.L.C. (1980). Pelletizing bagasse for fuel. *Sugar y Azucar* **75,** 22–26.

BROWN, L.R. (1980). *Food or Fuel? New Competition for the World's Cropland. Worldwatch Paper Number 35.* Worldwatch Institute, Washington.

BULL, T.A. AND GLASZIOU, K.T. (1963). The evolutionary significance of sugar accumulation in *Saccharum. Australian Journal of Biological Sciences* **16,** 737–742.

BURRIS, R.H., OKON, Y. AND ALBRECHT, S.L. (1978). Properties and reactions of *Spirillum lipoferum. Ecological Bulletin* **26,** 353–363.

CALLANDER, I.J. AND BARFORD, J.P. (1983). Recent advances in anaerobic digestion technology. *Process Biochemistry* **18,** 24–30.

CHEN, R. (1980). The development of biogas utilization in China. *Biomass* **1,** 39–44.

COOMBS, J. (1976). Interaction between chloroplasts and cytoplasm in C4 plants. In *The Intact Chloroplast* (J. Barber, Ed.), pp. 279–314. Elsevier, Amsterdam.

COOMBS, J. (1980). Renewable sources of energy (carbohydrates). *Outlook on Agriculture* **10,** 235–245.

COOMBS, J. (1981a). Biogas and power alcohol. *Chemistry and Industry* April, 223–226.

COOMBS, J. (1981b). Production of fuel alcohol from biomass. *Enzyme Microbial Technology* **3,** 171–173.

COOMBS, J. (1983). Aspects of carbon metabolism and prospects for enhancing yields. *Pesticide Science* **14,** 317–326.

COOMBS, J., HALL, D.O. AND CHARTIER, P. (1983). *Plants as Solar Collectors.* D. Reidel Publishing Company, Dordrecht.

CUSI, D.S. (1980). The storage and conversion of bagasse. *Sugar y Azucar* **75,** 39–51.

DA SILVA, J.G., SERRA, G.E., MOREIRA, J.R., CONCLAVES, J.C. AND GOLDBERG, J. (1978). Energy balance for ethyl alcohol production from crops. *Science* **201,** 903–906.

DOBEREINER, J. (1959). Influencia da cana-de-acucar na populacao de *Beijerinkia* do solo. *Review of Brazilian Biology* **19,** 251–258.

DOBEREINER, J. AND DAY, J.M. (1976). Associative symbiosis in tropical grasses: characterization of microorganisms and dinitrogen fixation sites. In *Proceedings of the First International Symposium on Nitrogen Fixation* (W.E. Newton and C.J. Nyman, Eds) pp. 518–538. Washington State University Press, Pullman.

DOBEREINER, J., DAY, J. AND DART, P.J. (1972). Nitrogenase activity in the rhizosphere of sugarcane and some other tropical grasses. *Plant and Soil* **37,** 191–196.

DONOVAN, P.A. (1978). A preliminary study of the energy inputs in the production of sugarcane. *Proceedings of the South African Sugar Technologists Association* **52,** 188–192.

ELLIS, R.J. (1976). Protein and nucleic acid synthesis by chloroplasts. In *The Intact Chloroplast* (J. Barber, Ed.) pp. 335–364. Elsevier, Amsterdam.

ESSIEN, D. AND PYLE, D.L. (1983). Energy conservation in ethanol production by fermentation. *Process Biochemistry* **18,** 31–37.

EVANS, H.J. AND BARBER, L.E. (1977). Biological nitrogen fixation for food and fibre production. *Science* **197,** 332–339.

FAO (1981). *Year Book of Statistics.* Food and Agriculture Organization, Rome.

FLAIM, S. AND HERTZMARK, D. (1981). Agricultural policies and biomass fuels. *Annual Reviews of Energy* **6,** 89–121.

GIAMALVA, M.J., CLARKE, S.J. AND BISCHOFF, K. (1981). Bagasse production from high fibre sugar cane hybrids. *Sugar Journal* August, 20–21.

GIBBS, M. AND LATZKO, E., EDS (1979). *Photosynthetic Carbon Metabolism and Related Processes. Encyclopaedia of Plant Physiology. New Series, volume 6.* Springer-Verlag, Berlin.

HALL, D.O., BARNARD, G.W. AND MOSS, P.A. (1981). *Energy from Biomass in the Developing Countries.* Pergamon Press, Oxford.

HARRIS, G., LEAMY, M.L., FRASER, T., DENT, J.B., BROWN, W.A.N., EARL, W.B., FOOKES,

T.W. AND GILBERT, J. (1979). *The Potential of Energy Farming for Transport Fuels in New Zealand. Report No. 46.* New Zealand Energy Research and Development Committee, Auckland.

HARTLEY, B.S., PAYTON, M.A., PYLE, D.L., MISTRY, P. AND SHAMA, G. (1983). Development and economics of a novel thermophilic ethanol fermentation. In *Biotech 83: Proceedings of the International Conference on the Commercial Applications and Implications of Biotechnology*, pp. 895–905. Online Publications, Northwood, London.

HATCH, M.D. AND SLACK, C.R. (1970). Photosynthetic CO_2 fixation pathways. *Annual Review of Plant Physiology* **21,** 141–162.

HEINZ, D.J. AND MEE, G.W.P. (1969). Plant differentiation from callus tissue of *Saccharum* species. *Crop Science* **9,** 346–348.

HEINZ, D.J., KRISHNAMURTHI, M., NICKELL, L.G. AND MARETZKI, A. (1977). Cell, tissue and organ culture in sugarcane improvement. In *Applied and Fundamental Aspects of Plant Cell, Tissue and Organ Culture* (J. Reinert and Y.P.S. Bajaj, Eds), pp. 3–17, 207–248. Springer-Verlag, Berlin.

HINNEN, A., HICKS, J.B. AND FINK, G.R. (1978). Transformation of yeast. *Proceedings of the National Academy of Sciences of the United States of America* **75,** 1929–1933.

HOLLENBERG, C.P., DEGEMANN, A. KUSTERMANN-KUHN, B. AND ROYER, H.D. (1976). Characterization of 2 μm DNA of *Saccharomyces cerevisiae* by restriction fragment, analysis and integration in an *Escherichia coli* plasmid. *Proceedings of the National Academy of Sciences of the United States of America* **73,** 2072–2076.

HOPKINSON, C.S. AND DAY, J.W. (1980). Net energy analysis of alcohol production from sugar cane. *Science* **207,** 302–303.

HUDSON, J.C. (1975). Sugarcane: energy relationships with fossil fuel. *Sugar Journal* **37,** 25–28.

JAMES, L.A. (1975). Sugar cane for livestock. *World Crops* **27,** 156.

JOHNSON, M.A. (1983). On gasohol and energy analysis. *Energy* **8,** 225–233.

KEIM, C.R. (1983). Technology and economics of fermentation alcohol—an update. *Enzyme Microbial Technology* **5,** 103–114.

KHAN, A.S. AND FOX, R.W. (1982). Net energy analyses of ethanol production from sugarcane in northeast Brazil. *Biomass* **2,** 213–222.

KORTSCHAK, H.P., HARTT, C.E. AND BURR, G.O. (1965). Carbon dioxide fixation in sugar cane leaves. *Plant Physiology* **40,** 209–212.

KOVARIK, B. (1982a). Third World fuel alcohol push shows mixed results. *Renewable Energy News* **6,** 31

KOVARIK, B. (1982b). *Fuel Alcohol: Energy and Environment in a Hungry World.* An Earthscan Publication, International Institute for Environment and Development, London.

LINK, G., BEDBROOK, J.R., BOGORAD, L., COHEN, C.M. AND RICH, A. (1978). The expression of the gene for the large subunit of ribulose 1,5-bisphosphate carboxylase in maize. In *Photosynthetic Carbon Assimilation* (H.W. Siegelman and G. Hind, Eds), pp. 349–361. Plenum Press, New York.

LADISCH, M.R., LIN, K.W., VOLOCH, M. AND TSAO, G.T. (1983). Process considerations in the enzymatic hydrolysis of biomass. *Enzyme and Microbial Technology* **5,** 82–102.

LAETSCH, W.M. (1974). The C4 syndrome—a structural analysis. *Annual Review of Plant Physiology* **24,** 27–52.

LEVINSON, A. (1982). The production of ethanol from crops in the United States: The food–energy conflict. *Resource Management and Optimization* **2,** 99–120.

LEVY, P.F., SANDERSON, J.E., KISPERT, R.G. AND WISE, D.L. (1981). Biorefining of biomass to liquid fuels and organic chemicals. *Enzyme Microbial Technology* **3,** 207–215.

LIPINSKY, E.S. (1981). Chemicals from biomass: Petrochemical Substitution Options. *Science* **212,** 1465–1471.

LIPINSKY, E.S. AND KRESOVICH, S. (1982). Sugar crops as a solar energy converter. *Experientia* **38,** 19–23.

Loomis, R.S. and Williams, W.A. (1963). Maximum crop productivity: an estimate. *Crop Science* **3,** 67.

Lorimer, G.H. (1981). The carboxylation and oxygenation of ribulose 1,5 bis phosphate. *Annual Reviews of Plant Physiology* **32,** 349–383.

Maiorella, B.L., Blanch, H.W. and Wilke, C.R. (1983). Distillery effluent treatment and by-product recovery. *Process Biochemistry* **18,** 5–8.

Mardon, C.J. (1981). *Methane from Biomass—Towards a Methane Economy. Research Review, Division of Chemical Technology*, pp. 39–55. Commonwealth Scientific Industrial Research Organisation, Melbourne.

Murata, D. (1979). Energy inventory for Hawaiian sugar factories—1978. *Hawaiian Planters Record* **59,** 177–185.

Mukherjee, S.K. (1957). Origin and distribution of *Saccharum. Botanical Gazette* **119,** 55–61.

Nathan, R.A. (1978). *Fuels from Sugar Crops. DOE Critical Review Series, No. TID*-22781. National Technical Information Service, Springfield, Virginia.

Ng, T.K., Busche, R.M., McDonald, C.C. and Hardy, R.W.F. (1983). Production of feedstock chemicals. *Science* **19,** 733–740.

Nickell, L.G. (1964). Tissue and cell cultures of sugar cane—another research tool. *Hawaiian Planters Record* **57,** 223–239.

Nickell, L.G. and Maretzki, A. (1969). Growth of suspension cultures of sugarcane cells in chemically defined media. *Physiologia Plantarum* **22,** 117–125.

OTA (1980). *Energy from Biological Processes.* Ballinger Publishing Company and Congressional Office of Technology Assessment, Cambridge, Massachusetts.

Paturau, J.M. (1982). *Byproducts of the Cane Sugar Industry*, 2nd edn. Elsevier, Amsterdam.

Pearce, J. (1981). *The Common Agricultural Policy. Chatham House Papers No. 13.* The Royal Institute for International Affairs. Routledge and Kegan Paul, London.

Pigden, W.J. (1974). Derinded sugarcane as an animal feed—a major breakthrough. *World Animal Review* **11,** 1–5.

Pimental, L.S. (1980). The Brazilian Ethanol Programme. *Biotechnology and Bioengineering* **22,** 1989–2012.

Ravno, A.B. (1979). Perspectives on ethanol manufacture. *South African Sugar Journal* **64,** 239–243.

Rolz, C., de Cabrea, S. and Garcia, R. (1979). Ethanol from sugar cane: EX-FERM Concept. *Biotechnology and Bioengineering* **21,** 2347–2349.

Rothman, H., Greenshields, R. and Calle, F.R. (1983). *The Alcohol Economy: Fuel Ethanol and the Brazilian Experience.* Frances Pinter Publishers, London.

Ruschel, A.P., Henis, Y. and Salati, E. (1975). Nitrogen-N-15 tracing of N-fixation with soil-grown sugarcane seedlings. *Soil Biology and Biochemistry* **7,** 181–182.

Ruschel, A.P., Victoria, R.L., Salati, E. and Henis, Y. (1978). Nitrogen fixation in sugar cane (*Saccharum officinarum*). *Ecological Bulletin* **26,** 297–303.

Simmonds, N.W., ed. (1976). *Evolution of Crop Plants.* Longman, London.

Somerville, C.R. and Ogren, W.L. (1982). Genetic modification of photorespiration. *Trends in Biochemical Sciences* May, 171–175.

Stevenson, G.C. (1965). *Genetics and Breeding of Sugar Cane.* Longman, London.

Stewart, G.A., Nix, H.A., Gartside, G., Rawlins, W.H.M., Gifford, R.M. and Siemon, J.R. (1979). *The Potential for Liquid Fuels from Agriculture and Forestry in Australia.* Commonwealth Scientific and Industrial Research Organization, Australia.

Stewart, G.G. and Russell, I. (1981). Yeast, the primary producer of fermentation ethanol. In *Gasohol, A Step To Energy Independence* (T.P. Lyons, Ed.), pp. 133–166. Alltech Technical Publications, Lexington.

Stone, J.E. and Marshall, H.B. (1980). *Analysis of Ethanol Production Potential from Cellulosic Feedstocks. Report no. 15SQ.23380-9-6463.* Department of Energy, Mines and Resources, Ottawa, Canada.

THOMPSON, G.D. (1978). The production of biomass by sugarcane. *Proceedings of the South African Sugar Technologists Association* **52,** 180–187.

TOLBERT, N.E. (1971). Microbodies, peroxisomes and glyoxysomes. *Annual Review of Plant Physiology* **22,** 45–74.

TSAO, G.T., LADISCH, M.R., VOLOCH, M. AND BIENKOWSKI, P. (1982). Production of ethanol and chemicals from cellulosic materials. *Process Biochemistry* **17,** 34–38.

WANG, D.I., COONEY, C.L., DEMAIN, A.L., GOMEZ, R.F. AND SINSKY, A.J. (1981). *Degradation of Cellulosic Biomass and its Subsequent Utilisation for the Production of Chemical Feedstock. Report number XR-9-8109-1, to USDE.* Department of Nutrition and Food Science, Massachusetts Institute of Technology.

WILKE, C.R., YANG, R.D., SCIAMANNA, A.F. AND PARK, Y.K. (1981). Raw materials evaluation and process development studies for conversion of biomass to sugars and ethanol. *Biotechnology and Bioengineering* **23,** 163–183.

WILKE, C.R., MAIORELLA, B., SCIAMANNA, A., TANGNU, K., WILEY, D. AND WONG, H. (1983). *Enzymatic Hydrolysis of Cellulose.* Noyes Data Corporation, Park Ridge, New Jersey.

12

Whey, a Potential Substrate for Biotechnology

G. MOULIN AND P. GALZY

Chaire de Génétique et Microbiologie, ENSA—INRA, 34060 Montpellier Cedex, France

Introduction

COMPOSITION

Whey is a by-product of the cheese-making industry following the separation of casein and butter fat as curd from milk. Whey composition varies according to its origin (ewe, goat or cow) and to the cheese-making technique employed. The composition of whey has been widely studied (Supplee, 1940; Olling, 1963; Guy, Vettel and Pallanche, 1966; Kosikowski, 1967; Mereo, 1971, Cerbulis, Woychik and Wodolowski, 1972; Mavropoulou and Kosikowski, 1973a, b; Kosikowski, 1975; Février and Bourdin, 1977). Cheese wheys can be divided into five groups (*Table 1*). The coagulation with rennet or rennet preparations yields sweet whey (pH 4·5–6·7), with high lipid contents. Coagulation by lactic

Table 1. Composition of different types of liquid whey (g/ℓ) according to Fèvrier and Bourdin. 1977).

	Cow			Ewe	Goat
Curd type	Rennet	Mixed	Lactic	Rennet	Lactic
Density	1·239	1·0247	1·0245	1·0234	1·0269
Dry matter	70·84	70·49	65·76	83·84	62·91
Lipids	5·06	3·38	0·85	6·46	0·40
Dry matter without lipids	65·78	67·11	64·91	77·38	62·51
Lactose	51·81	50·84	45·25	50·98	39·18
Total nitrogen	1·448	1·454	1·223	2·933	1·466
Non-protein nitrogen	0·368	0·414	0·536	0·796	0·669
Ammonium nitrogen	0·041	0·090	0·140	0·129	0·176
Urea nitrogen	0·141	0·095	0·070	0·139	0·122
Lactic acid	0·322	2·226	7·555	1·763	8·676
Citric acid	1·298	1·095	0·260	1·032	0·157
Ash	5·252	5·888	7·333	5·654	8·361
Phosphorus	0·412	0·470	0·649	0·545	0·703
Calcium	0·466	0·630	1·251	0·494	1·345
Potassium	1·455	1·491	1·485	1·281	1·812
Sodium	0·505	0·537	0·528	0·616	0·433
Chlorides (as NaCl)	2·195	2·208	2·092	2·368	3·287

Biotechnology and Genetic Engineering Reviews—Vol. 1, February 1984
0264–8725/84/01/347–28$10.00 + $00.00 © Intercept Ltd

Table 2. Average composition of defatted whey (% of dry matter) (according to Février and Bourdin, 1977).

Curd type	Cow			Ewe	Goat
	Rennet	Mixed	Lactic	Rennet	Lactic
Dry defatted extract	100	100	100	100	100
Lactose	78·76	75·76	69·71	65·88	62·68
Total nitrogen	2·20	2·17	1·88	3·79	2·35
Total protein (N × 6·38)	14·04	13·82	12·02	24·18	14·96
Total protein (N × 6·25)	13·75	13·56	11·75	23·69	14·69
Non-protein nitrogen	0·56	0·62	0·83	1·03	1·07
Ammonium nitrogen	0·06	0·13	0·22	0·17	0·28
Urea nitrogen	0·21	0·14	0·11	0·18	0·20
Lactic acid	0·49	3·32	11·64	2·28	13·38
Citric acid	1·97	1·63	0·40	1·33	0·25
Ash	7·98	8·77	11·30	7·31	13·37
Calcium	0·71	0·94	1·93	0·64	2·15
Phosphorus	0·63	0·70	1·00	0·70	1·12
Potassium	2·21	2·22	2·29	1·66	2·90
Sodium	0·77	0·80	0·81	0·80	0·69
Chlorides (as NaCl)	3·34	3·29	3·22	3·06	5·26

fermentation yields acid whey (pH 3·9–4·5), containing smaller quantities of lactose and proteins. It is noteworthy that whey obtained from ewe's milk contains more than twice the amount of protein found in whey from cow's milk. The comparison between the content of non-fat solids of wheys of different origins (*Table 2*), proposed by Février and Bourdin (1977), shows the importance of two factors: the high content of lactose (75%), and that of protein (N × 6·25) which varies from 12% to 14% of dry matter. Cheese whey and ultrafiltration permeate contain trace elements (*Table 3*) and the main vitamins (*Table 4*): whey is therefore extremely valuable nutritionally, and the permeates can be used for the industrial production of many micro-organisms capable of using lactose or its hydrolysis products.

QUANTITIES AVAILABLE

According to the Whey Product Institute of Chicago, Illinois, 11 804 000 t of whey were produced in the US in 1974 and 16 355 000 t were produced in 1979. In France, whey production rose from 4 800 000 t in 1974 to 8 000 000 t in 1979, and still continues to increase (SCEES, 1981). According to Coton (1983), world whey production had reached 118 261 000 t by 1981.

Initially, whey was considered to be a waste product to be disposed of, and was mainly redistributed to milk producers for animal feed. The increase in size of cheese plants, the necessity for reduction in the BOD and COD of the effluent (*see* Chapter 10, this volume) and the need to maximize returns on raw material have encouraged producers to seek new ways of using cheese whey. At first, cheese whey was concentrated or dried for animal feed. Since 1960, great efforts have been made to fractionate and use independently the main

Table 3. Average content of the main trace elements in cheese whey and ultrafiltration permeate (mg/100 g dry matter).

Element	Whey	Permeate
Iron	1–7	3–11
Copper	0·5–5	1–3
Zinc	5–9	30–33
Manganese	0·01–0·04	0·5–0·8

Table 4. Average vitamin content of cheese whey and ultrafiltration permeate (mg/100 g of dry matter).

Vitamin	Whey	Permeate
Vitamin A	100	80
Thiamin	4–6	5–6
Pyridoxin	6–10	5–10
Riboflavin	7–30	15–20
Calcium pantothenate	30–70	50–60
Biotin	0·2–0·3	0·1–0·3
Cobalamin	0·01–0·05	0·02–0·05
Vitamin C	30–50	20–40

constituents of whey, in particular whey proteins. Despite these efforts, only about 50% of world whey production has been turned to good account.

Utilization of cheese whey

LIQUID CHEESE WHEY

The use of liquid cheese whey for the purpose of feeding hogs is on the decline because of the constant increase in the output from cheese plants. Such large volumes are difficult to redistribute to farmers and would require the establishment of large feedlots nearby. However, in the case of small cheese plants, Février and Bourdin (1977) showed that this use remained of interest. Similarly, it is possible to incorporate into cattle and sheep feed rations up to 30% of dry matter as liquid whey (Schingoethe, 1976).

With regard to human food uses, the production of beverages from cheese whey has been attempted in several countries. Holsinger, Posati and De Vilbiss (1974), Holsinger *et al.* (1977) and Kosikowski (1981) reviewed different whey beverage processes. Several products, either protein-enriched or deproteinized, and either fermented or non-fermented, were tested. However, it is noteworthy that only the Rivella product has been marketed in Europe. A lactic fermentation was used for the production of this beverage.

CONCENTRATED CHEESE WHEY

The concentration processes for cheese whey were developed in order to reduce the costs of transport and storage and to improve the quality of the product

(Longuet, 1977). The technologies that were developed employed either the reverse osmosis technique (Pepper, 1981) or the multiple evaporation effect, with or without mechanical recompression of steam (Voisin, 1981). The reverse osmosis technique is well adapted to small production units whereas the concentration process through evaporation would require larger installations. Whey has sometimes been concentrated to 18–38% in order to facilitate transport. Concentrates at 40% and 60% could be used for animal feed. Molasses at 70% dry matter and cattle lick containing 75% dry matter have also been used in animal rations (Coton, 1980). Thivend, Vermorel, and Guilhermet (1977) showed that the use of whey concentrates was well adapted to cattle feeding. These authors were convinced that deproteinized permeate could also be used. Nitrogen could be provided by non-protein additives. Similarly, Barre *et al.* (1977) showed that concentration of whey yielded a microbiologically stable product at a lower production cost than that of drying.

WHEY POWDER

Although the production of whey powder is expensive and requires large amounts of energy, it is still a greatly used process because whey powder is easy to store and to transport. There is often a pre-concentration process by evaporation down to 50% dry matter, followed by drying in a spray tower (Chaput, 1977; Longuet, 1977; Hynd, 1980).

With regard to animal feed, whey powder is extensively used in hog rations and also in milk replacers for unweaned calves. This outlet is very important in Europe, and especially in France where the production of whey powder reached 285 000 t in 1979 (Lenoir, 1981). According to Toullec and Le Treut (1977) about 100 000 t of whey powder were incorporated into unweaned calves' rations in 1975 in France. However, it is noteworthy that present European legislation requires the incorporation (up to 60%) of skim milk powder into these rations: the development of this market for whey powder is thus becoming limited and unstable.

Many studies on the incorporation of whey powder into human foods (Guy, Wettel and Pallanche, 1966; de la Guérivière, 1977; Kosikowski, 1979; Mathur and Shahani, 1979; Grandadam, 1981; Mann, 1982) have shown that sweet whey can be added advantageously to many foods such as frozen desserts, cheese products, dried soup and gravy bases.

The main products containing whey powder have been described by Kosikowski (1979), Clark (1979), and Salmon (1981). The main industries using whey are infant food manufacturers, bakers, confectioners and meat processing plants. The amounts used vary very much from one country to another. Kosikowski (1979) estimated that about 30% of whey powder was incorporated into food products in the USA, whereas only about 1% was so incorporated in the UK.

The intended use of whey powder depends as much on its functional properties as on its nutritional value. These functional properties depend largely on the different treatments applied before the drying stage. Whey can be used as it is, partly or totally demineralized by electrodialysis or by ion exchange, deproteinized by ultrafiltration, or both demineralized and deproteinized. In fact, all these technologies have been employed in order to obtain a product with the best functional properties.

Protein recovery

From the nutritional point of view, cheese whey is unbalanced by its high mineral and lactose contents. Lactose is not well assimilated by many organisms. This had led to the early development of many processes for the recovery of proteins which constituted the most valuable part of whey. Recently, the ultrafiltration technique has allowed the retention of almost all milk proteins and has increased the yield in cheese manufacture (Maubois, Mocquot and Vassal, 1969). As a result it is now the permeates which need to be turned to better account.

PROTEIN PRECIPITATION PROCESSES

There are several cheap and easy processes for the recovery of proteins through precipitation at their isoelectric pH and at 90–95°C (Centriwhey, Bel Industrie). The industrial uses of these processes are, however, limited by the partial or total denaturation of the proteins thus obtained.

Moddler and Harwalkar (1981) proposed different heat treatments at acid pH (pH 2·5–3·5) at 95°C. The proteins thus obtained had interesting functional properties (Moddler and Emmons, 1977). Similarly, heat treatment at neutral pH (pH 6·8–7·5 at 80°C) increased the foaming power of the proteins obtained (Dewitt and Hontelez-Backx, 1981). Cold precipitation with salts, polymers or solvents has been reviewed recently by Mathur and Shahani (1979) and by Humbert and Alais (1981a). The main precipitation agents used were carboxymethyl cellulose, polyphosphates, ferriphosphate, polyacrylic acid, polyethylene glycol, chitosan, bentonite, lignosulphate and sodium laurylsulphate. Protein yields varied from 63% with polyacrylic acid to 91% for ferripolyphosphate. The precipitate contained a high proportion of ash, requiring further demineralization.

ULTRAFILTRATION

Ultrafiltration for the recovery of proteins was proposed in 1969 by Fallick, and this technology was rapidly developed by the dairy industry. According to Maubois and Brulé (1982), the area of filter membrane installed has increased from 300 m^2 in 1971 to 70 000 m^2 in 1981. In 1979, about 3% of available world whey production employed this technique, which reached 8–10% in some countries that are major producers of cheese.

Maubois (1980) and Maubois and Brulé (1982) have described the main types of membranes being used: flat sheet (proposed by Rhône-Poulenc, Pasilac and Dorr-Oliver), tube shape (by SFEC, Abcor, PCI, Wafilin), hollow fibre (by Romicon), spiral shape (by Abcor). According to Maubois (1980), the first-generation membrane made of cellulose acetate has now almost entirely been superseded. The membranes currently used are manufactured from synthetic organic polymers. These second-generation membranes have a better performance, and a better resistance to extremes of pH and temperature. More recently, mineral membranes (zirconium oxide on carbon graphite) have appeared with very high mechanical (2000 kPa), thermal (400°C) and physico-chemical resistances. These new membranes are of great interest for the ultrafiltration of solutions of very high protein content (Maubois, 1979; Goudefranche *et al* 1981).

The protein concentrates obtained by ultrafiltration have particularly interesting nutritional (Forsum, 1974) and functional properties (Dewitt and de Boer, 1975; Cheftel and Lorient, 1982), and could be incorporated into soft curd cheese. The protein concentrates are also useful animal-feed components, especially for unweaned calves (Stewart, Muller and Griffin, 1974; Toullec *et al.*, 1975). Their functional properties (solubility, viscosity, water-holding capacity, emulsifying power, foaming power) have encouraged attempts to use them in a great number of food products in replacing other more traditional additives such as milk powder and egg albumen (Richert, 1979; Paquet, 1981). The incorporation of these protein concentrates into new products such as gel-type foods (Humbert and Alais, 1981b) and comminuted meat products (Grandadam, 1981) has also been investigated.

The direct ultrafiltration of milk for preparing liquid 'pre-cheese' has been described by Maubois, Mocquot and Vassal (1969) (MMV process). This cheese-making technology produces, instead of whey, a deproteinized permeate which needs further processing. In 1981, 120 000 t of cheese were produced applying the ultrafiltration technique (Maubois and Brulé, 1982). Feta cheese production has been a major development in Denmark. Recent improvement in ultrafiltration equipment and, subsequently, in product quality (Mahaut *et al.*, 1982), should soon allow this technique to be used in the production of soft curd cheese.

ION-EXCHANGE SEPARATION

Two processes are being tested. The industrial process Vistec uses carboxymethyl cellulose to fix the proteins at pH 4, with elution at pH 9. The eluate has to be concentrated by ultrafiltration. The foaming property of the product obtained is similar to that of egg white (Palmer, 1977). The spherosil process (Mirabel 1978 a,b; 1981) uses controlled silica beads of a predetermined uniform size on which anionic or cationic active groups have previously been bonded. The protein retention capacity can reach 130 g/kg support. For treating acid whey, the process uses cation exchange beads at pH 4·6 whereas for treating sweet whey the column is filled with anion exchange beads at pH 6·6. The protein content of the eluate is 4·5%. This process leads to the production of undenatured, very pure proteins which can be separated from each other.

These processes may be used eventually for large-scale fractionation of proteins: in particular, it should be possible to produce immunoglobulins.

PROTEIN HYDROLYSATES

The production of protein hydrolysates with a membrane enzyme reactor was suggested by Roger, Brulé and Maubois (1981). This process comprised three stages: the preparation of a protein concentrate (70–90 g protein/ℓ); clarification of the concentrate, and hydrolysis in an enzyme reactor. The hydrolysis could be performed with a multienzyme system such as proteases or lipases, or with a single purified enzyme. This reactor functioned in the continuous state. When the protein concentrate was hydrolysed by the reactor with the enzyme pancreatin, Roger (1979) and Roger and Maubois (1981) showed that the hydrolysates obtained could have satisfied the nutritional nitrogen needs of post-operative patients. Using the same technique, Brulé *et al.* (1980) proposed the hydrolysis of a caseinate solution followed by the formation of phosphopeptidic complexes by the addition of calcium chloride and disodium phosphate; these phosphopeptide complexes were subsequently separated and purified. Phosphopeptides are potentially of interest because of their effect on the intestinal absorption of minerals (Mathan *et al.*, 1979; Mykkanen and Wasserman, 1980).

Utilization of permeates

All the processes described in the previous section yield a permeate with high contents of lactose (35–50 g/ℓ), minerals, vitamins, and sometimes lactic acid. The volume of permeates being produced is increasing. As there are several profitable uses at the preceding production step, either by incorporation of the proteins into cheese products or by recovery and fractionation of these proteins, the permeates have theoretically little or no value. However, the volumes produced are so large that some treatment at least is necessary before their disposal as effluent. Coton (1980) has shown that the permeate could be used *per se* or in a concentrated form for hog and cattle feed. There are also a few uses in the food industry; however, the enormous volumes of permeate to be treated demand other possible outlets, implying the use of fermentation techniques.

PRODUCTION OF FOOD YEAST

The nutritional qualities of yeasts are well recognized (Wasserman, 1961; Birolaud, 1971; Vrignaud, 1971). These qualities are related to their amino acid, sterol, fatty acid and vitamin composition. More recently the functional properties of yeast have been investigated, showing a high binding power, high water-retention capacity and a thickening power that was unaffected by heat treatment (Marzolf, 1977, 1981; de la Guérivière, 1981). The application of these functional properties will be linked with the development of new outlets in the food industry and with the use of yeast components in the pharmaceutical industry.

In a general review of food yeast, Meyrath and Bayer (1979) remarked that the production of yeast from whey was attempted as early as 1940 by Messrs Harmer in Spillern. Many other processes were subsequently proposed, such as the Polyvit process (Waeser, 1944), the Waldhoff process (Demmler, 1950), the Bel process (1958), and the Wheat process (Robe, 1964). Bel started its industrial process in 1958. The fermenters used corresponded to the air-lift-type Lefrançois–Marillier process (Lefrançois, 1964). The fermenters have not much changed to the present day: however, the biological aspects of the fermentations are better understood. Studies of the flora being used helped greatly to improve the performance of the strains. At present there are two possibilities: transformation of lactose thus making it available to the yeast strain chosen, or direct use of lactose by a yeast strain able to metabolize this substrate.

Among the first type of process was that proposed by Lembke and Baader (1975). Lactic bacteria were used to transform lactose into lactic acid, without aeration at pH 4·5 and 44°C. In a second step, lactic acid was used by *Candida utilis* and *Candida krusei* in a vigorously aerated fermenter. This technology involved the use of two successive fermenters employing different reaction conditions, and another drawback was that the medium was at a very vulnerable pH for part of the process. The process as a whole could be run in the semicontinuous or continuous fashion (Moebus and Kiesbye, 1975). Experiments with *Lactobacillus bulgaricus* and *Candida krusei* mixed cultures showed that it was possible to attain productivities of up to 2·65 g/ℓ/h under these conditions of continuous fermentation.

The direct transformation of lactose could be achieved by several yeast strains. The strain most often used has been *Kluyveromyces fragilis* (Porges, Pepinski and Jasewicz, 1951; Wasserman, 1960; Chapman, 1966; Admunson, 1967; Ladet *et al.*, 1972; Bernstein and Plantz, 1977; Moulin and Galzy, 1976; Moulin, Ratomahenina and Galzy, 1976; Giec and Kosikowski, 1982). *Candida intermedia* was also recommended by Meyrath and Bayer (1979) for the Vienna UKM process. The final choice depends as much on the strain as on the species, because many biological factors of great economic importance vary considerably within each species, for example phosphate requirements (Wasserman, 1961; Nour-el-Dien, Halasz and Lengyel, 1981) and vitamin requirements (Wasserman, 1961, Nour-el-Dien and Halasz, 1982a,b).

Studies with industrial-scale fermenters (Moulin and Malige, unpublished data) have shown that a progressive improvement of the strains was seen for the following factors: increase in growth rate; increase in protein and nucleic acid content; decrease in the relative content of storage compounds, especially glycogen; loss of sporulation and sexual characteristics. Within the fermenter the selection operated in favour of the features useful for survival in the absence of cell multiplication. We observed the same situation in other industrial systems using a continuous fermentation technique. It is possible to select yeast strains with rapid growth rate on permeates without any additive other than a source of nitrogen; however, industrial producers generally add small amounts of trace elements (Fe, Cu, Mn, Zn). Sterilization of the medium is not necessary because of the bacterial nature of the ultrafiltration permeates and the pH usually employed (pH 3·2). Oxygen transfer remains the limiting factor for the productivity of the fermenters. There are at present two processes for the production of food yeast by direct transformation of lactose in continuous culture—the Vienna process and the Bel process.

The Vienna process

The flow diagram of this process has been given by Meyrath and Bayer (1979). The fermentation was run continuously at 32–33°C and pH 3·4–3·6, with the addition of ammonia (0·8 kg/m^3) and ammonium sulphate (0·8 kg/m^3) as nitrogen source and for pH regulation. Under these conditions, for a lactose concentration input of 45 g/ℓ, the productivity obtained was 4·5 g/ℓ/h (yeast dry matter). Yield of dry matter for the amount of lactose used was 54%. The fermentation took place in a fermenter with an aeration system involving an external pump for rapid circulation of the reaction mixture; about 600 m^3/h of air were used for a 14 m^3 fermenter. The strain used was *Candida intermedia* which is characterized by an exclusively oxidative lactose metabolism (Lodder, 1970). This property enables undiluted permeate to be used because no alcoholic fermentation of lactose would be expected. However, this physiological advantage of the strain cannot, as yet, be exploited fully because of the limitations of the industrial fermenters currently available. The level of oxygen transfer obtained is not sufficient for the dilution rates which might be expected according to the strain and the amount of lactose used (45g/ℓ). Under laboratory conditions, the growth rate (expressed as $\mu = \log_2/T_g$, where T_g is the generation time of the yeast *Candida intermedia*) of this strain was 0·28/h. Such a growth rate could be reached in an industrial fermenter only by using diluted permeate.

The Bel process

The general scheme of this process has been described by Marzolf (1977, 1981), by de la Guérivière (1981) and by Moulin, Malige and Galzy (1983). This process is being employed, in France, in three industrial units producing about 10 000 t of food yeast annually. The fermentation has been run at pH 3·2 and 38°C ±1°C using as substrate deproteinized whey either from protein recovery units by ultrafiltration, or by protein precipitation at the isoelectric point (pH 4·5) at 90°C; in both cases the whey was stored before fermentation. Nutrients (nitrogen, trace elements) were added and the substrate was diluted before it was pumped into the fermenter. The whey solution thus prepared contained 20–25 g/ℓ of lactose and was continuously pumped into the fermentation tower; the dilution rate was constant at 0·33/h. Aeration was ensured with an air-lift system according to the scheme proposed by Lefrançois (1964) at 1·5 vol/vol/min. For a fermenter of 23 m^3 capacity, about 1800 m^3air was pumped per hour. Such a fermenter would produce 4·5 g yeast dry matter per litre per hour (55–60 g yeast per 100 g lactose utilized). Unlike the Vienna process, which uses a pure strain, the Bel process uses three species in equilibrium: Moulin, Malige and Galzy (1983) have shown that this equilibrium is between strains of the species *Kluyveromyces fragilis*, *Kluyveromyces lactis* and *Torulopsis bovina*, with *Kluyveromyces fragilis* strains representing 90% of the total flora. These strains grow well on lactose: they possess the general physiological feature of lack of glucose effect on their respiratory metabolism and a strong Pasteur effect (Chassang-Douillet *et al.*, 1973): they could, therefore, grow on a sugar substrate in the presence of air with a very weak fermentative metabolism. When oxygen was not the limiting factor, the production of ethanol was 0·06 mg/h per gram of yeast dry matter (Moulin, Legrand and Galzy, 1983). This ethanol production was sufficient to maintain the strains of *Torulopsis bovina* strains in the fermenter (Moulin, Malige and Galzy, 1983) (this species grows on ethanol but not on lactose); the strains of *Kluyveromyces lactis* species grew on lactic acid.

The equilibrium between the strains of these three species was remarkably stable. In fact, each species had an independent 'ecological niche' because each of them preferentially uses a different substrate. This equilibrium of the flora enables the various types of permeate to be treated under optimal conditions. The relative proportions of the three species may vary according to the relative lactose and lactic acid contents of the permeates, the lactose content of the fermenter being the most important factor. When this lactose input increased from 20 to 27 g/ℓ, the dry-matter yield was shown to fall from 63% to 48% (Moulin, Malige and Galzy, 1983). Again, in this process the limiting factor was oxygen transfer. It is noteworthy that the productivity of these two industrial processes was identical despite the technological differences and the different yeast strains used.

PRODUCTION OF PROTEIN-ENRICHED WHEY

In the studies described above, the aim was to produce yeast cells with separation of the micro-organisms from the medium. There have also been several studies on the production of a protein-enriched food, exploiting the growth of micro-organisms (yeast, filamentous fungi). At present, only the first two processes described below are used commercially.

The Devos process consists of growing *Saccharomyces cerevisiae* Hansen strain on the lactic acid in whey; the whey thus enriched is concentrated and dried. The yeast-enriched powder contains 15–18% protein.

The Société des Alcools du Vexin (SAV) process (1963) involves a two-stage fermentation followed by evaporation and drying. The substrate in the first fermenter is lactose, and lactic acid is used in the second fermenter. The productivity is 2 g/ℓ/h with a dry-matter yield from lactose of between 36% and 41%. This process can utilize whey from cheese plants or from casein plants as substrate.

The production of yeast-enriched milk proteins has also been studied by Molinaro, Hondermarck and Jacquot (1977). The principle of this process consisted of growing a *Kluyveromyces fragilis* strain on cheese whey. The whole culture mixture is subsequently concentrated or dried: the resulting powder contains up to 45% proteins (N × 6·25).

The enrichment of whey proteins has also been attempted using yeast-like moulds. Halter, Puhan and Kapelli (1981) proposed the enrichment of ultrafiltration permeates with a strain of *Trichosporon cutaneum*. The experiments performed with equipment of capacity 5000–7000 ℓ used a fed-batch culture technique. The ultrafiltration permeate with 46 g/ℓ lactose was first enriched with nitrogen (NH_4Cl 6·0 g/ℓ), trace elements and vitamins. Cell growth was limited by the restricted addition of ammonia so that only about 50% of lactose was consumed. The fermentation was run at pH 4 following pasteurization of the medium in order to avoid contamination by wild yeasts. The final product thus obtained contained 1·3–1·4% protein and 1·3–2% lactose as dry matter.

The Caliqua Sireb process (Fleury and Henriet, 1977) used a filamentous fungus of the species *Penicillium cyclopium* in a continuous culture. The medium contained amino acids and soluble peptides at the outlet of the fermenter. The biomass obtained could be dried and the protein content (N × 6·25) of the final product was about 40%.

PRODUCTION BY BACTERIAL FERMENTATION, OF A RUMINANT FEED SUPPLEMENT RICH IN CRUDE PROTEIN

Different processes have been described for the fermentation of lactose from whey into lactic acid (Whittier and Rogers, 1931; Johnson *et al.*, 1937; Jansen, 1945). Many authors have shown that ammonium salts of short-chain organic acids, in particular ammonium lactate, are better nitrogen sources for ruminants than urea or soybean meal (Varner and Woods, 1971; Allon and Henderson, 1972; Dutrow, Huber and Henderson, 1974; Crickenberger, Henderson and Reddy, 1981). The fermentation could be batch (Gerhardt and Reddy, 1978; Juengtst, 1979) or continuous (Keller and Gerhardt, 1975; Reddy, Henderson and Erdman, 1976), or continuous with dialysis (Coulman, Stierber and Gerhardt, 1977; Stierber and Gerhardt, 1979).

The continuous fermentation process was as follows: a strain of *Lactobacillus bulgaricus* was used; the growth medium was constituted with whey or deproteinized whey, to which corn steep liquor was added as a source of growth factor; the pH was maintained at 5·5 by the addition of ammonia; the temperature was maintained at 43°C. Sterilization of the medium was not necessary in view of the pH, temperature and anaerobic conditions used, and the high concentration of undissociated lactic acid. The fermentation lasted from 14 to 24 hours with a lactose–lactic acid transformation yield of 95% for lactose contents below 70 g/ℓ. After concentration to 70% dry matter, the final product contained 55% protein (N × 6·25): 76% of this was ammonium lactate, 17% was whey protein and 7% was bacterial cell protein. This product was termed 'fermented ammoniated condensed whey' (FACW).

The continuous fermentation process has been studied by Reddy, Henderson and Erdman (1976). Two fermenters in series were used at pH 5·5 with a retention time of 31 h; the transformation yield was 98%.

In order to increase the productivity, Coulman, Stieber and Gerhardt (1977) studied a continuous culture process with a dialysis stage. Under these conditions the retention time was reduced by 19 h with a transformation yield of 97%.

Comparison of the three processes showed that the rate of transformation of lactose (mg/ml/h) fell from 4·4 in the batch process to 1·1 in the continuous process. This rate increased to 12·1 in the process with dialysis. It is also noteworthy that this third process used a lactose input concentration of 230 g/ℓ compared with 70 g/ℓ for the batch process and 50 g/ℓ for the continuous culture process. The production of FACW appears to be of great potential interest: Gerhardt and Reddy (1978) have pointed out that, with appropriate adaptation, this could be an excellent process for the treatment of aqueous effluents from many food industries.

PRODUCTION OF BAKER'S YEAST

The production of baker's yeast from cheese whey with its lactose already transformed into lactic acid has been considered (Moebus and Kiesbye, 1975). *Lactobacillus bulgaricus* could transform lactose completely into lactic acid, which would imply the use of large quantities of a neutralization agent. In addition, Dion, Goulet and Lachance (1978) have shown that cell growth is strongly inhibited by high concentrations of lactic acid (above 10 g/ℓ). In order to reduce this inhibition problem, Champagne, Lachance and Goulet (1980) have proposed the use of a strain of *Streptococcus thermophilus*. This strain transforms only part of the lactose content into lactic acid and excretes galactose into the medium. The production of baker's yeast is based partly on lactic acid and partly on galactose as substrates. Under the operational conditions described by Champagne, Lachance and Goulet (1980) at pH 5·5, 30°C and 0·3 vol/vol/min, the dry-matter yield for lactose was 31%.

In 1977, the world production of baker's yeast was 177 000 t. The use of cheese whey to produce such a quantity would involve about 10% of the total world whey production.

PRODUCTION OF ETHANOL

The alcoholic fermentation of lactose was demonstrated in 1947 by Rogosa, Brown and Whittier. However, only a few species of yeast could ferment lactose. Several selection studies of strains capable of fermenting lactose directly have been undertaken (Gawel and Kosikowski 1978; Laham-Guillaume, Moulin and Galzy, 1979; Demott, Draughon and Herald, 1981; Friend, Cunningham and Shahani, 1982; Giec and Kosikowski, 1982; Izaguirre and Castillo, 1982). The strains selected by different authors belong to the species *Kluyveromyces fragilis* or *Candida pseudotropicalis*. In general, the fermentation of lactose up to a concentration of 200 g/ℓ could be performed with over 90% of the theoretical yield. Deproteinized whey can be fermented without any additive, and no variation in pH has been observed during the fermentation process. The fermentation period varied from 4 h for wheys containing 40 g/ℓ lactose to 20 h for wheys containing 200 g/ℓ lactose with recycling of yeast cells. In view of the inhibition mechanisms of alcoholic fermentation in the presence of lactose (Moulin, Boze and Galzy, 1980), the optimal lactose concentration was about 150 g/ℓ; this sugar concentration would yield 8% (v/v) ethanol.

Deproteinized cheese whey could thus be fermented into alcohol under favourable conditions. In 1979, the US Department of Energy in its report on the alcohol fuels policy, indicated that cheese whey is one of the cheapest feedstocks per gallon of ethanol. Other studies published recently (Reesen and Strube, 1978; Moulin and Galzy, 1981; Barry, 1982; Singh *et al.*, 1983) have shown the importance of prior recovery of proteins from cheese whey. These authors also demonstrated that ethanol production from cheese whey is competitive against chemical synthetic processes; in addition, there is a 30–40% net energy gain. At present, ethanol is being produced from deproteinized cheese whey in Ireland (Lyons and Cunningham, 1980; Barry, 1982), and in the USA (Walker, 1982).

Different projects are under investigation in the USA (Chen and Zall, 1982; Singh *et al.*, 1983), in New Zealand (Barry, 1982), and in France.

The production schemes given by Barry (1982) and by Reesen and Strube (1978) showed a continuous fermentation process in two stages, with a 12 h retention time. Yeast cells were partly recycled and some new cells were added. In the process described by Reesen and Strube (1978), the effluent from the distillation process was treated by anaerobic fermentation to produce methane. Following fermentation, the chemical oxygen demand of cheese whey was reduced by 90%: however it remained at a high level (5000–7000 mg/ℓ); further treatment of the effluent was thus required.

It is noteworthy that the development of lactose hydrolysis processes permitted the use of traditional distiller's strains (Kosikowski and Wzozeck, 1977; O'Leary *et al.* 1977a,b). However, the diauxy feature which appeared for glucose and galactose required the selection of derepressed mutants (Baily, Benitez and Woodward, 1982).

PRODUCTION OF GALACTOSE

The production of galactose by chemical means was studied by Clark (1921) and by the Préval Company (Société Préval) (1971). Clark used a chemical hydrolysis of lactose followed by a fractionation extraction of galactose with ethanol. The Préval process recommended the hydrolysis of lactose either by chemical or enzymatic means; glucose could then be eliminated by growing a yeast strain such as *Saccharomyces rosei*. Following fermentation and separation of yeast cells, the medium contained only galactose which could be extracted with ethanol.

The excretion of galactose was observed by Cooper *et al.* (1978) in an *Arthrobacter globiformis* culture growing on a medium containing lactose as carbon source. Similarly, Champagne, Lachance and Goulet (1980) showed that lactose was converted by *Lactobacillus thermophilus* partly into lactic acid and partly into galactose. Furthermore, yoghurt and kefir-type products resulting from lactose fermentation of milk always contain some galactose.

The production of galactose from whey permeate by biological means was proposed by Galzy and Moulin (1976) and Moulin, Varchon and Galzy (1977). *Kluyveromyces fragilis* and *Kluyveromyces lactis* yeast strains containing β-galactosidase are able to metabolize galactose. It should be possible to select mutants which have lost the ability to metabolize galactose while retaining the ability to hydrolyse lactose into glucose and galactose. Such a mutant could thus hydrolyse lactose, grow on the resulting glucose and secrete galactose. The genetics of lactose metabolism regulation in *Kluyveromyces lactis* has been studied by Dickson and Sheetz (1981), Sheetz and Dixon (1980), Carré (1978) and Fuentes (1981). The results showed that such mutants could be selected following multiple mutations. These mutant strains would have lost either their galactose permease galactokinase and epimerase, or their galactose permease galactokinase and transferase (Moulin *et al.*, unpublished results). These mutants could absorb lactose using a lactose permease, hydrolyse it with a β-galactosidase, and metabolize the resulting glucose with a growth rate of 0·3/h and a 0·45 growth yield (g dry weight per g glucose) from available glucose, with secretion of galactose into the medium. This fermentation was performed at pH 3·2, 32°C and the aeration was set at 1·5 vol/vol/min. The medium was supplemented with a nitrogen source and trace elements (Zn, Mn, Cu, Fe). The production of galactose from deproteinized whey containing up to 60 g/ℓ lactose reached 98% of the theoretical yield. The productivity obtained was 1·5 g/ℓ/h. The solution obtained following fermentation and separation of yeast cells contained only galactose as carbon substrate. Galactose could be extracted as in the processes described above for ethanol.

LACTOSE AND ITS DERIVATIVES

The manufacture of lactose from whey or from ultrafiltration permeates is a well-known process (Krevela, 1969; Nickerson, 1970, 1979). However, the amounts of lactose produced world-wide would require the use of only 5% of the whey available (Coton, 1980). The development of lactose chemistry would help to increase the amount of whey which could be utilized for lactose production. Nickerson (1979) in a review on lactose chemistry showed how this compound could be used for the absorption of flavours or as a complexing agent for metal ions. In addition, lactulose, lactitol and lactobionic acid are three important compounds derived from lactose.

Lactulose ((4-*O*-β-1) D-galacto-pyranosyl-D-fructose) is mainly used in pharmacology. This compound can be obtained by chemical means at 35°C under alkaline conditions (Hicks and Parrish, 1980), or by using an ion exchange resin (Demainay and Baron, 1978); these processes have relatively low yields. The production of lactulose from lactose with the enzyme β-galactosidase (EC 3.2.1.21) has been studied by Vaheri and Kauppinen (1978). The rates of galactosyl transfer and the biochemical nature of the products obtained vary according to the origin of the β-galactosidase enzyme. The β-galactosidase from *Kluyveromyces fragilis* could transform only 8% of lactose into lactulose, and other compounds (galactosyl-galactose and galactosyl-glucose) also were produced.

Lactitol can be produced by hydrogenation of lactose with nickel as catalyst (Van Velthuijsen, 1979). Hayashibara (1976) has demonstrated the possible use of lactitol as a sorbitol substitute. According to Van Velthuijsen (1979), lactitol has interesting possibilities in low-calorie diet foods. It is also noteworthy that the esterification of lactitol into lactitol palmitate would yield a compound with interesting emulsifying properties as an additive in foods or detergents (Zadow, 1979). Lactobionic acid, produced by enzymes, could be used as a food acid (Zadow, 1979).

The production of 'lactosylurea' is being much studied for ruminant feeding purposes. This compound is relatively slowly hydrolysed by the rumen microorganisms: this property would allow the increased use of urea as a non-protein nitrogen source in ruminant feeding without running the risk of reaching the toxicity level of ammonia (as when pure urea was included in the ration). This compound is obtained by reaction between urea and lactose.

The conversion of lactose into different products has been attempted mainly by chemical methods: enzymatic and biological techniques now offer new possibilities to be explored.

HYDROLYSIS OF LACTOSE

The partial or total hydrolysis of lactose is interesting from the viewpoint both of nutrition and of food technology. In addition, bearing in mind the development of ultrafiltration techniques, lactose hydrolysis would be an excellent way of furthering the use of permeates. According to Alam (1982), lactose intolerance is very widespread, affecting 70% of the non-white population of the USA and 95% of the populations of Asian and African countries. Harries (1982) puts the prevalence of hypolactasia at 50–90% in most non-Caucasian races and 2–30% in Caucasians. The limiting factor for using lactose as feed also seemed to depend on its hydrolysis by the animals following weaning (Février and Aumaitre, 1981). The hydrolysis of lactose would help to overcome the enzyme deficiency problem in some species.

From the technological point of view, the hydrolysis of lactose would improve two parameters important for the food industry, namely the sweetening power and the solubility of the sugars: the solubility of a glucose–galactose mixture is 60% whereas lactose alone would have a maximum solubility of only 22%; the solubility of sucrose is 68%. Similarly, the sweetening power of the glucose–galactose mixture is 70, whereas it is only 15 for lactose; again, sucrose is used as reference (sweetening power 100). In addition, hydrolysis of the lactose in the permeate would yield a mixture containing glucose and galactose, which can be metabolized by a greater number of micro-organisms.

Hydrolysis process

The hydrolysis of lactose can be performed either by chemical or by enzymatic means.

Hydrolysis of lactose with catalytic resin. Chemical hydrolysis is achieved by passing the substrate through a cationic resin, in the acidified state, at 90–100°C. Previous demineralization of whey (over 95%) would be necessary (Le Henaff, 1978; Coton, 1980). This process is still at the pilot plant stage and not all the process conditions have as yet been worked out. Furthermore, lactose solutions hydrolysed by this process contain (together with residual lactose, glucose and galactose) small amounts of other compounds such as lactulose, trisaccharides and high-molecular-weight sugars (Coton, 1980).

Several such processes have been put forward by different companies including Applexion Company (France), Portal Water Treatment Ltd (UK), Permutit Co. (UK), Technichem (USA) and Rohm and Hass (USA).

Enzymatic hydrolysis. The β-galactosidase enzyme has been isolated from animals, plants and micro-organisms, but only the latter have been used in the industrial production of β-galactosidase. Bacteria, yeast and fungi are all good sources of β-galactosidase: however there are important differences between the β-galactosidases from different sources. Their main features have been reviewed by Shukla (1975), Greenberg and Mahoney (1981) and Richmond, Gray and Stine (1981). In general, yeast β-galactosidases are characterized by a neutral pH and are less stable than those isolated from fungi. There are many published studies on bacterial β-galactosidases: however, the possibility of food poisoning due to coliforms precludes the use of bacterial β-galactosidase in food products and food processes. Only β-galactosidases isolated from yeast (*Kluyveromyces fragilis*, or *Kluyveromyces lactis*) or from fungi (*Aspergillus niger*) are used.

The hydrolysis can be carried out with soluble enzymes (Nipjels, 1978; Nipjels and Rheinlander, 1982; Giacin *et al.*, 1974): in this case the enzyme preparation is lost with each production batch. This process has often been used for the lactose hydrolysis of milk. Several such enzyme preparations have been developed in Italy, in Holland, and in the USA (Dahlquist *et al.*, 1977).

In addition, the hydrolysis of milk lactose before cheese manufacture could be of interest in the production of yoghurt, soft curd cheese and matured cheese (Thompson and Gyuriesek, 1974; Gyuriesek and Thompson, 1976; O'Leary and Woychik, 1976; Ramet, 1978), with important advantages to be gained either at the lactic fermentation stage or at the maturing stage.

Application of the hydrolysis process to the treatment of permeates would involve the recovery of soluble enzymes after treatment, because of the costs involved; such an attempt has been described by Brulé (1977).

The immobilization of β-galactosidase for lactose hydrolysis has been the subject of many papers. Richmond, Gray and Stine (1981) have reported that over 50 immobilization systems have been proposed: among these systems, immobilization processes by inclusion (Snam Progetti process) and by covalent bonding (Corning process) have yielded interesting results at the pre-industrial pilot plant stage (*see* Chapter 5 of this volume).

In the inclusion process, *Kluyveromyces fragilis* β-galactosidase was immobilized in triacetate cellulose fibres, and a reactor filled with these fibres was used for the pilot treatment of batches of 10 000 litres of milk: 70–80% of lactose was hydrolysed in 20 hours using 450 g fibres. Marconi *et al.* (1979) have extended this process to the treatment of whey.

The covalent bonding immobilization process has been described by Coton (1980) and by Dohan and Baret (1980). The β-galactosidase used was isolated from *Aspergillus niger*. The enzyme preparation was immobilized on a porous carrier using the glutaraldehyde technique described by Havewala and Weetall (1972). The immobilized lactase showed 500 activity (international) units per gram of carrier at 50°C and pH 3·2–4·3.

A preliminary pre-industrial pilot plant has been tested in England by the Milk Marketing Board: the flow diagram of this pilot plant has been given by Dohan and Baret (1980), and by Coton (1980). Cheese whey was pasteurized, ultrafiltered, demineralized and then hydrolysed by the reactor in a continuous system. The pilot plant achieved 80% hydrolysis at a flow rate of 360 litres/hour, and ran for six months, at 16 h/day and 5 days/week. After concentration, the resulting syrup contained 12% lactose, 22% glucose and 24% galactose (Dohan and Baret, 1980).

A second pilot plant which could treat 500 ℓ/hour was tested in France by the Union Laitière Normande according to the same process scheme. The results obtained for treating whey from casein manufacture confirmed the potential of the immobilization technology of β-galactosidase by covalent bonds (Dohan and Baret, 1980).

Main applications

Partially hydrolysed lactose syrups have been tested in many foods. They avoid the lactose intolerance problem, and the hydrolysis also increases the solubility and sweetness. In most cases, the best results were obtained with hydrolysed permeates which had already been partly or totally demineralized. De la Guérivière (1978), Lombrez (1978), Coton (1980), Salmon (1981) and Dicker (1982) have listed the main foods in which sucrose or corn syrup could be replaced successfully by hydrolysed lactose syrup: these include toffees, fudge, pectin, creams, jam and cakes. Development of these formulations would soon lead to a marked increase in the use of this technology for the treatment and utilization of permeates.

The availability of deproteinized whey containing hydrolysed lactose would also lead to new and greater uses of these products in fermentation processes. A limited number of micro-organisms are capable of hydrolysing lactose, whereas all micro-organisms are able to use glucose. However, the hydrolysate in fact contains a mixture of glucose, galactose and some residual lactose.

Permeates containing hydrolysed lactose permit the use of many strains for the production of food yeasts. Arnaud *et al.* (1978) have described a two-stage process. In an initial fermentation tower, glucose was used by a very rapidly growing yeast such as *Candida utilis*; *Kluyveromyces fragilis* could also be used for this purpose. These yeast cells were recovered by centrifugation and the medium was pumped into the second fermentation tower where its galactose and residual lactose could be used by flora identical to that described in the Bel process. The general operating conditions were a temperature of 35–38°C, pH 3·2, and aeration 1·5 vol/vol/min. Such a system was of interest in that the first tower could be fed with undiluted permeates (50g/ℓ total sugars).

Bakers' yeast may also be produced from permeates containing hydrolysed lactose, according to Arnaud *et al.* (1978). Bakers' yeast would grow on glucose then on galactose with a diauxy effect; it would thus be preferable to use a batch process. A continuous fermentation process would require an initial fermenter for the utilization of glucose, then a secondary fermenter for galactose and finally a third fermenter for the ethanol formed. Yeast cells also have to be recovered by centrifugation at each fermenter outlet.

Stineman (1978) described a process which preferentially used acid whey containing a mixture of glucose, galactose and lactic acid. The fermentation was performed at 30°C, pH 5·0 with an aeration of 1·5 vol/vol/min.

Various authors (O'Leary *et al.*, 1977 a, b; Baily, Benitez and Woodward, 1982) have considered the prior hydrolysis of lactose for alcoholic fermentation with classic distiller's strains. It would be necessary for this purpose to use derepressed strains which are not subject to the diauxy effect (Baily, Benitez and Woodward, 1982). These processes have not shown any obvious advantage compared with a direct fermentation process.

Hydrolysed lactose syrup could also be considered as a substitute for 'corn syrup' in beer brewing. According to Coton (1980) experiments have shown that it is possible to replace corn syrup with hydrolysed lactose syrup in beer without noticeable alteration in the taste of the beer.

Conclusions

The problem of cheese whey disposal appeared with the increase in size of cheese manufacture production units. Technological advances have continually encouraged the re-evaluation of suggested solutions of this problem.

The first change appeared at the raw material level. The development of protein recovery processes produced greater quantities of available permeates, with a concomitant reduction in the production of whey concentrate and whey powder for animal feedstuffs or human foods.

The possible uses of permeates are very varied. Some applications would require very large quantities of this product: ethanol production is one such process. Several production units are being developed in the USA (Fermentec, California and Milbrew, Wisconsin), and in Ireland (Carbery Milk). The economic interest of these units depends on their size and the economic situation (Delaney, 1981). If a country decided to convert ethanol into fuel, the volume of permeates available would provide only a small fraction of the necessary raw material.

Another example of permeate use is the production of ammonium lactate (fermented ammoniated condensed whey; FACW) for animal feed. This process is being developed by Color Agricultural Research Co. (Wisconsin) with the approval of the Food and Drug Administration. According to Delaney (1981), the daily production of 600 t of dry matter is expected. This product appears to be a competitor of soybeans for animal feed, which would ensure a very important outlet for permeates.

Some processes, however, call for only small volumes of the permeates available: such processes must lead to very valuable products in order to justify further development, and this is the case for the production of galactose and baker's yeast. Burrows (1979) has reported that fluctuations in the price, the availability and the quality of molasses would endanger the baker's yeast industry: the total replacement of molasses by permeates would require only 5–10% of the whey available. Similarly, the production of derivative compounds such as lactulose or lactobionic acid are excellent ways of using cheese whey: they call for only small volumes of whey and would never have raw material supply problems.

An intermediate situation is that of the production of single-cell proteins from permeates (30 000–35 000 t/year). The yeast cells produced are intended for two very different markets: human food applications are profitable but this market is quantitatively limited; animal feed uses would be of interest only if production costs were sufficiently low for the product to compete with other comparable protein sources. The further development of this product is difficult to predict, and will be linked with the possibilities of incorporating yeast into foods.

Another important area has opened up in the last few years following advances in the enzymatic hydrolysis of lactose. Hydrolysed whey and milk containing hydrolysed lactose are readily used in the food industry. Hydrolysed lactose syrup is another good example of the further use of permeates (Coton, 1980). The utilization conditions of these products are subject to government regulation in some countries. Luquet (1980) has reported that the use of hydrolysed lactose is unrestricted in Canada, is subject to a simple declaration in the UK and is authorized in Italy; in France, a temporary authorization has been given for some products. Hydrolysis of lactose in milk is necessary for access to some markets such as the Middle East, Mediterranean countries, and various African and Asian countries (Luquet, 1980; Alam, 1982). The novel functional properties of hydrolysed whey also opens up numerous industrial possibilities. The development of hydrolysis processes will depend on the current legislation of different cheese-producing countries.

The utilization of cheese whey and ultrafiltrate in the fermentation industries raises several problems:

1. These substrates with low dry-matter contents are excellent growth media for many micro-organisms. This feature, while constituting the essential advantage of these products, also provides the main problem with regard to their preservation and, subsequently, their transportation.
2. In many countries there is the problem of collecting whey over large geographical areas. Whereas in industrial centres, many large production units have been built often with a daily whey output of 500 000–1 500 000 litres, in other areas (particularly in mountainous regions) there are widely dispersed small units which produce about 5000 litres of whey daily. As on-the-spot treatment is impossible because of the small volumes involved, the cost of transportation of the raw material then becomes prohibitive.
3. In general, milk production is spread over the year, with some exceptions such as of ewe's milk for the production of Roquefort cheese. Thus, cheese whey is available throughout most of the year although sometimes there is a variation in the volume of production, depending on the season.

The factors listed above (dry-matter content, protein content, amounts available and geographical location) greatly influence the price of cheese whey. This dairy product can be either a source of income or a source of pollution. Cheese whey was formerly considered to be a waste product with no commercial value and for small production units it is still an expensive waste. On the other hand, for large production units, cheese whey is just one of the by-products to be obtained from milk and constitutes a source of income. The price of this raw material thus varies greatly depending on whether it is produced by small or large cheese manufacturers. From small plants, whey would be considered as having no value, however its transportation costs will be important. From large plants, the price would be superior to the existing by-products obtained from whey. In 1983, whey powder was sold at US $0·32/kg (September 1983 level); the cost of drying was about $0·27/kg, not counting transportation costs. These prices vary considerably depending on the market conditions. On this basis there is a return of 0·3 US cents for each litre of whey processed (15 ℓ of whey are needed to produce 1 kg of whey powder). Protein concentrates (35%) obtained by ultrafiltration are being sold at $0·76/kg, at a production cost of $0·63. On this basis, protein concentrates thus gave a return of 0·22 US cents per litre of whey treated (60 ℓ of whey are needed to produce 1 kg of protein concentrate). The resulting ultrafiltrate was considered to be valueless.

In comparison with other carbon substrates, whey lactose can be assimilated only by specific micro-organisms. In addition, some people cannot assimilate this sugar because of their inborn lack of β-galactosidase. It is evident that glucose and levulose are more universal substrates, permitting the multiplication of almost all kinds of organisms. Sucrose is a substrate which is not utilized by many micro-organisms, however more can grow on sucrose than on lactose. Other substrates such as starch or inulin permit the growth of only a very limited number of microbial species. It is noteworthy that cheese whey can be easily treated in order to transform lactose into lactic acid, ethanol, galactose or a mixture of glucose and galactose according to specific requirements. Cheese whey also contains sufficient proteins to constitute an excellent medium for proteolytic strains; in fact, cheese whey or whey ultrafiltrates are complete culture media. Whey and permeates contain all the minerals and trace elements required for the growth of micro-organisms. They also contain water-soluble vitamins (*Table 4*), and thus constitute an excellent basal medium for any type of culture required by the fermentation industry, whereas it appears to be difficult to produce a culture medium with wide applications from starchy materials without a preliminary treatment. For example, malt obtained from germinated barley is an expensive raw material. Beet or cane molasses currently are the most utilized substrates in biotechnology. Their supply is becoming limited and their quality varies according to the sugar extraction technology employed. These products have an advantage over cheese whey in that they are concentrated and easily transported. Other substrates such as chicory and Jerusalem artichoke juices can be used (Guiraud, 1981); these substrates are rich in inulin and proteins and constitute excellent growth media. Although Jerusalem artichoke can be a valuable substrate, depending on previous treatments such as biological hydrolysis of polyfructosan polymers, the production period is limited and storage of the raw material is necessary. Finally, the main advantage of cheese whey is its constant composition arising from its animal origin. Its competitors are plant products with composition varying according to the cultivar, the harvest date, the annual climatic variations, and pest attacks, as well as subsequent treatments and storage conditions. Moreover, cheese whey is an industrial by-product which is immediately available, and only molasses can be considered to be in the same class of raw material which does not require additional investment for production.

References

ADMUNSON, C.H. (1967). Increasing protein content of whey. *American Dairy Review* **29**, 22–23, 96–98.

ALAM, L. (1982). Effect of fermentation on lactose, glucose, and galactose content in milk and suitability of fermented milk products for lactose intolerant individuals. *Journal of Dairy Science* **65,** 346–352.

ALLON, C.K. AND HENDERSON, H.E. (1972). Ammonium salt as a source of crude protein for feedlot cattle. In *Report of Beef Cattle Research: Michigan State University Agricultural Experimental Station Report 174*, pp. 5–17. East Lansing, Michigan.

ARNAUD, M., MALIGE, B., GALZY, P. AND MOULIN, G. (1978). *Perfectionnement à la Fabrication de Levure Lactique*. French Patent No. 7830 229.

BAILY, R.B., BENITEZ, T. AND WOODWARD, A. (1982). *Saccharomyces cerevisiae* mutants resistant to catabolite repression: use in cheese whey hydrolysate fermentation. *Applied and Environmental Microbiology* **44,** 631–639.

BARRE, P., BEREAUX, J.C., LETOURNEUR, G. AND DELALANDE, P. (1977). La technologie du lactosérum en alimentation animale. In *Colloque: Les Lactosérums, une Richesse Alimentaire*, pp. 245–296. Association pour la Promotion Industrie–Agriculture, 35 rue du Général Foy, 75008, Paris.

BARRY, J.A. (1982). Alcohol production from cheese whey. *Dairy Industry International* **47,** 19–22.

BEL PROCESS (1958). *Procédé de Fabrication de Levures Alimentaires Lactiques.* French Patent No. 1 128–063.

BERNSTEIN, S. AND PLANTZ, P.E. (1977). Fermentation whey into yeast. *Food Engineering*, November, pp. 74–75.

BIROLAUD, P. (1971). Quelques données sur l'intérêt nutritif de la levure aliment et de son industrie. *Revue de l'Institut Pasteur de Lyon* **4,** 115–145.

BRULÉ, G. (1977). Hydrolyse du lactose. In *Les Lactosérums, une Richesse Alimentaire*, pp. 49–55. Association pour la Promotion Industrie–Agriculture, 35 rue du Général Foy, 75008, Paris

BRULÉ, G., ROGER, L., FAUQUANT, J. AND PIOT, M. (1980). *Procédé de Traitement d'une Matière à Base de Caséine Contenant des Phosphocaséinates de Cations Monovalents et leurs Dérivés. Produits Obtenus et Application.* French Patent No. 80.02.281

BURROWS, S. (1979). Baker's yeast. In *Economic Microbiology* (A. H. Rose, Ed.), volume 4, pp. 51–64. Academic Press, London.

CARRE, P. (1978). *Contrôle Génétique du Métabolisme du Lactose chez* Kluyveromyces *van der Walt.* Thèse Université des Sciences et Techniques du Languedoc. Montpellier, France.

CERBULIS, J.J., WOYCHIK, S.H. AND WODOLOWSKI, M.V. (1972). Composition of commercial whey. *Journal of Agricultural and Food Chemistry* **20,** 1957–1961.

CHAMPAGNE, C.P., LACHANCE, R.A. AND GOULET, J. (1980). Baker's yeast production from cheese whey. *Communication at the VIth International Symposium on yeast. London, Ontario, Canada.* (Reprints available from authors).

CHAPMAN, L.P.J. (1966). Food yeast from whey. *New Zealand Journal of Dairy Technology* **1,** 78–81.

CHAPUT, G. (1977). Le séchage des différents types de lactosérum. In *Colloque: Les Lactosérums, une Richesse Alimentaire*, pp. 25–33. Association pour la Promotion Industrie–Agriculture, 35 rue du Géneral Foy, 75008, Paris.

CHASSANG-DOUILLET, A., LADET, J., BOZE, H. AND GALZY, P. (1973). Remarques sur le métabolisme respiratoire de *Kluyveromyces fragilis* van der Walt. *Zeitschrift für allgemeine Mikrobiologie* **13,** 193–199.

CHEFTEL, J.C. AND LORIENT, D. (1982). Les propriétés fonctionnelles des protéines laitières et leur amélioration. *Le Lait* **62,** 435–483.

CHEN, H.C. AND ZALL, R.R. (1982). Continuous fermentation of whey into alcohol using an attached film expanded bed reactor. *Process Biochemistry* January, pp. 20–25.

CLARK, E. (1921). Preparation of galactose. *Journal of Biological Chemistry* **47,** 1–2

CLARK, W.S. (1979). Major whey product market. *Journal of Dairy Science* **62,** 96–98.

COOPER, D.G., KENNEDY, K.J., GERSON, D.F. AND ZAJIC, J.E. (1978). The production of extracellular galactose by *Arthrobacter globiformis* using lactose as substrate. *Journal of Fermentation Technology* **56,** 550–553.

COTON, S.G. (1980). Whey technology. The utilization of permeates from the ultra-filtration of whey and skim milk. *Journal of the Society of Dairy Technology* **33,** 89–94.

COTON, S.G. (1983). Utilization of whey and ultrafiltration permeate. In *Proceedings of the 1983/1984 Workshop on Production and Feeding of Single-Cell Protein* (Ferranti and Fiechter, Eds), pp. 135–146. Applied Science Publishers, New York.

COULMAN, G.A.R., STIERBER, W.S. AND GERHARDT, P. (1977). Dialysis continuous process for ammonium lactate fermentation of whey: mathematical model and computer simulation. *Applied and Environmental Microbiology* **34,** 725–732.

CRICKENBERGER, R.G., HENDERSON, H.E. AND REDDY, C.A. (1981). Fermented ammoniated condensed whey as a crude protein source for feedlot cattle. *Journal of Animal Science* **52,** 677–687.

DAHLQUIST, A.H., ASP, N.G., BURVALL, A. AND HAUSSING, H. (1977). Hydrolysis of lactose in milk and whey with minute amount of lactase. *Journal of Dairy Research* **44,** 541–548.

DE LA GUÉRIVIÈRE, J.F. (1977). Mise en oeuvre des lactosérums dans les industries de cuisson des produits céréaliers. In *Colloque: Les Lactosérums, une Richesse Alimentaire*, pp. 319–334. Association pour la Promotion Industrie–Agriculture, 35 rue de Général Foy, 75008, Paris.

DE LA GUÉRIVIÈRE, J.F. (1978). Acquis technologiques en vue de la valorisation des lactosérums hydrolysés dans certains secteurs de l'alimentation humaine. In *Journée d'étude—Hydrolyse du lactose—Technologie—Produits nouveaux*, pp. 111–112. Association pour la Promotion Industrie–Agriculture, 35 rue du Général Foy, 75008 Paris.

DE LA GUÉRIVÈRE, J.F. (1981). Les levures lactiques cultivées sur lactosérum déprotéiné. *La Technique Laitière* **952,** 89–92.

DELANEY R.A.M. (1981). Recent developments in the utilization of whey. *Cultured Dairy Products Journal* **16**, 11–17, 20–22.

DEMAINAY, M., AND BARON, C. (1978). Isomérisation du lactose en solution acqueuse et du lactosérum sur résine échangeuses d'ions. *Le Lait 575-579*, 234–245.

DEMMLER, G. (1950). Production of yeast from whey using the Waldhof method. *Milchwissenschaft* **4,** 11–17.

DEMOTT, B.J., DRAUGHON, F.A. AND HERALD, P.J. (1981). Fermentation of lactose in direct acid set cottage cheese whey. *Journal of Food Protection* **44,** 588–590.

DEWITT, J.N. AND DE BOER, R. (1975). Ultrafiltration of cheese whey and some functional properties of the resulting whey protein concentrate. *Netherland Milk Dairy Journal* **29,** 198–201.

DEWITT, J.N. AND HONTELEZ-BACKX, (1981). Les propriétés fonctionnelles des protéines du lactosérum: conséquences des traitements thermiques. *La Technique Laitière* **952,** 19–22.

DICKER, R. (1982). The uses of hydrolyzed whey in food products. *Food Trade Review*, June, pp. 295–297.

DICKSON, R. C. AND SHEETZ, M. (1981). LAC_4 is the structural gene for β-galactosidase in *Kluyveromyces lactis*. *Genetics* **98,** 729–745.

DION, P., GOULET, J. AND LACHANCE, R.A. (1978). Transformation du lactosérum déprotéine en milieu de culture pour la levure de boulangerie. *Journal de l'Institut Canadien de Science et Technologie Alimentaire* **11,** 78–81.

DOHAN, L.A. AND BARET, J.L. (1980). Lactose hydrolysis by immobilized lactase: semi industrial experience. *Enzyme Engineering* **5,** 279–293.

DUTROW, D.A., HUBER, S.T. AND HENDERSON, H.E. (1974). Comparison of ammonium salts and urea in rations for lactating dairy cows. *Journal of Animal Science* **38,** 1304–1308.

FALLICK, G.F. (1969). Industrial ultrafiltration. *Process Biochemistry* **4,** 29–34.

FÉVRIER, C., and AUMAITRE, A. (1981). Utilisation du lactose par les animaux d'élevage. In *Journée d'étude: Lactose–Galactose*, pp. 18–24. Centre National de la Recherche Scientifique, Paris.

FÉVRIER, C. AND BOURDIN, C. (1977). Utilisation du lactosérum et des produits lactosés par les porcins. In *Colloque: Les Lactosérums, une Richesse Alimentaire*, pp. 129–174. Association pour la Promotion Industrie–Agriculture, 35 rue du Général Foy, 75008 Paris.

FLEURY, H. AND HENRIET, P. (1977). Culture de champignons filamenteux. Traitement du lactosérum par le procédé Caliqua/Sireb. In *Colloque: Les Lactosérums, une Richesse Alimentaire,* pp. 101–116. Association pour la Promotion Industrie–Agriculture, 35 rue du General Foy, 75008, Paris.

FORSUM, E. (1974). Nutritional evaluation of whey protein concentrates and their fraction. *Journal of Dairy Science* **57,** 665–670.

FRIEND, B.A., CUNNINGHAM, M.L. AND SHAHANI, K.M. (1982). Industrial alcohol production via whey and grain fermentation. *Agricultural Wastes* **4,** 55–63.

FUENTES, J.L. (1981). *Etude Génétique et Physiologique de Mutants 'Galactose Négatif' de* Kluyveromyces lactis. Thèse Université des Sciences et Techniques du Languedoc, Montpellier, France.

GALZY, P., AND MOULIN, G. (1976). *Procédé de Préparation de Galactose et de Boisson à base de Galactose à partir de Solution contenant du Lactose.* US Patent No. 3, 98, 773.

GAWEL, J. AND KOSIKOWSKI, F.V. (1978). Improving alcohol fermentation in concentrated ultrafiltration permeate of cottage cheese whey. *Journal of Food Science* **43,** 1717–1719.

GERHARDT, P. AND REDDY, C.A. (1978). Conversion of agroindustrial wastes into ruminant feedstuff by ammoniated organic acid fermentation: a brief review and preview. *Developments in Industrial Microbiology* **19,** 71–78.

GIACIN, J.R., JAKUBOWSKI, J., JEEDER, J.G., GILBERT, S.G. AND KLEYN, D.H. (1974). Characterization of lactase immobilized on collagen: conversion of whey lactose by soluble and immobilized lactase. *Journal of Food Science* **39,** 751–756.

GIÉC, A. AND KOSIKOWSKI, F.V. (1982). Activity of lactose-fermenting yeast in producing biomass from concentrated whey permeate. *Journal of Food Science* **47,** 1992–1993/1907.

GOUDEFRANCHE, H., MAUBOIS, J.L., DUCRUET, P. AND MAHAUT, M. (1981). Utilisation de nouvelles membranes d'ultrafiltration pour la fabrication de fromages type St Paulin. *La Technique Laitière* **951,** 7–13.

GRANDADAM, Y. (1981). L'actuel et le futur de divers produits alimentaires utilisant le lactosérum. *La Technique Laitière* **952,** 99–101.

GREENBERG, N.A. AND MAHONEY, R.R. (1981). Immobilization of lactase (β-galactosidase) for use in dairy processing: a review. *Process Biochemistry*, Feb./March, pp. 2–8, 49.

GUIRAUD, J. (1981). *Utilization des Levures pour la Valorisation Industrielle des Polyfructosanes de Types Inulines.* Thesis USTL, Montpellier, France.

GUY, E.J., VETTEL, H.E. AND PALLANCHE, M.J. (1966). Utilization of dry cottage cheese whey. *Journal of Dairy Science* **49,** 694–699.

GYURIESEK, D.M. AND THOMPSON, M.P. (1976). Hydrolyzed lactose culture dairy product. Manufacture of yoghurt, buttermilk and cottage cheese. *Cultured Dairy Products Journal* August, pp. 12–13.

HALTER, N., PUHAN, Z. AND KAPPELI, O. (1981). Upgrading of milk U.F. permeate by yeast fermentation. In *Advances in Biotechnology* (M. Moo-Young, Ed.), Volume 2, pp. 351–356. Pergamon Press, New York.

HARRIES, J.T. (1982). Disorders of carbohydrate absorption. In *Familial Inherited Abnormalities.* (J.T. Harris, Ed.). *Clinics in Gastroenterology,* Volume 11, pp. 17–30. W.B.S. Saunders, London.

HAVEWALA, N.B. AND WEETALL, H.H. (1972). Continuous production of dextrose from corn-starch. A study of reactor parameters necessary for commercial application. In *Enzyme Engineering.* (L.B. Wingard, Ed.), pp. 244–266. Wiley, New York.

HAYASHIBARA, K. (1976). U.S. Patent 3,962,335.

HICKS, K.B. AND PARRISH, F.W. (1980). A new method for the preparation of lactulose from lactose. *Carbohydrate Résearch* **82,** 393–397.

HOLSINGER, V.H., POSATI, L.P. AND DE VILBISS, E.D. (1974). Whey beverages: A review. *Journal of Dairy Science* **57**, 849–859.

HOLSINGER, V.H., SUTTON, C.S., VETTEL, H.E., ALLEN, C. AND TALLEY, F.B. (1977). Acceptability of whey soy drink mix prepared with cottage whey. *Journal of Dairy Science* **60**, 1841–1845.

HUMBERT, G. AND ALAIS, C. (1981a). Possibilité d'application au lactosérum de nouveaux procédés de précipitation ou de fractionnement des protéines. Procédés de précipitation non thermique des protéines du lactosérum. *La Technique Laitière* **952**, 41–43.

HUMBERT, G. AND ALAIS, C. (1981b). Nouvelles voies de valorisation des protéines lactosériques. Produits gélifiés sucrés aux protéines lactosériques. *La Technique Laitière* **952**, 173–174.

HYND, J. (1980). Drying of whey. *Journal of the Society of Dairy Technology* **33**, 52–54.

IZAGUIRRE, M.E. AND CASTILLO, F.J. (1982). Selection of lactose fermenting yeast for ethanol production from whey. *Biotechnology Letters* **4**, 257–262

JANSEN, H.C. (1945). *Method for Preparation of Alkali Lactates, especially Ammonium Lactate, through Lactic Acid Fermentation of Sugar-Containing Solutions*. Dutch Patent 57, 848.

JOHNSON, S.M., WEISBERG, J., JOHNSON, J. AND PARKER, M.E. (1937). US Patent 207, 1346.

JUENGST, F.W., JR (1979). Use of total whey constituents for animal feed. *Journal of Dairy Science* **62**, 106–111.

KELLER, A.K. AND GERHARDT, P. (1975). Continuous lactic acid fermentation of whey to produce a ruminant feed supplement high in crude protein. *Biotechnology and Bioengineering* **17**, 997–1018.

KOSIKOWSKI, F.V. (1967). Greater utilization of whey powder for human consumption and nutrition. *Journal of Dairy Science* **50**, 1343–1348.

KOSIKOWSKI, F.V. (1975). A new-type acid whey concentrated product derived from ultrafiltration. *Journal of Dairy Science* **58**, 792–793.

KOSIKOWSKI, F.V. (1979). Whey utilization and whey products. *Journal of Dairy Science* **62**, 1149–1160.

KOSIKOWSKI, F.V. (1981). Boisson de lactosérum ayant une valeur potentielle. *La Technique Laitière* **952**, 93–97.

KOSIKOWSKI, F.V. AND WZOZECK, W. (1977). Whey wines from concentrates of reconstituted acid whey powder. *Journal of Dairy Science* **60**, 1982–1986.

KREVELA, A. (1969). Growth rates of lactose crystals in solutions of stable anhydro lactose. *Netherland Milk Dairy Journal* **23**, 258–275.

LADET, J., MOULIN, G., GALZY, P., JOUX, J.L. AND BIJU-DUVAL, F. (1972). Comparaison des rendements de croissance sur lactose de quelques *Kluyveromyces* van der Walt. *Le Lait 519-520*, 613–621.

LAHAM-GUILLAUME, M., MOULIN, G. AND GALZY P. (1979). Sélection de souches de levures en vue de la production d'alcool sur lactosérum. *Le Lait* **59**, 489–496.

LEFRANÇOIS, L. (1964). Problèmes de l'aération et de la circulation dans les cuves de fermentations aérobies. *Industries Alimentaires et Agricoles* **1**, 3–18.

LE HENAFF, L. (1978). Hydrolyse du lactose sur résine catalytique. In *Journée d'Étude. Hydrolyse du Lactose. Technologies–Produits Nouveaux*, pp. 29–39. Association pour la Promotion Industrie–Agriculture, 35 rue de Général Foy, 75008 Paris.

LEMBKE, A. AND BAADER, W. (1975). Sonderheft. *Berichte der Landwirtschaft* **192**, 903–907.

LENOIR, J. (1981). Le lactosérum, source de lactose. *Médecine et Nutrition* **17**, 201–206.

LODDER, J. (1970). *The Yeast: a Taxonomic Study*. North Holland Publishing Company. Amsterdam.

LOMBREZ, R. (1978). Aspect technico-économiques de l'emploi des dérivés laitiers à lactose hydrolysés. In *Journée d'Étude: Hydrolyse du Lactose—Technologies. Produits Nouveaux*, pp. 87–95. Association pour la Promotion Industrie–Agriculture, 35 rue du Général Foy, 75008, Paris.

LONGUET, R. (1977). Le transport et le stockage des lactosérums concentrés. In *Colloque: Les Lactosérums, une Richesse Alimentaire*, pp. 15–24. Association pour la Promotion Industrie–Agriculture, 35 rue du Général Foy, 75008, Paris.

LUQUET, J.F. (1980). L'emploi du lactosérum enfin autorisé. Aspect règlementaire. *La Technique Laitière* **950,** 23–25.

LYONS, T.P. and CUNNINGHAM, J.D. (1980). Fuel alcohol from whey. *American Dairy Review* **42,** 42A-42E.

MAHAUT, M., MAUBOIS, J.L., ZINCK, A., PANNETIER, R. AND VEYRE, R. (1982). Elément de fabrication de fromages frais par ultrafiltration sur membrane de coagulum de lait. *La Technique Laitière* **961,** 9–13.

MANN, E.J. (1982). Whey utilization in foods. *Dairy Industry International* **47,** 22–23.

MARCONI, W., BARTOLI, F., MORISI, F. AND MARIANI, A. (1979). Improved whey treatment by immobilized lactase. In *Enzyme Engineering* (H.H. Weetall and G.P. Royer, Eds), volume 5, pp. 269–278. Plenum Press, New York

MARZOLF, J.J. (1977). Point sur la situation actuelle de la production de levure cultivée sur lactosérum. In *Colloque: les Lactosérums, une Richesse Alimentaire*, pp. 73–82. Association pour la Promotion Industrie–Agriculture, 35 rue du Général Foy, 75008 Paris.

MARZOLF, J. J. (1981). La levure de lactosérum. In *Journée d'étude. Lactose–Galactose.* pp. 60–63. Centre National de la Recherche Scientifique, Paris.

MATHAN, V.I., BAKER, J.J., SOOD, S.K., RAMACHANDRAU, K. AND RAMATINGASWANI, V. (1979). The effect of ascorbic acid and protein supplementation on the reponse of pregnant women to iron, pteroylglutamic acid and cyanocobalamin therapy. *British Journal of Nutrition* **42,** 391–398.

MATHUR, B.N. AND SHAHANI, K.M. (1979). Use of total whey constituents for human food. *Journal of Dairy Science* **62,** 99–105.

MAUBOIS, J.L. (1979). Le procéde M.M.V. est désormais applicable à la fabrication industrielle du St Paulin. *Technicien du Lait* **12,** 934–939.

MAUBOIS J.L. (1980). Ultrafiltration of whey. *Journal of the Society of Dairy Technology* **35,** 55–58.

MAUBOIS, J.L. AND BRULÉ, G. (1982). Utilisation des techniques à membranes pour la séparation, la purification, la fragmentation des protéines laitières. *Le Lait* **62,** 484–510.

MAUBOIS, J.L., BRULÉ, G. AND GOURDON, P. (1981). L'ultrafiltration du lactosérum optimisation de la technologie du perméat. *La Technique Laitière* **952,** 29–33.

MAUBOIS, J.L., MOCQUOT, G. AND VASSAL, L. (1969). *Prétraitement du Lait et de Sous-produits Laitiers.* French Patent 2, 052, 121.

MAVROPOULOU, I.P. AND KOSIKOWSKI, F.V. (1973a). Composition, solubility and stability of whey powders. *Journal of Dairy Science* **56,** 1128–1134.

MAVROPOULOU, I.P. AND KOSIKOWSKI, F.V. (1973b). Free amino acids and soluble peptides of whey powders. *Journal of Dairy Science* **56,** 1135–1138.

MÉRÉO, I. (1971). Les utilisations industrielles de sérum de fromagerie. *Industries Alimentaires et Agricoles*, April, pp. 817–823.

MEYRATH, J. AND BAYER, K. (1979). Biomass from whey. In *Economic Microbiology* (A.H. Rose, Ed.), volume 4, pp. 207–269. Academic Press, London.

MIRABEL, B. (1978a). Nouveau procédé d'extraction des protéines du lactosérum. *Annales Nutrition Alimentation* **32,** 243–248

MIRABEL, B. (1978b). Nouveau procédé de valorisation du lactosérum. *Information Chimie* **175,** 105–109.

MIRABEL, B. (1981). Possibilité d'application au lactosérum de nouveaux procédés de précepitation ou de fractionnement des protéines. *La Technique Laitière* **952,** 37–40.

MODDLER, H.W. AND EMMONS, D.B. (1977). Properties of whey protein concentrate prepared under acidic conditions. *Journal of Dairy Science* **60,** 177–184.

MODDLER, H.W. AND HARWALKAR, V.R. (1981). Whey protein concentrate prepared under acidic conditions. *Milchwissenschaft* **36,** 537–542.

MOEBUS, O. AND KIESBYE, P. (1975). *Continuous Process for Producing Yeast Protein and Baker's Yeast.* G.F.R. Patent Application 2, 410, 349.

MOLINARO, R., HONDERMARCK, J.C. AND JACQUOT, L. (1977). Production de lacto-protéines levurées. In *Colloque: Les Lactosérums, une Richesse Alimentaire*, pp. 83–93. Association pour la Promotion Industrie–Agriculture, 35 rue du Général Foy, 75008 Paris.

MOULIN, G. AND GALZY, P. (1976). Une possibilité d'utilisation du lactosérum: La production de levure. *Industries Alimentaires et Agricoles* **11,** 1337–1343.

MOULIN, G. AND GALZY, P. (1981). Alcohol production from whey. In *Advances in Biotechnology* (M. Moo Young, Ed.) Volume 2, pp. 181–189. Pergamon Press, New York.

MOULIN, G., BOZE, H. AND GALZY, P. (1980). Inhibition of alcoholic fermentation by substrate and ethanol. *Biotechnology and Bioengineering* **22,** 2375–2381.

MOULIN, G., LEGRAND, M. AND GALZY, P. (1983). The importance of residual aerobic fermentation in aerated medium for the production of yeast from glucidic substrates. *Process Biochemistry* **18**(5), 5.

MOULIN, G., MALIGE, B. AND GALZY, P. (1981). Etude physiologique de *Kluyveromyces fragilis*: conséquence sur la production de levure sur lactosérum. *Le Lait* **61,** 323–332.

MOULIN, G., MALIGE, B. AND GALZY, P. (1983). Balanced flora of an industrial fermenter. Production of yeast from whey. *Journal of Dairy Science* **66,** 21–28.

MOULIN, G., RATOMAHENINA, R. AND GALZY, P. (1976). Sélection de levure en vue de la culture sur lactosérum. *Le Lait 553-554*, 135–142.

MOULIN, G., VARCHON, P. AND GALZY, P. (1977). Une nouvelle utilisation possible du lactosérum: La préparation de boisson à base de galactose. *Industries Alimentaires et Agricoles* **1,** 29–34.

MYKKANEN, H.M. AND WASSERMAN, R.H. (1980). Enhanced absorption of calcium by casein phosphopeptides in rachitic and normal chicks. *Journal of Nutrition* **110,** 2141–2148.

NICKERSON, T.A. (1970). Lactose. In *By-products from Milk* (B.M. Webb and O.Whittier, Eds), pp. 357–380. The Avi Publishing Co., Westport, Connecticut, USA.

NICKERSON, T.A. (1979). Lactose chemistry. *Journal of Agricultural and Food Chemistry* **27,** 672–677.

NIPJELS, H.N. AND RHEINLANDER, P.M. (1982). Laktose. Hydrolyse von Molke. *Deutsche milchwirtschaft* **33,** 529–538.

NIPJELS, J. (1978). Technique d'hydrolyse du lactose avec la lactase maxilac. In *Journée d'étude, Hydrolyse du Lactose—Technologie—Produits Nouveaux*, pp. 9–28. Association pour la Promotion Industrie–Agriculture, 35 rue du Général Foy, 75008 Paris.

NOUR-EL-DIEN, H. AND HALASZ, A. (1982a). Attempts to utilize whey for the production of yeast protein. Part II. Effect of biotin concentration and whey content at constant lactose concentration. *Acta Alimentaria* **11,** 11–19.

NOUR-EL-DIEN, H. AND HALASZ, A. (1982b). Attempts to utilize whey for the production of yeast protein. Part III. Effects of some vital growth factors. *Acta Alimentaria* **11,** 125–134.

NOUR-EL-DIEN, H., HALASZ, A. AND LENGYEL, Z. (1981). Attempts to utilize whey for the production of yeast protein. Part I. Effect of whey concentration of ammonium sulfate and of phosphate. *Acta Alimentaria* **10,** 11–25.

O'LEARY, V.S. AND WOYCHIK, J.H. (1976). Comparison of some chemical properties of yoghurt made from control and lactase treated milk. *Journal of Food Science* **41,** 891–893.

O'LEARY, V.S., GREEN, R., SULLIVAN, B.C. AND HOLSINGER, V.H. (1977a). Alcohol production by selected yeast strains in lactose hydrolyzed acid whey. *Biotechnology and Bioengineering* **19,** 1019–1035.

O'LEARY, V.S., SUTTON, C., BENCIVENGO, M., SULLIVAN, B. AND HOLSINGER, V.H. (1977b). Influence of lactose hydrolysis and solids concentration on alcohol production by yeast in whey ultrafiltrate. *Biotechnology and Bioengineering* **19,** 1689–1707.

OLLING, C.J. (1963). Composition of Friesian whey. *Netherlands Milk Dairy Journal* **17,** 177–181.

PALMER, D.E. (1977). High purity protein recovery. *Process Biochemistry* **12,** 24–28.

PAQUET, D. (1981). Nouvelles voies de valorisation des protéines lactosériques: Produits moussants succédanés du blanc d'oeuf. *La Technique Laitière* **952,** 69–71.

PEPPER, D. (1981). Whey concentration by reverse osmosis. *Dairy Industry International* **46,** 24–25.

PORGES, N., PEPINSKI, J.B. AND JASEWICZ, L.Z. (1951). Feedyeast from dairy by-product. *Journal of Dairy Science* **34,** 615–621.

RAMET, J.P. (1978). Application de l'hydrolyse enzymatique du lactose en fromagerie. In *Journée d'étude—Hydrolyse du Lactose—Technologie—Produits Nouveaux*, pp. 97–102. Association pour la Promotion Industrie–Agriculture, 35 rue du Général Foy, 75008, Paris.

REDDY, C.A., HENDERSON, H.E. AND ERDMAN, M.D. (1976). Bacterial fermentation of cheese whey for production of a ruminant feed supplement rich in crude protein. *Applied and Environmental Microbiology* **32,** 769–776.

REESEN, L. AND STRUBE, R. (1978). Complete utilization of whey for alcohol and methane production. *Process Biochemistry* November, 21–24.

RICHERT, H.S. (1979). Physico-chemical properties of whey protein foams. *Journal of Agricultural and Food Chemistry* **27,** 665–667.

RICHMOND, M.L., GRAY, J.I. AND STINE, C.M. (1981). β-galactosidase: a review of recent research related to technological application, nutritional concerns and immobilization. *Journal of Dairy Science* **64,** 1759–1771.

ROBE, K. (1964). Wheat Process. *Food Processing* **25,** 95–99.

ROGER, L. (1979). *Contribution à la Recherche d'une Meilleure Utilisation en Alimentation Humaine des Composants Glucidiques et Protéiques du Lactosérum grâce à l'Emploi des Techniques à Membranes*. Thèse, Ecole Nationale Supérieure Agronomique, Rennes, France.

ROGER, L. AND MAUBOIS, J.L. (1981). Actualité dans le domaine des technologies à membrane pour la préparation et la séparation des protéines laitières. *Revue Laitière Française* **400,** 67–75.

ROGER, L., BRULÉ, G. AND MAUBOIS, J.L. (1981). Nouvelles voies de valorisation des protéines lactosériques. Hydrolyse des protéines de lactosérum. Intérêt Thérapeutique. *La Technique Laitière* **952,** 65–67.

ROGOSA, M., BROWN, H.M. AND WHITTIER, E.O. (1947). Ethyl alcohol from whey. *Journal of Dairy Science* **30,** 263–269.

SALMON, M. (1981). Extraction et valorisation du lactose. Produits déminéralisés à lactose hydrolysé. *La Technique Laitière* **952,** 85–88.

SCEES (1981). *Industrie Laitière. Production, Collecte et Transformation, année 1979.* Service Central des Enquêtes et Études Statistiques. Avenue de Saint Maudé, 75570 Paris Cedex.

SCHINGOETHE, D.J. (1976). Whey utilization in animal feeding: a summary and evaluation. *Journal of Dairy Science* **59,** 556–570.

SHUKLA, T.P. (1975). Betagalactosidase technology: a solution to the lactose problem *C.R.C. Critical Reviews in Food Technology* **5,** 325–356.

SHEETZ, R.M. AND DICKSON, R.C. (1980). Mutation affecting synthesis of β-galactosidase activity in the yeast *Kluyveromyces lactis*. *Genetics* **95,** 877–890.

SINGH, V., HSU, C.C., CHEN, D.C. AND TZENG, C.H. (1983). Fermentation processes for dilute food and dairy wastes. *Process Biochemistry*, March/April, pp. 13–17/21.

STEWART, J.A., MULLER, U. AND GRIFFIN, A.T. (1974). Use of whey solids in calf feeding. *Australian Journal of Dairy Technology* **29,** 53–58.

STIERBER, R.W. AND GERHARD, I.P. (1979). Dialysis continuous process for ammonium lactate fermentation improved mathematical model and use of deproteinized whey. *Applied and Environmental Microbiology* **37,** 487–495.

STINEMAN, T.L. (1978). *Process for the Treatment of Acid Whey to Produce* Saccharomyces *Yeasts and Process for Growing* Saccharomyces *Yeasts on Treated Acid Whey*. US Patent 001, 2501.

SOCIÉTÉ DES ALCOOLS DU VEXIN (1963). *Méthode et Equipement pour le Traitement du Lactosérum*. French Patent 80, 198.

SOCIÉTÉ PRÉVAL (1971). *Procéde de Préparation du Galactose*. French Patent 71, 27, 592.

SUPPLEE, G.C. (1940). Whey as a source of vitamins and vitamin product. *Industrial Engineering* **32,** 238–243.

THIVEND, P., VERMOREL, D. AND GUILHERMET, R. (1977). Utilisation du lactosérum et de ses dérivés par les bovins et les ovins. In *Colloque: Les Lactosérums, une Richesse Alimentaire*, pp. 225–243. Association pour la Promotion Industrie–Agriculture, 35 rue de Général Foy, 75008, Paris.

THOMPSON, M.P. AND GYURIESEK, D.M. (1974). Manufacture of Cheddar cheese from hydrolysed lactose milk. *Journal of Dairy Science* **57,** 598–602.

TOULLEC, R. AND LE TREUT, J.H. (1977). Utilisation du lactosérum et de ses dérivés dans l'alimentation du veau préruminant. In *Colloque: Les Lactosérums, une Richesse Alimentaire*, pp. 187–210. Association pour la Promotion Industrie–Agriculture, 35 rue du Général Foy, 75008, Paris.

TOULLEC, R., FRAUTZEN, J.F., MAUBOIS, J.L. AND PION, R. (1975). Utilisation digestive par le veau préruminant des protéines de lactosérum traitées par ultrafiltration sur membrane. *Technicien du Lait* **828,** 15–21, 51.

VAHERI, M. AND KAUPPINEN, V. (1978). The formation of lactulose (4-*O*-β-galactopyranosyl fructose) by β-galactosidase. *Acta pharmaceutica fennica* **87,** 75–83.

VAN VELTHUIJSEN, J.A. (1979). Food additives derived from lactose: lactitol and lactitol palmitate. *Journal of Agricultural and Food Chemistry* **27,** 680–686.

VARNER, L.W. AND WOODS, W. (1971). Influence of ammonium salts of volatile fatty acid upon ration on digestibility rumen fermentation and nitrogen retention by steers. *Journal of Animal Science* **33,** 110–117.

VOISIN, M. (1981). Concentration de lactosérum: évaporation simple effet avec re-compression mécanique de vapeur. *La Technique Laitière* **952,** 47–49.

VRIGNAUD, Y. (1971). Levure lactique Bel. *Revue de l'Institut Pasteur Lyon* **2,** 1147–1165.

WAESER, B. (1944). *Chemiker Zeitung* **7,** 120–125.

WALKER, J. (1982). Production of fuel grade ethanol from soft drink bottling wastes. *Beverage Industry*, November, pp. 157–160, 163.

WASSERMAN, A.E. (1960). The rapid conversion of whey to yeast. *Dairy Engineering* **77,** 374–379.

WASSERMAN, A.E. (1961). Amino-acid and vitamin composition of *Saccharomyces fragilis* grown in whey. *Journal of Dairy Science* **44,** 379–386.

WHITTIER, E.O. AND ROGERS, L.A. (1931). Continuous fermentation in the production of lactic acid. *Industrial Engineering Chemistry* **23,** 532–534.

ZADOW, G. (1979). Modification of whey and whey component. *New Zealand Journal of Dairy Science and Technology* **14,** 131–138.

13
Viruses as Pest-Control Agents

J. B. CARTER

Department of Biology, Liverpool Polytechnic, Byrom Street, Liverpool L3 3AF, UK

Introduction

Pressures from a variety of sources are causing man to investigate alternatives to the chemical pesticides which have been used so widely during the past few decades. Pressures are brought to bear by environmentalists concerned about the effects of pesticides on wildlife, by pest-control experts concerned about the effects of these pesticides on parasites and predators of the pests and about the increasing resistance of the pests to the pesticides, by consumers concerned about toxic residues in food, and by public health officials concerned about human poisoning. In Sri Lanka, for example, more people die of pesticide poisoning than of malaria (Matthews, 1983). Furthermore, research, development and production costs for chemical pesticides have soared, making them expensive in the developed nations, while in the developing nations, if pesticides are used at all, farmers select the least expensive—which are usually the most toxic. 'Biological' control strategies, including the use of pathogens of pests, attempt to circumvent most of these problems.

Viruses have been used to control mites (Reed, 1981) and rabbits (Fenner, 1983) but this review is concerned principally with the insect-pathogenic baculoviruses (BVs). Insects are hosts to a wide variety of viruses, including picornaviruses, parvoviruses and poxviruses. Each of these groups also has representatives infecting vertebrate animals. Attention has been focused on the BVs as pesticidal agents because of their lack of similarity to any viruses of hosts other than invertebrates. There is some logic in this approach, but other groups of insect-pathogenic viruses should not be ignored. It may be that the host spectra of other virus groups, such as the cytoplasmic polyhedrosis viruses and the iridescent viruses, are restricted to invertebrates even though viruses with similar morphologies and biochemical characteristics infect vertebrates (and higher plants in the case of iridescent viruses). In fact, a cytoplasmic polyhedrosis virus is used in Japan against the pine caterpillar, *Dendrolimus*

Abbreviations: BV, baculovirus; DNA, deoxyribonucleic acid; ELISA, enzyme-linked immunosorbent assay; FP, few polyhedra; GV, granulosis virus; IB, inclusion body; IPM, integrated pest management; LD_{50}, median lethal dose; LT_{50}, median lethal time; MNPV, multiple nucleocapsids per virion envelope; MP, many polyhedra; NPV, nuclear polyhedrosis virus; REN, restriction endonuclease; RNA, ribonucleic acid; SDS-PAGE, sodium dodecyl sulphate–polyacrylamide gel electrophoresis; SNPV, single nucleocapsid per virion envelope; UV, ultra-violet.

Biotechnology and Genetic Engineering Reviews—Vol. 1, February 1984
0264–8725/84/01/375–45 \$10.00 + \$0.00 © Intercept Ltd

spectabilis (Aizawa, 1976), and an iridescent virus has been tested against leather-jackets (*Tipula* spp. larvae), although with disappointing results (Carter, 1978).

Insect viruses are in use, or are being considered for use, in forestry, horticulture and agriculture, including grassland. Entwistle (1983) listed 31 lepidopteran, 6 hymenopteran and one coleopteran pest species for which control with BVs has been demonstrated to be feasible or highly likely. There is little or no current effort to apply viruses for control of disease vectors or of stored product and timber pests.

If a virus is to be considered seriously as a pest-control agent then detailed knowledge of the virus, its host, and their interactions with the environment must be amassed. Information is required on the structural and biochemical characteristics of the virus, its host spectrum, and median lethal doses (LD_{50}s) and median lethal times (LT_{50}s) for different stages of the host(s). The habits and life cycle of the host, and the mechanisms whereby the virus persists and spreads in the field (epizootiology) must be understood. Techniques for mass production and purification of the virus must be developed and it must be shown to be safe for man and other non-target organisms. This chapter considers all these aspects and discusses the advantages and disadvantages of using insect-pathogenic viruses as pesticides. Industrial aspects are discussed only briefly as they will be considered in detail in a subsequent volume. Recent reviews on the use of viruses as pest-control agents include those of Falcon (1982), Payne (1982) and Entwistle (1983).

The baculoviruses

Only an outline of the structure and replication of BVs is given here. For more detailed accounts the reviews of Harrap and Payne (1979), Granados (1980a) and Kelly (1982) should be consulted.

The BVs have been classified into three subgroups according to whether or not the virions become embedded (occluded) in inclusion bodies (IBs), and, if so, on the size and shape of the IB (Matthews, R.E.F., 1982). Details of the subgroups are presented in *Figure 1* and *Table 1*. The rod-shaped virions are enveloped. Those of the granulosis viruses (GVs) become occluded in capsule-shaped IBs (granules), whereas the nuclear polyhedrosis virus (NPV) IBs (polyhedra) are polyhedral, cuboidal, or 'orange segment-shaped', depending on the virus. The NPVs are subdivided into those in which each occluded virion has a single nucleocapsid per virion envelope (SNPVs) and those in which the occluded virions have multiple nucleocapsids per virion envelope (MNPVs).

STRUCTURE

The virion

BV virions have been found to contain up to 35 proteins (Vlak, 1979), some of which are glycosylated (Dobos and Cochran, 1980) and/or phosphorylated (Tweeten, Bulla and Consigli, 1980). The nucleocapsid consists of a protein capsid containing DNA and further proteins (*Figure 1*). Its dimensions fall

Table 1. Baculovirus subgroups.

Subgroup	Inclusion body dimensions	Number of virions per inclusion body	Single or multiple nucleocapsids per virion envelope	Hosts
Nuclear polyhedrosis viruses	0·8–15 μm diameter	Many		
1. MNPVs			Multiple (1–5 usually; up to 39)	Lepidoptera
2. SNPVs			Single	Lepidoptera, Diptera, Hymenoptera, Trichoptera, Coleoptera, Neuroptera, Crustacea
Granulosis viruses	approx. 200 × 500 nm	1, usually	Single	Lepidoptera
Non-occluded baculoviruses	No inclusion bodies formed		Single	Coleoptera, Diptera (possibly) Mites, Crustacea

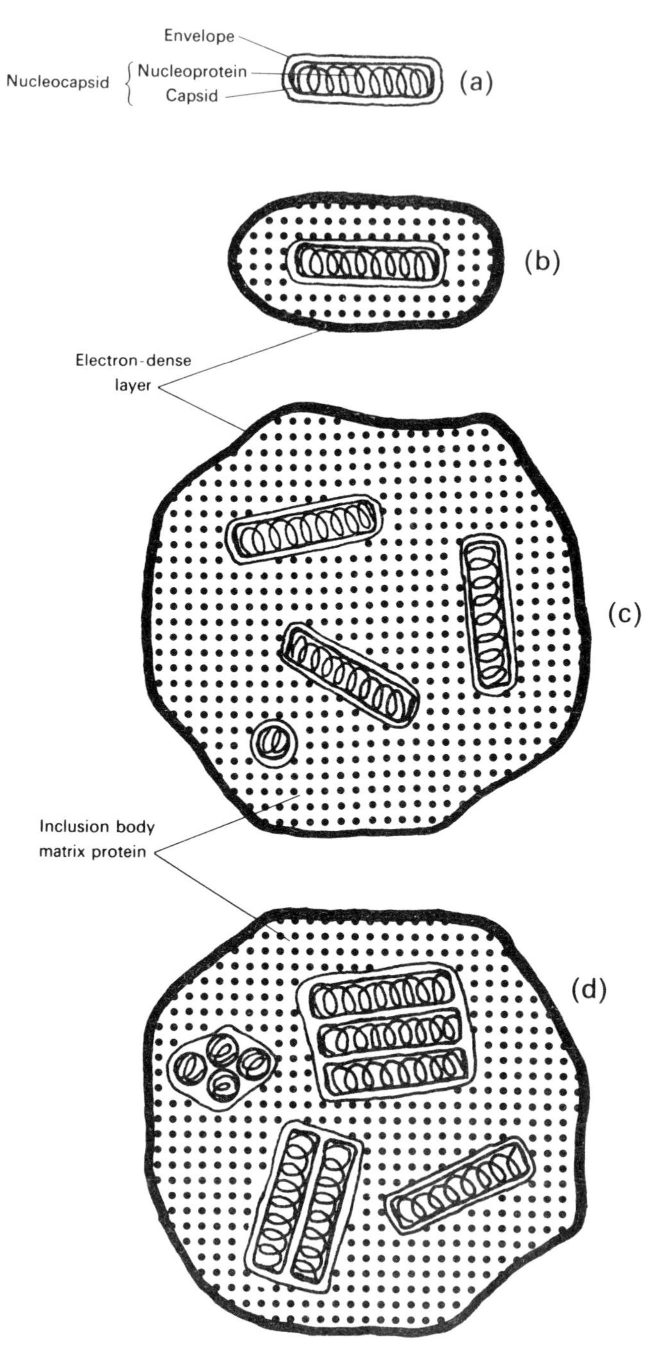

Figure 1. Baculovirus subgroups. (a) Non-occluded baculovirus virion. (b) Granulosis virus inclusion body. (c) Nuclear polyhedrosis virus inclusion body; singly enveloped nucleocapsids (SNPV). (d) Nuclear polyhedrosis virus inclusion body; multiple nucleocapsids per virion envelope (MNPV).

within the ranges 40–140 nm × 250–400 nm. The capsid is constructed from helically arranged subunits (Burley *et al.*, 1982), and structures described as claws and nipples (Kawanishi and Paschke, 1970) or caps (Federici, 1980) are present at its ends.

The double-stranded DNA molecule is a closed circle and is supercoiled. Most molecular weight estimates for BV DNAs fall between 70×10^6 and 120×10^6. Associated with the DNA is a highly basic protein (Tweeten, Bulla and Consigli, 1981) which may play a part in its condensation (Burley *et al.*, 1982).

Many of the larger virion proteins are associated with the lipid-containing membrane which forms the virion envelope. The virions of occluded BVs occur in two forms, each with a distinct envelope: the form which becomes occluded derives its envelope from membrane synthesized within the nucleus, while another form acquires its envelope by budding from the plasma membrane. The latter form normally has only one nucleocapsid per envelope, even if the virus is a MNPV, although Longworth and Singh (1980) observed that a few budded virions of *Epiphyas postvittana* MNPV had two nucleocapsids per virion. The occluded virions are specialized for infection of the host midgut cells, while the budded virions spread the infection to other cells and can readily infect susceptible cell cultures.

Some of the envelope proteins of occluded and budded forms of the same virus are distinct, while others are related (Volkman, 1983). At one end of a budded virion the envelope bears a number of spikes (Summers and Volkman, 1976), which are probably glycoproteins.

The polyamines spermidine and putrescine have been found in some NPVs. In *Heliothis zea* SNPV all of the spermidine and most of the putrescine was shown to be associated with the virion envelope (Elliott and Kelly, 1979).

The inclusion body

IBs are formed by the cytoplasmic polyhedrosis viruses and poxviruses of insects as well as by the occluded BVs. They afford protection to the virions outside the host, often for considerable periods between generations of larvae. Retention of occluded virus infectivity is far superior to that of viruses which do not form IBs, and is a further reason why interest has centred on the occluded viruses as microbial control agents.

The IB matrix is a paracrystalline lattice of protein subunits laid down to form an extremely stable structure which survives putrefaction of the dead host, but is broken down at low and high pH values. Reducing agents enhance the rate of IB dissolution in alkali (Croizier and Meynadier, 1972) and are essential for the dissolution of IBs of poxviruses and of the SNPV of *Tipula paludosa* at pH 10·5 (Bergoin, Guelpa and Meynadier, 1975).

The protein subunits are constructed from monomers, some of which have been reported to be glycosylated and phosphorylated (Kelly, 1981a). Proposals as to how the subunits are formed from the monomers include ionic, hydrophobic and disulphide bonding (Eppstein and Thoma, 1977).

There appears to be a high degree of similarity between the IB monomer proteins of different BVs. Their molecular weights all fall within the range

25 000–33 000. Several serological investigations with antisera have demonstrated relationships within and between the IB proteins of GVs, SNPVs and MNPVs, and these have been confirmed recently using monoclonal antibodies (Roberts and Naser, 1982a; Hohmann and Faulkner, 1983). The latter authors found stronger reactions within BV subgroups than between subgroups. The amino acid sequences of a few IB proteins have been determined and confirm that there is a high degree of similarity between them, especially between lepidopteran NPVs (Rohrmann *et al.*, 1981; Rohrmann, 1982).

Around each IB is a layer of material which appears electron-dense when sections are viewed in the electron microscope. It appears to be more resistant than the IB matrix to alkaline dissolution (Kawanishi, Egawa and Summers, 1972; Green, 1981) and may be composed of carbohydrate (Minion, Coons and Broome, 1979).

IBs from infected insects contain alkaline protease activity. This is displayed when IBs dissolve in alkali (Yamafuji, Yoshihara and Hirayama, 1957) and enhances their rate of dissolution (Summers and Smith, 1975). No such enzyme activity has been detected in IBs from infected cell cultures.

INFECTION OF THE HOST

Most infections are initiated by the ingestion of infective virus. The virions of occluded viruses are released by dissolution of the IBs in the alkali of the midgut. Gut enzymes (Faust and Adams, 1966) and the IB protease may also have roles. Granados and Lawler (1981) found that few *Autographa californica* MNPV IBs remained intact after 15 minutes in the larval midgut (pH 10·4) of the cabbage looper, *Trichoplusia ni.*

The virions must survive the harsh conditions of the midgut while they traverse the peritrophic membrane and attach to the microvilli of midgut cells. The virion envelope fuses with the microvillus membrane, releasing the nucleocapsid(s) into the cell (Granados, 1978).

The SNPVs of the most Diptera and Hymenoptera replicate only in the midgut cells, and IBs are shed into the gut lumen by lysis of infected cells. In the Lepidoptera, however, infection of the midgut is only the preliminary to infection of other tissues. Enveloped nucleocapsids develop and IB protein polymerization may occur in midgut cells, but virions are rarely occluded. Instead they bud into the haemocoel (Harrap, 1970) and are carried in the haemolymph to other susceptible tissues. It has also been suggested (Granados and Lawler, 1981) that some inoculum nucleocapsids may pass straight through the gut cells and bud into the haemocoel.

REPLICATION

BVs replicate in the nucleus, within which the DNA is released from NPV nucleocapsids, while GV nucleocapsids release their DNA into the nucleus via nuclear pores (Granados, 1980a). The infected nucleus hypertrophies and becomes the dominant feature of the cell. A 'virogenic stroma' is formed and nucleocapsids develop at its periphery. Nucleocapsids produced early enter the cytoplasm either by budding through the nuclear envelope (Injac *et al.*, 1971) or via ruptures in it (Adams, Goodwin and Wilcox, 1977). They then leave the cell by budding through a portion of modified plasma membrane which becomes the virion envelope (Hunter, Hoffmann and Collier, 1975). These virions spread the infection to other cells. Later in occluded BV infections most nucleocapsids are retained in the nucleus where they acquire envelopes and become occluded. Occlusion of naked nucleocapsids has never been observed, which suggests that there is an IB protein receptor on the virion envelope.

IBs develop randomly throughout the nucleus, except those of two dipteran SNPVs which develop in intimate association with the inner nuclear membrane (Smith and Xeros, 1954; Stoltz, Pavan and Da Cunha, 1973). The number of IBs produced per cell may vary from a few to several hundred, depending in part on the IB size. The yield per insect depends on many factors, including species and instar. Evans, Lomer and Kelly (1981) found maximum yields of $2{\cdot}7 \times 10^7$ IBs per first-instar larva and $3{\cdot}4 \times 10^9$ IBs per fifth-instar larva for *Mamestra brassicae* MNPV. IBs have been reported to constitute up to 40% of the insect dry weight (Bucher and Turnock, 1983).

Some progress has been made recently in understanding the biochemical events involved in BV replication. Kelly and Lescott (1981) identified four phases of virus protein synthesis in *Spodoptera frugiperda* cell cultures infected with *T. ni* MNPV. The phases were induced in a cascade fashion, with synthesis of one phase blocked if the proteins of the previous phase were rendered non-functional. The early proteins include enzymes such as thymidine kinase (Kelly, 1981b), while the virus structural proteins appear later. Synthesis of the later proteins is probably dependent on virus DNA synthesis, which reaches a high rate.

Nearly all of the late messenger RNA is virus-specific, with approximately 25% of that in *A. californica* MNPV-infected cells specific for IB protein (Adang and Miller, 1982). The control of IB formation is undoubtedly complex; studies with *A. californica* MNPV mutants led Potter and Miller (1980) to suggest that about half of the genome might be involved. Another small protein (molecular weight 10000) is produced late and in large quantities in *A. californica* MNPV-infected cells. It is present in the virion as a minor component, but its function is not known (Smith, Vlak and Summers, 1983).

Insect-cell culture

A brief account of insect-cell culture techniques is relevant because of their value in studies of BV replication (page 380), genetics (page 385) and safety testing (page 388). Furthermore, there are hopes that viruses used as insecticides might be mass produced in cell cultures (page 400). Recent reviews of insect-cell culture include those of Stockdale and Priston (1981), Vaughn and Dougherty (1981) and Grace (1982).

Increasing numbers of insect cell lines and culture media are becoming available (Hink, 1976, 1980). Most of the cell lines are from lepidopteran and dipteran insects; lines from several insect orders, including the Hymenoptera, have not yet been developed.

MNPVs, SNPVs and non-occluded BVs have been replicated in cell cultures.

All attempts to replicate GVs in cell lines have failed so far, but Vago and Bergoin (1963) and Rubinstein, Lawler and Granados (1982) have reported GV replication in primary cell cultures. Replication was incomplete in the latter case. As one of the preferred sites of GV replication is the fat body it will be of interest to see if a GV will replicate in any of the cell lines derived from fat bodies which are now becoming available (Mitsuhashi, 1981).

Susceptible cell cultures are readily infected with budded virions, so the haemolymph of an infected insect or medium from an infected cell culture provides effective inoculum. Virions released from IBs have much lower infectivity for cell cultures.

One of the most widely used cell lines is one derived by Hink (1970) from the ovaries of *T. ni* adults and designated TN-368. It has been used for plaque assays of *A. californica* MNPV (Hink and Strauss, 1977) and *Galleria mellonella* MNPV (Fraser and Hink, 1982). A plaque assay of *H. zea* SNPV in an *H. zea* cell line was described by Yamada and Maramorosch (1981).

A virus which will produce plaques in cell culture can be cloned by picking from single plaques, as carried out by Lee and Miller (1978) for *A. californica* MNPV in a *Spodoptera frugiperda* cell line.

A cell culture, like the whole organism, can harbour inapparent virus infections (Granados, Nguyen and Cato, 1978; Plus, 1978; Heine, Kelly and Avery, 1980). Plus (1980) stressed the importance of initiating cell lines from insects reared from surface-sterilized eggs as a precaution against virus contamination.

Baculovirus characterization and identification

ANALYSIS OF PROTEINS BY SODIUM DODECYL SULPHATE–POLYACRYLAMIDE GEL ELECTROPHORESIS

The technique of sodium dodecyl sulphate–polyacrylamide gel electrophoresis (SDS–PAGE) permits the number of virus proteins to be determined and their molecular weights to be estimated. It provides useful information, but suffers from a number of limitations and should not be used as the sole technique in virus identification (Allaway and Payne, 1983).

SEROLOGY

Serological methods are used to compare different viruses, and to diagnose infection in insects, especially in epizootiological studies. They are also used in safety testing (page 388) where they provide a means of detecting virus or virus components in non-target organisms and of detecting anti-viral antibodies in vertebrates exposed to the virus. Apparently the IB protein, the virion envelope and the nucleocapsid of an occluded BV each bears distinct antigenic determinants.

Prominent among several techniques which have been used is immunodiffusion, which is useful for investigating antigenic relationships, although it lacks sensitivity. The sensitive technique of enzyme-linked immunosorbent assay (ELISA) is becoming widely used. McCarthy and Henchal (1983) used an antiserum against nucleocapsids in an ELISA to detect *A. californica* MNPV virions in larvae and in cell cultures. Brown, Allen and Bignell (1982), investigating the relationships between four MNPVs of *Spodoptera* spp., used an indirect ELISA with enzyme-labelled protein A of *Staphylococcus aureus* in place of enzyme-labelled anti-immunoglobulin.

Monoclonal antibodies are increasing the specificity of serological techniques. Roberts and Naser (1982b) developed hybridomas secreting monoclonal antibodies against the IB protein and against a major virion protein of *A. californica* MNPV. These antibodies were used in several serological methods, and have recently been used in a protein-blotting technique incorporating ELISA (Naser and Miltenburger, 1983). Hohmann and Faulkner (1983) reported the application of a similar technique to investigate BV relationships. Volkman and Falcon (1982) used a monoclonal antibody against the IB protein of *T. ni* SNPV in an ELISA to diagnose infection in larvae. They found that host tissue caused interference, but concluded that the test was sensitive enough to be useful.

RESTRICTION ENDONUCLEASE ANALYSIS OF DNA

For definitive characterization and unequivocal identification of a BV it is preferable to analyse the genome rather than phenotypic characters. One of the techniques that discriminates best between double-stranded DNA viruses is restriction endonuclease (REN) analysis of their nucleic acids. Smith and Summers (1979) could differentiate five *A. californica* MNPV isolates by this technique, whereas the SDS–PAGE protein profiles of the isolates were identical.

BIOASSAYS

Precise bioassay techniques yield important information about the virus–host relationship. This information is vital for selecting virus strains with high infectivity and for estimating suitable rates for field application. Many factors can affect the dose–response relationship and/or the LD_{50} of an insect virus and each of these must be standardized. Larval instar (page 396), larval weight and/or age within instar (Burgerjon *et al.*, 1981; Evans, 1983) diet composition, IB purification technique (Baugher and Yendol, 1981) and incubation temperature (Boucias, Johnson and Allen, 1980) should all be carefully controlled.

Techniques in which a larva consumes only a portion of virus-inoculated diet are less preferable to those in which the whole of the dose is ingested on a leaf disc (Evans, 1981), a small piece of diet (Nordin, 1976), or in a small drop (Klein, 1978; Hughes and Wood, 1981). Laing and Jaques (1980) described a bioassay technique for larvae of boring species such as the codling moth, *Cydia pomonella.*

Estimates of LT_{50}s may also be useful, especially for predicting how rapidly insects will be killed in the field.

HOST RANGE

BV host ranges have been widely investigated but many of the results require confirmation, as viruses which replicated in inoculated hosts were not always identified, often because suitable techniques were not available when these experiments were carried out. Some of the cases of virus replication could have been due to latent virus activation (*see below*) rather than to cross-transmission. In many studies only gross effects of infection (e.g. IB formation, host death) were looked for, although a virus might infect a host sublethally or only some virus functions might be expressed without IB formation. Furthermore, an insect resistant to infection by ingestion of IBs might be susceptible if injected with budded virions.

With these provisos in mind it can be stated tentatively that the MNPVs are the least host-specific of the occluded BVs. *A. californica* MNPV has the widest known host range, infection having been reported in more than 30 insect species and in cell cultures from at least 13 species. Replication of *H. zea* SNPV, on the other hand, appears to be restricted to members of the genus *Heliothis* (Ignoffo and Couch, 1981). Some GVs have been transmitted to other species, e.g. *C. pomonella* GV to five closely related species (Huber, 1982), and *Heliothis armigera* GV to four other species including *T. ni* and two *Spodoptera* species (Hamm, 1982).

When selecting a virus for possible use against more than one pest species it is important to determine the dose–mortality relationship (page 383) for each host. A virus is not likely to control an insect if the LD_{50} is extremely high, as for *Agrotis segetum* GV in *Agrotis exclamationis* larvae in which the LD_{50} for neonate larvae was found to be $1{\cdot}2 \times 10^6$ IBs compared with $1{\cdot}1 \times 10^4$ IBs for the homologous host (Allaway and Payne, 1984).

LATENCY

There have been many reports of insects harbouring ‘latent’ viruses, especially BVs, but no firm conclusion can be drawn from many of them. The best-substantiated reports concern the development of a homologous NPV in an insect fed with IBs of a heterologous NPV, with both viruses being characterized (Longworth and Cunningham, 1968; Maleki-Milani, 1978; Jurkovíčová, 1979). McKinley *et al.* (1981) found that activation of a latent virus was more common than cross-infection after feeding four NPVs to heterologous hosts. Two aspects of their results are particularly interesting: first, there was a straight-line relationship between dose and mortality, i.e. there was no threshold dose of heterologous virus above which activation of homologous virus occurred; second, it appeared that each of the insects in their cultures carried a latent virus.

Because of the phenomenon of latency it is vital that all insect and cell-culture stocks used for virus studies are checked as closely as possible for the presence of latent viruses.

Baculovirus classification and nomenclature

The BV subgroups were described on page 376, and the reader will have gathered that an individual virus is identified by the name of the insect from which it was isolated, e.g. *Gilpinia hercyniae* SNPV, *Pieris rapae* GV. Some insects, e.g. *T. ni* and the Douglas fir tussock moth, *Orgyia pseudotsugata*, are host to both a SNPV and a MNPV; regrettably some authors do not specify which type of virus they have worked with.

The system of naming a BV after an insect host is far from satisfactory because many, if not most, of the BVs can infect several hosts. The wide host range of *A. californica* MNPV has been discussed (page 384), and DNA REN analyses indicate that this virus, *T. ni* and *G. mellonella* MNPVs (Smith and Summers, 1979) and an NPV from *Diparopsis watersi* (Croizier *et al.*, 1980) are very closely related. In fact many of the REN pattern differences between these viruses were no greater than the differences between strains of *A. californica* MNPV.

Sometimes a virus is found to be more infective for another host, e.g. *Pieris brassicae* GV is more infective for *P. rapae* than for the 'natural' host (Payne, Tatchell and Williams, 1981), and *M. brassicae* MNPV is more infective for *Plusia gamma* than for the 'natural' host (Allaway and Payne, 1984). Clearly, a more logical approach to BV classification and nomenclature is required.

Baculovirus genetics

There appears to be a multiplicity of genotypes for each of the BVs. A virus isolated from a single infected larva may contain a variety of genomes, as demonstrated by the regular presence of submolar fragments of DNA after REN digestion (e.g. Smith and Summers, 1978; McIntosh and Ignoffo, 1983). Even when no submolar fragments can be detected in REN analysis, a small proportion of the genomes may display variability which can be detected in plaque-purified strains (Smith and Summers, 1980).

There may be differences between virus isolates from members of the same host species collected from different geographical areas, e.g. isolates of *Spodoptera littoralis* MNPV (Kislev and Edelman, 1982), *Neodiprion sertifer* SNPV (Brown, 1982) and *Lacanobia oleracea* GV (Crook, Brown and Foster, 1982) differed in their DNA REN patterns. Heterogeneity in the genome of a single 'virus' is also reflected in variability of phenotypic characters. Isolates may differ serologically, e.g. *A. segetum* MNPV (Allaway and Payne, 1983), in their SDS–PAGE protein patterns, e.g. *N. sertifer* SNPV (Brown, 1982), and in biological characteristics of crucial importance in the use of these agents for pest control. Isolates of *Oryctes rhinoceros* non-occluded BV (Zelazny, 1979), *C. pomonella* GV (Harvey and Volkman, 1983) and *A. segetum* MNPV (Allaway and Payne, 1983) have been shown to differ in LD_{50} for their hosts.

For the reasons just outlined it is preferable that cloned virus strains be used in all investigations. Viruses which produce plaques in cell culture can be cloned from single plaques. For those viruses for which no plaque system is available, the next best approach is to inject groups of insects with serial dilutions of budded virions, and to select for virus isolation a single infected insect from a group injected with a dose smaller than the LD_{50}. Green (1981) used the latter approach with *T. paludosa* SNPV.

A. californica MNPV has been adopted for study by a number of laboratories and rapid progress is being made in mapping the genome of its dominant variant. Physical maps have been derived using RENs (Miller and Dawes, 1979;

Vlak, 1980; Cochran and Faulkner, 1983). *Eco*RI digestion yields 24 fragments, 21 of which have been cloned by Lübbert *et al.* (1981). It has been agreed that the map should start at *Eco*RI fragment I, which includes the IB protein gene. Smith and Summers (1982) found that DNAs from several NPVs, GVs and a non-occluded BV had sequences homologous with this fragment.

The locations on the physical map of the genes for several functions, including IB protein, have been found by marker rescue and by using Southern and Northern blotting techniques. The copy-DNA technique has been used to determine the relative amounts of virus messenger-RNA species in the infected cell, and to identify the proteins for which they code (Adang and Miller, 1982; Smith, Vlak and Summers, 1982; Erlandson and Carstens, 1983).

Recombination between MNPV genomes has been demonstrated. Croizier, Godse and Vlak (1980) inoculated *G. mellonella* larvae with MNPVs from *G. mellonella* and *A. californica*, and isolated recombinants. Smith and Summers (1980) plaque-purified recombinants between *A. californica* MNPV and *Rachiplusia ou* MNPV from wild-type *R. ou* MNPV, and suggested that recombination may be important in the evolution of BVs.

The genomes of BVs can now be manipulated using the techniques of genetic engineering. It may soon be possible to construct new virus strains with improved characteristics as microbial control agents.

Epizootiology

A common objective in pest control with a virus is the establishment of an epizootic in a pest population from which the virus is absent or in which it is only enzootic. In order to achieve this it is important that the mechanisms whereby the virus spreads from host to host within a generation and between generations are understood. Some knowledge of how well the virus persists in the field is also necessary.

Insects which feed at plant surfaces become infected with occluded viruses principally by ingesting IBs present on the plant, deposited there from the faeces or the cadavers of infected insects. In order to ensure virus persistence, large quantities of virus are produced, of which only a tiny proportion may be utilized as inoculum. During a SNPV epizootic in the European spruce sawfly, *G. hercyniae*, in Wales it was estimated that more than 10^{14} IBs/hectare were produced, of which only 0·00025% was utilized the following year (Evans and Harrap, 1982).

Virus may be disseminated by the movement of infected larvae, e.g. NPV-infected larvae of the cabbage moth, *M. brassicae* can move several metres in cabbage plots (Evans and Allaway, 1983). NPV-infected larvae of some species, e.g. the gypsy moth, *Lymantria dispar*, (Doane, 1970) and *M. brassicae* (Evans and Allaway, 1983) tend to climb to the tops of plants before they die, thus ensuring maximum contamination of the plants with their virus load. In many lepidopteran species the BV-killed cadaver hangs from the host plant while putrefaction occurs; then the skin bursts, shedding the liquefied contents together with the virus IBs. Soil-dwelling, plant-feeding insects, such as *Tipula* spp. larvae, are less likely to contaminate their food source with infective doses of virus.

The main mode of transmission for two viruses of *Tipula* spp. appears to be by cannibalism (Carter, 1973a, b; Green, 1981).

IBs may also be deposited on plants in the faeces of predators, or they may be transferred from the soil by rain-splash or by the activities of animals.

TRANSMISSION TO THE NEXT GENERATION

In a permanent ecosystem, such as a forest, IBs produced in one generation of larvae may persist on foliage until the next generation has hatched, as Entwistle and Adams (1977) showed for *G. hercyniae* SNPV. Virus may also contaminate the egg surfaces, and this may be ingested by the hatching larvae, as Doane (1975) demonstrated for *L. dispar* NPV.

In an annual crop, on the other hand, virus is transferred to the plants from a reservoir, usually the soil. There have been several investigations into the survival in soil of viruses of brassica pests. David and Gardiner (1967) reported good survival of *P. brassicae* GV in soil for at least two years, and Jaques (1969) found large amounts of an NPV of *T. ni* in soil 231 days after application with little or no evidence of leaching of IBs. Evans (1982), however, found a 98% loss of *M. brassicae* NPV IBs after 52 weeks: nevertheless, with sufficient IBs initially, enough could survive to infect the next generation.

Some larvae which receive small doses of virus, and/or which become infected in a late instar, may survive to produce infected adults which may disperse the virus and transmit it to their progeny. Entwistle (1976) considered that this was an important dispersal mechanism during an epizootic of *G. hercyniae* SNPV.

It has been claimed that some viruses are transmitted within the egg, but this has not yet been unequivocally demonstrated. It has been shown, however, that infected adults can contaminate the egg surface. Hamm and Young (1974) demonstrated transmission of *H. zea* NPV to the next generation in this way.

ROLES OF PARASITES AND PREDATORS

Hymenopteran parasites of insects can act as virus vectors when females oviposit in infected insects and subsequently in uninfected insects. The infective material probably consists of budded virions. Transmission by this mechanism has been shown for several viruses, including *P. rapae* GV (David, 1965), *Heliothis virescens* NPV (Irabagon and Brooks, 1974) and *L. dispar* NPV (Raimo, Reardon and Podgwaite, 1977).

Predators may disperse virus after feeding on infected insects. The following are a few examples of cases in which an infective BV, often in significant amounts, has been demonstrated in the faeces of predators: insects predatory upon *Heliothis punctiger* (Beekman, 1980) and *M. brassicae* (Evans and Allaway, 1983); birds predatory upon *G. hercyniae* (Entwistle, Adams and Evans, 1978) and *Wiseana* spp. (Kalmakoff and Crawford, 1982); and mammals and birds predatory upon *L. dispar* (Lautenschlager and Podgwaite, 1979).

Parasites and predators therefore have important roles in the transmission and dispersal of viruses, in addition to their more direct roles in regulating insect numbers. Integrated pest management (IPM) practices should therefore aim at maximum conservation of these animals.

EPIZOOTICS

The normal situation for most virus diseases is an enzootic, occasionally becoming epizootic when the host population density increases. Doane (1976) has described how an NPV epizootic develops in an *L. dispar* population, resulting in a spectacular reduction in population size, which is then likely to remain small for a number of years because of the high level of virus in the environment. Only when this declines is there likely to be a repeat of the cycle of resurgence in insect numbers followed by another epizootic. Briese (1981) proposed that climate, too, might influence the development of GV epizootics in the potato moth, *Phthorimaea operculella.*

Entwistle *et al.* (1983) described the patterns of virus dispersal in *G. hercyniae* SNPV epizootics. The spread of the disease from an initial focus became wave-like and then became random. These authors suggested that other insect viruses, e.g. *O. rhinoceros* non-occluded BV, might follow similar patterns of spread.

Safety

It has been argued (Burges, Croizier and Huber, 1980) that BVs are inherently safe for use as pesticides because man has been exposed to them throughout his evolution and no adverse effects are known. The presence of BVs can be demonstrated on marketed vegetables, some of which are eaten raw. There are, however, a number of potential hazards associated with the mass production and mass application of BVs, and these should be evaluated as fully as possible. It is better to use a pesticide with the confidence that it has passed a series of stringent safety tests than to risk an accident which could set back microbial control for decades.

The viruses which have been most exhaustively tested for safety to date are those registered for use in the US. In a large series of tests on *H. zea* SNPV no adverse effects have been found, except for possible enhancement of simian virus 40-transformation of human amnion cells (McIntosh and Maramorosch, 1973).

POTENTIAL HAZARDS

The virus, and other materials in the formulation, should be tested for infectivity, toxicity, carcinogenicity, teratogenicity and allergenicity in non-target organisms.

A change in the host specificity of a virus might occur by mutation, or by recombination with another virus or with cellular DNA. BV genomes resemble those of papovaviruses, many of which are oncogenic, in that they are both circular double-stranded DNA molecules. Tests for hybridization between BV and vertebrate virus DNAs, and between BV and cell DNAs, would provide an indication of the likelihood of recombination events.

The safety of humans is the prime concern, and it must be remembered that some highly susceptible individuals, i.e. those with hereditary immunodeficiency, those with acquired immune deficiency syndrome, and those receiving immunosuppressant therapy, could be exposed to virus-containing sprays and dusts. Persons involved in virus production and field application receive the greatest exposure, especially when the virus is disseminated as a spray.

Perhaps the most likely hazard is an allergic response in the skin or respiratory system. Repeated inhalation might lead to a pulmonary condition similar to farmers' lung disease. One worker involved in *H. zea* SNPV-production is reported to have developed an allergy (Rogoff, 1975).

The welfare of other organisms, including domestic animals, wild mammals, birds, fishes and beneficial insects, must also be safeguarded.

TESTS ON VERTEBRATE ANIMALS

Animals have been inoculated with BV IBs, virions and DNA via a variety of routes. In the vast majority of these tests, e.g. after feeding *H. zea* SNPV IBs to pregnant rats (Ignoffo, Anderson and Woodard, 1973) and after inoculating *O. rhinoceros* non-occluded BV into mice (Gourreau, Kaiser and Monsarrat, 1982), no harmful effects were found. *M. brassicae* NPV IBs and *A. californica* NPV virions were fed to rodents and no chromosomal aberrations were detected (Miltenburger, 1980).

There are two reports of adverse effects in BV-inoculated pigs: Gourreau *et al.* (1979) found an increased rate of liver lesions in pigs inoculated intraperitoneally with *O. rhinoceros* non-occluded BV; and Döller, Gröner and Straub (1983) found slight temperature increases in piglets fed *M. brassicae* NPV IBs.

G. Döller and co-workers have suggested that an antibody response in an animal is suggestive of virus replication, and have been unable to detect antibodies to IBs and virions in mammals exposed by feeding and inhalation (Döller and Huber, 1983; Döller, Gröner and Straub, 1983). Carey and Harrap (1980), however, found that some rats exposed to *Spodoptera* spp. NPVs developed antibodies to the virions and/or the IB protein and antibody responses have occurred in mice fed IBs (D. L. Knudson, in discussion after Granados, 1980b).

Workers involved in *H. zea* SNPV production (Ignoffo and Couch, 1981) and in field trials with *N. sertifer* SNPV (Entwistle *et al.*, 1978) have been monitored and no antibodies against BV components have been found in their sera.

Care is necessary when interpreting results of serum tests as a number of non-specific reactions have been detected between mammalian sera and IB proteins (Döller, 1980, 1981).

There is evidence both for the survival of BV IBs intact in the mammalian gut, and for their breakdown. Carey and Harrap (1980) recovered infective IBs 21 days after feeding to rats, while Döller, Gröner and Straub (1983) found evidence of IB breakdown in the piglet gut, but were unable to detect infectious virus in the organs.

To test for adverse effects on wildlife the approach of Lautenschlager, Rothenbacher and Podgwaite (1978) could be emulated. These authors monitored a variety of parameters in five species of caged and free-living mammals in a woodland after aerial application of *L. dispar* MNPV; they found no adverse effects. Döller and Enzmann (1982) showed that fish can mount a good antibody response to IB protein, and proposed that tests for immune responses in fish could form part of an environmental monitoring programme.

TESTS ON VERTEBRATE-CELL CULTURES

BVs have been inoculated into a wide variety of vertebrate-cell cultures and in the majority of cases no cytopathic effect occurred and no evidence of virus replication could be found, e.g. *H. zea* SNPV in primate cells (Ignoffo and Rafajko, 1972), *A. californica* MNPV in three mammalian cell lines (Miltenburger, 1980) and *O. rhinoceros* non-occluded BV in mammalian and fish cells (Gourreau, Kaiser and Montsarrat, 1981). Lack of IB production or other cytopathic effect should not be construed as lack of virus replication, but sensitive tests for a range of virus functions should be performed.

BV virions are readily taken up by vertebrate cells in culture. Granados (1980b) reported uptake of *A. californica* MNPV virions into cytoplasmic vacuoles in HeLa and fathead minnow cells, and similar observations were made by Volkman and Goldsmith (1983) and Miltenburger and Reimann (1980). The latter authors (Reimann and Miltenburger, 1983) also found evidence of some nucleocapsids breaking down in the vacuoles, and of others budding out of the cell. They could not detect virus in the cell nuclei, but Tjia, Zu Altenschildesche and Doerfler (1983), using a DNA hybridization technique, found DNA of *A. californica* MNPV in the nuclei of inoculated mammalian cells for at least 24 hours, after which it was rapidly lost. The limit for DNA detection by this technique is one viral genome per 5–10 cells (Miltenburger, 1980), so it is possible that it might have persisted undetected in a few cells. No evidence of transcription of the virus genome could be found.

McIntosh, Maramorosch and Riscoe (1979) found that *A. californica* MNPV virions were taken into cytoplasmic vacuoles in a viper cell line. There was no evidence of virus replication, but the cells grew more slowly, and there was a large increase in the number of C-type particles present in that cell line.

There have been a few reports of BV replication in mammalian cells. The first of these was by Himeno *et al.* (1967) who announced that IBs had developed in human cells inoculated with *Bombyx mori* NPV DNA. Aleshina *et al.* (1973a) subsequently reported replication of *B. mori* NPV in mouse fibroblasts. McIntosh and Shamy (1980) reported evidence of *A. californica* MNPV replication in a Chinese hamster cell line, but no evidence of replication was found by Volkman and Goldsmith (1983) in the same virus-cell system, or by Reimann and Miltenburger (1983) in another Chinese hamster cell line. One further report of a BV-induced change in mammalian cells is of an increase in nuclear size after inoculation with *L. dispar* NPV (Aleshina *et al.*, 1973b).

OTHER COMPONENTS OF BACULOVIRUS PREPARATIONS

Potential hazards from other materials present in a virus formulation must also be assessed. Insect fragments, insect diet and contaminant micro-organisms may be present, depending on the production method. Chemicals may be added to protect the virus from ultra-violet (UV)-light, to enhance adhesion to foliage or to stimulate larval feeding, and some viruses are applied in oil suspensions.

One cause for concern is the possible presence of contaminant viruses. Two small RNA viruses were found in a preparation of *Darna trima* GV (Harrap and Tinsley, 1978), and a small RNA virus has been found in *A. californica* MNPV preparations (Morris, Hess and Pinnock, 1979; Vail *et al.*, 1983). The latter virus has affinities with the mammalian caliciviruses, and is infective for *T. ni* larvae, in which small doses can initiate inapparent infections. *T. ni* larvae are used for *A. californica* MNPV production.

The risks posed by contaminant viruses are still largely unknown, but one small RNA virus (Nodamura virus) isolated from insects is lethal to mice when injected by various routes (Scherer, Verna and Richter, 1968). Until the risks can be shown to be negligible it would seem prudent for any virus which is to be applied as a spray to undergo a purification procedure sufficient, at least, to remove contaminant virions.

Most mass-produced insect virus preparations, however, consist of ground, lyophilized virus-infected larvae, and therefore contain insect material and contaminating micro-organisms. Podgwaite, Bruen and Shapiro (1983) found approximately 10^8–10^9 viable bacteria and fungi per gram of 'Gypchek' (*L. dispar* MNPV). Many of the organisms that they found are opportunistic human pathogens. Padhi and Maramorosch (1983) determined viable bacterial counts in commercial preparations of *H. zea* SNPV. They found 10^5 bacteria per gram in 'Elcar', whereas 'Viron/H' (now discontinued) contained 10^8 bacteria per gram, including *Bacillus cereus* which was pathogenic to silkworm larvae.

Dubois (1976) demonstrated that bacterial contaminants can be destroyed chemically. In the UK, field trials have been carried out with highly purified preparations of *N. sertifer* SNPV (Cunningham and Entwistle, 1981), *P. brassicae* GV (Tatchell and Payne, 1984) and *C. pomonella* GV (Glen and Payne, 1984).

'Gypchek' production also involves the hazard posed by the allergenic, urticarious setae of *L. dispar* larvae. Personnel are protected by filter masks, and a method has been devised for removing the setae during processing (Shapiro *et al.*, 1981).

Other components of BV preparations (e.g. UV-protectants, oils) should be tested for possible hazard, especially for carcinogenicity by inhalation.

REGISTRATION REQUIREMENTS AND GUIDE-LINES FOR SAFETY TESTING

Harrap (1982) has given a comprehensive account of registration requirements for viral (and other microbial) insecticides, and of the guide-lines produced by several national and international bodies for safety testing them. In the UK the controlling body is the Pesticides Safety Precautions Scheme of the Ministry of Agriculture, Fisheries and Food (Papworth, 1980), while in the US it is the Environmental Protection Agency (Rogoff, 1980). Many developing nations lack the facilities and resources for safety testing. Virus preparations which might be of value to those nations could be safety tested in laboratories in the developed nations as a contribution to their overseas aid programmes.

Too few BVs have been exhaustively safety tested to allow conclusions to be drawn about the safety of BVs in general, but the current impression is that these viruses appear to be safe for field use. However, the evidence that IBs can be dissolved in the mammalian gut and that virions can be taken into mammalian cells, together with the reports of replication in mammalian cells and of adverse effects in mammals, mean that several BVs will need to pass stringent safety tests before the group in general receives a blanket seal of approval.

Strategies for pest control with viruses

INTRODUCTION OF VIRUS

Many pests have been introduced into new areas of the world as a result of man's activities. It has been estimated that 30% of the most serious pests in the US are of foreign origin. The pests are often introduced without all of their natural enemies, including viruses. There have been several cases where a virus has subsequently been introduced, either deliberately or by accident, and has provided effective control of the pest. Two examples concern sawflies introduced into North American forests from Europe. In each case the subsequent release of an NPV from Europe initiated epizootics and controlled pests (Bird, 1953; Bird and Elgee, 1957).

This strategy was also applied to the non-occluded BV of the coconut palm rhinoceros beetle, *Oryctes rhinoceros*, which was discovered in Malaysia (Huger, 1966), but was apparently absent from the Pacific islands where *O. rhinoceros* causes serious damage to palms (pages 393 and 396).

SUPPLEMENTATION OF EXISTING DISEASE

Where a virus is present in an insect population, increasing the amount of virus in the environment may lead to a greater proportion of insects becoming infected. One virus application may be sufficient to reduce numbers of a pest to an economically acceptable level, especially in a forest. Cunningham and Entwistle (1981) stated that a single NPV application to young trees is likely to protect them from sawfly damage for their lifetime. In an agricultural situation it may be necessary to use a virus more like a chemical insecticide, with adequate protection provided only by several applications during the lifetime of the crop.

MANIPULATION OF EXISTING DISEASE

In some situations it is possible to increase the level of virus disease in a pest population by the adoption of certain management practices. An example concerns *Wiseana cervinata* which damages pasture in New Zealand. An NPV is widespread and can control this pest, but cultivation of the land buries the virus reservoir beyond the range of the larvae. Kalmakoff and Crawford (1982) therefore recommended oversowing damaged areas of pasture without cultivation. They also recommended the regular movement of stock over pastures to spread the virus.

Techniques for virus dissemination

RELEASE OF INFECTED/CONTAMINATED INSECTS

This dissemination technique has special attraction for viruses which survive poorly outside the host, e.g. *O. rhinoceros* non-occluded BV, which has been introduced into a number of South Pacific islands by releasing infected beetles (Bedford, 1981). Only the mid-gut cells are susceptible in the adult, which may survive for many weeks. The infection has a debilitative effect, however: the beetles stop boring into palms and females stop egg-laying. Monsarrat and Veyrunes (1976) estimated that an infected adult excretes about 300 ng virus per day. Some of this virus is transmitted during mating, and some serves as a source of infection for larvae, in which the infection rapidly becomes systemic and causes death.

Some insect viruses, e.g. *P. brassicae* GV (Tatchell, 1981), are transmitted to the progeny if the ovipositor of the female is contaminated, but this technique has not yet been widely applied to virus dissemination in the field.

SPRAYING

Most viruses are applied in aqueous sprays using equipment developed for spraying chemical insecticides. Morris (1980) and Smith and Bouse (1981) have argued for a research programme to design equipment specifically for the application of viruses and other microbes.

Equipment producing small droplets is preferred. Virus application in droplets with diameters of 100–150 μm usually results in higher insect mortality than in larger droplets (Smith and Bouse, 1981). Entwistle *et al.* (1978) used a microdroplet machine producing droplets with a mean diameter of 50 μm. Reed and Springett (1971) suggested that *P. operculella* GV might best be disseminated as a mist as the IBs would be more likely to enter the stomata, thereby becoming more accessible to the larvae within the leaves.

Virus dissemination in charged droplets from an electrostatic sprayer means that a larger proportion of IBs adhere to the leaves, especially the undersides which are the sides often favoured by insects, and which provide some protection from sunlight for the virus. A disadvantage of electrostatic sprayers is poorer spray penetration into the plant canopy (Matthews, G.A., 1982).

A modification of spray application was carried out by Hamm and Hare (1982) who introduced NPVs of *H. zea* and *S. frugiperda* on to corn via an overhead irrigation system. Instead of spraying the crop, Young and Yearian (1980) sprayed the soil with an NPV of the soybean looper, *Pseudoplusia includens*, at soybean planting time.

If a virus spreads rapidly, then blanket spraying may be unnecessary. This is the situation with the SNPV of the red-headed pine sawfly, *Neodiprion lecontei*, for which Cunningham (1982) has proposed spot introductions or 'zebra stripe spraying' from aircraft. Spot introductions into glasshouses of the GV of the tomato moth, *Lacanobia oleracea*, were suggested by Crook, Brown and Foster (1982).

BAITS

The application of insecticides in baits has the advantage that less insecticidal material is required, and the disadvantage of increased costs of field application. Baits are especially valuable if insects which have a burrowing or mining habit can be encouraged to spend longer at the plant surface and ingest larger doses of insecticidal material.

Most research into the application of viruses in baits has involved *H. zea* SNPV and baits based on cottonseed and soybean (page 395). Johnson and Lewis (1982) used wheat bran baits to apply two MNPVs to corn.

DIPPING SEEDLINGS

Ignoffo *et al.* (1980) suggested that IBs could be introduced on to cabbages by dipping them in an IB suspension at the time of transplanting.

Viruses undergoing trials and/or in use

Details of viruses registered for use in various countries are given in *Table 2*, and some of them are discussed more fully below.

1. *Heliothis zea* SNPV has been marketed for almost a decade in the US for the control of *H. zea* and *H. virescens* on cotton (Ignoffo and Couch, 1981). Some workers, e.g. Shieh and Bohmfalk (1980), have found it to be an effective

Table 2. Viruses registered for use.

Virus	Used on	Country	Product name
Heliothis zea SNPV	Cotton and other crops	US	Elcar
	Cotton, sorghum	Australia	
Orgyia pseudotsugata MNPV	Fir trees	US	TM Biocontrol-1
		Canada (temporary registration)	Virtuss
Lymantria dispar MNPV	Deciduous trees	US	Gypchek
		USSR	Virin–ENSh
Autographa californica MNPV	Several crops	US (experimental use permit)	SAN 404
Neodiprion sertifer SNPV	Pine trees	US	Neochek S
		USSR	Virin–Diprion
		Finland	none
Neodiprion lecontei SNPV	Pine trees	Canada (temporary registration)	Lecontvirus
Dendrolimus spectabilis cytoplasmic polyhedrosis virus	Pine trees	Japan	Matsukemin

insecticide, while others, e.g. Pfrimmer (1979), have obtained variable and sometimes disappointing results.

Much effort has been expended in attempts to achieve more consistent results. Some of this effort has involved the development of baits, and two in particular have been tested: 'Coax' based on cottonseed and 'Gustol' based on soybean. Many workers (e.g. Hostetter *et al.*, 1982, and Potter and Watson, 1983a) have shown in laboratory and field tests that applying the virus in a bait increases larval mortality. Some of the increased mortality may not be virus-induced, however, as treatment of cotton with 'Coax' alone results in increased mortality (Henry, 1982). This has been attributed to larvae spending longer at the surface before tunnelling into the bolls, thereby extending their exposure to parasites and predators.

Smith, Hostetter and Ignoffo (1978, 1979) compared different formulations, application rates, types of spray nozzle and nozzle pressures. They found that the efficiency of application was affected by nozzle type and droplet size.

H. zea SNPV can also control *Heliothis* spp. on other crops. Ignoffo *et al.* (1978) found that it reduced *H. zea* populations on soybeans by 92–100%, and Smith and Hostetter (1982) reported better control of *H. zea* on soybean and cabbage than on cotton. In Australia *H. zea* SNPV is undergoing tests for its ability to protect navybeans from *Heliothis* spp. (R. E. Teakle, personal communication).

2. *Autographa californica* MNPV, originally isolated from the alfalfa looper, is considered to have a potential commercial value because of its wide host range. It has been reported that it can control *T. ni* as effectively as chemicals on cabbage (Hostetter *et al.*, 1979) and lettuce (Vail, Seay and Debolt, 1980), and it is being assessed as an alternative to *Orgyia pseudotsugata* MNPV for the control of the Douglas-fir tussock moth, *O. pseudotsugata.* Although the latter virus can control its host effectively, the high cost of its production and its limited market mean that there is no commercial interest in it (Martignoni, Steltzer and Iwai, 1982).

3. *Neodiprion sertifer* SNPV has been extensively tested against its host, the European pine sawfly, in Eastern and Western Europe and in North America (Cunningham and Entwistle, 1981). Entwistle *et al.* (in press) have induced high larval mortality in pine forests in Scotland with applications of 5×10^9 to 2×10^{10} IBs/hectare. These quantities of virus can be produced in 20–50 larvae. This remarkable efficiency is attributed to the high larval susceptibility to this virus and to its rapid spread.

4. *Neodiprion lecontei* SNPV has shown promise in trials in Canada. Its host has been controlled with applications of 5×10^9 to 8×10^9 IBs/hectare (Cunningham, 1982).

5. *Galleria mellonella* MNPV was shown by Dougherty, Cantwell and Kuchinski (1982) to control wax moth larvae effectively in bee-hives. A non-hazardous insecticide is especially important for this pest of honeycomb.

6. *Panolis flammea* NPV has shown very promising results against its host, the pine beauty moth, which is a pest of lodgepole pines in Scotland (P. F. Entwistle, personal communication).

7. *Heliothis armigera* SNPV was shown to provide control of its host on sorghum in Botswana by Roome (1975) and is still under investigation in that country where *H. armigera* is a pest of many crops (Flattery, 1983).

8. *Choristoneura occidentalis* NPV and GV have shown promise for the control of their host, the western spruce budworm on Douglas fir. The impact of the NPV on the population size was still detectable one year after spraying (Shepherd, Gray and Cunningham, 1982), and a GV application rate of only 25 'larval equivalents'/acre resulted in 56% mortality (Cunningham, Kaupp and McPhee, 1983).

9. *Pieris brassicae* GV has been demonstrated by a number of workers, including Kelsey (1958), to provide control of larvae of the small cabbage white butterfly, *Pieris rapae*. Tatchell and Payne (1984) recently found that a spray containing 10^8 IBs/ml reduced the larval population by more than 90%. The virus is rapidly inactivated in the field, however, and regular spraying would be necessary to maintain satisfactory control.

10. *Cydia pomonella* GV has been tested in many countries for control of the codling moth in orchards. Huber and Dickler (1977) reported that four sprays resulted in good control, but there was no persistence of the disease into the next season. Much virus is probably removed from the orchard on the surface of the apples. Trials carried out by Glen and Payne (1984) led them to conclude that the use of *C. pomonella* GV effectively reduces the more severe forms of fruit damage, but the quantities of virus required to control less severe forms of damage would probably be uneconomic.

11. *Oryctes rhinoceros* non-occluded BV has been introduced into a number of South Pacific islands. In Tonga it was still infecting 84% of the beetle population after seven years (Young and Longworth, 1981). Control of the rhinoceros beetle has led to a revival of the copra industry in Western Samoa (Marschall and Ioane, 1982).

Timing of field applications

The timing of field applications of virus can be crucial in determining the level of pest control achieved. Significant pest damage is not usually noticed until the larvae are in the later instars, when larger doses of virus are necessary to infect them. This, coupled with the fact that most insect viruses kill their hosts more slowly than chemical insecticides, means that for many pests the virus must be applied before crop damage appears. Pest forecasting systems can be used to indicate when pest numbers are approaching damaging levels.

Increases in LD_{50} of 10^4-fold to 10^6-fold from early to late larval instars have been found for a number of lepidopteran BVs, including *P. brassicae* GV

(Payne, Tatchell and Williams, 1981) and *Mamestra configurata* MNPV (Bucher and Turnock, 1983), and LT_{50}s are often longer in later instars. In some cases the increased LD_{50} may be offset by the increased food consumption of larger larvae, as in *H. armigera* where the first three larval instars have a similar probability of becoming infected with *H. zea* SNPV in the field (R. E. Teakle and J. M. Jensen, personal communication).

Increases in resistance to NPVs in the later instars of sawflies appear to be small compared with those in the Lepidoptera. For *N. sertifer* and *G. hercyniae* the increases in LD_{50} are about tenfold to fiftyfold from the first to the fifth larval instar (Entwistle, Adams and Evans, unpublished work cited in Evans and Harrap, 1982; Entwistle *et al.*, in press). This means that these pests can be controlled if virus is applied after the first instar. Infection of the larvae when they are larger means that more virus is produced and is available to infect the next generation. This approach is more applicable in stable ecosystems, such as forests and pastures where some pest damage can be tolerated, than in annual crops.

Virus persistence in the field

A rapid loss of infective virus from plant surfaces can usually be detected after field application, which could be a physical loss of IBs from plants and/or a loss of infectivity in virus on the plants.

There have been many studies of rates of infectivity loss, but, as pointed out by Richards and Payne (1982), most of them have started with amounts of virus giving 100% mortality in bioassays and have not therefore achieved their objectives. These authors outlined a sound experimental approach which they applied to measure survival of infectivity of a *Pieris* sp. GV on cabbage in the UK. They found that the half-life varied from 0·35 day in June to 1·0 day in October. They showed, using ^{32}P-labelled IBs, that the IBs had not been lost from the cabbages.

If a virus can be protected from inactivation then smaller amounts need be applied and/or the timing of application becomes less critical. The extent to which virus inactivation can make timing of application critical was shown by Potter and Watson (1983b). If they sprayed *H. zea* SNPV against *H. virescens* just after the eggs were laid, 15% of the larvae died, whereas if they sprayed just before the eggs hatched, 80% died.

INACTIVATION BY ULTRA-VIOLET LIGHT

The main factor causing infectivity loss in the field appears to be the UV component of sunlight. Attempts are made to protect some IB preparations by adding a UV-absorbing substance. A polyflavinoid marketed as 'Shade' has been used with *H. zea* SNPV, and increased virus persistence and/or mortality of larvae has resulted. 'Shade' is incorporated into *L. dispar* MNPV preparations (Lewis, 1981), and has been shown to act as a UV-protectant for this virus in a laboratory test, although more protection was afforded by the feeding stimulant 'Coax' (Shapiro, Poh Agin and Bell, 1983).

INACTIVATION BY COTTON LEAF SECRETIONS

Cotton leaf secretions have a high pH due to substances secreted by epidermal glands (Elleman and Entwistle, 1982), and there is some evidence that IBs on the leaf surface can be affected (Andrews and Sikorowski, 1973). Richards (MSc thesis cited in Richards and Payne, 1982) found that an unbuffered suspension of *S. littoralis* NPV was completely inactivated 6 days after application to cotton leaves, whereas infectivity was preserved for much longer if the virus was applied in a phosphate buffer, pH 7. Some workers in the US have applied *H. zea* SNPV to cotton in buffered suspensions, but results have varied in different areas. Further investigations are necessary to determine whether there are advantages to be gained by applying viruses to cotton in buffered suspensions.

Virus production

Viral insecticides are currently produced in the host insect which is either collected in the field or reared in an insectary. For some species more than 10^8 larvae per year are produced. High standards of hygiene are vital to reduce the risk of infection by pathogens which could decimate the insect stocks and contaminate the product.

The production of *H. zea* SNPV ('Elcar') has been described by Ignoffo and Anderson (1979) and Ignoffo and Couch (1981). *H. zea* larvae are reared on a semi-synthetic diet in the wells of plastic trays. Each larva yields about $3{\cdot}5 \times 10^9$ IBs which are extracted, purified and spray dried. The final product contains 99·6% inert ingredients and is stored at −20°C. 'Elcar' is produced in the US by Sandoz Inc. who also produce smaller quantities of *A. californica* MNPV and an NPV of *T. ni* for experimental purposes. Both of these viruses are produced in *T. ni* larvae and the IB preparations are spray-dried (Yearian and Young, 1982).

A process for the mass production of *L. dispar* MNPV ('Gypchek') has been described by Shapiro *et al.* (1981) and Shapiro (1982), in which the IB yield represents a 5600-fold increase over the inoculum.

For the production of sawfly viruses, either field-collected larvae are infected and then maintained on host plant material, or infected larvae are collected in the field (Cunningham and Entwistle, 1981).

Production of viruses in insects is labour-intensive and therefore costly in the developed nations. The most time-consuming stage in 'Elcar' production is the introduction of larvae into the trays, while in 'Gypchek' production it is the removal of the infected larvae from their containers.

OPTIMIZING PRODUCTION

The host insect

The insect species from which a virus was isolated may not be the most susceptible (page 385). Use of a more susceptible host for virus production would mean a smaller inoculum requirement.

Alternative hosts might also be considered for insect species which have a long life cycle, which are small and produce a low yield, or which have allergenic and urticarious setae. Shapiro *et al.* (1982) suggested that *O. pseudotsugata* MNPV might be produced in the saltmarsh caterpillar, *Estigmene acrea*, which is more easily reared than the homologous host. It is important to check that virus produced in an alternative host does not have reduced virulence for the original host.

Insect diet

Diet may affect the growth rate of an insect, its susceptibility to virus infection, and the virus yield. Synthetic or semi-synthetic diets are used for most insects, and cost is an important factor. A diet rich in wheat germ was found to be the most cost effective for *L. dispar* MNPV production, although higher IB yields could be obtained using other, more expensive, diets (Shapiro, Bell and Owens, 1981). Shapiro (1982) found a substitute for agar which was 40% cheaper than agar and resulted in improved growth of *L. dispar* larvae with higher IB yields.

Glen and Payne (1984) increased the yield of *C. pomonella* GV by incorporating into the diet a juvenile hormone analogue (methoprene) which resulted in larger larvae.

Insect stage and virus dosage

The lower IB doses necessary to infect younger larvae must be balanced against the early deaths of these larvae with smaller IB yields. The optimum dosage must be determined. If it is too low, many larvae will not become infected, whereas if it is too high, inoculum will be wasted and larval growth will be retarded resulting in suboptimal yields.

Incubation environment

Temperature affects the rate of insect growth, the rate of virus replication and the virus yield. The optimum temperature for each of these may not be the same and it is necessary to determine the optimum for yield. Relative humidity and photoperiod must also be maintained at their optima.

Preservation of virus infectivity

Conditions which destroy infectivity (e.g. increased temperature, extremes of pH) must be avoided during harvesting, purification, formulation and storage of virus.

QUALITY CONTROL

Each batch of virus must be carefully bioassayed (page 383) and tested for the presence of harmful contaminants, especially human pathogens. Morris, Vail and Collier (1981) suggested that quality control procedures should include tests for contaminants such as small RNA viruses.

Future prospects

USE OF VIRAL INSECTICIDES IN INTEGRATED PEST MANAGEMENT

The relatively narrow host spectra of insect viruses may be environmentally attractive, but mean that markets for viral insecticides are restricted and that a virus alone is unlikely to afford protection against all the pests in a particular ecosystem. For example, if *C. pomonella* GV is used for codling moth control in orchards, other lepidopteran pests, especially tortrix moths, may resurge. On the other hand, use of the virus has the advantage that parasites and predators of the fruit-tree red spider mite, *Panonychus ulmi*, are not killed, so damaging numbers of this pest are not reached, which may occur if an organophosphorus insecticide is used to control codling moth (Glen *et al.*, 1984).

A virus may form a useful component of an IPM programme in which pests are controlled by husbandry practices, chemicals and biological agents. The most widely used microbial control agent is *Bacillus thuringiensis*, most strains of which have wide spectra of activity against lepidopteran insects. In fact, the existence of this microbial insecticide is one of the factors limiting the development of viral insecticides, although for some pests the two might be used together. *B. thuringiensis*, together with *P. rapae* GV and *A. californica* MNPV, have been reported to control *P. rapae* and *T. ni* on cabbage almost as effectively as chemical insecticides (Sears, Jaques and Laing, 1983).

IPM on cotton might include the use of *H. zea* SNPV and *B. thuringiensis* or chlordimeform against *Heliothis* spp., *A. californica* MNPV and *B. thuringiensis* against the cotton leaf-perforator, *Bucculatrix thurberiella* (Bell and Romine, 1982), and diflubenzuron against the cotton boll weevil, *Anthonomus grandis* (Bull *et al.*, 1979).

VIRUS PRODUCTION IN CELL CULTURE

Several laboratories are attempting to develop reliable and economic cell-culture systems for the mass production of insect viruses as alternatives to production in insects, which has a number of associated problems (pages 398–399), and because a high-purity product is more feasible from cell cultures. Much progress has been made, but several problems remain to be solved. An outline of insect-cell culture techniques was given on pages 381–382. The possible application of those techniques to virus production will now be discussed.

Production systems

Insect cells can be grown in fermenters of the type used for vaccine production and Vaughn (1981) has suggested that the slack periods of such plants could be used for insect-virus production. When *A. californica* MNPV was produced in TN-368 cells in fermenters 2–3 litres in volume it was found (Hink and Strauss, 1980) that more vigorous aeration was required than in small volumes, and this resulted in foaming and cell damage. Antifoam was added and the concentration of methylcellulose, already present to inhibit cell clumping, was increased to protect the cells.

Attempts have been made to avoid the stresses imposed on cells in traditional fermenters by using alternative systems. Miltenburger and David (1980) blew air through silicone rubber tubing coiled inside a fermenter. Oxygen diffused through the silicone rubber into the medium. Hilwig and Alapatt (1981) and Vaughn and Dougherty (1981) have worked on roller bottle systems, but Stockdale and Priston (1981) believe that they are too bulky and labour-intensive for adoption by industry. Vaughn and Dougherty (1981) are also developing a 'perfusion culture system': this consists of vessels containing coils which provide a large surface area for cell attachment; pH and oxygen concentration are adjusted outside the vessel. Pollard and Khosrovi (1978) presented a design for a continuous-flow tubular fermenter.

Optimizing production

Some of the factors which can affect IB yield were investigated by Gardiner, Priston and Stockdale (1976) for *A. californica* MNPV in TN-368 cells. They found an optimum temperature of 27°C, an optimum pH range of 5·5–6·5 and an optimum osmotic pressure range of 250–500 milliosmoles. For the same virus–cell system Hink (1982) reported production of 10^8 IBs/ml medium and suggested that this must be increased twentyfold before the system becomes economic.

The IB yield can be affected by the growth phase of the cells at the time of virus inoculation (Lynn and Hink, 1978), and by their concentration. The cell concentration giving maximum IB yield per ml of medium was higher than that at which the maximum number of IBs per cell was produced (Hink, Strauss and Ramoska, 1977; Stockdale and Gardiner, 1977). The latter authors suggested that the reduced IB production at higher cell densities might be due to depletion of a vital precursor. Wood, Johnston and Burand (1982) reported a 98% reduction in virus production in high-density attached cultures compared with low-density cultures. Inhibition of virus production did not occur unless there was cell-to-cell contact. Further investigations are necessary into the mechanisms of, and ways of overcoming, inhibition of IB production at high cell concentrations.

Virus strains and cell strains should be selected to give a high-yielding system. Cells should be cloned and the clones screened for desirable properties, e.g. more rapid growth rate (McIntosh and Rechtoris, 1974). The quality of IBs produced in a cell line must also be checked. Lynn and Hink (1980) found that *A. californica* MNPV IBs produced in cells from four insect species were less infective for *T. ni* larvae than IBs produced in *T. ni* cells.

Most insect-cell culture media are expensive, principally because most of them contain foetal bovine serum. Dougherty, Cantwell and Kuchinski (1982) calculated that, for *G. mellonella* MNPV production in cell culture, half the cost, including labour, was for serum. Serum-free media are now being developed and have been reported for a *S. frugiperda* cell line with replication of *A. californica* MNPV (Wilkie, Stockdale and Pirt, 1980), an *L. dispar* cell line with replication of *L. dispar* NPV (Goodwin and Adams, 1980) and for several other cell lines (Mitsuhashi, 1982). Weiss *et al.* (1981) reduced the cost of their medium for *S. frugiperda* cells by omitting antibiotics. They encountered no contamination problems.

Changes in virus on passage

Several studies, e.g. Faulkner and Henderson (1972), have demonstrated that IBs produced during the first few passages in cell culture are as infective as IBs produced in insects. Upon repeated passage, however, the quality and yield of IBs have been found to decline. Hirumi, Hirumi and McIntosh (1975) reported that passage of *A. californica* MNPV in *T. ni* cells led to the production of aberrant virions and a reduction in IB yield. After 40 passages the yield had dropped a hundredfold, with IBs developing in only 4% or less of the cells (McIntosh, Shamy and Ilsley, 1979). MacKinnon *et al.* (1974) found a reduction in average yield of *T. ni* MNPV IBs from 28 per cell initially to 2·5 per cell after 50 passages, with extensive production of abnormal capsids. Knudson and Harrap (1976) found that passage of *S. frugiperda* NPV in *S. frugiperda* cells led to the production of IBs containing few or no virions, and Yamada, Sherman and Maramorosch (1982) reported reduced yields of *H. zea* SNPV IBs after 20 passages in *H. zea* cells.

Hink and Strauss (1976) described two plaque morphologies after passage of *A. californica* MNPV *in vitro*. In one type of plaque there were between 81 and 352 IBs per nucleus, while in the other there were only 2–13 IBs per nucleus. The plaque types were named many-polyhedra (MP) and few-polyhedra (FP) plaques respectively. The FP plaques became increasingly dominant on passage. IBs from MP plaques contained normal virions (multiple nucleocapsids per virion) and were much more infective for *T. ni* larvae than IBs from FP plaques, which either contained only a few virions (each with only a single nucleocapsid per virion) or appeared devoid of virions. Similar phenomena have been reported for *T. ni* MNPV (Potter, Faulkner and MacKinnon, 1976), *G. mellonella* MNPV (Fraser and Hink, 1982) and for *H. zea* SNPV (M. J. Fraser and W. J. McCarthy, unpublished, in Fraser, Smith and Summers, 1983).

Most FP forms are genetically stable. They have a selective advantage *in vitro* as FP-infected cells produce higher titres of budded virions than MP-infected cells (Potter, Jaques and Faulkner, 1978). Wood (1980) suggested that FP forms might be deletion mutants of MP forms, but several FP forms have been found to contain insertions of host DNA (Miller and Miller, 1982; Fraser, Smith and Summers, 1983).

One way of avoiding the FP form becoming dominant in cell cultures would be to return regularly for inoculum to haemolymph from insects infected by ingestion of IBs. Insect haemolymph, however, is unlikely to supply the quantities of inoculum that an industrial-scale process would demand, so some means will have to be found of preventing the development of FP and other aberrant virus forms *in vitro*.

VIRUS PURIFICATION

The debate concerning the degree of purification necessary before a virus is sprayed in the field has not yet been resolved. Of primary concern are the potential hazards posed by the presence of contaminants (pages 390–391). Other considerations are the costs involved and the possible effects of purification on virus infectivity and persistence.

There have been several reports (e.g. Magnoler, 1968; Carner, Hudson and Barnett, 1979; Evans and Harrap, 1982) of laboratory and field tests in which purified IB preparations had lower infectivity and/or poorer environmental persistence than IB preparations contaminated with insect fragments, gut contents and micro-organisms. It is well known that proteins can protect viruses from inactivation, so the contaminants may afford some protection, especially from UV light. *N. sertifer* SNPV, however, controls its host effectively when applied as a highly purified IB preparation (Entwistle *et al.*, in press). It is interesting that *L. dispar* larvae were deterred from feeding on foliage contaminated with decayed cadavers or extracts from healthy larvae, but were not deterred by foliage treated with purified NPV IBs (Capinera, Kirouac and Barbosa, 1976).

The most efficient way of purifying IBs is by some form of gradient centrifugation, but this can make the final product prohibitively expensive in the developed nations, let alone the developing nations. In the future, if viruses are produced in cell cultures they should be free from contaminating micro-organisms and minimal purification should be necessary. In the meantime, tests should be carried out to evaluate the hazards posed by contaminants in virus preparations produced in insects, which should be subjected to rigorous quality control procedures before field application.

NEW VIRUS STRAINS

Ideal attributes in a viral pesticide are high infectivity and high virulence for a broad range of pest species, rapid replication with high yields, and good field persistence with high resistance to UV inactivation. No known virus is endowed with all of these attributes, but progress towards the development of such an agent should be possible by two approaches, i.e. by searching in nature for new virus strains and by the genetic manipulation of existing isolates.

There can be no doubt that the number of insect virus strains isolated to date is only a tiny fraction of the total in nature. Virus strains with desirable properties will undoubtedly be found among future isolates.

A virus might be genetically improved by selecting for desired traits, or by using the techniques of genetic engineering. The former approach was used by Brassel and Benz (1979) who selected a strain of *C. pomonella* GV which was 5·6 times more resistant to UV light than the original isolate, and remained infective for twice as long in the field. Wood *et al.* (1981) induced mutations in *A. californica* MNPV, then isolated a mutant with increased virulence for *T. ni* larvae, demonstrated by a significantly reduced LT_{50}. Among several possibilities for genetically engineering virus strains is the suggestion by Miller, Lingg and Bulla (1983) that the gene for an insect-specific toxin might be incorporated into the virus genome to kill the host more rapidly.

PATENTS

A search by Stockdale (in press) revealed that 13 patents had been filed for processes or formulations involving insect viruses. It is not possible to patent the viruses, however, and this is one of the reasons why viral pesticides have not been developed more rapidly. There are some hopes that this situation may change and that it may become possible to patent an organism if it is the product of a biotechnological process (Crespi, 1980) or if it has undergone genetic manipulation (Kayton, 1983). If these hopes are realized then there will be more incentive for commercial concerns to invest in microbial pesticide development.

DEVELOPMENT OF PEST RESISTANCE

There are not yet any reports of selection of an insect strain with high resistance to a virus, as occurred in the rabbit to myxoma virus (Fenner, 1983). It is a possible outcome, however, if a virus is widely used over a long period. Genetic variability, upon which selection could operate, has been demonstrated in a number of insect species, e.g. varying levels of susceptibility to a GV in the Indian meal moth, *Plodia interpunctella* (Hunter and Hoffmann, 1973) and to an NPV in the light brown apple moth, *Epiphyas postvittana* (Briese *et al.*, 1980).

Some workers have attempted to select for virus resistance. Ignoffo and Allen (1972) failed to select for increased resistance to an NPV in *H. zea* after inoculating 25 generations of larvae with doses at, or greater than, the LD_{50} and breeding from the survivors. Briese and Mende (1983), however, selected for resistance to a GV in *P. operculella* within six generations of insects from the wild, but they were able to select for only a slight increase in resistance in a laboratory strain which was already highly resistant.

CONCLUSION

Advances in BV research are providing greater insight into the viruses themselves, e.g. their genetics, and into their interactions with their hosts, e.g. their epizootiology. This information means that the viruses can be used as pesticides on a more rational basis.

A number of insect viruses are currently used as pesticides and the potential of others has been demonstrated. Entwistle (1983) is optimistic that BVs will become the principal means of regulating lepidopteran and sawfly pests of forests. Viral pesticides can have an important role in Third World countries if they can be produced locally and if it can be shown that they are safe to disseminate in an unpurified or semi-purified state.

The potential for viral insecticides in the agriculture and horticulture of the Developed World is more limited at present. Consumers demand fruit and vegetables free from blemishes, so growers look for products which provide a quick and virtually complete kill of pests. Often this cannot be achieved with viruses, so until there is a change in consumer attitude, chemicals are likely to remain the main tools for pest control. As the integrated pest management approach gains ground, however, viruses of pests should have increasingly useful roles.

Acknowledgements

I am grateful to Drs J. C. Cunningham, P. F. Entwistle, C. C. Payne, H. Stockdale and R. E. Teakle for helpful discussions and for allowing me access to their unpublished work, and to Karen Bernard and Christine Guy for typing the manuscript.

References

ADAMS, J.R., GOODWIN, R.H. AND WILCOX, T.A. (1977). Electron microscopic investigations on invasion and replication of insect baculoviruses *in vivo* and *in vitro*. *Biologie Cellulaire* **28,** 261–268.

ADANG, M.J. AND MILLER, L.K. (1982). Molecular cloning of DNA complementary to mRNA of the baculovirus *Autographa californica* nuclear polyhedrosis virus: location and gene products of RNA transcripts found late in infection. *Journal of Virology* **44,** 782–793.

AIZAWA, K. (1976). Recent development in the production and utilisation of microbial insecticides in Japan. In *Proceedings of the First International Colloquium on Invertebrate Pathology, Kingston, Canada,* (T.A. Angus, P. Faulkner and A. Rosenfield, Eds), pp. 59–63. Society for Invertebrate Pathology, Queen's University at Kingston.

ALESHINA, O.A., EGIAZARYAN, L.A., SOLDATOVA, N.V. AND MARTYNOVA, G.S. (1973a). The infection of culture L of mouse cells with nuclear polyhedrosis virus (first communication. [English abstract.] *Review of Applied Entomology, Series A* **61,** 812–813.

ALESHINA, O.A., SOLDATOVA, N.V., MARTYNOVA, G.S. AND EGIAZARYAN, L.A. (1973b). The caryometrical study of the cytopathogenic action of the nuclear polyhedrosis virus on transplantable cultures of mammalian cells. [English abstract.] *Review of Applied Entomology, Series A* **61,** 812.

ALLAWAY, G.P. AND PAYNE, C.C. (1983). A biochemical and biological comparison of three European isolates of nuclear polyhedrosis viruses from *Agrotis segetum. Archives of Virology* **75,** 43–54.

ALLAWAY, G.P. AND PAYNE, C.C. (1984). Host range and virulence of five baculoviruses from lepidopterous hosts. *Annals of Applied Biology,* in press.

ANDREWS, G.L. AND SIKOROWSKI, P.P. (1973). Effects of cotton leaf surfaces on the nuclear polyhedrosis virus of *Heliothis zea* and *Heliothis virescens* (Lepidoptera : Noctuidae). *Journal of Invertebrate Pathology* **22,** 290–291.

BAUGHER, D.G. AND YENDOL, W.G. (1981). Virulence of *Autographa californica* baculovirus preparations fed with different food sources to cabbage loopers. *Journal of Economic Entomology* **74,** 309–313.

BEDFORD, G.O. (1981). Control of the rhinoceros beetle by baculovirus. In *Microbial Control of Pests and Plant Diseases 1970–1980* (H.D. Burges, Ed.), pp. 409–426. Academic Press, London.

BEEKMAN, A.G.B. (1980). The infectivity of polyhedra of nuclear polyhedrosis virus (N.P.V.) after passage through gut of an insect-predator. *Experientia* **36,** 858–859.

BELL, M.R. AND ROMINE, C.L. (1982). Cotton leafperforator (Lepidoptera : Lyonetiidae): effect of two microbial insecticides on field populations. *Journal of Economic Entomology* **75,** 1140–1142.

BERGOIN, M., GUELPA, B. AND MEYNADIER, G. (1975). Ultrastructure du virus de la polyédrose nucléaire du Diptère *Tipula paludosa* Meig. *Journal de Microscopie et de Biologie Cellulaire* **23,** 9a–10a.

BIRD, F.T. (1953). The use of a virus disease in the biological control of the European pine sawfly, *Neodiprion sertifer* (Geoffr.). *Canadian Entomologist* **85,** 437–446.

BIRD, F.T. AND ELGEE, D.E. (1957). A virus disease and introduced parasites as factors controlling the European spruce sawfly, *Diprion hercyniae* (Htg.) in central New Brunswick. *Canadian Entomologist* **89,** 371–378.

BOUCIAS, D.G., JOHNSON, D.W. AND ALLEN, G.E. (1980). Effects of host age, virus dosage, and temperature on the infectivity of a nucleopolyhedrosis virus against velvetbean caterpillar, *Anticarsia gemmatalis*, larvae. *Environmental Entomology* **9,** 59–61.

BRASSEL, J. AND BENZ, G. (1979). Selection of a strain of the granulosis virus of the codling moth with improved resistance against artificial ultraviolet radiation and sunlight. *Journal of Invertebrate Pathology* **33,** 358–363.

BRIESE, D.T. (1981). The incidence of parasitism and disease in field populations of the potato moth *Phthorimaea operculella* (Zeller) in Australia. *Journal of the Australian Entomological Society* **20,** 319–326.

BRIESE, T.D. AND MENDE, H.A. (1983). Selection for increased resistance to a granulosis virus in the potato moth, *Phthorimaea operculella* (Zeller) (Lepidoptera: Gelechiidae). *Bulletin of Entomological Research* **73,** 1–9.

BRIESE, D.T., MENDE, H.A., GRACE, T.D.C. AND GEIER, P.W. (1980). Resistance to a nuclear polyhedrosis virus in the light-brown apple moth *Epiphyas postvittana* (Lepidoptera: Tortricidae). *Journal of Invertebrate Pathology* **36,** 211–215.

BROWN, D.A. (1982). Two naturally occurring nuclear polyhedrosis virus variants of *Neodiprion sertifer* Geoffr. (Hymenoptera: Diprionidae). *Applied and Environmental Microbiology* **43,** 65–69.

BROWN, D.A., ALLEN, C.J. AND BIGNELL, G.N. (1982). The use of a protein A conjugate in an indirect enzyme-linked immunosorbent assay (ELISA) of four closely related baculoviruses from *Spodoptera* species. *Journal of General Virology* **62,** 375–378.

BUCHER, G.E. AND TURNOCK, W.J. (1983). Dosage responses of the larval instars of the bertha armyworm, *Mamestra configurata* (Lepidoptera: Noctuidae), to a native nuclear polyhedrosis. *Canadian Entomologist* **115,** 341–349.

BULL, D.L., HOUSE, V.S., ABLES, J.R. AND MORRISON, R.K. (1979). Selective methods for managing insect pests of cotton. *Journal of Economic Entomology* **72,** 841–846.

BURGERJON, A., BIACHE, G., CHAUFAUX, J. AND PETRÉ, Z. (1981). Sensibilité comparée, en fonction de leur âge, des chenilles de *Lymantria dispar*, *Mamestra brassicae* et *Spodoptera littoralis* aux virus de la polyédrose nucléaire. *Entomophaga* **26,** 47–58.

BURGES, H.D., CROIZIER, G. AND HUBER, J. (1980). A review of safety tests on baculoviruses. *Entomophaga* **25,** 329–340.

BURLEY, S.K., MILLER, A., HARRAP, K.A. AND KELLY, D.C. (1982). Structure of the *Baculovirus* nucleocapsid. *Virology* **120,** 433–440.

CAPINERA, J.L., KIROUAC, S.P. AND BARBOSA, P. (1976). Phagodeterrency of cadaver components to gypsy moth larvae, *Lymantria dispar*. *Journal of Invertebrate Pathology* **28,** 277–279.

CAREY, D. AND HARRAP, K.A. (1980). Safety tests on the nuclear polyhedrosis viruses of *Spodoptera littoralis* and *Spodoptera exempta*. In *Invertebrate Systems In Vitro* (E. Kurstak, K. Maramorosch and A. Dübendorfer, Eds), pp. 441–450. Fifth International Conference of Invertebrate Tissue Culture, Rigi-Kaltbad, 1979. Elsevier/North-Holland, Amsterdam.

CARNER, G.R., HUDSON, J.S. AND BARNETT, O.W. (1979). The infectivity of a nuclear polyhedrosis virus of the velvetbean caterpillar for eight noctuid hosts. *Journal of Invertebrate Pathology* **33,** 211–216.

CARTER, J.B. (1973a). The mode of transmission of *Tipula* iridescent virus I. Source of infection. *Journal of Invertebrate Pathology* **21,** 123–130.

CARTER, J.B. (1973b). The mode of transmission of *Tipula* iridescent virus II. Route of infection. *Journal of Invertebrate Pathology* **21,** 136–143.

CARTER, J.B. (1978). Field trials with *Tipula* iridescent virus against *Tipula* spp. larvae in grassland. *Entomophaga* **23,** 169–174.

COCHRAN, M.A. AND FAULKNER, P. (1983). Location of homologous DNA sequences interspersed at five regions in the baculovirus *Ac*NPV genome. *Journal of Virology* **45,** 961–970.

CRESPI, S. (1980). Patenting nature's secrets and protecting microbiologists' interests. *Nature* **284,** 590–591.

CROIZIER, G. AND MEYNADIER, G. (1972). Les protéines des corps d'inclusion des *Baculovirus* 1. Etude de leur solubilisation. *Entomophaga* **17,** 231–239.

CROIZIER, G., GODSE, D. AND VLAK, J. (1980). Sélection de types viraux dans les infections doubles à *Baculovirus* chez les larves de Lépidoptère. *Comptes rendus des séances de l'Académie des sciences, Série D* **290,** 579–582.

CROIZIER, G., AMARGIER, A., GODSE, D.-B., JACQUEMARD, P. AND DUTHOIT, J.-L. (1980). Un virus de polyédrose nucléaire découvert chez le lépidoptère Noctuidae *Diparopsis watersi* (Roth.) nouveau variant du *Baculovirus* d'*Autographa californica* (Speyer). *Coton et Fibres Tropicales* **35,** 415–423.

CROOK, N.E., BROWN, J.D. AND FOSTER, G.N. (1982). Isolation and characterization of a granulosis virus from the tomato moth, *Lacanobia oleracea*, and its potential as a control agent. *Journal of Invertebrate Pathology* **40,** 221–227.

CUNNINGHAM, J.C. (1982). Field trials with baculoviruses: control of forest insect pests. In *Microbial and Viral Pesticides* (E. Kurstak, Ed.), pp. 335–386. Marcel Dekker, New York.

CUNNINGHAM, J.C. AND ENTWISTLE, P.F. (1981). Control of sawflies by baculovirus. In *Microbial Control of Pests and Plant Diseases 1970–1980* (H.D. Burges, Ed.), pp. 379–407. Academic Press, London.

CUNNINGHAM, J.C., KAUPP, W.J. AND MCPHEE, J.R. (1983). Ground spray trials with two baculoviruses on western spruce budworm. *Canadian Forestry Service Research Notes* **3,** 10–11.

DAVID, W.A.L. (1965). The granulosis virus of *Pieris brassicae* L. in relation to natural limitation and ecological control. *Annals of Applied Biology* **56,** 331–334.

DAVID, W.A.L. AND GARDINER, B.O.C. (1967). The persistence of a granulosis virus of *Pieris brassicae* in soil and sand. *Journal of Invertebrate Pathology* **9,** 342–347.

DOANE, C.C. (1970). Primary pathogens and their role in the development of an epizootic in the gypsy moth. *Journal of Invertebrate Pathology* **15,** 21–33.

DOANE, C.C. (1975). Infectious sources of nuclear polyhedrosis virus persisting in natural habitats of the gypsy moth. *Environmental Entomology* **4,** 392–394.

DOANE, C.C. (1976). Epizootiology of diseases of the gypsy moth. In *Proceedings of the First International Colloquium on Invertebrate Pathology, Kingston, Canada,* (T.A. Angus, P. Faulkner and A. Rosenfield, Eds), pp. 161–165. Society for Invertebrate Pathology, Queen's University at Kingston.

DOBOS, P. AND COCHRAN, M.A. (1980). Protein synthesis in cells infected by *Autographa californica* nuclear polyhedrosis virus (*Ac*–NPV): the effect of cytosine arabinoside. *Virology* **103,** 446–464.

DÖLLER, G. (1980). Solid phase radioimmunoassay for the detection of polyhedrin antibodies. In *Safety Aspects of Baculoviruses as Biological Insecticides* (H.G. Miltenburger, Ed.), pp. 203–210. Symposium Proceedings, Jülich, 1978, Bundesministerium für Forschung und Technologie, Bonn, Federal Republic of Germany.

DÖLLER G. (1981). Unspecific interaction between granulosis virus and mammalian immunoglobulins. *Naturwissenschaften* **68,** 573–574.

DÖLLER, G. AND ENZMANN, P.-J. (1982). Induction of baculovirus specific antibodies in rainbow trout and carp. *Bulletin of the European Association of Fish Pathologists* **2,** 53–55.

DÖLLER, G. AND HUBER, J. (1983). Sicherheitsstudie zur Prüfung einer Vermehrung des Granulosevirus aus *Laspeyresia pomonella* in Säugern. *Zeitschrift für angewandte Entomologie* **95,** 64–69.

DÖLLER, G., GRÖNER, A. AND STRAUB, O.C. (1983). Safety evaluation of nuclear polyhedrosis virus replication in pigs. *Applied and Environmental Microbiology* **45,** 1229–1233.

DOUGHERTY, E.M., CANTWELL, G.E. AND KUCHINSKI, M. (1982). Biological control of the greater wax moth (Lepidoptera: Pyralidae), utilising in vivo- and in vitro-propagated baculovirus. *Journal of Economic Entomology* **75,** 675–679.

DUBOIS, N. (1976). Effectiveness of chemically decontaminated *Neodiprion sertifer* polyhedral inclusion body suspensions. *Journal of Economic Entomology* **69,** 93–95.

ELLEMAN, C.J. AND ENTWISTLE, P.F. (1982). A study of glands on cotton responsible for the high pH and cation concentration of the leaf surface. *Annals of Applied Biology* **100,** 553–558.

ELLIOTT, R.M. AND KELLY, D.C. (1979). Compartmentalization of the polyamines contained by a nuclear polyhedrosis virus from *Heliothis zea. Microbiologica* **2,** 409–413.

ENTWISTLE, P.F. (1976). The development of an epizootic of a nuclear polyhedrosis virus disease in European spruce sawfly, *Gilpinia hercyniae.* In *Proceedings of the First International Colloquium on Invertebrate Pathology, Kingston, Canada* (T.A. Angus, P. Faulkner and A. Rosenfield, Eds), pp. 184–188. Society for Invertebrate Pathology, Queen's University at Kingston.

ENTWISTLE, P.F. (1983). Viruses for insect pest control. *Span* **26,** 59–62.

ENTWISTLE, P.F. AND ADAMS, P.H.W. (1977). Prolonged retention of infectivity in the nuclear polyhedrosis virus of *Gilpinia hercyniae* (Hymenoptera: Diprionidae) on foliage of spruce species. *Journal of Invertebrate Pathology* **29,** 392–394.

ENTWISTLE, P.F., ADAMS, P.H.W. AND EVANS, H.F. (1978). Epizootiology of a nuclear polyhedrosis virus in European spruce sawfly (*Gilpinia hercyniae*): the rate of passage of infective virus through the gut of birds during cage tests. *Journal of Invertebrate Pathology* **31,** 307–312.

ENTWISTLE, P.F., ADAMS, P.H.W., EVANS, H.F. AND RIVERS, C.F. (1983). Epizootiology of a nuclear polyhedrosis virus (Baculoviridae) in European spruce sawfly (*Gilpinia hercyniae*): spread of disease from small epicentres in comparison with spread of baculovirus diseases in other hosts. *Journal of Applied Ecology* **20,** 473–487.

ENTWISTLE, P.F., EVANS, H.F., HARRAP, K.A. AND ROBERTSON, J.S. (1978). *Field Trials on the Control of Pine Sawfly* (Neodiprion sertifer) *using Purified Nuclear Polyhedrosis Virus. First series 1977, Technical Report No. 1.* Unit of Invertebrate Virology, Oxford, UK.

ENTWISTLE, P.F., EVANS, H.F., HARRAP, K.A. AND ROBERTSON, J.S. (in press). Control of European pine sawfly (*Neodiprion sertifer*) (Geoffr.) with its nuclear polyhedrosis virus in Scotland. In *Population Dynamics of Forest Pests* (D. Bevan, Ed.), Proceedings of IUFRO Meeting, Dornoch, Scotland, 1980.

EPPSTEIN, D.A. AND THOMA, J.A. (1977). Characterization and serology of the matrix protein from a nuclear-polyhedrosis virus of *Trichoplusia ni* before and after degradation by an endogenous proteinase. *Biochemical Journal* **167,** 321–332.

ERLANDSON, M.A. AND CARSTENS, E.B. (1983). Mapping early transcription products of *Autographa californica* nuclear polyhedrosis virus. *Virology* **126,** 398–402.

EVANS, H.F. (1981). Quantitative assessment of the relationships between dosage and response of the nuclear polyhedrosis virus of *Mamestra brassicae. Journal of Invertebrate Pathology* **37,** 101–109.

EVANS, H.F. (1982). The ecology of *Mamestra brassicae* NPV in soil. In *Invertebrate Pathology and Microbial Control, Proceedings of the Third International Colloquium on Invertebrate Pathology, Brighton, UK* (C.C. Payne and H.D. Burges, Eds), pp. 307–312. Society for Invertebrate Pathology, Glasshouse Crops Research Institute, Littlehampton.

EVANS, H.F. (1983). The influence of larval maturation on responses of *Mamestra brassicae* L. (Lepidoptera: Noctuidae) to nuclear polyhedrosis virus infection. *Archives of Virology* **75,** 163–170.

EVANS, H.F. AND ALLAWAY, G.P. (1983). Dynamics of baculovirus growth and dispersal in *Mamestra brassicae* L. (Lepidoptera: Noctuidae) larval populations introduced into small cabbage plots. *Applied and Environmental Microbiology* **45,** 493–501.

EVANS, H.F. AND HARRAP, K.A. (1982). Persistence of insect viruses. In *Virus Persistence, 33rd Symposium of the Society for General Microbiology* (B.W.J. Mahy, A.C. Minson and G.K. Darby, Eds), pp. 57–96. Cambridge University Press.

EVANS, H.F., LOMER, C.J. AND KELLY, D.C. (1981). Growth of nuclear polyhedrosis virus in larvae of the cabbage moth, *Mamestra brassicae* L. *Archives of Virology* **70,** 207–214.

FALCON, L.A. (1982). Use of pathogenic viruses as agents for the biological control of insect pests. In *Population Biology of Infectious Diseases* (R.M. Anderson and R.M. May, Eds), pp. 191–210. Springer-Verlag, Berlin.

FAULKNER, P. AND HENDERSON, J.F. (1972). Serial passage of a nuclear polyhedrosis disease virus of the cabbage looper (*Trichoplusia ni*) in a continuous tissue culture cell line. *Virology* **50,** 920–924.

FAUST, R.M. AND ADAMS, J.R. (1966). The silicon content of nuclear and cytoplasmic viral inclusion bodies causing polyhedrosis in Lepidoptera. *Journal of Invertebrate Pathology* **8,** 526–530.

FEDERICI, B.A. (1980). Mosquito baculovirus: sequence of morphogenesis and ultrastructure of the virion. *Virology* **100,** 1–9.

FENNER, F. (1983). Biological control, as exemplified by smallpox eradication and myxomatosis. *Proceedings of the Royal Society of London, B* **218**, 259–285.

FLATTERY, K.E. (1983). Bioassay of a purified nuclear polyhedrosis virus against *Heliothis armigera. Annals of Applied Biology* **102,** 301–304.

FRASER, M.J. AND HINK, W.F. (1982). The isolation and characterization of the MP and FP plaque variants of *Galleria mellonella* nuclear polyhedrosis virus. *Virology* **117,** 366–378.

FRASER, M.J., SMITH, G.E. AND SUMMERS, M.D. (1983). Acquisition of host cell DNA sequences by baculoviruses: relationship between host DNA insertions and FP mutants of *Autographa californica* and *Galleria mellonella* nuclear polyhedrosis viruses. *Journal of Virology* **47,** 287–300.

GARDINER, G.R., PRISTON, R.A.J. AND STOCKDALE, H. (1976). Studies on the production of baculoviruses in insect tissue culture. In *Proceedings of the First International Colloquium on Invertebrate Pathology, Kingston, Canada* (T. A. Angus, P. Faulkner and A. Rosenfield, Eds), pp. 99–103. Society for Invertebrate Pathology, Queen's University at Kingston.

GLEN, D.M. AND PAYNE, C.C. (1984). Production and field evaluation of codling moth granulosis virus against *Cydia pomonella* in the United Kingdom. *Annals of Applied Biology* **104,** in press.

GLEN, D.M., WILTSHIRE, C.W., MILSOM, N.F. AND BRAIN, P. (1984). Codling moth granulosis virus: effects of its use on other orchard fauna. *Annals of Applied Biology*, in press.

GOODWIN, R.H. AND ADAMS, J.R. (1980). Liposome incorporation of factors permitting serial passage of insect viruses in Lepidopteran cells grown in serum-free medium. *In Vitro* **16,** 222.

GOURREAU, J.M., KAISER, C. AND MONTSARRAT, P. (1981). Étude de l'action pathogène éventuelle du baculovirus d'*Oryctes* sur cultures cellulaires de vertébrés en lignée continue. *Annales de Virologie (Institut Pasteur)* **132E,** 347–355.

GOURREAU, J.M., KAISER, C. AND MONSARRAT, P. (1982). Study of the possible pathogenic action of the *Oryctes* baculovirus in the white mouse. *Annales de Virologie (Institut Pasteur)* **133E,** 423–428.

GOURREAU, J.-M., KAISER, C., LAHELLEC, M., CHEVRIER, L. AND MONSARRAT, P. (1979). Étude de l'action pathogène éventuelle du *Baculovirus* d'*Oryctes* pour le porc. *Entomophaga* **24,** 213–219.

GRACE, T.D.C. (1982). Development of insect cell culture. In *Invertebrate Cell Culture Applications* (K. Maramorosch and J. Mitsuhashi, Eds), pp. 1–8. Academic Press, London.

GRANADOS, R.R. (1978). Early events in the infection of *Heliothis zea* midgut cells by a baculovirus. *Virology* **90,** 170–174.

GRANADOS, R.R. (1980a). Infectivity and mode of action of baculoviruses. *Biotechnology and Bioengineering* **22,** 1377–1405.

GRANADOS, R.R. (1980b). Replication phenomena of insect viruses *in vivo* and *in vitro*. In *Safety Aspects of Baculoviruses as Biological Insecticides, Symposium Proceedings, Jülich, 1978* (H.G. Miltenburger, Ed.), pp. 163–184. Bundesministerium für Forschung und Technologie, Bonn, Federal Republic of Germany.

GRANADOS, R.R. AND LAWLER, K.A. (1981). *In vivo* pathway of *Autographa californica* baculovirus invasion and infection. *Virology* **108,** 297–308.

GRANADOS, R.R., NGUYEN, T. AND CATO, B. (1978). An insect cell line persistently infected with a baculovirus-like particle. *Intervirology* **10,** 309–317.

GREEN, E.I. (1981). *Interactions Between a Baculovirus and its Host,* Tipula paludosa *(Meigen)*. PhD thesis, Liverpool Polytechnic.

HAMM, J.J. (1982). Extension of the host range for a granulosis virus from *Heliothis armiger* from South Africa. *Environmental Entomology* **11,** 159–160.

HAMM, J.J. AND HARE, W.W. (1982). Application of entomopathogens in irrigation water for control of fall armyworms and corn earworms (Lepidoptera: Noctuidae) on corn. *Journal of Economic Entomology* **75,** 1074–1079.

HAMM, J.J. AND YOUNG, J.R. (1974). Mode of transmission of nuclear-polyhedrosis virus to progeny of adult *Heliothis zea*. *Journal of Invertebrate Pathology* **24,** 70–81.

HARRAP, K.A. (1970). Cell infection by a nuclear polyhedrosis virus. *Virology* **42,** 311–318.

HARRAP, K.A. (1982). Assessment of the human and ecological hazards of microbial insecticides. *Parasitology* **84,** 269–296.

HARRAP, K.A. AND PAYNE, C.C. (1979). The structural properties and identification of insect viruses. *Advances in Virus Research* **25,** 273–355.

HARRAP, K.A. AND TINSLEY, T.W. (1978). The international scope of invertebrate virus research in controlling pests. In *Viral Pesticides: Present Knowledge and Potential Effects on Public and Environmental Health, Environmental Protection Agency Symposium, Myrtle Beach, South Carolina, US, 1977* (M.D. Summers and C.Y. Kawanishi, Eds), pp. 27–42. EPA Health Effects Research Laboratory, Research Triangle Park.

HARVEY, J.P. AND VOLKMAN, L.E. (1983). Biochemical and biological variation of *Cydia pomenella* (codling moth) granulosis virus. *Virology* **124,** 21–34.

HEINE, C.W., KELLY, D.C. AND AVERY, R.J. (1980). The detection of intracellular retrovirus-like entities in *Drosophila melanogaster* cell cultures. *Journal of General Virology* **49,** 385–395.

HENRY, J.E. (1982). Use of baits in microbial control of insects. In *Invertebrate Pathology and Microbial Control, Proceedings of the Third International Colloquium on Invertebrate Pathology, Brighton, UK* (C.C. Payne and H.D. Burges, Eds), pp. 45–48. Society for Invertebrate Pathology, Glasshouse Crops Research Institute, Littlehampton.

HILWIG, I. AND ALAPATT, F. (1981). Insect cells lines in suspension, cultivated in roller bottles. *Zeitschrift für angewandte Entomologie* **91,** 1–7.

HIMENO, M., SAKAI, F., ONODERA, K., NAKAI, H., FUKUDA, T. AND KAWADE, Y. (1967). Formation of nuclear polyhedral bodies and nuclear polyhedrosis virus of silkworm in mammalian cells infected with viral DNA. *Virology* **33,** 507–512.

HINK, W.F. (1970). Established insect cell line from the cabbage looper, *Trichoplusia ni*. *Nature* **226,** 466–467.

HINK, W.F. (1976). A compilation of invertebrate cell lines and culture media. In *Invertebrate Tissue Culture Research Applications* (K. Maramorosch, Ed.), pp. 319–369. Academic Press, New York.

HINK, W.F. (1980). The 1979 compilation of invertebrate cell lines and culture media. In *Invertebrate Systems In Vitro, Fifth International Conference on Invertebrate Tissue Culture, Rigi-Kaltbad, 1979* (E. Kurstak, K. Maramorosch and A. Dübendorfer, Eds), pp. 553–578. Elsevier/North-Holland, Amsterdam.

HINK, W.F. (1982). Production of *Autographa californica* nuclear polyhedrosis virus in cells from large-scale suspension cultures. In *Microbial and Viral Pesticides* (E. Kurstak, Ed.), pp. 493–506. Marcel Dekker, New York.

HINK, W.F. AND STRAUSS, E. (1976). Replication and passage of alfalfa looper nuclear polyhedrosis virus plaque variants in cloned cell cultures and larval stages of four host species. *Journal of Invertebrate Pathology* **27,** 49–55.

HINK, W.F. AND STRAUSS, E.M. (1977). An improved technique for plaque assay of *Autographa californica* nuclear polyhedrosis virus on TN–368 cells. *Journal of Invertebrate Pathology* **29,** 390–391.

HINK, W.F. AND STRAUSS, E.M. (1980). Semi-continuous culture of the TN–368 cell line in fermentors with virus production in harvested cells. In *Invertebrate Systems In Vitro, Fifth International Conference on Invertebrate Tissue Culture, Rigi-Kaltbad, 1979* (E. Kurstak, K. Maramorosch and A. Dübendorfer, Eds), pp. 27–33. Elsevier/North-Holland, Amsterdam.

HINK, W.F., STRAUSS, E.M. AND RAMOSKA, W.A. (1977). Propagation of *Autographa californica* nuclear polyhedrosis virus in cell culture: methods for infecting cells. *Journal of Invertebrate Pathology* **30,** 185–191.

HIRUMI, H., HIRUMI, K. AND McINTOSH, A.H. (1975). Morphogenesis of a nuclear polyhedrosis virus of the alfalfa looper in a continuous cabbage looper cell line. *Annals of the New York Academy of Sciences* **266,** 302–326.

HOHMANN, A.W. AND FAULKNER, P. (1983). Monoclonal antibodies to baculovirus structural proteins: determination of specificities by Western blot analysis. *Virology* **125,** 432–444.

HOSTETTER, D.L., BIEVER, K.D., HEIMPEL, A.M. AND IGNOFFO, C.M. (1979). Efficacy of the nuclear polyhedrosis virus of the alfalfa looper against cabbage looper larvae on cabbage in Missouri. *Journal of Economic Entomology* **72,** 371–373.

HOSTETTER, D.L., SMITH, D.B., PINNELL, R.E., IGNOFFO, C.M. AND McKIBBEN, G.H. (1982). Laboratory evaluation of adjuvants for use with *Baculovirus heliothis* virus. *Journal of Economic Entomology* **75,** 1114–1119.

HUBER, J. (1982). The baculoviruses of *Cydia pomonella* and other tortricids. In *Invertebrate Pathology and Microbial Control, Proceedings of the Third International Colloquium on Invertebrate Pathology, Brighton, UK* (C.C. Payne and H.D. Burges, Eds), pp. 119–124. Society for Invertebrate Pathology, Glasshouse Crops Research Institute, Littlehampton.

HUBER, J. AND DICKLER, E. (1977). Codling moth granulosis virus: its efficiency in the field in comparison with organophosphorus insecticides. *Journal of Economic Entomology* **70,** 557–561.

HUGER, A.M. (1966). A virus disease of the Indian rhinoceros beetle, *Oryctes rhinoceros* (Linnaeus), caused by a new type of insect virus, *Rhabdionvirus oryctes* gen.n., sp.n. *Journal of Invertebrate Pathology* **8,** 38–51.

HUGHES, P.R. AND WOOD, H.A. (1981). A synchronous peroral technique for the bioassay of insect viruses. *Journal of Invertebrate Pathology* **37,** 154–159.

HUNTER, D.K. AND HOFFMANN, D.F. (1973). Susceptibility of two strains of Indian meal moth to a granulosis virus. *Journal of Invertebrate Pathology* **21,** 114–115.

HUNTER, D.K., HOFFMANN, D.F. AND COLLIER, S.J. (1975). Observations on a granulosis virus of the potato tuberworm, *Phthorimaea operculella. Journal of Invertebrate Pathology* **26,** 397–400.

IGNOFFO, C.M. AND ALLEN, G.E. (1972). Selection for resistance to a nucleopolyhedrosis virus in laboratory populations of the cotton bollworm, *Heliothis zea. Journal of Invertebrate Pathology* **20,** 187–192.

IGNOFFO, C.M. AND ANDERSON, R.F. (1979). Bioinsecticides. In *Microbial Technology* (H.J. Peppler and D. Perlman, Eds), Volume 1, 2nd edn, pp. 1–28. Academic Press, New York.

IGNOFFO, C.M. AND COUCH, T.L. (1981). The nucleopolyhedrosis virus of *Heliothis* species as a microbial insecticide. In *Microbial Control of Pests and Plant Diseases 1970–1980* (H.D. Burges, Ed.), pp. 329–362. Academic Press, London.

IGNOFFO, C.M. AND RAFAJKO, R.R. (1972). *In vitro* attempts to infect primate cells with the nucleopolyhedrosis virus of *Heliothis*. *Journal of Invertebrate Pathology* **20,** 321–325.

IGNOFFO, C.M., ANDERSON, R.F. AND WOODARD, G. (1973). Teratogenic potential in rats fed the nuclear polyhedrosis virus of *Heliothis*. *Environmental Entomology* **2,** 337–338.

IGNOFFO, C.M., GARCIA, C., HOSTETTER, D.L. AND PINNELL, R.E. (1980). Transplanting: a method of introducing an insect virus into an ecosystem. *Environmental Entomology* **9,** 153–154.

IGNOFFO, C.M., HOSTETTER, D.L., BIEVER, K.D., GARCIA, C., THOMAS, G.D., DICKERSON, W.A. AND PINNELL, R. (1978). Evaluation of an entomopathogenic bacterium, fungus and virus for control of *Heliothis zea* on soybeans. *Journal of Economic Entomology* **71,** 165–168.

INJAC, M., VAGO, C., DUTHOIT, J.-L. AND VEYRUNES, J.-C. (1971). Libération (release) des virions dans les polyédroses nucléaires. *Comptes rendus des séances de l'Académie des sciences, Série D* **273**, 439–441.

IRABAGON, T.A. AND BROOKS, W.M. (1974). Interaction of *Campoletis sonorensis* and a nuclear polyhedrosis virus in larvae of *Heliothis virescens*. *Journal of Economic Entomology* **67,** 229–231.

JAQUES, R.P. (1969). Leaching of the nuclear polyhedrosis virus of *Trichoplusia ni* from soil. *Journal of Invertebrate Pathology* **13,** 256–263.

JOHNSON, T.B. AND LEWIS, L.C. (1982). Evaluation of *Rachiplusia ou* and *Autographa californica* nuclear polyhedrosis viruses in suppressing black cutworm damage to seedling corn in greenhouse and field. *Journal of Economic Entomology* **75,** 401–404.

JURKOVÍČOVÁ, M. (1979). Activation of latent virus infections in larvae of *Adoxophyes orana* (Lepidoptera: Tortricidae) and *Barathra brassicae* (Lepidoptera: Noctuidae) by foreign polyhedra. *Journal of Invertebrate Pathology* **34,** 213–223.

KALMAKOFF, J. AND CRAWFORD, A.M. (1982). Enzootic virus control of *Wiseana* spp. in the pasture environment. In *Microbial and Viral Pesticides* (E. Kurstak, Ed.), pp. 435–448. Marcel Dekker, New York.

KAWANISHI, C.Y. AND PASCHKE, J.D. (1970). The relationship of buffer pH and ionic strength on the yield of virions and nucleocapsids obtained by the dissolution of Rachiplusia ou nuclear polyhedra. In *Proceedings of the Fourth International Colloquium on Insect Pathology, Maryland, US* (A.M. Heimpel, Ed.), pp. 127–146. Society for Invertebrate Pathology, Agricultural Research Service, Beltsville, Maryland, USA.

KAWANISHI, C.Y., EGAWA, K. AND SUMMERS, M.D. (1972). Solubilization of *Trichoplusia ni* granulosis virus proteinic crystal. II. Ultrastructure. *Journal of Invertebrate Pathology* **20,** 95–100.

KAYTON, I. (1983). Does copyright law apply to genetically engineered cells? *Trends in Biotechnology* **1,** 2–3.

KELLY, D.C. (1981a). Baculovirus replication: electron microscopy of the sequence of infection of *Trichoplusia ni* nuclear polyhedrosis virus in *Spodoptera frugiperda* cells. *Journal of General Virology* **52,** 209–219.

KELLY, D.C. (1981b). Baculovirus replication: stimulation of thymidine kinase and DNA polymerase activities in *Spodoptera frugiperda* cells infected with *Trichoplusia ni* nuclear polyhedrosis virus. *Journal of General Virology* **52,** 313–319.

KELLY, D.C. (1982). Baculovirus replication. *Journal of General Virology* **63,** 1–13.

KELLY, D.C. AND LESCOTT, T. (1981). Baculovirus replication: protein synthesis in *Spodoptera frugiperda* cells infected with *Trichoplusia ni* nuclear polyhedrosis virus. *Microbiologica* **4,** 35–57.

KELSEY, J.M. (1958). Control of *Pieris rapae* by granulosis viruses. *New Zealand Journal of Agricultural Research* **1,** 778–782.

KISLEV, N. AND EDELMAN, M. (1982). DNA restriction-pattern differences from geographic isolates of *Spodoptera littoralis* nuclear polyhedrosis virus. *Virology* **119,** 219–222.

KLEIN, M. (1978). An improved peroral administration technique for bioassay of nucleopolyhedrosis viruses against Egyptian cotton worm, *Spodoptera littoralis*. *Journal of Invertebrate Pathology* **31,** 134–136.

KNUDSON, D.L. AND HARRAP, K.A. (1976). Replication of a nuclear polyhedrosis virus in a continuous cell culture of *Spodoptera frugiperda*: microscopy study of the sequence of events of the virus infection. *Journal of Virology* **17,** 254–268.

LAING, D.R. AND JAQUES, R.P. (1980). Codling moth: techniques for rearing larvae and bioassaying granulosis virus. *Journal of Economic Entomology* **73,** 851–853.

LAUTENSCHLAGER, R.A. AND PODGWAITE, J.D. (1979). Passage of nucleopolyhedrosis virus by avian and mammalian predators of the gypsy moth, *Lymantria dispar*. *Environmental Entomology* **8,** 210–214.

LAUTENSCHLAGER, R.A., ROTHENBACHER, H. AND PODGWAITE, J.D. (1978). Response of small mammals to aerial applications of the nucleopolyhedrosis virus of the gypsy moth, *Lymantria dispar*. *Environmental Entomology* **7,** 676–684.

LEE, H.H. AND MILLER, L.K. (1978). Isolation of genotypic variants of *Autographa californica* nuclear polyhedrosis virus. *Journal of Virology* **27,** 754–767.

LEWIS, F.B. (1981). Control of the gypsy moth by a baculovirus. In *Microbial Control of Pests and Plant Diseases 1970–1980* (H.D. Burges, Ed.), pp. 363–377. Academic Press, London.

LONGWORTH, J.F. AND CUNNINGHAM, J.C. (1968). The activation of occult nuclear-polyhedrosis viruses by foreign nuclear polyhedra. *Journal of Invertebrate Pathology* **10,** 361–367.

LONGWORTH, J.F. AND SINGH, P. (1980). A nuclear polyhedrosis virus of the light brown apple moth, *Epiphyas postvittana* (Lepidoptera: Tortricidae). *Journal of Invertebrate Pathology* **35,** 84–87.

LÜBBERT, H., KRUCZEK, I., TJIA, S. AND DOERFLER, W. (1981). The cloned *Eco*RI fragments of *Autographa californica* nuclear polyhedrosis virus DNA. *Gene* **16,** 343–345.

LYNN, D.E. AND HINK, W.F. (1978). Infection of synchronized TN–368 cell cultures with alfalfa looper nuclear polyhedrosis virus. *Journal of Invertebrate Pathology* **32,** 1–5.

LYNN, D.E. AND HINK, W.F. (1980). Comparison of nuclear polyhedrosis virus replication in five lepidopteran cell lines. *Journal of Invertebrate Pathology* **35,** 234–240.

MCCARTHY, W.J. AND HENCHAL, L.S. (1983). Detection of *Autographa californica* baculovirus nonoccluded virions in vitro and in vivo by enzyme-linked immunosorbent assay. *Journal of Invertebrate Pathology* **41,** 401–404.

MCINTOSH, A.H. AND IGNOFFO, C.M. (1983). Restriction endonuclease patterns of three baculoviruses isolated from species of *Heliothis*. *Journal of Invertebrate Pathology* **41,** 27–32.

MCINTOSH, A.H. AND MARAMOROSCH, K. (1973). Retention of insect virus infectivity in mammalian cell cultures. *Journal of the New York Entomological Society* **81,** 175–182.

MCINTOSH, A.H. AND RECHTORIS, C. (1974). Insect cells: colony formation and cloning in agar medium. *In Vitro* **10,** 1–5.

MCINTOSH, A.H. AND SHAMY, R. (1980). Biological studies of a baculovirus in a mammalian cell line. *Intervirology* **13,** 331–341.

MCINTOSH, A.H., MARAMOROSCH, K. AND RISCOE, R. (1979). *Autographa californica* nuclear polyhedrosis virus (NPV) in a vertebrate cell line: localization by electron microscopy. *Journal of the New York Entomological Society* **87,** 55–58.

MCINTOSH, A.H., SHAMY, R. AND ILSLEY, C. (1979). Interference with polyhedral inclusion body (PIB) production in *Trichoplusia ni* cells infected with a high passage strain of *Autographa californica* nuclear polyhedrosis virus (NPV). *Archives of Virology* **60,** 353–358.

MCKINLEY, D.J., BROWN, D.A., PAYNE, C.C. AND HARRAP, K.A. (1981). Cross-infectivity and activation studies with four baculoviruses. *Entomophaga* **26,** 79–90.

MACKINNON, E.A., HENDERSON, J.F., STOLTZ, D.B. AND FAULKNER, P. (1974). Morphogenesis of nuclear polyhedrosis virus under conditions of prolonged passage *in vitro*. *Journal of Ultrastructure Research* **49,** 419–435.

MAGNOLER, A. (1968). The differing effectiveness of purified and non purified suspensions of the nuclear-polyhedrosis virus of *Porthetria dispar*. *Journal of Invertebrate Pathology* **11,** 326–328.

Maleki-Milani, H. (1978). Influence de passages répétés du virus de la polyèdrose nucléaire de *Autographa californica* chez *Spodoptera littoralis* [Lep.: Noctuidae]. *Entomophaga* **23,** 217–224.

Marschall, K.J. and Ioane, I. (1982). The effect of re-release of *Oryctes rhinoceros* baculovirus in the biological control of rhinoceros beetles in Western Samoa. *Journal of Invertebrate Pathology* **39,** 267–276.

Martignoni, M.E., Stelzer, M.J. and Iwai, P.J. (1982). *Baculovirus* of *Autographa californica* (Lepidoptera: Noctuidae): a candidate biological control agent for Douglas-fir tussock moth (Lepidoptera: Lymantriidae). *Journal of Economic Entomology* **75,** 1120–1124.

Matthews, G.A. (1982). Prospects of better deposition of microbial pesticides using electrostatic sprayers. In *Invertebrate Pathology and Microbial Control, Proceedings of the Third International Colloquium on Invertebrate Pathology, Brighton, UK* (C.C. Payne and H.D. Burges, Eds), pp. 55–59. Society for Invertebrate Pathology, Glasshouse Crops Research Institute, Littlehampton.

Matthews, G.A. (1983). Can we control insect pests? *New Scientist* **98,** 368–372.

Matthews, R.E.F. (1982). Classification and nomenclature of viruses. *Intervirology* **17,** 1–199.

Miller, D.W. and Miller, L.K. (1982). A virus mutant with an insertion of a *copia*-like transposable element. *Nature* **299,** 562–564.

Miller, L.K. and Dawes, K. (1979). Physical map of the DNA genome of *Autographa californica* nuclear polyhedrosis virus. *Journal of Virology* **29,** 1044–1055.

Miller, L.K., Lingg, A.J. and Bulla, L.A. (1983). Bacterial, viral and fungal insecticides. *Science* **219,** 715–721.

Miltenburger, H.G. (1980). Viral pesticides: hazard evaluation for non-target organisms and safety testing. In *Environmental Protection and Biological Forms of Control of Pest Organisms* (B. Lundholm and M. Stackerud, Eds), *Ecological Bulletins* **31,** 57–74. Swedish Natural Science Research Council, Stockholm.

Miltenburger, H.G. and David, P. (1980). Mass production of insect cells in suspension. In *Proceedings of the Third General Meeting of the European Society of Animal Cell Technology 1979, Developments in Biological Standardization* volume 46, pp. 183–186. S. Karger, Basel.

Miltenburger, H.G. and Reimann, R. (1980). Viral pesticides: biohazard evaluation on the cytogenetic level. In *Proceedings of the Third General Meeting of the European Society of Animal Cell Technology 1979, Developments in Biological Standardization* volume 46, pp. 217–222. S. Karger, Basel.

Minion, F.C., Coons, L.B. and Broome, J.R. (1979). Characterization of the polyhedral envelope of the nuclear polyhedrosis virus of *Heliothis virescens*. *Journal of Invertebrate Pathology* **34,** 303–307.

Mitsuhashi, J. (1981). Establishment and some characteristics of a continuous cell line derived from fat bodys of the cabbage armyworm (Lepidoptera, Noctuidae). *Development, Growth and Differentiation* **23,** 63–72.

Mitsuhashi, J. (1982). Media for insect cell cultures. *Advances in Cell Culture* **2,** 133–196.

Monsarrat, P. and Veyrunes, J.C. (1976). Evidence of *Oryctes* virus in adult feces and new data for virus characterization. *Journal of Invertebrate Pathology* **27,** 387–389.

Morris, O.N. (1980). Entomopathogenic viruses: strategies for use in forest insect pest management. *Canadian Entomologist* **112,** 573–584.

Morris, T.J., Hess, R.T. and Pinnock, D.E. (1979). Physicochemical characterization of a small RNA virus associated with baculovirus infection in *Trichoplusia ni*. *Intervirology* **11,** 238–247.

Morris, T.J., Vail, P.V. and Collier, S.S. (1981). An RNA virus in *Autographa californica* nuclear polyhedrosis preparations: detection and identification. *Journal of Invertebrate Pathology* **38,** 201–208.

Naser, W.L. and Miltenburger, H.G. (1983). Rapid baculovirus detection, identification, and serological classification by Western blotting—ELISA using a monoclonal antibody. *Journal of General Virology* **64,** 639–647.

NORDIN, G.L. (1976). Microsporidian bioassay technique for third-instar *Pseudaletia unipuncta* larvae. *Journal of Invertebrate Pathology* **27,** 397–398.

PADHI, S.B. AND MARAMOROSCH, K. (1983). *Heliothis zea* baculovirus and *Bombyx mori:* safety considerations. *Applied Entomology and Zoology* **18,** 136–138.

PAPWORTH, D.S. (1980). Registration requirements in the UK for bacteria, fungi and viruses used as pesticides. In *Environmental Protection and Biological Forms of Control of Pest Organisms* (B. Lundholm and M. Stackerud, Eds), *Ecological Bulletins* **31,** 135–143. Swedish Natural Science Research Council, Stockholm.

PAYNE, C.C. (1982). Insect viruses as control agents. *Parasitology* **84,** 35–77.

PAYNE, C.C., TATCHELL, G.M. AND WILLIAMS, C.F. (1981). The comparative susceptibilities of *Pieris brassicae* and *P. rapae* to a granulosis virus from *P. brassicae. Journal of Invertebrate Pathology* **38,** 273–280.

PFRIMMER, T.R. (1979). *Heliothis* spp.: control on cotton with pyrethroids, carbamates, organophosphates, and biological insecticides. *Journal of Economic Entomology* **72,** 593–598.

PLUS, N. (1978). Endogenous viruses of *Drosophila melanogaster* cell lines: their frequency, identification and origin. *In Vitro* **14,** 1015–1021.

PLUS, N. (1980). Further studies on the origin of the endogenous viruses of *Drosophila melanogaster* cell lines. In *Invertebrate Systems In Vitro, Fifth International Conference on Invertebrate Tissue Culture, Rigi-Kaltbad, 1979* (E. Kurstak, K. Maramorosch and A. Dübendorfer, Eds), pp. 435–439. Elsevier/North-Holland, Amsterdam.

PODGWAITE, J.D., BRUEN, R.B. AND SHAPIRO, M. (1983). Microorganisms associated with production lots of the nucleopolyhedrosis virus of the gypsy moth, *Lymantria-dispar* [*Lep.: Lymantriidae*]. *Entomophaga* **28,** 9–16.

POLLARD, R. AND KHOSROVI, B. (1978). Reactor design for fermentation of fragile tissue cells. *Process Biochemistry* **13,** 31–37.

POTTER, K.N. AND MILLER, L.K. (1980). Correlating genetic mutations of a baculovirus with the physical map of the DNA genome. In *Animal Virus Genetics, ICN–UCLA Symposia on Molecular and Cellular Biology* (B.N. Fields, R. Jaenisch and C.F. Fox, Eds), volume 18, pp. 71–80. Academic Press, New York.

POTTER, K.N., FAULKNER, P. AND MACKINNON, E.A. (1976). Strain selection during serial passage of *Trichoplusia ni* nuclear polyhedrosis virus. *Journal of Virology* **18,** 1040–1050.

POTTER, K.N., JAQUES, R.P. AND FAULKNER, P. (1978). Modification of *Trichoplusia ni* nuclear polyhedrosis virus passaged *in vivo. Intervirology* **9,** 76–85.

POTTER, M.F. AND WATSON, T.F. (1983a). Laboratory and greenhouse performance of *Baculovirus heliothis,* combined with feeding stimulants for control of neonate tobacco budworm. *Protection Ecology* **5,** 161–165.

POTTER, M.F. AND WATSON, T.F. (1983b). Timing of nuclear polyhedrosis virus-bait spray combinations for control of egg and larval stages of tobacco budworm (Lepidoptera: Noctuidae). *Journal of Economic Entomology* **76,** 446–448.

RAIMO, B., REARDON, R.C. AND PODGWAITE, J.D. (1977). Vectoring gypsy moth nuclear polyhedrosis virus by *Apanteles melanoscelus* [*Hym.: Braconidae*]. *Entomophaga* **22,** 207–215.

REED, D.K. (1981). Control of mites by non-occluded viruses. In *Microbial Control of Pests and Plant Diseases 1970–1980* (H.D. Burges, Ed.), pp. 427–432. Academic Press, London.

REED, E.M. AND SPRINGETT, B.P. (1971). Large-scale field testing of a granulosis virus for the control of the potato moth (*Phthorimaea operculella* (Zell.) (Lep., Gelechiidae)). *Bulletin of Entomological Research* **61,** 223–233.

REIMANN, R. AND MILTENBURGER, H.G. (1983). Cytogenetic studies in mammalian cells after treatment with insect pathogenic viruses [*Baculoviridae*]. II *In vitro* studies with mammalian cell lines. *Entomophaga* **28,** 33–44.

RICHARDS, M.G. AND PAYNE, C.C. (1982). Persistence of baculoviruses on leaf surfaces. In *Invertebrate Pathology and Microbial Control, Proceedings of the Third International Colloquium on Invertebrate Pathology, Brighton, UK* (C.C. Payne and H.D. Burges, Eds), pp. 296–301. Society for Invertebrate Pathology, Glasshouse Crops Research Institute, Littlehampton.

ROBERTS, P.L. AND NASER, W. (1982a). Characterization of monoclonal antibodies to the *Autographa californica* nuclear polyhedrosis virus. *Virology* **122,** 424–430.

ROBERTS, P.L. AND NASER, W. (1982b). Preparation of monoclonal antibodies to a baculovirus. *FEMS Microbiology Letters* **14,** 79–83.

ROGOFF, M.H. (1975). Exposure of humans to nuclear polyhedrosis virus during industrial production. In *Baculoviruses for Insect Pest Control: Safety Considerations, EPA–USDA Symposium, Bethesda, Maryland, USA* (M. Summers, R. Engler, L.A. Falcon and P.V. Vail, Eds), pp. 102–105. American Society for Microbiology, Washington D.C.

ROGOFF, M.H. (1980). Testing requirements for registering biological pesticides in the United States—current status. In *Environmental Protection and Biological Forms of Control of Pest Organisms* (B. Lundholm and M. Stackerud, Eds), *Ecological Bulletins* **31,** 111–134. Swedish Natural Science Research Council, Stockholm.

ROHRMANN, G.F. (1982). Genetic characterization and sequence analysis of polyhedrins. In *Invertebrate Pathology and Microbial Control, Proceedings of the Third International Colloquium on Invertebrate Pathology, Brighton, UK* (C.C. Payne and H.D. Burges, Eds), pp. 226–232. Society for Invertebrate Pathology, Glasshouse Crops Research Institute, Littlehampton.

ROHRMANN, G.F., PEARSON, M.N., BAILEY, T.J., BECKER, R.R. AND BEAUDREAU, G.S. (1981). N-terminal polyhedrin sequences and occluded *Baculovirus* evolution. *Journal of Molecular Evolution* **17,** 329–333.

ROOME, R.E. (1975). Field trials with a nuclear polyhedrosis virus and *Bacillus thuringiensis* against larvae of *Heliothis armigera* on sorghum and cotton in Botswana. *Bulletin of Entomological Research* **65,** 507–514.

RUBINSTEIN, R., LAWLER, K.A. AND GRANADOS, R.R. (1982). Use of primary fat body cultures for the study of baculovirus replication. *Journal of Invertebrate Pathology* **40,** 266–273.

SCHERER, W.F., VERNA, J.E. AND RICHTER, G.W. (1968). Nodamura virus, an ether- and chloroform-resistant arbovirus from Japan. *American Journal of Tropical Medicine and Hygiene* **17,** 120–128.

SEARS, M.K., JAQUES, R.P. AND LAING, J.E. (1983). Utilization of action thresholds for microbial and chemical control of lepidopterous pests (Lepidoptera: Noctuidae, Pieridae) on cabbage. *Journal of Economic Entomology* **76,** 368–374.

SHAPIRO, M. (1982). In vivo mass production of insect viruses for use as pesticides. In *Microbial and Viral Pesticides* (E. Kurstak, Ed.), pp. 463–492. Marcel Dekker, New York.

SHAPIRO, M., BELL, R.A. AND OWENS, C.D. (1981). Evaluation of various artificial diets for in vivo production of the gypsy moth nucleopolyhedrosis virus. *Journal of Economic Entomology* **74,** 110–111.

SHAPIRO, M., POH AGIN, P. AND BELL, R.A. (1983). Ultraviolet protectants of the gypsy moth (Lepidoptera: Lymantriidae) nucleopolyhedrosis virus. *Environmental Entomology* **12,** 982–985.

SHAPIRO, M., MARTIGNONI, M.E., CUNNINGHAM, J.C. AND GOODWIN, R.H. (1982). Potential use of the saltmarsh caterpillar as a production host for nucleopolyhedrosis viruses. *Journal of Economic Entomology* **75,** 69–71.

SHAPIRO, M., OWENS, C.D., BELL, R.A. AND WOOD, H.A. (1981). Simplified, efficient system for in vivo mass production of gypsy moth nucleopolyhedrosis virus. *Journal of Economic Entomology* **74,** 341–343.

SHEPHERD, R.F., GRAY, T.G. AND CUNNINGHAM, J.C. (1982). Effects of nuclear polyhedrosis virus and *Bacillus thuringiensis* on western spruce budworm (Lepidoptera:

Tortricidae) 1 and 2 years after aerial application. *Canadian Entomologist* **114,** 281–282.

SHIEH, T.R. AND BOHMFALK, G.T. (1980). Production and efficacy of baculoviruses. *Biotechnology and Bioengineering* **22,** 1357–1375.

SMITH, D.B. AND BOUSE, L.F. (1981). Machinery and factors that affect the application of pathogens. In *Microbial Control of Pests and Plant Diseases 1970–1980* (H.D. Burges, Ed.), pp. 635–653. Academic Press, London.

SMITH, D.B. AND HOSTETTER, D.L. (1982). Laboratory and field evaluations of pathogen-adjuvant treatments. *Journal of Economic Entomology* **75,** 472–476.

SMITH, D.B., HOSTETTER, D.L. AND IGNOFFO, C.M. (1978). Formulation and equipment effects on application of a viral (*Baculovirus heliothis*) insecticide. *Journal of Economic Entomology* **71,** 814–817.

SMITH, D.B., HOSTETTER, D.L. AND IGNOFFO, C.M. (1979). Nozzle size-pressure and concentration combinations for *Heliothis zea* control with an aqueous suspension of polyvinyl alcohol and *Baculovirus heliothis*. *Journal of Economic Entomology* **72,** 920–923.

SMITH, G.E. AND SUMMERS, M.D. (1978). Analysis of baculovirus genomes with restriction endonucleases. *Virology* **89,** 517–527.

SMITH, G.E. AND SUMMERS, M.D. (1979). Restriction maps of five *Autographa californica* *M*NPV variants, *Trichoplusia ni* *M*NPV and *Galleria mellonella* *M*NPV DNAs with endonucleases *Sma*I, *Kpn*I, *Bam*HI, *Sac*I, *Xho*I, and *Eco*RI. *Journal of Virology* **30,** 828–838.

SMITH, G.E. AND SUMMERS, M.D. (1980). Restriction map of *Rachiplusia ou* and *Rachiplusia ou–Autographa californica* baculovirus recombinants. *Journal of Virology* **33,** 311–319.

SMITH, G.E. AND SUMMERS, M.D. (1982). DNA homology among subgroup A, B and C baculoviruses. *Virology* **123,** 393–406.

SMITH, G.E., VLAK, J.M. AND SUMMERS, M.D. (1982). In vitro translation of *Autographa californica* nuclear polyhedrosis virus early and late mRNAs. *Journal of Virology* **44,** 199–208.

SMITH, G.E., VLAK, J.M. AND SUMMERS, M.D. (1983). Physical analysis of *Autographa californica* nuclear polyhedrosis virus transcripts for polyhedrin and 10,000-molecular-weight protein. *Journal of Virology* **45,** 215–225.

SMITH, K.M. AND XEROS, N. (1954). An unusual virus disease of a dipterous larva. *Nature* **173,** 866–867.

STOCKDALE, H. (in press). Microbial insecticides. In *The Practice of Biotechnology: Commodity Products, Comprehensive Biotechnology, Vol. 2*, Pergamon Press, Oxford.

STOCKDALE, H. AND GARDINER, G.R. (1977). The influence of the condition of cells and medium on production of polyhedra of *Autographa californica* nuclear polyhedrosis virus in vitro. *Journal of Invertebrate Pathology* **30,** 330–336.

STOCKDALE, H. AND PRISTON, R.A.J. (1981). Production of insect viruses in cell culture. In *Microbial Control of Pests and Plant Diseases 1970–1980* (H.D. Burges, Ed.), pp. 313–328. Academic Press, London.

STOLTZ, D.B., PAVAN, C. AND DA CUNHA, A.B. (1973). Nuclear polyhedrosis virus: a possible example of *de novo* intranuclear membrane morphogenesis. *Journal of General Virology* **19,** 145–150.

SUMMERS, M.D. AND SMITH, G.E. (1975). *Trichoplusia ni* granulosis virus granulin: a phenol-soluble phosphorylated protein. *Journal of Virology* **16,** 1108–1116.

SUMMERS, M.D. AND VOLKMAN, L.E. (1976). Comparison of biophysical and morphological properties of occluded and extracellular nonoccluded baculovirus from in vivo and in vitro host systems. *Journal of Virology* **17,** 962–972.

TATCHELL, G.M. (1981). The transmission of a granulosis virus following the contamination of *Pieris brassicae* adults. *Journal of Invertebrate Pathology* **37,** 210–213.

TATCHELL, G.M. AND PAYNE, C.C. (1984). Field evaluation of a granulosis virus for control of *Pieris rapae* in the United Kingdom. *Entomophaga*, in press.

TJIA, S.T., ZU ALTENSCHILDESCHE, G.M. AND DOERFLER, W. (1983). Autographa californica nuclear polyhedrosis virus (AcNPV) DNA does not persist in mass cultures of mammalian cells. *Virology* **125,** 107–117.

TWEETEN, K.A., BULLA, L.A. AND CONSIGLI, R.A. (1980). Structural polypeptides of the granulosis virus of *Plodia interpunctella*. *Journal of Virology* **33,** 877–886.

TWEETEN, K.A., BULLA, L.A. AND CONSIGLI, R.A. (1981). Applied and molecular aspects of insect granulosis viruses. *Microbiological Reviews* **45,** 379–408.

VAGO, C. AND BERGOIN, M. (1963). Dévelopment des virus à corps d'inclusion du Lepidoptère *Lymantria dispar* en cultures cellulaires. *Entomophaga* **8,** 253–261.

VAIL, P.V., SEAY, R.E. AND DEBOLT, J. (1980). Microbial and chemical control of the cabbage looper on fall lettuce. *Journal of Economic Entomology* **73,** 72–75.

VAIL, P.V., MORRIS, T.J., COLLIER, S.S. AND MACKEY, B. (1983). An RNA virus in *Autographa californica* nuclear polyhedrosis virus preparations: incidence and influence on baculovirus activity. *Journal of Invertebrate Pathology* **41,** 171–178.

VAUGHN, J.L. (1981). Insect cells for insect virus production. *Advances in Cell Culture* **1,** 281–295.

VAUGHN, J.L. AND DOUGHERTY, E.M. (1981). Recent progress in *in vitro* studies of baculoviruses. In *Beltsville Symposia in Agricultural Research,* [*5*] *Biological Control in Crop Production* (G.C. Papavizas, Ed.), pp. 249–258. Allanheld, Osmun, New Jersey.

VLAK, J.M. (1979). The proteins of nonoccluded *Autographa californica* nuclear polyhedrosis virus produced in an established cell line of *Spodoptera frugiperda*. *Journal of Invertebrate Pathology* **34,** 110–118.

VLAK, J.M. (1980). Mapping of *Bam*HI and *Sma*I DNA restriction sites on the genome of the nuclear polyhedrosis virus of the alfalfa looper, *Autographa californica*. *Journal of Invertebrate Pathology* **36,** 409–414.

VOLKMAN, L.E. (1983). Occluded and budded *Autographa californica* nuclear polyhedrosis virus: immunological relatedness of structural proteins. *Journal of Virology* **46,** 221–229.

VOLKMAN, L.E. AND FALCON, L.A. (1982). Use of monoclonal antibody in an enzyme-linked immunosorbent assay to detect the presence of *Trichoplusia ni* (Lepidoptera: Noctuidae) S nuclear polyhedrosis virus polyhedrin in *T. ni* larvae. *Journal of Economic Entomology* **75,** 868–871.

VOLKMAN, L.E. AND GOLDSMITH, P.A. (1983). In vitro survey of *Autographa californica* nuclear polyhedrosis virus interaction with nontarget vertebrate host cells. *Applied and Environmental Microbiology* **45,** 1085–1093.

WEISS, S.A., SMITH, G.C., KALTER, S.S. AND VAUGHN, J.L. (1981). Improved method for the production of insect cell cultures in large volume. *In Vitro* **17,** 495–502.

WILKIE, G.E.I., STOCKDALE, H. AND PIRT, S.V. (1980). Chemically-defined media for production of insect cells and viruses *in vitro*. In *Proceedings of the Third General Meeting of the European Society of Animal Cell Technology 1979, Developments in Biological Standardization, Vol. 46*, pp. 29–37. S. Karger, Basel.

WOOD, H.A. (1980). Isolation and replication of an occlusion body-deficient mutant of the *Autographa californica* nuclear polyhedrosis virus. *Virology* **105,** 338–344.

WOOD, H.A., JOHNSTON, L.B. AND BURAND, J.P. (1982). Inhibition of *Autographa californica* nuclear polyhedrosis virus replication in high-density *Trichoplusia ni* cell cultures. *Virology* **119,** 245–254.

WOOD, H.A., HUGHES, P.R., JOHNSTON, L.B. AND LANGRIDGE, W.H.R. (1981). Increased virulence of *Autographa californica* nuclear polyhedrosis virus by mutagenesis. *Journal of Invertebrate Pathology* **38,** 236–241.

YAMADA, K. AND MARAMOROSCH, K. (1981). Plaque assay of *Heliothis zea* baculovirus employing a mixed agarose overlay. *Archives of Virology* **67,** 187–189.

YAMADA, K., SHERMAN, K.E. AND MARAMOROSCH, K. (1982). Serial passage of *Heliothis zea* singly embedded nuclear polyhedrosis virus in a homologous cell line. *Journal of Invertebrate Pathology* **39,** 185–191.

YAMAFUJI, K., YOSHIHARA, F. AND HIRAYAMA, K. (1957). Protease and desoxyribonuclease in viral polyhedral crystal. *Enzymologia* **19,** 53–58.

YEARIAN, W.C. AND YOUNG, S.Y. (1982). Control of insect pests of agricultural importance by viral insecticides. In *Microbial and Viral Pesticides* (E. Kurstak, Ed.), pp. 387–423. Marcel Dekker, New York.

YOUNG, E.C. AND LONGWORTH, J.F. (1981). The epizootiology of the baculovirus of the coconut palm rhinoceros beetle (*Oryctes rhinoceros*) in Tonga. *Journal of Invertebrate Pathology* **38,** 362–369.

YOUNG, S.Y. AND YEARIAN, W.C. (1980). Soil application of *Pseudoplusia* NPV: persistence and incidence of infection in soybean looper caged on soybean. *Environmental Entomology* **8,** 860–864.

ZELAZNY, B. (1979). Virulence of the baculovirus of *Oryctes rhinoceros* from ten locations in the Philippines and in Western Samoa. *Journal of Invertebrate Pathology* **33,** 106–107.

Index